LONDON MATHEMATICAL SOCIETY LECTURE NOTE SERIES

Managing Editor: Professor M. Reid, Mathematics Institute,
University of Warwick, Coventry CV4 7AL, United Kingdom

The titles below are available from booksellers, or from Cambridge University Press at
http://www.cambridge.org/mathematics

London Mathematical Society Lecture Note Series: 415

Automorphic Forms and Galois Representations

Volume 2

Edited by

FRED DIAMOND
King's College London

PAYMAN L. KASSAEI
McGill University, Montréal

MINHYONG KIM
University of Oxford

CAMBRIDGE
UNIVERSITY PRESS

CAMBRIDGE
UNIVERSITY PRESS

University Printing House, Cambridge CB2 8BS, United Kingdom

One Liberty Plaza, 20th Floor, New York, NY 10006, USA

477 Williamstown Road, Port Melbourne, VIC 3207, Australia

314-321, 3rd Floor, Plot 3, Splendor Forum, Jasola District Centre, New Delhi - 110025, India

79 Anson Road, #06-04/06, Singapore 079906

Cambridge University Press is part of the University of Cambridge.

It furthers the University's mission by disseminating knowledge in the pursuit of
education, learning and research at the highest international levels of excellence.

www.cambridge.org
Information on this title: www.cambridge.org/9781107693630

© Cambridge University Press 2014

First published 2014

A catalogue record for this publication is available from the British Library

Library of Congress Cataloging in Publication data
Automorphic forms and galois representations / edited by Fred Diamond, King's College
London, Payman L. Kassaei, McGill University, Montréal, Minhyong Kim, University of
Oxford.
volumes <1–2> cm. – (London Mathematical Society lecture note series ; 414, 415)
Papers presented at the London Mathematical Society, and EPSRC (Great Britain
Engineering and Physical Sciences Research Council), Symposium on Galois
Representations and Automorphic Forms, held at the University of Durham from
July 18–28, 2011.
ISBN 978-1-107-69192-6 (v. 1) – ISBN 978-1-107-69363-0 (v. 2)
1. Automorphic forms–Congresses.   2. Automorphic functions–Congresses.   3. Forms
(Mathematics)–Congresses.   4. Galois theory–Congresses.   I. Diamond, Fred, editor of
compilation.   II. Kassaei, Payman L., 1973– editor of compilation.   III. Kim, Minhyong,
editor of compilation.   IV. Symposium on Galois Representations and Automorphic
Forms (2011 : Durham, England)
QA353.A9A925   2014
515´.9–dc23
2014001841

ISBN 978-1-107-69363-0 Paperback

Contents

Contributors

Mladen Dimitrov, UFR Mathématiques, Université Lille 1, Lille, 59655 Villeneuve d'Ascq Cedex, France.

Laurent Fargues, CNRS, Institut de Mathématiques de Jussieu, Paris, 75252, France.

Jean-Marc Fontaine, Laboratoire de Mathématiques d'Orsay, Université Paris Sud, Paris 91405 Orsay, France.

Hidekazu Furusho, Graduate School of Mathematics, Nagoya University, Nagoya, 464-8602, Japan.

Thomas J. Haines, Department of Mathematics, Univeristy of Maryland, College Park, MD 20742-4015, USA.

Yuichiro Hoshi, Research Institute for Mathematical Sciences, Kyoto University, Kyoto, 606-8502, Japan.

Chandrashekhar Khare, Department of Mathematics, UCLA, Los Angeles, CA 90095-1555, USA.

Vytautas Paškūnas, Fakultät für Mathematik, Universität Duisburg–Essen, Essen 45127, Germany.

Peter Schneider, Mathematischen Instituts, Westfälische Wilhelms-Universität Münster, 48149 Germany.

Marie-France Vignéras, Institut de Mathématiques de Jussieu, Université Paris Diderot (Paris 7), 775205 Paris Cedex 13, France.

Jean-Pierre Wintenberger, Département de Mathématiques, Université de Strasbourg, 67084 Strasbourg Cedex, France.

Gergely Zabradi, Institute of Mathematics, Eötvös Loránd University, H-1518 Budapest, Pf. 120, Hungary.

Preface

The London Mathematical Society Symposium – EPSRC Symposium on Galois Representations and Automorphic Forms was held at the University of Durham from 18th July until 28th July 2011. These topics have been playing an important role in present-day number theory, especially via the Langlands program and the connections it entails. The meeting brought together researchers from around the world on these and related topics, with lectures on a variety of recent major developments in the area. Roughly half of these talks were individual lectures, while the rest constituted series on the following themes:

- p-adic local Langlands
- Curves and vector bundles in p-adic Hodge theory
- The fundamental lemma and trace formula
- Anabelian geometry
- Potential automorphy

These Proceedings present much of the progress described in those lectures. The organizers are very grateful to all the speakers and to others who contributed articles. We also wish to thank the London Mathematical Society and EPSRC for the financial support that made the meeting possible. We warmly appreciate the assistance and hospitality provided by the University of Durham's Department of Mathematics and Grey College. These institutions have helped to make the Symposia such a well established and highly valued event in the number theory community.

Fred Diamond
Payman Kassaei
Minhyong Kim

1

On the local structure of ordinary Hecke algebras at classical weight one points

Mladen Dimitrov

Abstract

The aim of this chapter is to explain how one can obtain information regarding the membership of a classical weight one eigenform in a Hida family from the geometry of the Eigencurve at the corresponding point. We show, in passing, that all classical members of a Hida family, including those of weight one, share the same local type at all primes dividing the level.

1. Introduction

Classical weight one eigenforms occupy a special place in the correspondence between Automorphic Forms and Galois Representations since they yield two dimensional Artin representations with odd determinant. The construction of those representations by Deligne and Serre [5] uses congruences with modular forms of higher weight. The systematic study of congruences between modular forms has culminated in the construction of the p-adic Eigencurve by Coleman and Mazur [4]. A p-stabilized classical weight one eigenform corresponds then to a point on the ordinary component of the Eigencurve, which is closely related to Hida theory.

An important result of Hida [11] states that an ordinary cuspform of weight at least two is a specialization of a unique, up to Galois conjugacy, primitive Hida family. Geometrically this translates into the smoothness of the Eigencurve at that point (in fact, Hida proves more, namely that the map

The author is partially supported by Agence Nationale de la Recherche grants
ANR-10-BLAN-0114 and ANR-11-LABX-0007-01.

Automorphic Forms and Galois Representations, ed. Fred Diamond, Payman L. Kassaei and Minhyong Kim. Published by Cambridge University Press. © Cambridge University Press 2014.

to the weight space is etale at that point). Whereas Hida's result continues to hold at all non-critical classical points of weight two or more [13], there are examples where this fails in weight one [6]. The purely quantitive question of how many Hida families specialize to a given classical p-stabilized weight one eigenform, can be reformulated geometrically as to describe the local structure of the Eigencurve at the corresponding point. An advantage of the new formulation is that it provides group theoretic and homological tools for the study of the original question thanks to Mazur's theory of deformations of Galois representations. Moreover, this method gives more qualitative answers, since the local structure of the Eigencurve at a given point contains more information than the collection of all Hida families passing through that point.

The local structure at weight one forms with RM was first investigated by Cho and Vatsal [3] in the context of studying universal deformation rings, who showed that in many cases the Eigencurve is smooth, but not etale over the weight space, at those points. The main result of a joint work with Joël Bellaïche [1] states that the p-adic Eigencurve is smooth at all classical weight one points which are regular at p and gives a precise criterion for etalness over the weight space at those points. The author has learned recently that the work [10] of Greenberg and Vatsal contains a slightly weaker version of this result. It would be interesting to describe the local structure at irregular points, to which we hope to come back in a future work.

The chapter is organized as follows. Section 2 describes some p-adic aspects in the theory of weight one eigenforms. Sections 3 and 4 introduce, respectively, the ordinary Hecke algebras and primitive Hida families, which are central objects in Hida theory [12]. In Section 5 various Galois representations are studied with emphasis on stable lattices, leading to the construction of a representation (1.10) which is a bridge between a primitive Hida family and its classical members. This is used in Section 6 to establish the rigidity of the local type in a Hida family, including in weight one (see Proposition 1.8). Section 7 quotes the main results of [1] and describes their consequences in classical Hida theory (see Corollary 1.15). The latter would have been rather straightforward, should the Eigencurve have been primitive, in the sense that the irreducible component of its ordinary locus would have corresponded (after inverting p) to primitive Hida families. Lacking a reference for the construction of such an Eigencurve, we establish a local isomorphism, at the points of interest, between the reduced Hecke algebra, used in the definition of the Eigencurve, and the new quotient of the full Hecke algebra, used in the definition of primitive Hida families (see Corollary 1.14).

Acknowledgements. The author would like to thank Joël Bellaïche for many helpful discussions, as well as the referee for his careful reading of the manuscript and for pointing out some useful references.

2. Artin modular forms and the Eigencurve

We let $\bar{\mathbb{Q}} \subset \mathbb{C}$ be the field of algebraic numbers, and denote by $\mathrm{Gal}(\bar{\mathbb{Q}}/\mathbb{Q})$ the absolute Galois group of $\mathbb{Q}$. For a prime ℓ we fix a decomposition subgroup G_ℓ of $\mathrm{Gal}(\bar{\mathbb{Q}}/\mathbb{Q})$ and denote by I_ℓ its inertia subgroup and by Frob_ℓ the arithmetic Frobenius in G_ℓ/I_ℓ.

We fix a prime number p and an embedding $\bar{\mathbb{Q}} \hookrightarrow \bar{\mathbb{Q}}_p$.

Let $f(z) = \sum_{n \geq 1} a_n q^n$ be a newform of weight one, level M and central character ϵ. Thus $a_1 = 1$ and for every prime $\ell \nmid M$ (resp. $\ell \mid M$) f is an eigenvector with eigenvalue a_ℓ for the Hecke operator T_ℓ (resp. U_ℓ). By a theorem of Deligne and Serre [5] there exists a unique continuous irreducible representation:

$$\rho_f : \mathrm{Gal}(\bar{\mathbb{Q}}/\mathbb{Q}) \to \mathrm{GL}_2(\mathbb{C}), \tag{1.1}$$

such that its Artin L-function $L(\rho_f, s)$ equals

$$L(f, s) = \sum_n \frac{a_n}{n^s} = \prod_{\ell \nmid M} (1 - a_\ell \ell^{-s} + \epsilon(\ell)\ell^{-2s})^{-1} \prod_{\ell \mid M} (1 - a_\ell \ell^{-s})^{-1}.$$

It follows that if $a_\ell \neq 0$ for $\ell \mid M$, then a_ℓ is the eigenvalue of $\rho_f(\mathrm{Frob}_\ell)$ acting on the unique line fixed by I_ℓ. Since ρ_f has finite image, a_ℓ is an eigenvalue of a finite order matrix, hence it is a root of unity.

Similarly, for $\ell \nmid M$ the characteristic polynomial $X^2 - a_\ell X + \epsilon(\ell)$ of $\rho_f(\mathrm{Frob}_\ell)$ has two (possibly equal) roots α_ℓ and β_ℓ which are both roots of unity.

In order to deform f p-adically, one should first choose a p-stabilization of f with finite slope, that is an eigenform of level $\Gamma_1(M) \cap \Gamma_0(p)$ sharing the same eigenvalues as f away from p and having a non-zero U_p-eigenvalue. By the above discussion if such a stabilization exists, then it should necessarily be ordinary. We distinguish two cases:

If p does not divide M, then f has two p-stabilizations $f_\alpha(z) = f(z) - \beta_p f(pz)$ and $f_\beta(z) = f(z) - \alpha_p f(pz)$ with U_p-eigenvalue α_p and β_p, respectively.

If p divides M and $a_p \neq 0$, then f is already p-stabilized. We let then $\alpha_p = a_p$ and $f_\alpha = f$.

Denote by N the prime to p-part of M.

Definition 1.1. We say that f_α is *regular* at p if either p divides M and $a_p \neq 0$, or p does not divide M and $\alpha_p \neq \beta_p$.

The Eigencurve $\mathcal{C}$ of tame level $\Gamma_1(N)$ is a rigid analytic curve over $\mathbb{Q}_p$ parametrizing systems of eigenvalues for the Hecke operators T_ℓ ($\ell \nmid Np$) and U_p appearing in the space of finite slope overconvergent modular forms of tame level dividing N. We refer to the original article of Coleman and Mazur [4] for the case $N = 1$ and $p > 2$, and to Buzzard [2] for the general case. Recall that $\mathcal{C}$ is reduced and endowed with a flat and locally finite weight map $\kappa : \mathcal{C} \to \mathcal{W}$, where $\mathcal{W}$ is the rigid space over $\mathbb{Q}_p$ representing homomorphisms $\mathbb{Z}_p^\times \times (\mathbb{Z}/N\mathbb{Z})^\times \to \mathbb{G}_m$.

The p-stabilized newform f_α defines a point on the ordinary component of $\mathcal{C}$, whose image by κ is a character of finite order.

Theorem 1.2. [1] *Let f be a classical weight one cuspidal eigenform form which is regular at p. Then the Eigencurve $\mathcal{C}$ is smooth at the point defined by f_α, so there is a unique irreducible component of $\mathcal{C}$ containing that point. In particular, if f has CM by a quadratic field in which p splits, then all classical points of that component also have CM by the same field.*

Moreover, $\mathcal{C}$ is etale over the weight space $\mathcal{W}$ at the point defined by f_α, unless f has RM by a quadratic field in which p splits.

In Section 7 we will revisit this theorem from the perspective of Hida families.

3. Ordinary Hecke algebras

The results in this and the following two sections are due to Hida [11, 12] when p is odd and have been completed for $p = 2$ by Wiles [18] and Ghate–Kumar [8].

Let $\Lambda = \mathbb{Z}_p[[\mathrm{Gal}(\mathbb{Q}_\infty/\mathbb{Q})]] \simeq \mathbb{Z}_p[[1 + p^\nu \mathbb{Z}_p]]$ be the Iwasawa algebra of the cyclotomic $\mathbb{Z}_p$ extension $\mathbb{Q}_\infty$ of $\mathbb{Q}$, where $\nu = 2$ if $p = 2$ and $\nu = 1$ otherwise. It is a complete local $\mathbb{Z}_p$-algebra which is an integral domain of Krull dimension 2. Let χ_{cyc} be the universal Λ-adic cyclotomic character obtained by composing $\mathrm{Gal}(\bar{\mathbb{Q}}/\mathbb{Q}) \twoheadrightarrow \mathrm{Gal}(\mathbb{Q}_\infty/\mathbb{Q})$ with the canonical continuous group homomorphism from $\mathrm{Gal}(\mathbb{Q}_\infty/\mathbb{Q})$ to the units of its completed group ring Λ.

We say that a height one prime ideal $\mathfrak{p}$ of a finite Λ-algebra $\mathbb{T}$ is of weight k (an integer ≥ 1) if $P = \mathfrak{p} \cap \Lambda$ is the kernel of a $\mathbb{Z}_p$-algebra homomorphism $\Lambda \to \bar{\mathbb{Q}}_p$ whose restriction to a finite index subgroup of $1 + p^\nu \mathbb{Z}_p$ is given by $x \mapsto x^{k-1}$. Such an ideal $\mathfrak{p}$ induces a Galois orbit of $\mathbb{Z}_p$-algebra homomorphisms $\mathbb{T} \to \mathbb{T}/\mathfrak{p} \hookrightarrow \bar{\mathbb{Q}}_p$ called specializations in weight k.

By definition a Λ-adic ordinary cuspform of level N (a positive integer not divisible by p) is a formal q-expansion with coefficients in the integral closure of Λ in some finite extension of its fraction field, whose specialization in any weight $k \geq 2$ yield the q-expansion of a p-stabilized, ordinary, normalized cuspform of tame level N and weight k. However, specializations in weight one are not always classical.

The ordinary Hecke algebra $\mathbb{T}_N$ of tame level N is defined as the Λ-algebra generated by the Hecke operators U_ℓ (resp. T_ℓ, $\langle \ell \rangle$) for primes ℓ dividing Np (resp. not dividing Np) acting on the space of Λ-adic ordinary cuspforms of tame level N. Hida proved that $\mathbb{T}_N$ is free of finite rank over Λ and its height one primes of weight $k \geq 2$ are in bijection with the (Galois orbits of) classical ordinary eigenforms of weight k and tame level dividing N.

A Λ-adic ordinary cuspform of level N is said to be N-new if all specializations in weights ≥ 2 are p-stabilized, ordinary cuspforms of tame level N which are N-new.

Define $\mathbb{T}_N^{\text{new}}$ as the quotient of $\mathbb{T}_N$ acting faithfully on the space of Λ-adic ordinary cuspforms of level N, which are N-new. A result of Hida (see [12, Corollaries 3.3 and 3.7]) states that $\mathbb{T}_N^{\text{new}}$ is a finite, reduced, torsion free Λ-algebra, whose height one primes of weight $k \geq 2$ are in bijection with the Galois orbits of classical ordinary eigenforms of weight k and tame level N which are N-new.

4. Primitive Hida families

A primitive Hida family $F = \sum_{n \geq 1} A_n q^n$ of tame level N is by definition a Λ-adic ordinary cuspform, new of level N and which is a normalized eigenform for all the Hecke operators, i.e., a common eigenvector of the operators U_ℓ, T_ℓ and $\langle \ell \rangle$ as above. The relations between coefficients and eigenvalues for the Hecke operators are the usual ones for newforms. One can see from [12, p. 265] that primitive Hida families can be used to write down a basis of the space of Λ-adic ordinary cuspforms in the same fashion as classically newforms can be used to write down a basis of the space of cuspforms.

The central character $\psi_F : (\mathbb{Z}/Np^\nu)^\times \to \mathbb{C}^\times$ of the family is defined by $\psi_F(\ell) = \text{eigenvalue of } \langle \ell \rangle$.

Galois orbits of primitive Hida families of level N are in bijection with the minimal primes of $\mathbb{T} = \mathbb{T}_N^{\text{new}}$. More precisely, a primitive Hida family determines and is uniquely determined by a Λ-algebra homomorphism $\mathbb{T} \to \overline{\text{Frac}(\Lambda)}$, sending each Hecke operator to its eigenvalue on F, whose kernel is a minimal prime $\mathfrak{a} \subset \mathbb{T}$. Since $\mathbb{T}$ is a finite and reduced Λ-algebra, its localization $\mathbb{T}_\mathfrak{a}$ is a finite field extension of $\text{Frac}(\Lambda)$. Hence, we obtain the following homomorphisms of Λ-algebras:

$$\mathbb{T} \twoheadrightarrow \mathbb{T}/\mathfrak{a} \hookrightarrow \widetilde{\mathbb{T}/\mathfrak{a}} \hookrightarrow \mathbb{T}_\mathfrak{a} \xrightarrow{\sim} K_F \subset \overline{\text{Frac}(\Lambda)}, \qquad (1.2)$$

where $\widetilde{\mathbb{T}/\mathfrak{a}}$ denotes the integral closure of the domain $\mathbb{T}/\mathfrak{a}$ in its field of fractions $\mathbb{T}_\mathfrak{a}$. In particular, the image K_F of $\mathbb{T}_\mathfrak{a}$ in $\overline{\mathrm{Frac}(\Lambda)}$ is a finite extension of $\mathrm{Frac}(\Lambda)$ generated by the coefficients of F.

By definition all specializations of F in weight $k \geq 2$ yield p-stabilized, ordinary newforms of tame level N and weight k. In weight one, there are only finitely many classical specializations, unless F has CM by a quadratic field in which p splits (see [9] and [6]). Nevertheless, a theorem of Wiles [18] asserts that any p-stabilized newform of weight one occurs as a specialization of a primitive Hida family.

Given a primitive Hida family $F = \sum_{n \geq 1} A_n q^n$ of level N, Hida constructed in [11, Theorem 2.1] an absolutely irreducible continuous representation:

$$\rho_F : \mathrm{Gal}(\bar{\mathbb{Q}}/\mathbb{Q}) \to \mathrm{GL}_2(K_F), \tag{1.3}$$

unramified outside Np, such that for all ℓ not dividing Np the trace of the image of Frob_ℓ equals A_ℓ. Moreover $\det \rho_F = \psi_F \chi_{\mathrm{cyc}}$. Finally by Wiles [18, Theorem 2.2.2] the space of I_p-coinvariants is a line on which Frob_p acts by A_p.

5. Galois representations

5.1. Minimal primes

The total quotient field of $\mathbb{T}$ is given by $\mathbb{T} \otimes_\Lambda \mathrm{Frac}(\Lambda) \simeq \prod_\mathfrak{a} \mathbb{T}_\mathfrak{a}$ where the product is taken over all minimal primes of $\mathbb{T}$. The representation (1.3) can be rewritten as

$$\rho_\mathfrak{a} : \mathrm{Gal}(\bar{\mathbb{Q}}/\mathbb{Q}) \to \mathrm{GL}_2(\mathbb{T}_\mathfrak{a}) \tag{1.4}$$

and by putting those together we obtain a continuous representation

$$\rho_\mathbb{T} : \mathrm{Gal}(\bar{\mathbb{Q}}/\mathbb{Q}) \to \mathrm{GL}_2(\mathbb{T} \otimes_\Lambda \mathrm{Frac}(\Lambda)) \tag{1.5}$$

unramified outside Np, such that for all ℓ not dividing Np the trace of the image of Frob_ℓ equals T_ℓ. Moreover the space of I_p-coinvariants is free of rank one and Frob_p acts on it as U_p.

5.2. Maximal primes

Since $\mathbb{T}$ is a finite Λ-algebra, it is semi-local, and is isomorphic to the direct product $\prod_\mathfrak{m} \mathbb{T}_\mathfrak{m}$ where the product is taken over all maximal primes. By composing (1.5) with the canonical projection, one obtains:

$$\rho_{\mathfrak{m}} : \mathrm{Gal}(\bar{\mathbb{Q}}/\mathbb{Q}) \to \mathrm{GL}_2(\mathbb{T}_{\mathfrak{m}} \otimes_{\Lambda} \mathrm{Frac}(\Lambda)). \tag{1.6}$$

The composition:

$$\mathrm{Gal}(\bar{\mathbb{Q}}/\mathbb{Q}) \xrightarrow{\mathrm{Tr}(\rho_{\mathfrak{m}})} \mathbb{T}_{\mathfrak{m}} \to \mathbb{T}/\mathfrak{m} \tag{1.7}$$

is a pseudo-character taking values in a field and sending the complex conjugation to 0. By a result of Wiles [18, §2.2] it is the trace of a unique semi-simple representation:

$$\bar{\rho}_{\mathfrak{m}} : \mathrm{Gal}(\bar{\mathbb{Q}}/\mathbb{Q}) \to \mathrm{GL}_2(\mathbb{T}/\mathfrak{m}). \tag{1.8}$$

Note that whereas each minimal prime $\mathfrak{a} \subset \mathbb{T}$ is contained in a unique maximal prime, there may be several minimal primes contained in a given maximal prime $\mathfrak{m}$, those corresponding to primitive Hida families sharing the same residual Galois representation $\bar{\rho}_{\mathfrak{m}}$.

5.3. Galois stable lattices

A lattice over a noetherian domain R (or R-lattice) is a finitely generated R-submodule of a finite dimensional $\mathrm{Frac}(R)$-vector space which spans the latter. This definition extends to a noetherian reduced ring R and its total quotient field $\prod_{\mathfrak{a}} R_{\mathfrak{a}}$, where $\mathfrak{a}$ runs over the (finitely many) minimal primes of R.

The continuity of $\rho_{\mathfrak{a}}$ implies the existence of a Galois stable $\widetilde{\mathbb{T}/\mathfrak{a}}$-lattice in $\mathbb{T}_{\mathfrak{a}}^2$, and similar statements hold for ρ_F, $\rho_{\mathbb{T}}$ and $\rho_{\mathfrak{m}}$. It is worth mentioning that $\rho_{\mathbb{T}}$ cannot necessarily be defined over the normalization of $\mathbb{T}$ in $\prod_{\mathfrak{a}} \mathbb{T}_{\mathfrak{a}}$. In other words $\rho_{\mathfrak{a}}$ does not necessarily stabilize a *free* $\widetilde{\mathbb{T}/\mathfrak{a}}$-lattice. There is an exception: if $K_F = \mathrm{Frac}(\Lambda)$ and $p > 2$ the regularity of Λ implies that ρ_F always admits a Galois stable free Λ-lattice (see [11, §2]).

If $\mathfrak{m}$ is a maximal prime such that the residual Galois representation $\bar{\rho}_{\mathfrak{m}}$ is absolutely irreducible, then by a result of Nyssen [15] and Rouquier [16] $\rho_{\mathfrak{m}}$ stabilizes a free $\mathbb{T}_{\mathfrak{m}}$-lattice. It follows that for every minimal prime $\mathfrak{a} \subset \mathfrak{m}$, the representation $\rho_{\mathfrak{a}}$ stabilizes a free lattice over $\mathbb{T}_{\mathfrak{m}}/\mathfrak{a} = \mathbb{T}/\mathfrak{a}$.

5.4. Height one primes

Let f be a p-stabilized, ordinary, newform of tame level N and weight k. It determines uniquely a height one prime $\mathfrak{p} \subset \mathbb{T}$ and an embedding of $\mathbb{T}_{\mathfrak{p}}/\mathfrak{p}$ into $\bar{\mathbb{Q}}_p$, although not every height one prime of $\mathbb{T}$ of weight one is obtained in this way. Our main interest is in the structure of the Λ_P-algebra $\mathbb{T}_{\mathfrak{p}}$, where $P = \mathfrak{p} \cap \Lambda$. The ring $\mathbb{T}_{\mathfrak{p}}$ is local, noetherian, reduced of Krull dimension 1, but is not necessarily integrally closed. It might even not be a domain, since f

could be a specialization of several, non Galois conjugate, Hida families (see [6, §7.4]), hence there may be several minimal primes $\mathfrak{a}$ of $\mathbb{T}$ contained in $\mathfrak{p}$.

Let $\rho_f : \mathrm{Gal}(\bar{\mathbb{Q}}/\mathbb{Q}) \to \mathrm{GL}_2(\bar{\mathbb{Q}}_p)$ be the continuous irreducible representation attached to f by Deligne when $k \geq 2$ and by (1.1) when $k = 1$ via the fixed embeddings $\mathbb{C} \supset \bar{\mathbb{Q}} \hookrightarrow \bar{\mathbb{Q}}_p$. Since ρ_f is odd, it can be defined over the ring of integers of the subfield of $\bar{\mathbb{Q}}_p$ generated by its coefficients, and hence defines an isomorphic representation:

$$\bar{\rho}_{\mathfrak{p}} : \mathrm{Gal}(\bar{\mathbb{Q}}/\mathbb{Q}) \to \mathrm{GL}_2(\mathbb{T}_{\mathfrak{p}}/\mathfrak{p}), \qquad (1.9)$$

admitting a model over the integral closure of $\mathbb{T}/\mathfrak{p}$ in its field of fractions $\mathbb{T}_{\mathfrak{p}}/\mathfrak{p}$.

The normalization of $\mathbb{T}_{\mathfrak{p}}$ in its total quotient field $\prod_{\mathfrak{a} \subset \mathfrak{p}} \mathbb{T}_{\mathfrak{a}}$ is given by $\prod_{\mathfrak{a} \subset \mathfrak{p}} \widetilde{\mathbb{T}_{\mathfrak{p}}/\mathfrak{a}}$, where $\widetilde{\mathbb{T}_{\mathfrak{p}}/\mathfrak{a}} \simeq \widetilde{(\mathbb{T}/\mathfrak{a})}_{\mathfrak{p}}$ is the integral closure of $\mathbb{T}_{\mathfrak{p}}/\mathfrak{a} \simeq (\mathbb{T}/\mathfrak{a})_{\mathfrak{p}}$ in $\mathbb{T}_{\mathfrak{a}}$.

Denote by $\widehat{\mathbb{T}_{\mathfrak{p}}/\mathfrak{a}}$ the completion of the discrete valuation ring $\widetilde{\mathbb{T}_{\mathfrak{p}}/\mathfrak{a}}$. Note that they share the same residue field which is a finite extension of $\mathbb{T}_{\mathfrak{p}}/\mathfrak{p}$ and that there is a natural bijection between the set of $\widehat{\mathbb{T}_{\mathfrak{p}}/\mathfrak{a}}$-lattices in a given $\mathbb{T}_{\mathfrak{a}}$-vector space V and the set of $\widehat{\mathbb{T}_{\mathfrak{p}}/\mathfrak{a}}$-lattices in $V \otimes_{\mathbb{T}_{\mathfrak{a}}} \mathrm{Frac}(\widehat{\mathbb{T}_{\mathfrak{p}}/\mathfrak{a}})$. Since $\bar{\rho}_{\mathfrak{p}}$ is absolutely irreducible and $\widehat{\mathbb{T}_{\mathfrak{p}}/\mathfrak{a}}$ is local and complete, by a result of Nyssen [15] and Rouquier [16] the representation $\rho_{\mathfrak{a}} \otimes_{\widehat{\mathbb{T}_{\mathfrak{a}}}} \mathrm{Frac}(\widehat{\mathbb{T}_{\mathfrak{p}}/\mathfrak{a}})$ stabilizes a free $\widehat{\mathbb{T}_{\mathfrak{p}}/\mathfrak{a}}$-lattice. The latter lattice yields (by intersection) a free $\widetilde{\mathbb{T}_{\mathfrak{p}}/\mathfrak{a}}$-lattice stable by $\rho_{\mathfrak{a}}$. In other terms there exists a unique, up to conjugacy, continuous representation:

$$\rho_{\mathfrak{p}}^{\mathfrak{a}} : \mathrm{Gal}(\bar{\mathbb{Q}}/\mathbb{Q}) \to \mathrm{GL}_2(\widetilde{\mathbb{T}_{\mathfrak{p}}/\mathfrak{a}}), \qquad (1.10)$$

such that $\rho_{\mathfrak{p}}^{\mathfrak{a}} \otimes_{\widetilde{\mathbb{T}_{\mathfrak{p}}/\mathfrak{a}}} \mathbb{T}_{\mathfrak{a}} \simeq \rho_{\mathfrak{a}}$ and $\rho_{\mathfrak{p}}^{\mathfrak{a}} \mod \mathfrak{p} \simeq \bar{\rho}_{\mathfrak{p}}$.

This representation is a bridge between a form and a family and will be used in Section 6 to transfer properties in both directions.

The exact control theorem for ordinary Hecke algebras, proved by Hida for $p > 2$ and by Ghate–Kumar [8] for $p = 2$, has the following consequence:

Theorem 1.3. [11, Corollary 1.4] *Assume that $k \geq 2$. Then the local algebra $\mathbb{T}_{\mathfrak{p}}$ is etale over the discrete valuation ring Λ_P. In particular, f is a specialization of a unique, up to Galois conjugacy, Hida family corresponding to a minimal prime $\mathfrak{a}$.*

Assume for the rest of this section that $\mathbb{T}_{\mathfrak{p}}$ is a domain. Then the field of fractions of $\mathbb{T}_{\mathfrak{p}}$ is isomorphic to $\mathbb{T}_{\mathfrak{a}}$, where $\mathfrak{a}$ is the unique minimal prime of $\mathbb{T}$ contained in $\mathfrak{p}$. Since normalization and localization commute, we have

$\widetilde{(\mathbb{T}/\mathfrak{a})}_\mathfrak{p} \simeq \widetilde{(\mathbb{T}/\mathfrak{a})}_\mathfrak{p} \simeq \widetilde{\mathbb{T}}_\mathfrak{p}$. Therefore, the collection of representations (1.10) are replaced by a unique, up to conjugacy, continuous representation:

$$\rho_\mathfrak{p} : \mathrm{Gal}(\bar{\mathbb{Q}}/\mathbb{Q}) \to \mathrm{GL}_2(\widetilde{\mathbb{T}}_\mathfrak{p}), \qquad (1.11)$$

such that $\rho_\mathfrak{p} \otimes_{\widetilde{\mathbb{T}}_\mathfrak{p}} \mathbb{T}_\mathfrak{a} \simeq \rho_\mathfrak{a}$ and $\rho_\mathfrak{p} \mod \mathfrak{p} \simeq \bar{\rho}_\mathfrak{p}$.

If we further assume that $\mathbb{T}_\mathfrak{p}$ is etale over Λ_P, then $\mathbb{T}_\mathfrak{p}$ is itself a discrete valuation ring, hence $\mathbb{T}_\mathfrak{p} \simeq \widetilde{\mathbb{T}}_\mathfrak{p}$.

6. Rigidity of the automorphic type in a Hida family

By definition, all specializations in weight at least two of a primitive Hida family F of level N share the same tame level. Also, by [7, Proposition 2.2.4], the tame conductor of ρ_F equals N. The aim of this section is to show that the tame level of all classical weight one specializations of F is also N, and to show that all classical specializations of F (including those of weight one) share the same automorphic type at all primes dividing N.

6.1. Minimally ramified Hida families

Recall that a newform f is said to be minimally ramified if it has minimal level amongst the underlying newforms of all its twists by Dirichlet characters.

Lemma 1.4. *Let F be a primitive Hida family and let χ be a Dirichlet character of conductor prime to p. There exists a unique primitive Hida family F_χ underlying $F \otimes \chi$, in the sense that the p-stabilized, ordinary newform underlying a given specialization of $F \otimes \chi$ can be obtained by specializing F_χ.*

Proof. By [12, p. 250] one can write any Λ-adic ordinary cuspform as a linear combination of translates of primitive Hida families of lower or equal level. Since $F \otimes \chi$ is an eigenform for all but finitely many Hecke operators, it is necessarily a linear combination of translates of the same primitive Hida family, denoted F_χ. It follows that any specialization of F_χ in weight at least two is the p-stabilized, ordinary newform underlying the corresponding specialization of $F \otimes \chi$. $\qquad\square$

Definition 1.5. We say that a primitive Hida family F of level N is minimally ramified if for every Dirichlet character χ of conductor prime to p, the level of F_χ is a multiple of N.

As for newforms, it is clear that any primitive Hida family admits a unique twist which is minimally ramified.

Lemma 1.4 implies that being minimally ramified is pure with respect to specializations in weight at least two, that is to say, all specializations of a minimally ramified primitive Hida family are minimally ramified, and a primitive Hida family admitting a minimally ramified specialization is minimally ramified. This observation together with the classification of the admissible representations of $GL_2(\mathbb{Q}_\ell)$, easily implies:

Lemma 1.6. *Let $F = \sum_{n \geq 1} A_n q^n$ be a minimally ramified, primitive Hida family of level N and let ℓ be a prime dividing N. Denote by $\mathrm{unr}(C)$ the unramified character of G_ℓ sending Frob_ℓ to C.*

(i) *If ψ_F is unramified at ℓ and ℓ^2 does not divide N, then every specialization in weight at least two corresponds to an automorphic form which is special at ℓ. In particular $A_\ell \neq 0$ and the restriction of ρ_F to G_ℓ is an unramified twist of an extension of 1 by $\mathrm{unr}(\ell)$.*

(ii) *If the conductor of ψ_F and N share the same ℓ-part, then every specialization in weight at least two corresponds to an automorphic form which is a ramified principal series at ℓ. In particular $A_\ell \neq 0$ and the restriction of ρ_F to G_ℓ equals $\mathrm{unr}(A_\ell) \oplus \mathrm{unr}(B_\ell)\psi_F$, for some $B_\ell \in K_F$.*

(iii) *In all other cases, every specialization in weight at least two corresponds to an automorphic form which is supercuspidal at ℓ. In particular $A_\ell = 0$ and the restriction of ρ_F to G_ℓ is irreducible.*

6.2. General case

Definition 1.7. Let F be a primitive Hida family of level N and let ℓ be a prime dividing N. We say that F is special (resp. ramified principal series or supercuspidal) at ℓ, if a minimally ramified twist of F falls in case (i) (resp. (ii) or (iii)) of Lemma 1.6.

It follows from Lemma 1.6, that being special, principal series or supercuspidal is pure with respect to specializations, that is to say, all specializations in weight at least two are of the same type. We will now describe the local automorphy type in greater detail and deduce information about classical weight one specializations.

Proposition 1.8. *Let F be a primitive Hida family of level N and let ℓ be a prime dividing N. If F is special at ℓ, so are all its specializations in weight at least two and F does not admit any classical weight one specialization. Otherwise, $\rho_F(I_\ell)$ is a finite group invariant under any classical specialization, including in weight one. More precisely*

(i) *If F is a ramified principal series at ℓ, then the restriction of ρ_F to G_ℓ is isomorphic to $\varphi_\ell \oplus \varphi'_\ell$, where φ_ℓ and φ'_ℓ are characters whose restrictions to inertia have finite order.*

(ii) *If F is supercuspidal at ℓ, then either the restriction of ρ_F to G_ℓ is induced from a character Φ_ℓ of an index two subgroup of G_ℓ whose restriction to inertia has finite order, or $\ell = 2$ and all classical specializations of F are extraordinary supercuspidal representations at 2.*

In particular, all classical weight one specializations of F have tame level N.

Proof. Although parts of the proposition seem to be well-known to experts, for the benefit of the reader, we will give a complete proof.

If F is special at ℓ, then the claim about specializations in weight at least two follows directly from Lemma 1.6(i). Moreover in this case $\rho_F|_{G_\ell}$ is by definition reducible and the quotient of the two characters occurring in its semi-simplification equals $\mathrm{unr}(\ell)$. Since ℓ is not a root of unity, F does not admit any classical weight one specializations.

Suppose now that F is *not* special at ℓ. Since $\ell \neq p$ and ρ_F is continuous, Grothendieck's ℓ-adic monodromy theorem implies that $\rho_F(I_\ell)$ is finite. Let $\mathfrak{p}$ be a height one prime of $\mathbb{T}$ corresponding to a classical cusp form f of weight k, containing the minimal prime $\mathfrak{a}$ defined by F. Denote by L the free rank two $\widetilde{\mathbb{T}_\mathfrak{p}} / \mathfrak{a}$-lattice on which $\rho^\mathfrak{a}_\mathfrak{p}$ acts (see (1.10)). Recall that $\rho^\mathfrak{a}_\mathfrak{p} \mod \mathfrak{p} \simeq \bar{\rho}_\mathfrak{p}$ and consider the natural projection:

$$\rho_F(I_\ell) \simeq \rho^\mathfrak{a}_\mathfrak{p}(I_\ell) \twoheadrightarrow \bar{\rho}_\mathfrak{p}(I_\ell) \simeq \rho_f(I_\ell), \qquad (1.12)$$

which we claim is an isomorphism. In fact, an eigenvalue ζ of an element of the kernel has to be a root of unity since the latter is a finite group, in particular $\zeta \in \bar{\mathbb{Q}}_p$. Since by assumption $(\zeta - 1)^2 \in \mathfrak{p}$, the product of all its G_p-conjugates belongs to $\mathfrak{p} \cap \mathbb{Q}_p$, which is $\{0\}$ because $\mathbb{T}_\mathfrak{p}$ is a $\mathbb{Q}_p$-algebra. Hence $\zeta = 1$ which implies that the kernel is trivial.

The claim (i) follows directly from 1.6(ii), so we can assume for the rest of the proof that F is supercuspidal at ℓ. Denote by W_ℓ the wild inertia subgroup of I_ℓ. Suppose first that $\rho_\mathfrak{a}|_{W_\ell}$ is reducible, isomorphic to $\Phi_\ell \oplus \Phi'_\ell$ with $\Phi_\ell \neq \Phi'_\ell$. Since W_ℓ is normal in G_ℓ, it follows easily that Φ_ℓ extends to an index two subgroup of G_ℓ, as claimed. If $\rho_\mathfrak{a}|_{W_\ell}$ is reducible and isotropic, then by taking an eigenvector for a topological generator of I_ℓ / W_ℓ one sees that $\rho_\mathfrak{a}|_{I_\ell}$ is reducible too, which allows us to conclude as in the previous case. Suppose finally that $\rho_\mathfrak{a}|_{W_\ell}$ is irreducible. Then by a classical result on hyper-solvable groups its image is a dihedral group, hence $\ell = 2$. Assume further $\rho_\mathfrak{a}|_{G_\ell}$ is not dihedral, since this case can be handled as above. Then, any specialization in weight at least two of F yields an eigenform f which is an extraordinary supercuspidal representation at $\ell = 2$. The isomorphism (1.12) implies that all other classical specializations of F are also extraordinary supercuspidal representations at $\ell = 2$ (we refer to [17] and [14, §5.1] for a detailed analysis of this case). □

7. Local structure of the ordinary Hecke algebras at classical weight one points

7.1. A deformation problem

Let f_α be a weight one p-stabilized newform of tame level N as in Section 2. Assume that f is regular at p. By ordinarity the restriction of ρ_f to G_p is a sum of two characters ψ_1 and ψ_2, and by regularity exactly one of those characters, say ψ_1, is the unramified character sending Frob_p to α_p. By (1.9) the Galois representation ρ_f is defined over a finite extension $E = \mathbb{T}_\mathfrak{p}/\mathfrak{p}$ of $\mathbb{Q}_p$, where $\mathfrak{p}$ denotes the height one prime of $\mathbb{T}$ determined by f.

Consider the functor $\mathcal{D}$ sending a local Artinian ring A with maximal ideal $\mathfrak{m}_A$ and residue field $A/\mathfrak{m}_A = E$ to the set of strict equivalence classes of representations $\widetilde{\rho} : \mathrm{Gal}(\bar{\mathbb{Q}}/\mathbb{Q}) \to \mathrm{GL}_2(A)$ such that $\widetilde{\rho} \mod \mathfrak{m}_A \simeq \rho_f$ and fitting in an exact sequence

$$0 \to \widetilde{\psi}_2 \to \widetilde{\rho} \to \widetilde{\psi}_1 \to 0$$

of $A[G_p]$-modules, free over A, and such that $\widetilde{\psi}_1$ is an unramified character whose reduction modulo $\mathfrak{m}_A$ equals ψ_1. We define $\mathcal{D}'$ as the subfunctor of $\mathcal{D}$ consisting of deformation with constant determinant. Finally, define $\mathcal{D}_{\min}$ (resp. $\mathcal{D}'_{\min}$) as the subfunctor of $\mathcal{D}$ (resp. $\mathcal{D}'$) of deformations $\widetilde{\rho}$ such that for all ℓ dividing N such that $a_\ell \neq 0$, the I_ℓ-invariants in $\widetilde{\rho}$ are a free A-module of rank one.

The functors $\mathcal{D}$ and $\mathcal{D}_{\min}$ are pro-representable by local noetherian complete E-algebras $\mathcal{R}$ and $\mathcal{R}_{\min}$, while $\mathcal{D}'$ and $\mathcal{D}'_{\min}$ are representable by local Artinian E-algebras $\mathcal{R}'$ and $\mathcal{R}'_{\min}$. Denote by $t_\mathcal{D}$ the tangent space of $\mathcal{D}$, etc.

Using the interpretation of $t_\mathcal{D}$ and $t_{\mathcal{D}'}$ in terms of Galois cohomology groups, the main technical result of [1] states:

Theorem 1.9. *If f is regular at p, then $\dim t_\mathcal{D} = 1$. If we further assume that f does not have RM by a quadratic field in which p splits, then $t_{\mathcal{D}'} = 0$.*

7.2. Modular deformations

Denote by $\widehat{\mathbb{T}_\mathfrak{p}}$ the completion of $\mathbb{T}_\mathfrak{p}$ The composition:

$$\mathrm{Gal}(\bar{\mathbb{Q}}/\mathbb{Q}) \xrightarrow{\mathrm{Tr}(\rho_\mathbb{T})} \mathbb{T} \to \mathbb{T}_\mathfrak{p} \to \widehat{\mathbb{T}_\mathfrak{p}} \tag{1.13}$$

is a two dimensional pseudo-character taking values in a complete local ring and whose reduction modulo the maximal ideal is the trace of the absolutely irreducible representation $\bar{\rho}_{\mathfrak{p}}$. By a result of Nyssen [15] and Rouquier [16] the pseudo-character (1.13) is the trace of a two dimensional irreducible representation:

$$\widehat{\rho}_{\mathfrak{p}} : \operatorname{Gal}(\bar{\mathbb{Q}}/\mathbb{Q}) \to \operatorname{GL}_2(\widehat{\mathbb{T}_{\mathfrak{p}}}). \qquad (1.14)$$

This representation contains more information than the collection $(\rho_{\mathfrak{p}}^{\mathfrak{a}})_{\mathfrak{a} \subset \mathfrak{p}}$ and plays a central role in the analysis of the p-adic deformations of f.

Note that $\widehat{\Lambda_P}$ is formal power series over its residue field $\widehat{\Lambda_P}/P \simeq \Lambda_P/P$ which is a finite extension of $\mathbb{Q}_p$. Consider the local Artinian $\mathbb{Q}_p$-algebra

$$\mathcal{T} = \widehat{\mathbb{T}_{\mathfrak{p}}} \otimes_{\widehat{\Lambda_P}} \widehat{\Lambda_P}/P. \qquad (1.15)$$

Reducing (1.14) modulo P yields a continuous representation:

$$\rho_{\mathcal{T}} : \operatorname{Gal}(\bar{\mathbb{Q}}/\mathbb{Q}) \to \operatorname{GL}_2(\mathcal{T}), \qquad (1.16)$$

such that $\det(\rho_{\mathcal{T}}) = \det(\rho_f)$. In fact, for all $\ell \in (\mathbb{Z}/Np)^\times$, the image of $\langle \ell \rangle$ in $\mathcal{T}$ is given by the henselian lift of its image in $\mathcal{T}/\mathfrak{p}$, hence is fixed.

As in [1], one can describe the local behavior of $\widehat{\rho}_{\mathfrak{p}}$ at bad primes.

Proposition 1.10. (i) *For all ℓ dividing N such that $a_\ell \neq 0$, the space of I_ℓ-invariants in $\widehat{\rho}_{\mathfrak{p}}$ is free of rank one and Frob_ℓ acts on it as U_ℓ.*
(ii) *Assume that f is regular at p. Then $\widehat{\rho}_{\mathfrak{p}}$ is ordinary, in the sense that the space of I_p-coinvariants is free of rank one and Frob_p acts on it as U_p.*

By Proposition 1.10 the Galois representation (1.14) (resp. (1.16)) defines a point of $\mathcal{D}_{\min}$ (resp. of $\mathcal{D}'_{\min}$). One deduces the following surjective homomorphisms of local reduced Λ_P-algebras (resp. local Artinian E-algebras):

$$\begin{aligned} \mathcal{R} \twoheadrightarrow \mathcal{R}_{\min} \twoheadrightarrow \widehat{\mathbb{T}_{\mathfrak{p}}}, \text{ and} \\ \mathcal{R}' \twoheadrightarrow \mathcal{R}'_{\min} \twoheadrightarrow \mathcal{T}. \end{aligned} \qquad (1.17)$$

7.3. Smoothness and etaleness

Lemma 1.11. *The Λ_P-algebra $\mathbb{T}_{\mathfrak{p}}$ is etale if, and only if, $\mathcal{T}$ is a field.*

Proof. Since $\mathbb{T}$ is flat over Λ, so is $\mathbb{T} \otimes_\Lambda \Lambda_P$ over Λ_P. The algebra $\mathbb{T} \otimes_\Lambda \Lambda_P$ is unramified over Λ_P if, and only if, $\mathbb{T} \otimes_\Lambda \Lambda_P/P$ is unramified over Λ_P/P, that is to say is a product of fields. Since $\mathbb{T} \otimes_\Lambda \Lambda_P = \prod_{\mathfrak{p} \cap \Lambda = P} \mathbb{T}_{\mathfrak{p}}$, we have

$$\prod_{\mathfrak{p} \cap \Lambda = P} \mathbb{T}_{\mathfrak{p}} \otimes_{\Lambda_P} \Lambda_P/P = \mathbb{T} \otimes_\Lambda \Lambda_P/P \simeq \mathbb{T} \otimes_\Lambda \widehat{\Lambda_P}/P = \prod_{\mathfrak{p} \cap \Lambda = P} \widehat{\mathbb{T}_{\mathfrak{p}}} \otimes_{\widehat{\Lambda_P}} \widehat{\Lambda_P}/P.$$

One deduces that $\mathbb{T}_\mathfrak{p}$ is unramified over Λ_P if, and only if, $\widehat{\mathbb{T}_\mathfrak{p}}$ is unramified over $\widehat{\Lambda_P}$ if, and only if, $\mathcal{T} = \widehat{\mathbb{T}_\mathfrak{p}} \otimes_{\widehat{\Lambda_P}} \widehat{\Lambda_P}/P$ is a field. $\square$

Proposition 1.12. *Suppose that f is regular at p. Then $\mathbb{T}_\mathfrak{p}$ is a discrete valuation ring and the homomorphisms in (1.17) are isomorphisms. Moreover, if f does not have RM by a quadratic field in which p splits, then $\mathbb{T}_\mathfrak{p}$ is etale over Λ_P, and otherwise, under the additional assumption that $\dim_E \mathcal{R}' \leq 2$, the ramification index of $\mathbb{T}_\mathfrak{p}$ over Λ_P equals 2.*

Proof. Since f is regular at p, Theorem 1.9 implies that $\dim t_\mathcal{D} = 1$. Since $\dim \widehat{\mathbb{T}_\mathfrak{p}} > 0$, one deduces that the natural surjective homomorphism $\mathcal{R} \twoheadrightarrow \widehat{\mathbb{T}_\mathfrak{p}}$ is an isomorphism of discrete valuation rings, hence $\mathbb{T}_\mathfrak{p}$ is a discrete valuation ring too. By reducing the isomorphism $\mathcal{R} \simeq \widehat{\mathbb{T}_\mathfrak{p}}$ modulo P we obtain the isomorphism $\mathcal{R}' \simeq \mathcal{T}$.

If $t_{\mathcal{D}'} = 0$, by Nakayama's lemma the structural homomorphisms $E \to \mathcal{R}'$ and $\widehat{\Lambda_P} \to \mathcal{R}$ are isomorphisms. By Lemma 1.11, it follows that $\mathbb{T}_\mathfrak{p}$ is etale over Λ_P, as claimed.

Assume now that $\dim t_{\mathcal{D}'} = 1$, in which case Theorem 1.9 implies that f has RM by a quadratic field in which p splits. Since $\dim_E \mathcal{R}' \leq 2$ by assumption and $\dim_E \mathcal{T} \geq 2$ by [6, Proposition 2.2.4], we deduce that the ramification index is 2. $\square$

Remark 1.13. Cho and Vatsal [3] have proved that $\dim_E \mathcal{R}' \leq 2$ under some additional assumptions, and their method is expected to continue to work under the only assumption of regularity at p.

7.4. Reduced Hecke algebras

Define the reduced ordinary Hecke algebra $\mathbb{T}' = \mathbb{T}'_N$ of tame level N as the subalgebra of $\mathbb{T}_N$ generated over Λ by the Hecke operators U_p, T_ℓ and $\langle \ell \rangle$ for primes ℓ not dividing Np. By the theory of newforms, the natural composition:

$$\mathbb{T}'_N \to \mathbb{T}_N \to \prod_{N'|N} \mathbb{T}^{\mathrm{new}}_{N'}, \tag{1.18}$$

is injective, in particular $\mathbb{T}'_N$ is reduced.

A classical result from Hida theory says that the localization of (1.18) at any height one prime of weight at least two yields an isomorphism. Let $\mathfrak{p}$ be a height one prime $\mathbb{T}$ corresponding to a p-stabilized classical weight one eigenform f_α and denote by $\mathfrak{p}'$ the corresponding height one prime of $\mathbb{T}'$.

Corollary 1.14. *Suppose that f is regular at p. Then the localization of (1.18) yields an isomorphism $\mathbb{T}'_{\mathfrak{p}'} \simeq \mathbb{T}_\mathfrak{p}$.*

Proof. Proposition 1.8 implies $\mathbb{T}^{\text{new}}_{N',\mathfrak{p}} = \{0\}$ for all $N' < N$, hence localizing (1.18) at $\mathfrak{p}$ yields an injective homomomorphism $\mathbb{T}'_{\mathfrak{p}'} \hookrightarrow \mathbb{T}_{\mathfrak{p}}$. Since $\mathbb{T}'_{\mathfrak{p}}$ and $\mathbb{T}_{\mathfrak{p}}$ are finite over the Zariski ring Λ_P, it is enough to check the surjectivity after completion. This follows from Proposition 1.12, since the surjective homomorphism $\mathcal{R} \twoheadrightarrow \widehat{\mathbb{T}_{\mathfrak{p}}}$ factors through $\widehat{\mathbb{T}'_{\mathfrak{p}'}}$. $\qquad\qquad\square$

We will conclude this chapter, by giving a partial answer to the original question that motivated this research.

Corollary 1.15. *Let f be a classical weight one cuspidal eigenform form which is regular at p. Then there exists a unique Hida family specializing to f_α. In particular, if f has CM by a quadratic field in which p splits, then the family has also CM by the same field.*

References

[1] J. BELLAÏCHE AND M. DIMITROV. On the Eigencurve at classical weight one points, arXiv:1301.0712, submitted.

[2] K. BUZZARD, Eigenvarieties, in *L-functions and Galois Representations*, vol. 320 of London Math. Soc. Lecture Note Ser., Cambridge University Press, Cambridge, 2007, pp. 59–120.

[3] S. CHO AND V. VATSAL, Deformations of induced Galois representations, *J. Reine Angew. Math.*, 556 (2003), 79–98.

[4] R. COLEMAN AND B. MAZUR, The eigencurve, in *Galois Representations in Arithmetic Algebraic Geometry* (Durham, 1996), vol. 254 of London Math. Soc. Lecture Note Ser., Cambridge University Press, Cambridge, 1998, pp. 1–113.

[5] P. DELIGNE AND J.-P. SERRE, Formes modulaires de poids 1, *Ann. Sci. École Norm. Sup.* 7 (4), (1974), 507–530.

[6] M. DIMITROV AND E. GHATE, On classical weight one forms in Hida families, *J. Théor. Nombres Bordeaux*, 24, 3 (2012), 669–690

[7] M. EMERTON, R. POLLACK AND T. WESTON, Variation of Iwasawa invariants in Hida families, *Invent. Math.*, 163 (2006), 523–580.

[8] E. GHATE AND N. KUMAR, Control theorems for ordinary 2-adic families of modular forms, to appear in the *Proceedings of the International Colloquium on Automorphic Representations and L-functions*, TIFR, 2012.

[9] E. GHATE AND V. VATSAL, On the local behaviour of ordinary Λ-adic representations, *Ann. Inst. Fourier, Grenoble*, 54, 7 (2004), 2143–2162.

[10] R. GREENBERG AND V. VATSAL, Iwasawa theory for Artin representations, in preparation.

[11] H. HIDA, Galois representations into $GL_2(\mathbf{Z}_p[[X]])$ attached to ordinary cusp forms, *Invent. Math.*, 85 (1986), 545–613.

[12] ———, Iwasawa modules attached to congruences of cusp forms, *Ann. Sci. Ecole Norm. Sup.* 19 (4), (1986), 231–273.

[13] M. KISIN, Overconvergent modular forms and the Fontaine-Mazur conjecture, *Invent. Math.*, 153 (2003), 373–454.

[14] P. KUTZKO, The Langlands conjecture for Gl_2 of a local field, *Ann. of Math.* 112
      (2), (1980), pp. 381–412.
[15] L. NYSSEN, Pseudo-représentations, *Math. Ann.*, 306 (1996), 257–283.
[16] R. ROUQUIER, Caractérisation des caractères et pseudo-caractères, *J. Algebra*,
      180 (1996), 571–586.
[17] A. WEIL, Exercices dyadiques, *Invent. Math.*, 27 (1974), 1–22.
[18] A. WILES, On ordinary λ-adic representations associated to modular forms,
      Invent. Math., 94 (1988), 529–573.

2

Vector bundles on curves and p-adic Hodge theory

Laurent Fargues and Jean-Marc Fontaine

Contents

Introduction

This text is an introduction to our work [12] on curves and vector bundles in p-adic Hodge theory. This is a more elaborate version of the reports [13] and [14]. We give a detailed construction of the "fundamental curve of p-adic Hodge theory" together with sketches of proofs of the main properties of the objects showing up in the theory. Moreover, we explain thoroughly the classification theorem for vector bundles on the curve, giving a complete proof for rank two vector bundles. The applications to p-adic Hodge theory, the theorem "weakly admissible implies admissible" and the p-adic monodromy theorem, are not given here but can be found in [13] and [14].

We would like to thank the organizers of the EPSRC Durham Symposium "Automorphic forms and Galois representations" for giving us the opportunity to talk about this subject.

Automorphic Forms and Galois Representations, ed. Fred Diamond, Payman L. Kassaei and Minhyong Kim. Published by Cambridge University Press. © Cambridge University Press 2014.

17

1. Holomorphic functions of the variable π

1.1. Background on holomorphic functions in a p-adic punctured disk after Lazard ([25])

1.1.1. The Frechet algebra B

Let F be a complete non-archimedean field for a non trivial valuation

$$v : F \longrightarrow \mathbb{R} \cup \{+\infty\},$$

with characteristic p residue field. We note $|\cdot| = p^{-v(\cdot)}$ the associated absolute value. Consider the open punctured disk

$$\mathbb{D}^* = \{0 < |z| < 1\} \subset \mathbb{A}_F^1$$

as a rigid analytic space over F, where z is the coordinate on the affine line. If $I \subset]0, 1[$ is a compact interval set

$$\mathbb{D}_I = \{|z| \in I\} \subset \mathbb{D}^*,$$

an annulus that is an affinoïd domain in $\mathbb{D}^*$ if $I = [\rho_1, \rho_2]$ with $\rho_1, \rho_2 \in \sqrt{|F^\times|}$. There is an admissible affinoïd covering

$$\mathbb{D}^* = \bigcup_{I \subset]0,1[} \mathbb{D}_I$$

where I goes through the preceding type of compact intervals.

Set now

$$B = \mathcal{O}(\mathbb{D}^*)$$
$$= \left\{ \sum_{n \in \mathbb{Z}} a_n z^n \mid a_n \in F, \ \forall \rho \in]0, 1[\ \lim_{|n| \to +\infty} |a_n| \rho^n = 0 \right\}$$

the ring of holomorphic functions on $\mathbb{D}^*$. In the preceding description of B, one checks the infinite set of convergence conditions associated to each $\rho \in]0, 1[$ can be rephrased in the following two conditions

$$\begin{cases} \liminf\limits_{n \to +\infty} \frac{v(a_n)}{n} \geq 0 \\ \lim\limits_{n \to +\infty} \frac{v(a_{-n})}{n} = +\infty. \end{cases} \tag{2.1}$$

For $\rho \in]0, 1[$ and $f = \sum_n a_n z^n \in B$ set

$$|f|_\rho = \sup_{n \in \mathbb{Z}} \{|a_n| \rho^n\}.$$

If $\rho = p^{-r}$ with $r > 0$ one has $|f|_\rho = p^{-v_r(f)}$ with

$$v_r(f) = \inf_{n \in \mathbb{Z}} \{v(a_n) + nr\}.$$

Then $|\cdot|_\rho$ is the Gauss supremum norm on the annulus $\{|z| = \rho\}$. It is in fact a *multiplicative* norm, that is to say v_r *is a valuation*. Equipped with the set of norms $(|\cdot|_\rho)_{\rho\in]0,1[}$, B is a Frechet algebra. The induced topology is the one of uniform convergence on compact subsets of the Berkovich space associated to $\mathbb{D}^*$. If $I \subset]0, 1[$ is a compact interval then

$$\mathrm{B}_I = \mathcal{O}(\mathbb{D}_I)$$

equipped with the set of norms $(|\cdot|_\rho)_{\rho\in I}$ is a Banach algebra. In fact if $I = [\rho_1, \rho_2]$ then by the maximum modulus principle, for $f \in \mathrm{B}_I$

$$\sup_{\rho\in I} |f_\rho| = \sup\{|f|_{\rho_1}, |f|_{\rho_2}\}.$$

One then has

$$\mathrm{B} = \varprojlim_{I\subset]0,1[} \mathrm{B}_I$$

as a Frechet algebra written as a projective limit of Banach algebras.

For $f = \sum_n a_n z^n \in \mathrm{B}$ one has

$$|f|_1 = \lim_{\rho\to 1} |f|_\rho = \sup_{n\in\mathbb{Z}} |a_n| \in [0, +\infty].$$

We will later consider the following closed sub-$\mathcal{O}_F$-algebra of B

$$\begin{aligned}
\mathrm{B}^+ &= \{f \in \mathrm{B} \mid \|f\|_1 \le 1\} \\
&= \Big\{ \sum_{n\in\mathbb{Z}} a_n z^n \in \mathrm{B} \mid a_n \in \mathcal{O}_F \Big\} \\
&= \Big\{ \sum_{n\in\mathbb{Z}} a_n z^n \mid a_n \in \mathcal{O}_F, \ \lim_{n\to+\infty} \frac{v(a_{-n})}{n} = +\infty \Big\}.
\end{aligned}$$

Set now

$$\begin{aligned}
\mathrm{B}^b &= \{f \in \mathrm{B} \mid \exists N \in \mathbb{N}, \ z^N f \in \mathcal{O}(\mathbb{D}) \text{ and is bounded on } \mathbb{D}\} \\
&= \Big\{ \sum_{n\gg-\infty} a_n z^n \mid a_n \in F, \ \exists C \ \forall n \ |a_n| \le C \Big\}.
\end{aligned}$$

This is a dense sub-algebra of B. In particular *one can find back* B *from* B^b *via completion with respect to the norms* $(|\cdot|_\rho)_{\rho\in]0,1[}$. In the same way

$$\begin{aligned}
\mathrm{B}^{b,+} &= \mathrm{B}^b \cap \mathrm{B}^+ \\
&= \Big\{ \sum_{n\gg-\infty} a_n z^n \mid a_n \in \mathcal{O}_F \Big\} \\
&= \mathcal{O}_F[\![z]\!][\tfrac{1}{z}]
\end{aligned}$$

is dense in B^+ and thus *one can find back* B^+ *from* $\mathrm{B}^{b,+}$ *via completion*.

1.1.2. Zeros and growth of holomorphic functions

Recall the following properties of holomorphic functions overs $\mathbb{C}$. Let f be holomorphic on the open disk of radius 1 (one could consider the punctured disk but we restrict to this case to simplify the exposition). For $\rho \in [0, 1[$ set

$$M(\rho) = \sup_{|z|=\rho} |f(z)|.$$

The following properties are verified:

- the function $\rho \mapsto -\log M(\rho)$ is a concave function of $\log \rho$ (Hadamard),
- if $f(0) \neq 0$, f has no zeros on the circle of radius ρ and $(a_1, \ldots, a_n)$ are its zeros counted with multiplicity in the disk of radius ρ, then as a consequence of Jensen's formula

$$-\log |f(0)| \geq \sum_{i=1}^{n} (-\log |a_i|) - n\rho - \log M(\rho).$$

In the non-archimedean setting we have an exact formula linking the growth of an holomorphic function and its zeros. For this, recall the formalism of the Legendre transform. Let

$$\varphi : \mathbb{R} \longrightarrow]-\infty, +\infty]$$

be a convex decreasing function. Define the Legendre transform of φ as the concave function (see Figure 1)

$$\mathscr{L}(\varphi) :]0, +\infty[\longrightarrow [-\infty, +\infty[$$
$$\lambda \longmapsto \inf_{x \in \mathbb{R}} \{\varphi(x) + \lambda x\}.$$

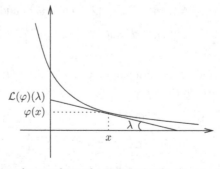

Figure 1 The Legendre transform of φ evaluated at the slope λ where by definition the slope is the opposite of the derivative (we want the slopes to be the valuations of the roots for Newton polygons).

If φ_1, φ_2 are convex decreasing functions as before define $\varphi_1 * \varphi_2$ as the convex decreasing function defined by

$$(\varphi_1 * \varphi_2)(x) = \inf_{a+b=x} \{\varphi_1(a) + \varphi_2(b)\}.$$

We have the formula

$$\mathscr{L}(\varphi_1 * \varphi_2) = \mathscr{L}(\varphi_1) + \mathscr{L}(\varphi_2).$$

One can think of the Legendre transform as being a "tropicalization" of the Laplace transform:

$$(\mathbb{R}, +, \times) \xrightarrow{\text{tropicalization}} (\mathbb{R}, \inf, +)$$

Laplace transform $\rightsquigarrow$ Legendre transform

usual convolution $*$ $\rightsquigarrow$ tropical $*$ just defined.

The function φ is a polygon, that is to say piecewise linear, if and only if $\mathscr{L}(\varphi)$ is a polygon. Moreover in this case:

- the slopes of $\mathscr{L}(\varphi)$ are the x-coordinates of the breakpoints of φ,
- the x-coordinates of the breakpoints of $\mathscr{L}(\varphi)$ are the slopes of φ.

Thus $\mathscr{L}$ and its inverse give a duality

slopes $\longleftrightarrow$ x-coordinates of breakpoints.

From these considerations one deduces that if φ_1 and φ_2 are convex decreasing polygons such that $\forall i = 1, 2, \ \forall \lambda \in]0, +\infty[, \ \mathscr{L}(\varphi_i)(\lambda) \neq -\infty$ then the slopes of $\varphi_1 * \varphi_2$ are obtained by concatenation from the slopes of φ_1 and the ones of φ_2.

For $f = \sum_{n \in \mathbb{Z}} a_n z^n \in B$ set now

$$\text{Newt}(f) = \text{decreasing convex hull of } \{(n, v(a_n))\}_{n \in \mathbb{Z}}.$$

This is a polygon with integral x-coordinate breakpoints. Moreover the function $r \mapsto v_r(f)$ defined on $]0, +\infty[$ is the Legendre transform of Newt(f).

Then, the statement of the "p-adic Jensen formula" is the following: the slopes of Newt(f) are the valuations of the zeros of f (with multiplicity).

Example 2.1. Take $f \in \mathcal{O}(\mathbb{D})$, $f(0) \neq 0$. Let $\rho = p^{-r} \in]0, 1[$ and $(a_1, \ldots, a_n)$ be the zeros of f in the ball $\{|z| \leq \rho\}$ counted with multiplicity. Then, as a consequence of the fact that $r \mapsto v_r(f)$ is the Legendre transform

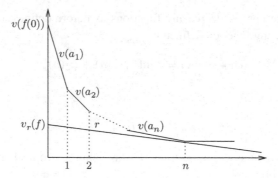

Figure 2 An illustration of the "p-adic Jensen formula". The numbers over each line are their slopes.

of a polygon whose slopes are the valuations of the roots of f, one has the formula

$$v(f(0)) = v_r(f) - nr + \sum_{i=1}^{n} v(a_i)$$

(see Figure 2).

Finally, remark that B^+ is characterized in terms of Newton polygons:

$$B^+ = \{f \in B \mid \text{Newt(f)} \subset \text{upper half plane } y \geq 0\}.$$

1.1.3. Weierstrass products

For a compact interval $I = [\rho_1, \rho_2] \subset]0, 1[$ with $\rho_1, \rho_2 \in \sqrt{|F^{\times}|}$ the ring B_I is a P.I.D. with $\text{Spm}(B_I) = |\mathbb{D}_I|$. In particular $\text{Pic}(\mathbb{D}_I) = 0$. Now let's look at $\text{Pic}(\mathbb{D}^*)$. In fact in the following we will only be interested in the submonoid of effective line bundles

$$\text{Pic}^+(\mathbb{D}^*) = \{[\mathscr{L}] \in \text{Pic}(\mathbb{D}^*) \mid H^0(\mathbb{D}^*, \mathscr{L}) \neq 0\}.$$

Set

$$\text{Div}^+(\mathbb{D}^*) = \left\{ D = \sum_{x \in |\mathbb{D}^*|} m_x[x] \mid m_x \in \mathbb{N}, \ \forall I \subset]0, 1[\right.$$

$$\left. \text{compact } \text{supp}(D) \cap \mathbb{D}_I \text{ is finite} \right\}$$

the monoid of effective divisors on $\mathbb{D}^*$. There is an exact sequence

$$0 \to B \setminus \{0\}/B^{\times} \xrightarrow{\text{div}} \text{Div}^+(\mathbb{D}^*) \longrightarrow \text{Pic}^+(\mathbb{D}^*) \longrightarrow 0.$$

We are thus led to the question: for $D \in \text{Div}^+(\mathbb{D}^*)$, does there exist $f \in B \setminus \{0\}$ such that $\text{div}(f) = D$?

This is of course the case if supp(D) is finite. Suppose thus it is infinite. We will suppose moreover F is algebraically closed (the discrete valuation case is easier but this is not the case we are interested in) and thus $|\mathbb{D}^*| = \mathfrak{m}_F \setminus \{0\}$ where $\mathfrak{m}_F$ is the maximal ideal of $\mathcal{O}_F$.

Suppose first there exists $\rho_0 \in]0, 1[$ such that supp(D) $\subset \{0 < |z| \leq \rho_0\}$. Then we can write

$$D = \sum_{i \geq 0}[a_i], \ a_i \in \mathfrak{m}_F \setminus \{0\}, \ \lim_{i \to +\infty} |a_i| = 0.$$

The infinite product

$$\prod_{i=0}^{+\infty} \left(1 - \frac{a_i}{z}\right),$$

converges in the Frechet algebra B and its divisor is D.

We are thus reduced to the case supp(D) $\subset \{\rho_0 < |z| < 1\}$ for some $\rho_0 \in]0, 1[$. But if we write

$$D = \sum_{i \geq 0}[a_i], \ \lim_{i \to +\infty} |a_i| = 1$$

then neither of the infinite products "$\prod_{i>0}\left(1 - \frac{a_i}{z}\right)$" or "$\prod_{i \geq 0} 1 - \frac{z}{a_i}$" converges. Recall that over $\mathbb{C}$ this type of problem is solved by introducing renormalization factors. Typically, if we are looking for a holomorphic function f on $\mathbb{C}$ such that $\operatorname{div}(f) = \sum_{n \in \mathbb{N}}[-n]$ then the product "$z \prod_{n \in \mathbb{N}}\left(1 + \frac{z}{n}\right)$" does not converge but $z \prod_{n \in \mathbb{N}}\left[\left(1 + \frac{z}{n}\right)e^{-\frac{z}{n}}\right] = \frac{1}{e^{\gamma z}\Gamma(z)}$ does. In our non-archimedean setting this problem has been solved by Lazard.

Theorem 2.2 (Lazard [25]). *If F is spherically complete then there exists a sequence $(h_i)_{i \geq 0}$ of elements of $B^\times$ such that the product $\prod_{i \geq 0}[(z - a_i).h_i]$ converges.*

Thus, if F is spherically complete $\operatorname{Pic}^+(\mathbb{D}^*) = 0$ (and in fact $\operatorname{Pic}(\mathbb{D}^*) = 0$). In the preceding problem, it is easy to verify that for any F there always exist renormalization factors $h_i \in B \setminus \{0\}$ such that $\prod_{i \geq 0}[(z - a_i).h_i]$ converges and thus an $f \in B \setminus \{0\}$ such that $\operatorname{div}(f) \geq D$. The difficulty is thus to introduce renormalization factors that do not add any new zero.

Let's conclude this section with a trick that sometimes allows us not to introduce any renormalization factors. Over $\mathbb{C}$ this is the following. Suppose we are looking for a holomorphic function on $\mathbb{C}$ whose divisor is $\sum_{n \in \mathbb{Z}}[n]$. The infinite product "$z \prod_{n \in \mathbb{Z}\setminus\{0\}}\left(1 - \frac{z}{n}\right)$" does not converge. Nevertheless, regrouping the terms, the infinite product $z \prod_{n \geq 1}\left[\left(1 - \frac{z}{n}\right)\left(1 + \frac{z}{n}\right)\right] = \frac{\sin \pi z}{\pi}$ converges.

In the non-archimedean setting there is a case where this trick works. This is the following. Suppose $E|\mathbb{Q}_p$ is a finite extension and $\overline{E}$ is an algebraic closure of E. Let $\mathcal{LT}$ be a Lubin–Tate group law over $\mathcal{O}_E$. Its logarithm $\log_{\mathcal{LT}}$ is a rigid analytic function on the open disk $\mathbb{D}$ with zeros the torsion points of the Lubin–Tate group law,

$$\mathcal{LT}[\pi^\infty] = \{x \in \mathfrak{m}_{\overline{E}} \mid \exists n \geq 1, \; [\pi^n]_{\mathcal{LT}}(x) = 0\}.$$

The infinite product

$$z \prod_{\zeta \in \mathcal{LT}[\pi^\infty]\setminus\{0\}} 1 - \frac{z}{\zeta}$$

does not converge since in this formula $|\zeta| \to 1$. Nevertheless, the infinite product

$$z \prod_{n \geq 1} \left[\prod_{\zeta \in \mathcal{LT}[\pi^n]\setminus\mathcal{LT}[\pi^{n-1}]} \left(1 - \frac{z}{\zeta} \right) \right]$$

converges in the Frechet algebra of holomorphic functions on $\mathbb{D}$ and equals $\log_{\mathcal{LT}}$. This is just a reformulation of the classical formula

$$\log_{\mathcal{LT}} = \lim_{n \to +\infty} \pi^{-n} [\pi^n]_{\mathcal{LT}}.$$

1.2. Analytic functions in mixed characteristic

1.2.1. The rings B and B⁺

Let E be a local field with uniformizing element π and finite residue field $\mathbb{F}_q$. Thus, either E is a finite extension of $\mathbb{Q}_p$ or $E = \mathbb{F}_q((\pi))$. Let $F|\mathbb{F}_q$ be a valued complete extension for a non trivial valuation $v : F \to \mathbb{R} \cup \{+\infty\}$. Suppose moreover F is perfect (in particular v is not discrete).

Let $\mathcal{E}|E$ be the unique complete unramified extension of E inducing the extension $F|\mathbb{F}_q$ on the residue fields, $\mathcal{O}_{\mathcal{E}}/\pi\mathcal{O}_{\mathcal{E}} = F$. There is a Teichmüller lift $[\cdot] : F \to \mathcal{O}_{\mathcal{E}}$ and

$$\mathcal{E} = \left\{ \sum_{n \gg -\infty} [x_n]\pi^n \mid x_n \in F \right\} \quad \text{(unique writing)}.$$

If $\mathrm{char}\, E = p$ then $[\cdot]$ is additive, $\mathcal{E}|F$, and $\mathcal{E} = F((\pi))$. If $E|\mathbb{Q}_p$ then

$$\mathcal{E} = W_{\mathcal{O}_E}(F)[\tfrac{1}{\pi}] = W(F) \otimes_{W(\mathbb{F}_q)} E$$

the ramified Witt vectors of F. There is a Frobenius φ acting on $\mathcal{E}$,

$$\varphi\left(\sum_n [x_n]\pi^n \right) = \sum_n [x_n^q]\pi^n.$$

If $E|\mathbb{Q}_p$ then on $W(F) \otimes_{W(\mathbb{F}_q)} E$ one has $\varphi = \varphi_{\mathbb{Q}_p}^f \otimes \mathrm{Id}$ where in this formula $q = p^f$ and $\varphi_{\mathbb{Q}_p}$ is the usual Frobenius of the Witt vectors. In this case the addition law of $W_{\mathcal{O}_E}(F)$ is given by

$$\sum_{n \geq 0} [x_n]\pi^n + \sum_{n \geq 0} [y_n]\pi^n = \sum_{n \geq 0} [P_n(x_0, \ldots, x_n, y_0, \ldots, y_n)]\pi^n \quad (2.2)$$

where $P_n \in \mathbb{F}_q[X_i^{q^{i-n}}, Y_j^{q^{j-n}}]_{0 \leq i, j \leq n}$ are generalized polynomials. The multiplication law is given in the same way by such kind of generalized polynomials.

Definition 2.3. (1) Define

$$B^b = \left\{ \sum_{n \gg -\infty} [x_n]\pi^n \in \mathcal{E} \mid \exists C, \forall n \ |x_n| \leq C \right\}$$

$$B^{b,+} = \left\{ \sum_{n \gg -\infty} [x_n]\pi^n \in \mathcal{E} \mid x_n \in \mathcal{O}_F \right\}$$

$$= W_{\mathcal{O}_E}(\mathcal{O}_F)[\tfrac{1}{\pi}] \text{ if } E|\mathbb{Q}_p$$

$$= \mathcal{O}_F[\![\pi]\!][\tfrac{1}{\pi}] \text{ if } E = \mathbb{F}_q((\pi)).$$

(2) For $x = \sum_n [x_n]\pi^n \in B^b$ and $r \geq 0$ set

$$v_r(x) = \inf_{n \in \mathbb{Z}} \{v(x_n) + nr\}.$$

If $\rho = q^{-r} \in]0, 1]$ set $|x|_\rho = q^{-v_r(x)}$.
(3) For $x = \sum_n [x_n]\pi^n \in B^b$ set

$$\mathrm{Newt}(x) = \text{decreasing convex hull of } \{(n, v(x_n))\}_{n \in \mathbb{Z}}.$$

In the preceding definition one can check that the function v_r does not depend on the choice of a uniformizing element π. In the equal characteristic case, that is to say $E = \mathbb{F}_q((\pi))$, setting $z = \pi$ one finds back the rings defined in Section 1.1.

For $x \in B^b$ the function $r \mapsto v_r(x)$ defined on $]0, +\infty[$ is the Legendre transform of $\mathrm{Newt}(x)$. One has $v_0(x) = \lim_{r \to 0} v_r(x)$. The Newton polygon of x is $+\infty$ exactly on $]-\infty, v_\pi(x)[$ and moreover $\lim_{+\infty} \mathrm{Newt}(x) = v_0(x)$. One has to be careful that since the valuation of F is not discrete, this limit is not always reached, that is to say $\mathrm{Newt}(x)$ may have an infinite number of strictly positive slopes going to zero. One key point is the following proposition whose proof is not very difficult but needs some work.

Proposition 2.4. *For $r \geq 0$, v_r is a valuation on B^b.*

Thus, for all $\rho \in]0, 1]$, $|\cdot|_\rho$ is a multiplicative norm. One deduces from this that for all $x, y \in B^b$,

$$\text{Newt}(xy) = \text{Newt}(x) * \text{Newt}(y)$$

(see 1.1.2). In particular the slopes of $\text{Newt}(xy)$ are obtained by concatenation from the slopes of $\text{Newt}(x)$ and the ones of $\text{Newt}(y)$. For example, as a consequence, if $a_1, \ldots, a_n \in \mathfrak{m}_F \setminus \{0\}$, then

$$\text{Newt}\big((\pi - [a_1]) \ldots (\pi - [a_n])\big)$$

is $+\infty$ on $] - \infty, 0[$, 0 on $[n, +\infty[$ and has non-zero slopes $v(a_1), \ldots, v(a_n)$.

Definition 2.5. Define

- $B = $ completion of B^b with respect to $(|\cdot|_\rho)_{\rho \in]0, 1[}$,
- $B^+ = $ completion of $B^{b,+}$ with respect to $(|\cdot|_\rho)_{\rho \in]0, 1[}$,
- for $I \subset]0, 1[$ a compact interval $B_I = $ completion of B^b with respect to $(|\cdot|_\rho)_{\rho \in I}$.

The rings B and B^+ are E-Fréchet algebras and B^+ is the closure of $B^{b,+}$ in B. Moreover, if $I = [\rho_1, \rho_2] \subset]0, 1[$, for all $f \in B$

$$\sup_{\rho \in I} |f|_\rho = \sup\{|f|_{\rho_1}, |f|_{\rho_2}\}$$

because the function $r \mapsto v_r(f)$ is concave. Thus, B_I is an E-Banach algebra. As a consequence, the formula

$$B = \varprojlim_{I \subset]0, 1[} B_I$$

expresses the Fréchet algebra B as a projective limit of Banach algebras.

Remark 2.6. Of course, the preceding rings are not new and appeared for example under different names in the work of Berger ([2]) and Kedlaya ([21]). The new point of view here is to see them as rings of holomorphic functions of the variable π. In particular, the fact that v_r is a valuation (Proposition 2.4) had never been noticed before.

The Frobenius φ extends by continuity to automorphisms of B and B^+, and for $[\rho_1, \rho_2] \subset]0, 1[$ to an isomorphism $\varphi : B_{[\rho_1, \rho_2]} \xrightarrow{\sim} B_{[\rho_1^q, \rho_2^q]}$.

Remark 2.7. In the case $E = \mathbb{F}_q((\pi))$, setting $z = \pi$ as in Section 1.1, the Frobenius φ just defined is given by $\varphi\big(\sum_n x_n z^n\big) = \sum_n x_n^q z^n$. This is thus an *arithmetic Frobenius*, the geometric one being $\sum_n x_n z^n \mapsto \sum_n x_n z^{qn}$.

The ring B^+ satisfies a particular property. In fact, if $x \in B^{b,+}$ and $r \geq r' > 0$ then

$$v_{r'}(x) \geq \frac{r'}{r} v_r(x). \qquad (2.3)$$

Thus, if $0 < \rho \leq \rho' < 1$ we have

$$B^+_{[\rho,\rho']} = B^+_\rho \subset B^+_{\rho'}$$

where for a compact interval $I \subset]0, 1[$ we note B^+_I for the completion of $B^{b,+}$ with respect to the $(|\cdot|_\rho)_{\rho \in I}$, and $B^+_\rho := B^+_{\{\rho\}}$. One deduces that for any $\rho_0 \in]0, 1[$, $B^+_{\rho_0}$ is stable under φ and

$$B^+ = \bigcap_{n \geq 0} \varphi^n(B^+_{\rho_0})$$

the biggest sub-algebra of $B^+_{\rho_0}$ on which φ is bijective.

Suppose $E = \mathbb{Q}_p$ and choose $\rho \in |F^\times| \cap]0, 1[$. Let $a \in F$ such that $|a| = \rho$. Define

$$B^+_{cris,\rho} = \big(p\text{-adic completion of the P.D. hull of the ideal}$$

$$W(\mathcal{O}_F)[a] \text{ of } W(\mathcal{O}_F)\big) \otimes_{\mathbb{Z}_p} \mathbb{Q}_p.$$

This depends only on ρ since $W(\mathcal{O}_F)[a] = \{x \in W(\mathcal{O}_F) \mid |x|_0 \geq \rho\}$. We thus have

$$B^+_{cris,\rho} = \left(W(\mathcal{O}_F)\left[\frac{[a^n]}{n!}\right]_{n \geq 1}\right)\left[\frac{1}{p}\right].$$

One has

$$B^+_{\rho^p} \subset B^+_{cris,\rho} \subset B^+_{\rho^{p-1}}.$$

From this one deduces

$$B^+ = \bigcap_{n \geq 0} \varphi^n(B^+_{cris,\rho}),$$

the ring usually denoted "B^+_{rig}" in p-adic Hodge theory. The ring $B^+_{cris,\rho}$ appears naturally in comparison theorems where it has a natural interpretation in terms of crystalline cohomology. But the structure of the ring B^+ is simpler as we will see. Moreover, if $k \subset \mathcal{O}_F$ is a perfect sub-field, $K_0 = W(k)_\mathbb{Q}$ and (D, φ) a k-isocrystal, then

$$\big(D \otimes_{K_0} B^+_{cris,\rho}\big)^{\varphi=\mathrm{Id}} = \big(D \otimes_{K_0} B^+\big)^{\varphi=\mathrm{Id}}$$

because φ is bijective on D. Replacing $B^+_{cris,\rho}$ by B^+ is thus harmless most of the time.

Remark 2.8. Suppose $E|\mathbb{Q}_p$ and let $(x_n)_{n\in\mathbb{Z}}$ be a sequence of $\mathcal{O}_F$ such that $\lim\limits_{n\to+\infty} \frac{v(x_{-n})}{n} = +\infty$. Then, the series

$$\sum_{n\in\mathbb{Z}} [x_n]\pi^n$$

converges in B^+. But:

- we don't know if each element of B^+ is of this form,
- for such an element of B^+ we don't know if such a writing is unique,
- we don't know if the sum or product of two element of this form is again of this form.

The same remark applies to B.

Nevertheless, there is a sub E-vector space of B^+ where the preceding remark does not apply. One can define for any $\mathcal{O}_E$-algebra R the group of (ramified) Witt bivectors $BW_{\mathcal{O}_E}(R)$. Elements of $BW_{\mathcal{O}_E}(R)$ have a Teichmüller expansion that is infinite on the left and on the right. One has

$$BW_{\mathcal{O}_E}(\mathcal{O}_F) := \varprojlim_{\substack{\mathfrak{a}\subset\mathcal{O}_F \\ \text{non zero ideal}}} BW_{\mathcal{O}_E}(\mathcal{O}_F/\mathfrak{a})$$

$$= \left\{ \sum_{n\in\mathbb{Z}} V_\pi^n[x_n] \mid x_n \in \mathcal{O}_F, \ \liminf_{n\to-\infty} v(x_n) > 0 \right\}$$

$$= \left\{ \sum_{n\in\mathbb{Z}} [y_n]\pi^n \mid y_n \in \mathcal{O}_F, \ \liminf_{n\to-\infty} q^n v(x_n) > 0 \right\} \subset B^+.$$

The point is that in $BW_{\mathcal{O}_E}(\mathcal{O}_F)$

$$\sum_{n\in\mathbb{Z}} [x_n]\pi^n + \sum_{n\in\mathbb{Z}} [y_n]\pi^n = \sum_{n\in\mathbb{Z}} \left[\lim_{k\to+\infty} P_k(x_{n-k}, \ldots, x_n, y_{n-k}, \ldots, y_n) \right] \pi^n$$

where the generalized polynomials

$$P_k \in \mathbb{F}_q\left[X_i^{q^{i-k}}, Y_j^{q^{j-k}} \right]_{1\leq i,j\leq k}$$

give the addition law of the Witt vectors as in formula (2.2) and the limits in the preceding formulas exist thanks to [15], prop.1.1 chap. II, and the convergence condition appearing in the definition of the bivectors.

In fact, periods of π-divisible $\mathcal{O}_E$-modules (that is to say p-divisible groups when $E = \mathbb{Q}_p$) lie in $BW_{\mathcal{O}_E}(\mathcal{O}_F)$, and $BW_{\mathcal{O}_E}(\mathcal{O}_F)$ contains all periods whose Dieudonné–Manin slopes lie in $[0, 1]$. In equal characteristic, when $E = \mathbb{F}_q((\pi))$, there is no restriction on Dieudonné–Manin slopes of formal $\mathcal{O}_E$-modules (what we call here a formal $\mathcal{O}_E$-module is a Drinfeld module in dimension 1). This gives a meta-explanation to the fact that in equal characteristic all elements of B have a unique power series expansion and the fact that this may not be the case in unequal characteristic.

1.2.2. Newton polygons

Since the elements of B may not be written uniquely as a power series $\sum_{n \in \mathbb{Z}} [x_n] \pi^n$, we need a trick to define the Newton polygon of such elements. The following proposition is an easy consequence of the following Dini type theorem: if a sequence of concave functions on $]0, +\infty[$ converges point-wise then the convergence is uniform on all compact subsets of $]0, +\infty[$ (but not on all $]0, +\infty[$ in general).

Proposition 2.9. *If* $(x_n)_{n \geq 0}$ *is a sequence of* B^b *that converges to* $x \in B \setminus \{0\}$ *then for all* $I \subset]0, 1[$ *compact,*

$$\exists N, \ n \geq N \text{ and } q^{-r} \in I \implies v_r(x_n) = v_r(x).$$

One deduces immediately:

Corollary 2.10. *For* $x \in B$ *the function* $r \mapsto v_r(x)$ *is a concave polygon with integral slopes.*

This leads us to the following definition.

Definition 2.11. For $x \in B$, define $\mathrm{Newt}(x)$ as the inverse Legendre transform of the function $r \mapsto v_r(x)$.

Thus, $\mathrm{Newt}(x)$ is a polygon with integral x-coordinate breakpoints. Moreover, if $(\lambda_i)_{i \in \mathbb{Z}}$ are its slopes, where λ_i is the slope on the segment $[i, i+1]$ (we set $\lambda_i = +\infty$ if $\mathrm{Newt}(x)$ is $+\infty$ on this segment), then

$$\lim_{i \to +\infty} \lambda_i = 0 \text{ and } \lim_{i \to -\infty} \lambda_i = +\infty.$$

In particular $\lim_{-\infty} \mathrm{Newt}(x) = +\infty$. Those properties of $\mathrm{Newt}(x)$ are the only restrictions on such polygons.

Remark 2.12. If $x_n \xrightarrow[n \to +\infty]{} x$ in B with $x_n \in B^b$ and $x \neq 0$ then one checks using Proposition 2.9 that in fact for any compact subset K of $\mathbb{R}$, there exists an integer N such that for $n \geq N$, $\mathrm{Newt}(x_n)_{|K} = \mathrm{Newt}(x)_{|K}$. The advantage of Definition 2.11 is that it makes it clear that $\mathrm{Newt}(x)$ does not depend on the choice of a sequence of B^b going to x.

Example 2.13.
(1) If $(x_n)_{n \in \mathbb{Z}}$ is a sequence of F satisfying the two conditions of formula (2.1) then the polygon $\mathrm{Newt}\left(\sum_{n \in \mathbb{Z}} [x_n] \pi^n\right)$ is the decreasing convex hull of $\{(n, v(x_n))\}_{n \in \mathbb{N}}$.

(2) If $(a_n)_{n\geq 0}$ is a sequence of $F^\times$ going to zero then the infinite product $\prod_{n\geq 0}\left(1-\frac{[a_n]}{\pi}\right)$ converges in B and its Newton polygon is zero on $[0, +\infty[$ and has slopes the $(v(a_n))_{n\geq 0}$ on $]-\infty, 0]$.

Of course the Newton polygon of x does not give more information than the polygon $r \mapsto v_r(x)$. But it is much easier to visualize and its interest lies in the fact that we can appeal to our geometric intuition from the usual case of holomorphic functions recalled in Section 1.1 to guess and prove results. Here is a typical application: the proof of the following proposition is not very difficult once you have convinced yourself it has to be true by analogy with the usual case of holomorphic functions.

Proposition 2.14. *We have the following characterizations:*

(1) $B^+ = \{x \in B \mid \text{Newt}(x) \geq 0\}$.
(2) $B^b = \{x \in B \mid \text{Newt}(x) \text{ is bounded below and } \exists A, \text{ Newt}(x)_{|]-\infty, A]} = +\infty\}$.
(3) *The algebra* $\{x \in B \mid \exists A, \text{ Newt}(x)_{|]-\infty, A]} = +\infty\}$ *is a subalgebra of* $W_{\mathcal{O}_E}(F)[\frac{1}{\pi}]$ *equal to* $\left\{ \sum_{n\gg -\infty}[x_n]\pi^n \mid \liminf_{n\to+\infty} \frac{v(x_n)}{n} \geq 0\right\}$.

This has powerful applications that would be difficult to obtain without Newton polygons. For example one obtains the following.

Corollary 2.15.
(1) $B^\times = \left(B^b\right)^\times = \{x \in B^b \mid \text{Newt}(x) \text{ has 0 as its only non infinite slope}\}$.
(2) *One has* $B^{\varphi=\pi^d} = 0$ *for* $d < 0$, $B^{\varphi=\text{Id}} = E$ *and for* $d \geq 0$,

$$B^{\varphi=\pi^d} = \left(B^+\right)^{\varphi=\pi^d}.$$

Typically, the second point is obtained in the following way. If $x \in B$ satisfies $\varphi(x) = \pi^d x$ then $\text{Newt}(\varphi(x)) = \text{Newt}(\pi^d x)$ that is to say $\text{Newt}(x)$ satisfies the functional equation $q\text{Newt}(x)(t) = \text{Newt}(x)(t - d)$. By solving this functional equation and applying Proposition 2.14 one finds the results.

2. The space $|Y|$

2.1. Primitive elements

We would like to see the Frechet algebra B defined in the preceding section as an algebra of holomorphic functions on a "rigid analytic space" Y. This is of course the case if $E = \mathbb{F}_q((\pi))$ since we can take $Y = \mathbb{D}^*$ a punctured disk as in Section 1.1. This is not the case anymore when $E|\mathbb{Q}_p$, at least as a Tate rigid space. But nevertheless we can still define a topological space $|Y|$ that embeds in the Berkovich space $\mathcal{M}(B)$ of rank 1 continuous valuations on B. It should be thought of as the set of classical points of this "space" Y that would remain to construct.

To simplify the exposition, in the following we always assume $E|\mathbb{Q}_p$, that is to say we concentrate on the most difficult case. When $E = \mathbb{F}_q((\pi))$, all stated results are easy to obtain by elementary manipulation and are more or less already contained in the backgrounds of Section 1.1.

Definition 2.16.

(1) An element $x = \sum_{n\geq 0}[x_n]\pi^n \in W_{\mathcal{O}_E}(\mathcal{O}_F)$ is primitive if $x_0 \neq 0$ and there exists an integer n such that $x_n \in \mathcal{O}_F^\times$. For such a primitive element x we define $\deg(x)$ as the smallest such integer n.

(2) A primitive element of strictly positive degree is irreducible if it can not be written as a product of two primitive elements of strictly lower degree.

If k_F is the residue field of $\mathcal{O}_F$ there is a projection

$$W_{\mathcal{O}_E}(\mathcal{O}_F) \twoheadrightarrow W_{\mathcal{O}_E}(k_F).$$

Then, x is primitive if and only if $x \notin \pi W_{\mathcal{O}_E}(\mathcal{O}_F)$ and its projection $\tilde{x} \in W_{\mathcal{O}_E}(k_F)$ is non-zero. For such an x, $\deg(x) = v_\pi(\tilde{x})$. We deduce from this that the product of a degree d by a degree d' primitive element is a degree $d+d'$ primitive element. Degree 0 primitive elements are the units of $W_{\mathcal{O}_E}(\mathcal{O}_F)$. Any primitive degree 1 element is irreducible.

In terms of Newton polygons, $x \in W_{\mathcal{O}_E}(\mathcal{O}_F)$ is primitive if and only if $\mathrm{Newt}(x)(0) \neq +\infty$ and $\mathrm{Newt}(x)(t) = 0$ for $t \gg 0$.

Definition 2.17. Define $|Y|$ to be the set of primitive irreducible elements modulo multiplication by an element of $W_{\mathcal{O}_E}(\mathcal{O}_F)^\times$.

There is a degree function

$$\deg : |Y| \to \mathbb{N}_{\geq 1}$$

given by the degree of any representative of a class in $|Y|$. If x is primitive note $\bar{x} \in \mathcal{O}_F$ for its reduction modulo π. We have $|\bar{x}| = |\bar{y}|$ if $y \in W_{\mathcal{O}_E}(\mathcal{O}_F)^\times.x$. There is thus a function

$$\|\cdot\| : |Y| \longrightarrow\;]0,1[$$
$$W_{\mathcal{O}_E}(\mathcal{O}_F)^\times.x \longmapsto |\bar{x}|^{1/\deg(x)}.$$

A primitive element $x \in W_{\mathcal{O}_E}(\mathcal{O}_F)$ of strictly positive degree is irreducible if and only if the ideal generated by x is prime. In fact, if $x = yz$ with $y,z \in W_{\mathcal{O}_E}(\mathcal{O}_F)$ and x primitive then projecting to $W_{\mathcal{O}_E}(k_F)$ and $\mathcal{O}_F$ the preceding equality one obtains that y and z are primitive. There is thus an embedding

$$|Y| \subset \mathrm{Spec}(W_{\mathcal{O}_E}(\mathcal{O}_F)).$$

The Frobenius φ induces a bijection

$$\varphi : |Y| \overset{\sim}{\to} |Y|$$

that leaves invariant the degree and satisfies

$$\|\varphi(y)\| = \|y\|^q.$$

Remark 2.18. When $E = \mathbb{F}_q((\pi))$, replacing $W_{\mathcal{O}_E}(\mathcal{O}_F)$ by $\mathcal{O}_F[\![z]\!]$ in the preceding definitions (we set $z = \pi$) there is an identification $|Y| = |\mathbb{D}^*|$. In fact, according to Weierstrass, any irreducible primitive $f \in \mathcal{O}_F[\![z]\!]$ has a unique irreducible unitary polynomial $P \in \mathcal{O}_F[z]$ in its $\mathcal{O}_F[\![z]\!]^\times$-orbit satisfying: $P(0) \neq 0$ and the roots of P have absolute value < 1. Then for $y \in |\mathbb{D}^*|$, $\deg(y) = [k(y) : F]$ and $\|y\|$ is the distance from y to the origin of the disk $\mathbb{D}$.

2.2. Background on the ring $\mathscr{R}$

For an $\mathcal{O}_E$-algebra A set

$$\mathscr{R}(A) = \left\{ \left(x^{(n)}\right)_{n \geq 0} \mid x^{(n)} \in A, \ \left(x^{(n+1)}\right)^q = x^{(n)} \right\}.$$

If A is π-adic, I is a closed ideal of A such that A is $I + (\pi)$-adic, then the reduction map induces a bijection

$$\mathscr{R}(A) \overset{\sim}{\to} \mathscr{R}(A/I)$$

with inverse given by

$$\left(x^{(n)}\right)_{n \geq 0} \longmapsto \left(\lim_{k \to +\infty} \left(x^{(n+k)}\right)^{q^k} \right)_{n \geq 0},$$

where $x^{(n+k)} \in A$ is any lift of $x^{(n+k)} \in A/I$, and the preceding limit is for the $I + (\pi)$-adic topology. In particular, applying this for $I = (\pi)$, we deduce that the set valued functor $\mathscr{R}$ factorizes canonically as a functor

$$\mathscr{R} : \pi\text{-adic } \mathcal{O}_E\text{-algebras} \longrightarrow \text{perfect } \mathbb{F}_q\text{-algebras}.$$

If $W_{\mathcal{O}_E}$ stands for the (ramified) Witt vectors there is then a couple of adjoint functors

$$\pi\text{-adic } \mathcal{O}_E\text{-algebras} \underset{W_{\mathcal{O}_E}}{\overset{\mathscr{R}}{\rightleftarrows}} \text{perfect } \mathbb{F}_q\text{-algebras}$$

where $W_{\mathcal{O}_E}$ is left adjoint to $\mathcal{R}$ and the adjunction morphisms are given by:

$$A \xrightarrow{\sim} \mathcal{R}(W_{\mathcal{O}_E}(A))$$
$$a \longmapsto \left(\left[a^{q^{-n}}\right]\right)_{n \geq 0}$$

and

$$\theta : W_{\mathcal{O}_E}(\mathcal{R}(A)) \longrightarrow A$$
$$\sum_{n \geq 0} [x_n] \pi^n \longmapsto \sum_{n \geq 0} x_n^{(0)} \pi^n.$$

If $L|\mathbb{Q}_p$ is a complete valued extension for a valuation $w : L \to \mathbb{R} \cup \{+\infty\}$ extending a multiple of the p-adic valuation, then $\mathcal{R}(L)$ equipped with the valuation

$$x \longmapsto w(x^{(0)})$$

is a characteristic p perfect complete valued field with ring of integers $\mathcal{R}(\mathcal{O}_L)$ (one has to be careful that the valuation on $\mathcal{R}(L)$ may be trivial). It is not very difficult to prove that if L is algebraically closed then $\mathcal{R}(L)$ is too. A reciprocal to this statement will be stated in the following sections.

2.3. The case when is F algebraically closed

Theorem 2.19. *Suppose F is algebraically closed. Let $\mathfrak{p} \in \mathrm{Spec}(W_{\mathcal{O}_E}(\mathcal{O}_F))$ generated by a degree one primitive element. Set*

$$A = W_{\mathcal{O}_E}(\mathcal{O}_F)/\mathfrak{p}$$

and $\theta : W_{\mathcal{O}_E}(\mathcal{O}_F) \twoheadrightarrow A$ the projection. The following properties are satisfied:

(1) *There is an isomorphism*

$$\mathcal{O}_F \xrightarrow{\sim} \mathcal{R}(A)$$
$$x \longmapsto \left(\theta\left(\left[x^{q^{-n}}\right]\right)\right)_{n \geq 0}.$$

(2) *The map*

$$\mathcal{O}_F \longrightarrow A$$
$$x \longmapsto \theta([x])$$

  is surjective.

(3) *There is a unique valuation w on A such that for all $x \in \mathcal{O}_F$,*

$$w(\theta([x])) = v(x).$$

Moreover, $A[\frac{1}{\pi}]$ equipped with the valuation w is a complete algebraically closed extension of E with ring of integers A. There is an identification of valued fields $F = \mathscr{R}(A[\frac{1}{\pi}])$.

(4) *If $\underline{\pi} \in \mathscr{R}(A)$ is such that $\underline{\pi}^{(0)} = \pi$ then*

$$\mathfrak{p} = ([\underline{\pi}] - \pi).$$

Remark 2.20. One can reinterpret points (2) and (4) of the preceding theorem in the following way. Let x be primitive of degree 1. Then:

- we have a *Weierstrass division* in $W_{\mathcal{O}_E}(\mathcal{O}_F)$: given $y \in W_{\mathcal{O}_E}(\mathcal{O}_F)$, there exists $z \in W_{\mathcal{O}_E}(\mathcal{O}_F)$ and $a \in \mathcal{O}_F$ such that

$$y = zx + [a],$$

- we have a Weierstrass factorization of x: there exists $u \in W_{\mathcal{O}_E}(\mathcal{O}_F)^\times$ and $b \in \mathcal{O}_F$ such that

$$x = u.(\pi - [b]).$$

One has to be careful that, contrary to the classical situation, the remainder term a is not unique in such a Weierstrass division. Similarly, b is not uniquely determined by x in the Weierstrass factorization.

Indications on the proof of Theorem 2.19. Statement (1) is an easy consequence of general facts about the ring $\mathscr{R}$ recalled in Section 2.2: if $\mathfrak{p} = (x)$

$$\mathscr{R}(A) \xrightarrow{\sim} \mathscr{R}(A/\pi A) = \mathscr{R}(\mathcal{O}_F/\bar{x}) = \mathcal{O}_F$$

since $\mathcal{O}_F$ is perfect.

According to point (1), point (2) is reduced to proving that any element of A has a q-th root. If $E^0 = W(\mathbb{F}_q)_{\mathbb{Q}}$, the norm map N_{E/E^0} induces a norm map $W_{\mathcal{O}_E}(\mathcal{O}_F) \to W(\mathcal{O}_F)$ sending a primitive element of degree 1 to a primitive element of degree 1. Using this one can reduce the problem to the case $E = \mathbb{Q}_p$. Suppose $p \neq 2$ to simplify. Then any element of $1 + p^2 W(\mathcal{O}_F)$ has a p-th root. Using this fact plus some elementary manipulations one is reduced to solving some explicit equations in the truncated Witt vectors of length 2, $W_2(\mathcal{O}_F)$. One checks this is possible, using the fact that F is algebraically closed (here this hypothesis is essential, the hypothesis F perfect is not sufficient).

In fact, the preceding proof gives that for any integer n, any element of A has an n-th root, the case when n is prime to p being easier than the case $n = p$ we just explained (since then any element of $1 + p W_{\mathcal{O}_E}(\mathcal{O}_F)$ has an n-th root).

Point (4) is an easy consequence of the following classical characterization of $\ker \theta$: an element $y = \sum_{n \geq 0}[y_n]\pi^n \in \ker \theta$ such that $y_0^{(0)} \in \pi A^\times$ is a generator of $\ker \theta$. In fact, if y is such an element then $\ker \theta = (y) + \pi \ker \theta$ and one concludes $\ker \theta = (y)$ by applying the π-adic Nakayama lemma ($\ker \theta$ is π-adically closed).

In point (3), the difficulty is to prove that the complete valued field $L = A[\frac{1}{\pi}]$ is algebraically closed; other points following easily from point (2). Using the fields of norms theory one verifies that L contains an algebraic closure of $\mathbb{Q}_p$. More precisely, one can suppose thanks to point (4) that $\mathfrak{p} = (\pi - [\underline{\pi}])$. There is then an embedding $\mathbb{F}_q((\underline{\pi})) \subset F$ that induces $F' := \overline{\mathbb{F}_q((\underline{\pi}))} \subset F$. This induces a morphism $L' = W_{O_E}(O_{F'})[\frac{1}{\pi}]/(\pi - [\underline{\pi}]) \to L$. But thanks to the fields of norms theory, L' is the completion of an algebraic closure of E. In particular, L contains all roots of unity. Let us notice that since $O_L/\pi O_L = O_F/\underline{\pi}O_F$, the residue field of L is the same as the one of F and is thus algebraically closed. Now, we use the following proposition that is well known in the discrete valuation case thanks to the theory of ramification groups (those ramification groups do not exist in the non discrete valuation case, but one can define some ad hoc one to obtain the proposition).

Proposition 2.21. *Let K be a complete valued field for a rank 1 valuation and $K'|K$ a finite degree Galois extension inducing a trivial extension on the residue fields. Then the group $\mathrm{Gal}(K'|K)$ is solvable.*

Since for any integer n any element of L has an n-th root, one concludes L is algebraically closed using Kümmer theory. $\square$

Note now $O_C = W_{O_E}(O_F)/\mathfrak{p}$ with fraction field C. One has to be careful that the valuation w on C extends only a multiple of the π-adic valuation of E: $q^{-w(\pi)} = \|\mathfrak{p}\|$. The quotient morphism $B^{b,+} \longrightarrow C$ extends in fact by continuity to surjective morphisms

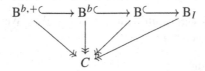

where $I \subset]0, 1[$ is such that $\|\mathfrak{p}\| \in I$. This is a consequence of the inequality

$$q^{-w(f)} \leq |f|_\rho$$

for $f \in B^b$ and $\rho = \|\mathfrak{p}\|$. If $\mathfrak{p} = (x)$ then all kernels of those surjections are the principal ideals generated by x in those rings.

Convention: we will now see $|Y|^{\deg=1}$ *as a subset of* Spm(B). *For* $\mathfrak{m} \in |Y|^{\deg=1}$, *we note*

$$C_\mathfrak{m} = B/\mathfrak{m}, \quad \theta_\mathfrak{m} : B \twoheadrightarrow C_\mathfrak{m}$$

and $v_\mathfrak{m}$ *the valuation such that*

$$v_\mathfrak{m}(\theta_\mathfrak{m}([x])) = v(x).$$

Theorem 2.22. *If* F *is algebraically closed then the primitive irreducible elements are of degree one.*

Remark 2.23. One can reinterpret the preceding theorem as a factorization statement: if $x \in W_{\mathcal{O}_E}(\mathcal{O}_F)$ is primitive of degree d then

$$x = u.(\pi - [a_1])\ldots(\pi - [a_d]), \quad u \in W_{\mathcal{O}_E}(\mathcal{O}_F)^\times, \; a_1, \ldots, a_d \in \mathfrak{m}_F \setminus \{0\}.$$

Indications on the proof of theorem 2.22. Let $f \in W_{\mathcal{O}_E}(\mathcal{O}_F)$ be primitive. The theorem is equivalent to saying that f "has a zero in $|Y|^{\deg=1}$" that is to say there exists $\mathfrak{m} \in |Y|^{\deg=1}$ such that $\theta_\mathfrak{m}(f) = 0$. The method to construct such a zero is a Newton type method by successive approximations. To make it work we need to know it converges in a sense that has to be specified. We begin by proving the following.

Proposition 2.24. *For* $\mathfrak{m}_1, \mathfrak{m}_2 \in |Y|^{\deg=1}$ *set*

$$d(\mathfrak{m}_1, \mathfrak{m}_2) = q^{-v_{\mathfrak{m}_1}(a)} \; \text{if } \theta_{\mathfrak{m}_1}(\mathfrak{m}_2) = \mathcal{O}_{C_{\mathfrak{m}_1}} a.$$

Then:

(1) *This defines an ultrametric distance on* $|Y|^{\deg=1}$.
(2) *For any* $\rho \in]0, 1[$, $\left(|Y|^{\deg=1, \|\cdot\| \geq \rho}, d\right)$ *is a complete metric space.*

Remark 2.25. In equal characteristic, if $E = \mathbb{F}_q((\pi))$, then $|Y| = |\mathbb{D}^*| = \mathfrak{m}_F \setminus \{0\}$ and this distance is the usual one induced by the absolute value $|\cdot|$ of F.

The approximation algorithm then works like this. We define a sequence $(\mathfrak{m}_n)_{n \geq 1}$ of $|Y|^{\deg=1}$ such that:

- $(\|\mathfrak{m}_n\|)_{n \geq 1}$ is constant,
- it is a Cauchy sequence,
- $\lim\limits_{n \to +\infty} v_{\mathfrak{m}_n}(f) = +\infty$.

Write $f = \sum_{k \geq 0}[x_k]\pi^k$. The Newton polygon of f as defined in Section 1.2.2 is the same as Newton polygon of $g(T) = \sum_{k \geq 0} x_k T^k \in \mathcal{O}_F[\![T]\!]$. Let $z \in \mathfrak{m}_F$ be a root of $g(T)$ with valuation the smallest one among the valuations of the roots of $g(T)$ (that is to say the smallest non-zero slope of Newt(f)). Start with $\mathfrak{m}_1 = (\pi - [z]) \in |Y|^{\deg=1}$. If $\mathfrak{m}_n$ is defined, $\mathfrak{m}_n = (\xi)$ with ξ primitive of degree 1, we can write

$$f = \sum_{k \geq 0} [a_k] \xi^k$$

in $W_{\mathcal{O}_E}(\mathcal{O}_F)$ (this is a consequence of point (2) of Theorem 2.19 and the fact that $W_{\mathcal{O}_E}(\mathcal{O}_F)$ is ξ-adic). We check the power series $h(T) = \sum_{k \geq 0} a_k T^k \in \mathcal{O}_F[\![T]\!]$ is primitive of degree d. Let z be a root of $h(T)$ of maximal valuation. Then $\xi - [z]$ is primitive of degree 1 and we set $\mathfrak{m}_{n+1} = (\xi - [z])$.

We then prove the sequence $(\mathfrak{m}_n)_{n \geq 1}$ satisfies the required properties. $\qquad\square$

2.4. Parametrization of $|Y|$ when F is algebraically closed

Suppose F is algebraically closed. We see $|Y|$ as a subset of Spm(B). As we saw, for any $\mathfrak{m} \in |Y|$ there exists $a \in \mathfrak{m}_F \setminus \{0\}$ such that $\mathfrak{m} = (\pi - [a])$. The problem is that such an a is not unique. Moreover, given $a, b \in \mathfrak{m}_F \setminus \{0\}$, there is no simple rule to decide whether $(\pi - [a]) = (\pi - [b])$ or not.

Here is a solution to this problem. Let $\mathcal{LT}$ be a Lubin–Tate group law over $\mathcal{O}_E$. We note

$$Q = [\pi]_{\mathcal{LT}} \in \mathcal{O}_E[\![T]\!]$$

and $\mathcal{G}$ the associated formal group on $\mathrm{Spf}(\mathcal{O}_E)$. We have

$$\mathcal{G}(\mathcal{O}_F) = \left(\mathfrak{m}_F, \underset{\mathcal{LT}}{+} \right).$$

The topology induced by the norms $(| \cdot |_\rho)_{\rho \in]0,1[}$ on $W_{\mathcal{O}_E}(\mathcal{O}_F)$ is the "weak topology" on the coefficients of the Teichmüller expansion, that is to say the product topology via the bijection

$$\mathcal{O}_F^{\mathbb{N}} \xrightarrow{\sim} W_{\mathcal{O}_E}(\mathcal{O}_F)$$

$$(x_n)_{n \geq 0} \longmapsto \sum_{n \geq 0} [x_n] \pi^n.$$

If $a \in \mathfrak{m}_F \setminus \{0\}$, this coincides with the $([a], \pi)$-adic topology. Moreover $W_{\mathcal{O}_E}(\mathcal{O}_F)$ is complete, that is to say closed in B. If

$$W_{\mathcal{O}_E}(\mathcal{O}_F)^{00} = \left\{ \sum_{n \geq 0} [x_n] \pi^n \mid x_0 \in \mathfrak{m}_F \right\} = \{ x \in W_{\mathcal{O}_E}(\mathcal{O}_F) \mid x \bmod \pi \in \mathfrak{m}_F \},$$

then

$$\mathcal{G}\left(W_{\mathcal{O}_E}(\mathcal{O}_F) \right) = \left(W_{\mathcal{O}_E}(\mathcal{O}_F)^{00}, \underset{\mathcal{LT}}{+} \right).$$

One verifies the following proposition.

Proposition 2.26. *For $x \in \mathfrak{m}_F$, the limit*

$$[x]_Q := \lim_{n \to +\infty} [\pi^n]_{\mathcal{LT}}\left(\left[x^{q^{-n}}\right]\right)$$

exists in $W_{\mathcal{O}_E}(\mathcal{O}_F)$, reduces to x modulo π, and defines a "lift"

$$[\cdot]_Q : \mathcal{G}(\mathcal{O}_F) \hookrightarrow \mathcal{G}(W_{\mathcal{O}_E}(\mathcal{O}_F)).$$

The usual Teichmüller lift $[\cdot]$ is well adapted to the multiplicative group law: $[xy] = [x].[y]$. The advantage of the twisted Teichmüller lift $[\cdot]_Q$ is that it is more adapted to the Lubin–Tate one:

$$[x]_Q \underset{\mathcal{LT}}{+} [y]_Q = \left[x \underset{\mathcal{LT}}{+} y\right]_Q.$$

When $E = \mathbb{Q}_p$ and $\mathcal{LT} = \mathbb{G}_m$ one has $[x]_Q = [1 + x] - 1$.

Definition 2.27. For $\epsilon \in \mathfrak{m}_F \setminus \{0\}$ define $u_\epsilon = \dfrac{[\epsilon]_Q}{[\epsilon^{1/q}]_Q} \in W_{\mathcal{O}_E}(\mathcal{O}_F)$.

This is a primitive degree one element since it is equal to the power series $\dfrac{Q(T)}{T}$ evaluated at $[\epsilon^{1/q}]_Q$. For example, if $E = \mathbb{Q}_p$ and $\mathcal{LT} = \mathbb{G}_m$, setting $\epsilon' = 1 + \epsilon$ one has

$$u_\epsilon = 1 + \left[\epsilon'^{\frac{1}{p}}\right] + \cdots + \left[\epsilon'^{\frac{p-1}{p}}\right].$$

Proposition 2.28. *There is a bijection*

$$(\mathcal{G}(\mathcal{O}_F) \setminus \{0\})/\mathcal{O}_E^\times \xrightarrow{\sim} |Y|$$
$$\mathcal{O}_E^\times.\epsilon \longmapsto (u_\epsilon).$$

The inverse of this bijection is given by the following rule. For $\mathfrak{m} \in |Y|$, define

$$X(\mathcal{G})(\mathcal{O}_{C_{\mathfrak{m}}}) = \left\{(x_n)_{n \geq 0} \mid x_n \in \mathcal{G}(\mathcal{O}_{C_{\mathfrak{m}}}), \ \pi.x_{n+1} = x_n\right\}.$$

More generally, $X(\mathcal{G})$ will stand for the projective limit "$\varprojlim\limits_{n \geq 0} \mathcal{G}$" where the transition mappings are multiplication by π (one can give a precise geometric meaning to this but this is not our task here, see [12] for more details). The reduction modulo π map induces a bijection

$$X(\mathcal{G})(\mathcal{O}_{C_{\mathfrak{m}}}) \xrightarrow{\sim} X(\mathcal{G})(\mathcal{O}_{C_{\mathfrak{m}}}/\pi\mathcal{O}_{C_{\mathfrak{m}}})$$

with inverse given by

$$(x_n)_n \mapsto \left(\lim_{k \to +\infty} \pi^k \hat{x}_{n+k}\right)_n$$

where $\hat{x}_{n+k}$ is any lift of x_{n+k}. But $[\pi]_{\mathcal{LT}}$ modulo π is the Frobenius Frob_q. We thus have

$$X(\mathcal{G})(\mathcal{O}_{C_{\mathfrak{m}}}/\pi\mathcal{O}_{C_{\mathfrak{m}}}) = \mathcal{G}\big(\mathcal{R}(\mathcal{O}_{C_{\mathfrak{m}}}/\pi\mathcal{O}_{C_{\mathfrak{m}}})\big) = \mathcal{G}(\mathcal{O}_F)$$

since

$$\mathcal{O}_F \xrightarrow[\theta_{\mathfrak{m}} \circ [\cdot]]{\sim} \mathcal{R}(\mathcal{O}_{C_{\mathfrak{m}}}) \xrightarrow{\sim} \mathcal{R}(\mathcal{O}_{C_{\mathfrak{m}}}/\pi\mathcal{O}_{C_{\mathfrak{m}}}).$$

The Tate module

$$T_\pi(\mathcal{G}) = \{(x_n)_{n>0} \in X(\mathcal{G})(\mathcal{O}_{C_{\mathfrak{m}}}) \mid x_0 = 1\} \subset X(\mathcal{G})(\mathcal{O}_{C_{\mathfrak{m}}})$$

embeds thus in $\mathcal{G}(\mathcal{O}_F)$. This is a rank 1 sub-$\mathcal{O}_E$-module. The inverse of the bijection of Proposition 2.28 sends $\mathfrak{m}$ to this $\mathcal{O}_E$-line in $\mathcal{G}(\mathcal{O}_F)$.

2.5. The general case: Galois descent

Let now F be general (but still perfect), that is to say not necessarily algebraically closed. Let $\overline{F}$ be an algebraic closure of F with Galois group $G_F = \mathrm{Gal}(\overline{F}|F)$. Since the field F will vary we now put a subscript in the preceding notation to indicate this variation. The set $|Y_{\widehat{\overline{F}}}|$ is equipped with an action of G_F.

Theorem 2.29. *For* $\mathfrak{p} \in |Y_F| \subset \mathrm{Spec}(W_{\mathcal{O}_E}(\mathcal{O}_F))$ *set*

$$L = \big(W_{\mathcal{O}_E}(\mathcal{O}_F)/\mathfrak{p}\big)\big[\tfrac{1}{\pi}\big]$$

and

$$\theta : \mathbf{B}_F^{b,+} \twoheadrightarrow L.$$

Then:

(1) *There is a unique valuation* w *on* L *such that for* $x \in \mathcal{O}_F$,

$$w(\theta([x])) = v(x).$$

(2) (L, w) *is a complete valued extension of* E.
(3) (L, w) *is perfectoid in the sense that the Frobenius of* $\mathcal{O}_L/\pi\mathcal{O}_L$ *is surjective.*
(4) *Via the embedding*

$$\mathcal{O}_F \hookrightarrow \mathcal{R}(L)$$
$$a \longmapsto \big(\theta([a^{q^{-n}}])\big)_{n\geq 0}$$

one has $\mathcal{R}(L)|F$ *and this extension is of finite degree*

$$[\mathcal{R}(L) : F] = \deg \mathfrak{p}.$$

Remark 2.30. What we call here perfectoid is what we called "strictly p-perfect" in [13] and [14]. The authors decided to change their terminology because meanwhile the work [29] appeared.

Indications on the proof. The proof is based on Theorems 2.19 and 2.22 via a Galois descent argument from $|Y_{\widehat{\overline{F}}}|$ to $|Y_F|$. For this we need the following.

Theorem 2.31. *One has* $\mathfrak{m}_F.H^1(G_F, \mathcal{O}_{\widehat{\overline{F}}}) = 0$.

Using Tate's method ([31]) this theorem is a consequence of the following "almost etalness" statement whose proof is much easier than in characteristic 0.

Proposition 2.32. *If $L|F$ is a finite degree extension then $\mathfrak{m}_F \subset \mathrm{tr}_{L/F}(\mathcal{O}_L)$.*

Sketch of proof. The trace $\mathrm{tr}_{L/F}$ commutes with the Frobenius $\varphi = \mathrm{Frob}_q$. Choosing $x \in \mathcal{O}_L$ such that $\mathrm{tr}_{L/F}(x) \neq 0$ one deduces that

$$\lim_{n \to +\infty} |\mathrm{tr}_{L/F}(\varphi^{-n}(x))| = 1.$$

$\square$

From Theorem 2.31 one deduces that $H^1(G_F, \mathfrak{m}_{\widehat{\overline{F}}}) = 0$ which implies

$$H^1(G_F, 1 + W_{\mathcal{O}_E}(\mathfrak{m}_{\widehat{\overline{F}}})) = 0 \tag{2.4}$$

where

$$1 + W_{\mathcal{O}_E}(\mathfrak{m}_{\widehat{\overline{F}}}) = \ker\left(W_{\mathcal{O}_E}(\mathcal{O}_{\widehat{\overline{F}}})^{\times} \longrightarrow W_{\mathcal{O}_E}(k_{\overline{F}})^{\times}\right).$$

Finally, let us notice that thanks to Ax's theorem (that can be easily deduced from 2.32) one has

$$W_{\mathcal{O}_E}(\mathcal{O}_{\widehat{\overline{F}}})^{G_F} = W_{\mathcal{O}_E}(\mathcal{O}_F).$$

Proposition 2.33. *Let $x \in W_{\mathcal{O}_E}(\mathcal{O}_{\widehat{\overline{F}}})$ be a primitive element such that $\forall \sigma \in G_F$, $(\sigma(x)) = (x)$ as an ideal of $W_{\mathcal{O}_E}(\mathcal{O}_{\widehat{\overline{F}}})$. Then, there exists $y \in W_{\mathcal{O}_E}(\mathcal{O}_F)$ such that $(x) = (y)$.*

Proof. If $\tilde{x} \in W_{\mathcal{O}_E}(k_{\overline{F}})$ is the projection of x via $W_{\mathcal{O}_E}(\mathcal{O}_{\widehat{\overline{F}}}) \to W_{\mathcal{O}_E}(k_{\overline{F}})$, up to multiplying x by a unit, one can suppose $\tilde{x} = \pi^{\deg x}$. Looking at the cocycle $\sigma \mapsto \frac{\sigma(x)}{x}$, the proposition is then a consequence of the vanishing (2.4). $\square$

Let now $x \in W_{\mathcal{O}_E}(\mathcal{O}_F)$ be primitive irreducible of degree d. There exist $y_1, \ldots, y_r \in W_{\mathcal{O}_E}(\mathcal{O}_{\widehat{\overline{F}}})$ primitive of degree 1 satisfying $(y_i) \neq (y_j)$ for $i \neq j$, $a_1, \ldots, a_r \in \mathbb{N}_{\geq 1}$ and $u \in W_{\mathcal{O}_E}(\mathcal{O}_{\widehat{\overline{F}}})^{\times}$ such that

$$x = u.\prod_{i=1}^{r} y_i^{a_i}.$$

The finite subset

$$\{(y_i)\}_{1 \le i \le r} \subset |Y_{\widehat{\overline{F}}}|$$

is stable under G_F. Using Proposition 2.33 and the irreducibility of x one verifies that this action is transitive and $a_1 = \cdots = a_r = 1$. In particular one has $r = d$. Note $\mathfrak{m}_i = (y_i)$ and $C_{\mathfrak{m}_i}$ the associated algebraically closed residue field. Let $K|F$ be the degree d extension of F in $\overline{F}$ such that

$$G_K = \operatorname{Stab}_{G_F}(\mathfrak{m}_1).$$

One has

$$B_{\widehat{\overline{F}}}^{b,+}/(x) = \prod_{i=1}^{d} C_{\mathfrak{m}_i}$$

and thus

$$\left(B_{\widehat{\overline{F}}}^{b,+}/(x)\right)^{G_F} = C_{\mathfrak{m}_1}^{G_K}.$$

Now, one verifies using that x is primitive that if $a \in \mathfrak{m}_F \setminus \{0\}$ then

$$\left(W_{\mathcal{O}_E}(\mathcal{O}_F)/(x)\right)\left[\tfrac{1}{\pi}\right] = \left(W_{\mathcal{O}_E}(\mathcal{O}_F)/(x)\right)\left[\tfrac{1}{[a]}\right].$$

Theorem 2.31 implies that for such an a,

$$[a].H^1\left(G_F, W_{\mathcal{O}_E}(\mathcal{O}_{\widehat{\overline{F}}})\right) = 0.$$

One deduces from this that

$$L = B_F^{b,+}/(x) = \left(B_{\widehat{\overline{F}}}^{b,+}/(x)\right)^{G_F}$$

that is thus a complete valued field. Moreover

$$\mathscr{R}(L) = \mathscr{R}\left(C_{\mathfrak{m}_1}^{G_K}\right) = \mathscr{R}(C_{\mathfrak{m}_1})^{G_K} = \widehat{\overline{F}}^{G_K} = K.$$

Other statements of Theorem 2.29 are easily deduced in the same way.   □

In the preceding theorem the quotient morphism $\theta : B_F^{b,+} \twoheadrightarrow L$ extends by continuity to a surjection $B_F \twoheadrightarrow L$ with kernel the principal ideal $B_F\mathfrak{p}$. *From now on, we will see $|Y_F|$ as a subset of $Spm(B)$. If $\mathfrak{m} \in |Y_F|$ we note*

$$L_{\mathfrak{m}} = B_F/\mathfrak{m}, \quad \theta_{\mathfrak{m}} : B \twoheadrightarrow L_{\mathfrak{m}}.$$

The preceding arguments give the following.

Theorem 2.34. *Let* $\left|Y_{\widehat{\overline{F}}}\right|^{G_F-\text{fin}}$ *be the elements of* $\left|Y_{\widehat{\overline{F}}}\right|$ *whose* G_F-*orbit is finite. There is a surjection*

$$\beta : \left|Y_{\widehat{\overline{F}}}\right|^{G_F-\text{fin}} \longrightarrow |Y_F|$$

whose fibers are the G_F-orbits:

$$\left|Y_{\widehat{\overline{F}}}\right|^{G_F-\mathrm{fin}}/G_F \xrightarrow{\sim} |Y_F|.$$

Moreover:

- *for $\mathfrak{m} \in |Y_F|$, one has*

$$\#\beta^{-1}(\mathfrak{m}) = [\mathscr{R}(L_\mathfrak{m}) : F] = \deg \mathfrak{m},$$

- *for $\mathfrak{n} \in \left|Y_{\widehat{\overline{F}}}\right|^{G_F-\mathrm{fin}}$, if $\mathfrak{m} = \beta(\mathfrak{n})$, one has an extension $C_\mathfrak{n}|L_\mathfrak{m}$ that identifies $C_\mathfrak{n}$ with the completion of the algebraic closure $\overline{L}_\mathfrak{m}$ of $L_\mathfrak{m}$ and*

$$\mathrm{Gal}\left(\overline{F}|\mathscr{R}(L_\mathfrak{m})\right) \xrightarrow{\sim} \mathrm{Gal}\left(\overline{L}_\mathfrak{m}|L_\mathfrak{m}\right).$$

Remark 2.35.

(1) One has to be careful that $\theta_\mathfrak{m}(W_{\mathcal{O}_E}(\mathcal{O}_F)) \subset \mathcal{O}_{L_\mathfrak{m}}$ is only an order. It is equal to $\mathcal{O}_{L_\mathfrak{m}}$ if and only if $\deg \mathfrak{m} = 1$. Using Theorem 2.31 one can verify that this order contains the maximal ideal of $\mathcal{O}_{L_\mathfrak{m}}$.
(2) Contrary to the case when F is algebraically closed, in general an $\mathfrak{m} \in |Y_F|$ of degree 1 is not generated by an element of the form $\pi - [a]$, $a \in \mathfrak{m}_F \setminus \{0\}$.

2.6. Application to perfectoid fields

Reciprocally, given a complete valued field $L|E$ for a rank 1 valuation, it is perfectoid if and only if the morphism

$$\theta : W_{\mathcal{O}_E}(\mathscr{R}(\mathcal{O}_L)) \longrightarrow \mathcal{O}_L$$

is surjective. In this case one can check that the kernel of θ is generated by a primitive degree 1 element. The preceding considerations thus give the following.

Theorem 2.36.

(1) *There is an equivalence of categories between perfectoid fields $L|E$ and the category of couples $(F, \mathfrak{m})$ where $F|\mathbb{F}_q$ is perfectoid and $\mathfrak{m} \in |Y_F|$ is of degree 1.*
(2) *In the preceding equivalence, if L corresponds to $(F, \mathfrak{m})$, the functor $\mathscr{R}$ induces an equivalence between finite étale L-algebras and finite étale F-algebras. The inverse equivalence sends the finite extension $F'|F$ to $B_{F'}/B_{F'}\mathfrak{m}$.*

Example 2.37.

(1) The perfectoid field $L = \mathbb{Q}_p(\zeta_{p^\infty})$ corresponds to $F = \mathbb{F}_p((T))^{\mathrm{perf}}$ and
$$\mathfrak{m} = \left(1 + [T^{1/p} + 1] + \cdots + [T^{\frac{p-1}{p}} + 1]\right).$$

(2) Choose $\underline{p} \in \mathscr{R}(\overline{\mathbb{Q}}_p)$ such that $\underline{p}^{(0)} = p$. The perfectoid field $L = \widehat{M}$ with $M = \cup_{n \geq 0} \mathbb{Q}_p(\underline{p}^{(n)})$ corresponds to $F = \mathbb{F}_p((T))^{\mathrm{perf}}$ and $\mathfrak{m} = ([T] - p)$.

Remark 2.38. In the preceding correspondence, F is maximally complete if and only if L is. In particular, one finds back the formula given in [26] for p-adic maximally complete fields: they are of the form $W(\mathcal{O}_F)[\frac{1}{p}]/([x] - p)$ where F is maximally complete of characteristic p and $x \in F^\times$ satisfies $v(x) > 0$.

Remark 2.39. Suppose $F = \overline{\mathbb{F}_p((T))}$. One can ask what are the algebraically closed residue fields $C_\mathfrak{m}$ up to isomorphism when $\mathfrak{m}$ goes through $|Y_F|$. Let us note $\mathbb{C}_p$ the completion of an algebraic closure of $\mathbb{Q}_p$. Thanks to the fields of norms theory it appears as a residue field $C_\mathfrak{m}$ for some $\mathfrak{m} \in |Y_F|$. The question is: is it true that for all $\mathfrak{m} \in |Y_F|$, $C_\mathfrak{m} \simeq \mathbb{C}_p$? The authors do not know the answer to this question. They know that for each integer $n \geq 1$, $\mathcal{O}_{C_\mathfrak{m}}/p^n\mathcal{O}_{C_\mathfrak{m}} \simeq \mathcal{O}_{\mathbb{C}_p}/p^n\mathcal{O}_{\mathbb{C}_p}$ but in a non canonical way.

As a consequence of the preceding theorem we deduce almost étalness for characteristic 0 perfectoïd fields.

Corollary 2.40. *For $L|E$ a perfectoïd field and $L'|L$ a finite extension we have*
$$\mathfrak{m}_L \subset tr_{L'|L}(\mathcal{O}_{L'}).$$

Proof. Set $F' = \mathscr{R}(L')$ and $F = \mathscr{R}(L)$. If L corresponds to $\mathfrak{m} \in |Y_F|$, $\mathfrak{a} = \{x \in \mathcal{O}_F \mid |x| \leq \|\mathfrak{m}\|\}$ and $\mathfrak{a}' = \{x \in \mathcal{O}_{F'} \mid |x| \leq \|\mathfrak{m}\|\}$ we have identifications
$$\mathcal{O}_L/\pi\mathcal{O}_L = \mathcal{O}_F/\mathfrak{a}$$
$$\mathcal{O}_{L'}/\pi\mathcal{O}_{L'} = \mathcal{O}_{F'}/\mathfrak{a}'.$$

According to point (2) of the preceding theorem, with respect to those identifications the map $tr_{L'|L}$ modulo π is induced by the map $tr_{F'|F}$. The result is thus a consequence of Proposition 2.32. $\qquad\Box$

Remark 2.41.

(1) Let K be a complete valued extension of $\mathbb{Q}_p$ with discrete valuation and perfect residue field and $M|K$ be an algebraic infinite degree arithmetically profinite extension. Then, by the fields of norms theory ([32]), $L = \widehat{M}$ is perfectoid and point (2) of the preceding theorem is already contained in [32].

(2) In [29] Scholze has obtained a different proof of point (2) of Theorem 2.36 and of Corollary 2.40.

Using Sen's method ([30]), Corollary 2.40 implies the following.

Theorem 2.42. *Let $L|E$ be a perfectoid field with algebraic closure $\overline{L}$. Then the functor $V \mapsto V \otimes_L \widehat{\overline{L}}$ induces an equivalence of categories between finite dimensional L-vector spaces and finite dimensional $\widehat{\overline{L}}$-vector spaces equipped with a continuous semi-linear action of $\mathrm{Gal}(\overline{L}|L)$. An inverse is given by the functor $W \mapsto W^{\mathrm{Gal}(\overline{L}|L)}$.*

Using this theorem one deduces by dévissage the following that we will use later.

Theorem 2.43. *Let $\mathfrak{m} \in |Y_F|$ and note*

$$B^+_{\widehat{\overline{F}},dR,\mathfrak{m}} = \prod_{\substack{\mathfrak{m}' \in |Y_{\widehat{\overline{F}}}| \\ \beta(\mathfrak{m}')=\mathfrak{m}}} B^+_{\widehat{\overline{F}},dR,\mathfrak{m}'}$$

the $B_{\widehat{\overline{F}}}\mathfrak{m}$-adic completion of $B_{\widehat{\overline{F}}}$ (see Definition 2.44). Then the functor

$$M \longmapsto M \otimes_{B^+_{F,dR,\mathfrak{m}}} B^+_{\widehat{\overline{F}},dR,\mathfrak{m}}$$

induces an equivalence of categories between finite type $B^+_{F,dR,\mathfrak{m}}$-modules and finite type $B^+_{\widehat{\overline{F}},dR,\mathfrak{m}}$-modules equipped with a continuous semi-linear action of $\mathrm{Gal}(\overline{F}|F)$. An inverse is given by the functor $W \mapsto W^{\mathrm{Gal}(\overline{F}|F)}$.

3. Divisors on Y

3.1. Zeros of elements of B

We see $|Y|$ as a subset of $\mathrm{Spm}(B)$. For $\mathfrak{m} \in |Y|$, we set

$$L_\mathfrak{m} = B/\mathfrak{m} \text{ and } \theta_\mathfrak{m} : B \twoheadrightarrow L_\mathfrak{m}.$$

We note $v_\mathfrak{m}$ the valuation on $L_\mathfrak{m}$ such that

$$v_\mathfrak{m}(\theta_\mathfrak{m}([a])) = v(a).$$

One has

$$\|\mathfrak{m}\| = q^{-v_\mathfrak{m}(\pi)/\deg \mathfrak{m}}$$

where $\|.\|$ was defined after Definition 2.17.

Definition 2.44. For $\mathfrak{m} \in |Y|$ define $B^+_{dR,\mathfrak{m}}$ as the $\mathfrak{m}$-adic completion of B.

The ring $B^+_{dR,\mathfrak{m}}$ is a discrete valuation ring with residue field $L_\mathfrak{m}$ and the natural map $B \to B^+_{dR,\mathfrak{m}}$ is injective. We note again $\theta_\mathfrak{m} : B^+_{dR,\mathfrak{m}} \to L_\mathfrak{m}$. We note

$$\operatorname{ord}_\mathfrak{m} : B^+_{dR,\mathfrak{m}} \longrightarrow \mathbb{N} \cup \{+\infty\}$$

its normalized valuation.

Example 2.45. If $E = \mathbb{F}_q((\pi))$ then $|Y| = |\mathbb{D}^*|$ and if $\mathfrak{m} \in |Y|$ corresponds to $x \in |\mathbb{D}^*|$ then $B^+_{dR,\mathfrak{m}} = \mathcal{O}_{\mathbb{D}^*,x}$.

Theorem 2.46. *For* $f \in B$, *the non-zero finite slopes of* Newt(f) *are the* $-\log_q \|\mathfrak{m}\|$ *with multiplicity* $\operatorname{ord}_\mathfrak{m}(f) \deg(\mathfrak{m})$ *where* $\mathfrak{m}$ *goes through the elements of* $|Y|$ *such that* $\theta_\mathfrak{m}(f) = 0$.

Indications on the proof. It suffices to prove that for any finite non-zero slope λ of Newt(f) there exists $\mathfrak{m} \in |Y|$ such that

$$q^{-\lambda} = \|\mathfrak{m}\| \text{ and } \theta_\mathfrak{m}(f) = 0.$$

As in Proposition 2.24 there is a metric d on $|Y|$ such that for all $\rho \in]0, 1[$, $\{\|\mathfrak{m}\| \geq \rho\}$ is complete. For $\mathfrak{m}_1, \mathfrak{m}_2 \in |Y|$, if $\theta_{\mathfrak{m}_1}(\mathfrak{m}_2) = \mathcal{O}_{L_{\mathfrak{m}_1}} x$ then

$$d(\mathfrak{m}_1, \mathfrak{m}_2) = q^{-v_{\mathfrak{m}_1}(x)/\deg \mathfrak{m}_2}.$$

We begin with the case when

$$f = \sum_{n \geq 0} [x_n] \pi^n \in W_{\mathcal{O}_E}(\mathcal{O}_F).$$

If $d \geq 0$ set

$$f_d = \sum_{n=0}^{d} [x_n] \pi^n.$$

For $d \gg 0$, λ appears as a slope of Newt(f_d) with the same multiplicity as in Newt(f). For each d, $f_d = [a_d].g_d$ for some $a_d \in \mathcal{O}_F$ and $g_d \in W_{\mathcal{O}_E}(\mathcal{O}_F)$ primitive. Thanks to the preceding results, we already know the result for each g_d. Thus, setting

$$X_d = \{\mathfrak{m} \in |Y| \mid \|\mathfrak{m}\| = q^{-\lambda} \text{ and } \theta_\mathfrak{m}(f_d) = 0\},$$

we know that for $d \gg 0$, $X_d \neq \emptyset$. Moreover $\#X_d$ is bounded when d varies. Now, if $\mathfrak{m} \in X_d$, looking at $v_\mathfrak{m}(\theta_\mathfrak{m}(f_{d+1}))$, one verifies that there exists $\mathfrak{m}' \in X_{d+1}$ such that

$$d(\mathfrak{m}, \mathfrak{m}') \leq q^{-\frac{(d+1)\lambda - v(x_0)}{\#X_{d+1}}}.$$

From this one deduces there exists a Cauchy sequence $(\mathfrak{m}_d)_{d \gg 0}$ where $\mathfrak{m}_d \in X_d$. If $\mathfrak{m} = \lim\limits_{d \to +\infty} \mathfrak{m}_d$ then $\theta_\mathfrak{m}(f) = 0$. This proves the theorem when $f \in W_{\mathcal{O}_E}(\mathcal{O}_F)$ and thus when $f \in B^b$.

The general case is now obtained in the same way by approximating $f \in B$ by a converging sequence of elements of B^b. $\qquad\square$

Example 2.47. As a corollary of the preceding theorem and Proposition 2.14, for $f \in B \setminus \{0\}$ one has $f \in B^\times$ if and only if for all $\mathfrak{m} \in |Y|$, $\theta_\mathfrak{m}(f) \neq 0$.

3.2. A factorization of elements of B when F is algebraically closed

Set

$$B_{[0,1[} = \{ f \in B \mid \exists A, \ \mathrm{Newt}(f)_{|]-\infty,A]} = +\infty \}$$
$$= \Big\{ \sum_{n \gg -\infty} [x_n]\pi^n \in W_{\mathcal{O}_E}(F)\big[\tfrac{1}{\pi}\big] \ \Big| \ \liminf_{n \to +\infty} \frac{v(x_n)}{n} \geq 0 \Big\}$$

(see Proposition 2.14 for this equality). Suppose F is algebraically closed. Given $f \in B$, applying Theorem 2.46, if $\lambda_1, \ldots, \lambda_r > 0$ are some slopes of $\mathrm{Newt}(f)$ one can write

$$f = g \cdot \prod_{i=1}^{r} \Big(1 - \frac{[a_i]}{\pi}\Big)$$

where $v(a_i) = \lambda_i$. Using this one proves the following.

Theorem 2.48. *Suppose F is algebraically closed. For $f \in B$ there exists a sequence $(a_i)_{i \geq 0}$ of elements of $\mathfrak{m}_F$ going to zero and $g \in B_{[0,1[}$ such that*

$$f = g \cdot \prod_{i=0}^{+\infty} \Big(1 - \frac{[a_i]}{\pi}\Big).$$

If moreover $f \in B^+$ there exists such a factorization with $g \in W_{\mathcal{O}_E}(\mathcal{O}_F)$.

3.3. Divisors and closed ideals of B

Definition 2.49. Define

$$\mathrm{Div}^+(Y) = \Big\{ \sum_{\mathfrak{m} \in |Y|} a_\mathfrak{m}[\mathfrak{m}] \ \Big| \ \forall I \subset]0,1[$$

$$\text{compact } \{\mathfrak{m} \mid a_\mathfrak{m} \neq 0 \text{ and } \|\mathfrak{m}\| \in I\} \text{ is finite} \Big\}.$$

For $f \in B \setminus \{0\}$ set

$$\mathrm{div}(f) = \sum_{\mathfrak{m} \in |Y|} \mathrm{ord}_{\mathfrak{m}}(f)[\mathfrak{m}] \in \mathrm{Div}^+(Y).$$

Definition 2.50. For $D \in \mathrm{Div}^+(Y)$ set $\mathfrak{a}_{-D} = \{f \in B \mid \mathrm{div}(f) \geq D\}$, an ideal of B.

For each $\mathfrak{m} \in |Y|$, the function $\mathrm{ord}_{\mathfrak{m}} : B \to \mathbb{N} \cup \{+\infty\}$ is upper semi-continuous. From this one deduces that the ideal $\mathfrak{a}_{-D}$ is closed in B.

Theorem 2.51. *The map $D \mapsto \mathfrak{a}_{-D}$ induces an isomorphism of monoids between $\mathrm{Div}^+(Y)$ and the monoid of closed non-zero ideals of B. Moreover, $D \leq D'$ if and only if $\mathfrak{a}_{-D'} \subset \mathfrak{a}_{-D}$.*

If $\mathfrak{a}$ is a closed ideal of B then

$$\mathfrak{a} \xrightarrow{\sim} \varprojlim_{I \subset]0,1[} B_I \mathfrak{a}.$$

The result is thus a consequence of the following.

Theorem 2.52. *For a compact non-empty interval $I \subset]0, 1[$:*

- *if $I = \{\rho\}$ with $\rho \notin |F^\times|$ then B_I is a field*
- *if not then the ring B_I is a principal ideal domain with maximal ideals $\{B_I \mathfrak{m} \mid \mathfrak{m} \in |Y|, \|\mathfrak{m}\| \in I\}$.*

Sketch of proof. The proof of this theorem goes as follows. First, given $f \in B_I$, one can define a bounded Newton polygon

$$\mathrm{Newt}_I(f).$$

If $f \in B^b$ then $\mathrm{Newt}_I(f)$ is obtained from $\mathrm{Newt}(f)$ by removing the slopes λ such that $q^{-\lambda} \notin I$ (if there is no such slope we define $\mathrm{Newt}_I(f)$ as the empty polygon). Now, if $f \in B$ the method used in Definition 2.11 to define the Newton polygon does not apply immediately. For example, if $I = \{\rho\}$ and $f \in B^b$ then $|f|_\rho$ does not determine $\mathrm{Newt}_I(f)$ (more generally, if $I = [q^{-\lambda_1}, q^{-\lambda_2}]$ one has the same problem with the definition of the pieces of $\mathrm{Newt}_I(f)$ where the slopes are λ_1 and λ_2). But one verifies that if $f_n \xrightarrow[n \to +\infty]{} f$ in B_I with $f_n \in B^b$ then for $n \gg 0$, $\mathrm{Newt}_I(f_n)$ is constant and does not depend on the sequence of B^b going to f.

Then, if $f \in B_I$ we prove a theorem that is analogous to Theorem 2.46: the slopes of $\mathrm{Newt}_I(f)$ are the $-\log_q \|\mathfrak{m}\|$ with multiplicity $\mathrm{ord}_{\mathfrak{m}}(f) \deg \mathfrak{m}$ where

$\|m\| \in I$ and $\theta_m(f) = 0$. This gives a factorization of any $f \in B_I$ as a product

$$f = g \cdot \prod_{i=1}^{r} \xi_i$$

where the ξ_i are irreducible primitive, $\|\xi_i\| \in I$, and $g \in B_I$ satisfies $\mathrm{Newt}_I(g) = \emptyset$.

Finally, we prove that if $f \in B_I$ satisfies $\mathrm{Newt}_I(f) = \emptyset$ then $f \in B_I^\times$. For $f \in B^b$ this is verified by elementary manipulations. Then if $f_n \xrightarrow[n \to +\infty]{} f$ in B_I with $f_n \in B^b$, since $\mathrm{Newt}_I(f) = \emptyset$ for $n \gg 0$ one has $\mathrm{Newt}_I(f_n) = \emptyset$. But then for $n \gg 0$ and $\rho \in I$,

$$|f_{n+1}^{-1} - f_n^{-1}|_\rho = |f_{n+1}|_\rho^{-1} \cdot |f_n|_\rho^{-1} \cdot |f_{n+1} - f_n|_\rho \xrightarrow[n \to +\infty]{} 0.$$

Thus the sequence $(f_n^{-1})_{n \gg 0}$ of B_I converges towards an inverse of f. □

Example 2.53. For $f, g \in B \setminus \{0\}$, f is a multiple of g if and only if $\mathrm{div}(f) \geq \mathrm{div}(g)$. In particular there is an injection of monoids

$$\mathrm{div} : B \setminus \{0\}/B^\times \hookrightarrow \mathrm{Div}^+(Y).$$

Corollary 2.54. *The set $|Y|$ is the set of closed maximal ideals of* B.

Remark 2.55. Even when F is spherically complete we do not know whether $\mathrm{div} : B^\times \to \mathrm{Div}^+(Y)$ is surjective or not (see 2.2).

4. Divisors on $Y/\varphi^{\mathbb{Z}}$

4.1. Motivation

Suppose we want to classify φ-modules over B, that is to say free B-modules equipped with a φ-semi-linear automorphism. This should be the same as vector bundles on

$$Y/\varphi^{\mathbb{Z}}$$

where Y is this "rigid" space we did not really define but that should satisfy

- $\Gamma(Y, \mathcal{O}_Y) = B$
- $|Y|$ is the set of "classical points" of Y.

Whatever this space Y is, since $\|\varphi(m)\| = \|m\|^q$, φ acts in a proper discontinuous way without fixed point on it. Thus, $Y/\varphi^{\mathbb{Z}}$ should have a sense as a "rigid" space. Let's look in more details at what this space $Y/\varphi^{\mathbb{Z}}$ should be.

It is easy to classify rank 1 φ-modules over B. They are parametrized by $\mathbb{Z}$: to $n \in \mathbb{Z}$ one associates the φ-module with basis e such that $\varphi(e) = \pi^n e$. We thus should have

$$\mathbb{Z} \xrightarrow{\sim} \mathrm{Pic}(Y/\varphi^{\mathbb{Z}})$$

$$n \longmapsto \mathscr{L}^{\otimes n}$$

where $\mathscr{L}$ is a line bundle such that for all $d \in \mathbb{Z}$,

$$H^0(Y/\varphi^{\mathbb{Z}}, \mathscr{L}^{\otimes d}) = \mathrm{B}^{\varphi = \pi^d}.$$

If $E = \mathbb{F}_q((\pi))$ and F is algebraically closed Hartl and Pink classified in [20] the φ-modules over B, that is to say φ-equivariant vector bundles on $\mathbb{D}^*$. The first step in the proof of their classification ([20] theo.4.3) is that if (M, φ) is such a φ-module then for $d \gg 0$

$$M^{\varphi = \pi^d} \neq 0.$$

The same type of result appears in the context of φ-modules over the Robba ring in the work of Kedlaya (see for example [22] prop. 2.1.5). From this one deduces that the line bundle $\mathscr{L}$ should be ample. We are thus led to study the scheme

$$\mathrm{Proj}\left(\bigoplus_{d \geq 0} \mathrm{B}^{\varphi = \pi^d} \right)$$

for which one hopes it is "uniformized by Y" and allows us to study φ-modules over B. In fact if (M, φ) is such a φ-module, we hope the quasi-coherent sheaf

$$\left(\bigoplus_{d \geq 0} M^{\varphi = \pi^d} \right)$$

is the vector bundle associated to the φ-equivariant vector bundle on Y attached to (M, φ).

4.2. Multiplicative structure of the graded algebra P

Definition 2.56. Define

$$P = \bigoplus_{d \geq 0} \mathrm{B}^{\varphi = \pi^d}$$

as a graded E-algebra. We note $P_d = \mathrm{B}^{\varphi = \pi^d}$ the degree d homogeneous elements.

In fact we could replace B by B^+ in the preceding definition since

$$\mathrm{B}^{\varphi = \pi^d} = (\mathrm{B}^+)^{\varphi = \pi^d}$$

(Corollary 2.15). One has $P_0 = E$.

Definition 2.57. Define

$$\text{Div}^+(Y/\varphi^{\mathbb{Z}}) = \{D \in \text{Div}^+(Y) \mid \varphi^* D = D\}.$$

There is an injection

$$|Y|/\varphi^{\mathbb{Z}} \hookrightarrow \text{Div}^+(Y/\varphi^{\mathbb{Z}})$$
$$\mathfrak{m} \longrightarrow \sum_{n\in\mathbb{Z}} [\varphi^n(\mathfrak{m})]$$

that makes $\text{Div}^+(Y/\varphi^{\mathbb{Z}})$ *a free abelian monoid on* $|Y|/\varphi^{\mathbb{Z}}$. *If* $x \in \text{B}^{\varphi=\pi^d} \setminus \{0\}$ *then* $\text{div}(x) \in \text{Div}^+(Y/\varphi^{\mathbb{Z}})$.

Theorem 2.58. *If* F *is algebraically closed the morphism of monoids*

$$\text{div} : \left(\bigcup_{d\geq 0} P_d \setminus \{0\} \right) / E^\times \longrightarrow \text{Div}^+(Y/\varphi^{\mathbb{Z}})$$

is an isomorphism.

Let us note the following important corollary.

Corollary 2.59. *If* F *is algebraically closed the graded algebra* P *is graded factorial with irreducible elements of degree* 1.

In the preceding theorem, the injectivity is an easy application of Theorem 2.51. In fact, if $x \in P_d$ and $y \in P_{d'}$ are non-zero elements such that $\text{div}(x) = \text{div}(y)$ then $x = uy$ with $u \in \text{B}^\times$. But $\text{B}^\times = (\text{B}^b)^\times$ (see the comment after Proposition 2.14). Thus,

$$u \in (\text{B}^b)^{\varphi=\pi^{d-d'}} = \begin{cases} 0 \text{ if } d \neq d' \\ E \text{ if } d = d'. \end{cases}$$

The surjectivity uses Weierstrass products. For this, let $x \in W_{\mathcal{O}_E}(\mathcal{O}_F)$ be a primitive degree d element and $D = \text{div}(x)$ its divisor. We are looking for $f \in P_d \setminus \{0\}$ satisfying

$$\text{div}(f) = \sum_{n\in\mathbb{Z}} \varphi^n(D).$$

Up to multiplying x by a unit we can suppose

$$x \in \pi^d + W_{\mathcal{O}_E}(\mathfrak{m}_F).$$

Then the infinite product

$$\Pi^+(x) = \prod_{n\geq 0} \left(\frac{\varphi^n(x)}{\pi^d} \right)$$

converges. For example, if $x = \pi - [a]$,

$$\Pi^+(\pi - [a]) = \prod_{n \geq 0} \left(1 - \frac{[a^{q^n}]}{\pi}\right).$$

One has

$$\text{div}(\Pi^+(x)) = \sum_{n \geq 0} \varphi^n(D).$$

We then would like to define

$$\text{``}\Pi^-(x) = \prod_{n < 0} \varphi^n(x)\text{''}$$

and then set

$$\Pi(x) = \Pi^+(x).\Pi^-(x)$$

that would satisfy $\Pi(x) \in P_d$ and $\text{div}(\Pi(x)) = \sum_{n \in \mathbb{Z}} \varphi^n(D)$. But the infinite product defining $\Pi^-(x)$ does not converge. Nevertheless let us remark it satisfies formally the functional equation

$$\varphi(\Pi^-(x)) = x\Pi^-(x).$$

Moreover, if $a = x \mod \pi$, $a \in \mathfrak{m}_F$, if we are trying to define $\Pi^-(x)$ modulo π, one should have formally

$$\prod_{n<0} \varphi^n(x) \mod \pi = \prod_{n<0} a^{q^n} = a^{\sum_{n<0} q^n} = a^{\frac{1}{q-1}}.$$

This means that up to an $\mathbb{F}_q^\times$-multiple one would like to define $\Pi^-(x)$ modulo π as a solution of the Kümmer equation $X^{q-1} - a = 0$. Similarly, for an element $y \in 1 + \pi^k W_{\mathcal{O}_E}(\mathcal{O}_F)$ where $k \geq 1$, via the identification

$$1 + \pi^k W_{\mathcal{O}_E}(\mathcal{O}_F)/1 + \pi^{k+1} W_{\mathcal{O}_E}(\mathcal{O}_F) \xrightarrow{\sim} \mathcal{O}_F$$

if

$$y \mod 1 + \pi^k W_{\mathcal{O}_E}(\mathcal{O}_F) \longmapsto b$$

one would have formally

$$\prod_{n<0} \varphi^n(y) \mod 1 + \pi^{k+1} W_{\mathcal{O}_E}(\mathcal{O}_F) \longmapsto \sum_{n<0} b^{q^{-n}}$$

that is formally a solution of the Artin–Schreier equation $X^q - X - b = 0$ (the remark that one can write solutions of Artin–Schreier equations in F as such non-converging series is due to Abhyankar, see [26]).

In fact, we have the following easy proposition whose proof is by successive approximations, solving first a Kümmer and then Artin–Schreier equations.

Proposition 2.60. *Suppose F is algebraically closed and let $z \in W_{\mathcal{O}_E}(\mathcal{O}_F)$ be a primitive element. Up to an $E^{\times}$-multiple there is a unique $\Pi^-(z) \in B^{b,+} \setminus \{0\}$ such that $\varphi(\Pi^-(z)) = z\Pi^-(z)$.*

Define $\Pi^-(x)$ using the preceding proposition. It is well defined up to an $E^{\times}$-multiple. Moreover

$$\varphi(\Pi^-(x)) = x\Pi^-(x) \implies \varphi\big(\mathrm{div}(\Pi^-(x))\big) = \mathrm{div}(\Pi^-(x)) + \underbrace{\mathrm{div}(x)}_{D}$$

$$\implies \mathrm{div}(\Pi^-(x)) = \sum_{n<0} \varphi^n(D).$$

Setting $\Pi(x) = \Pi^+(x)\Pi^-(x)$, this is a solution to our problem:

- $\Pi(x) \in P_d \setminus \{0\}$
- $\mathrm{div}(\Pi(x)) = \sum_{n\in\mathbb{Z}} \varphi^n(D)$.

4.3. Weierstrass products and the logarithm of a Lubin–Tate group

We use the notation from Section 2.4. For $\epsilon \in \mathfrak{m}_F \setminus \{0\}$ and $u_\epsilon = \dfrac{[\epsilon]_Q}{[\epsilon^{1/q}]_Q}$ one has

$$\varphi^n(u_\epsilon) = \frac{[\pi^n]_{\mathcal{LT}}([\epsilon]_Q)}{[\pi^{n-1}]_{\mathcal{LT}}([\epsilon]_Q)}$$

and thus

$$\Pi^+(u_\epsilon) = \prod_{n\geq 0} \left(\frac{\varphi^n(u_\epsilon)}{\pi}\right)$$

$$= \frac{1}{\pi[\epsilon^{1/q}]_Q} \cdot \lim_{n\to+\infty} \pi^{-n}[\pi^n]_{\mathcal{LT}}\big([\epsilon]_Q\big)$$

$$= \frac{1}{\pi[\epsilon^{1/q}]_Q} \log_{\mathcal{LT}}\big([\epsilon]_Q\big)$$

where $\log_{\mathcal{LT}}$ is the logarithm of the Lubin–Tate group law $\mathcal{LT}$. Moreover, one can take

$$\Pi^-(u_\epsilon) = \pi[\epsilon^{1/q}]_Q$$

and thus

$$\Pi(u_\epsilon) = \log_{\mathcal{LT}}([\epsilon]_Q).$$

Thus, the Weierstrass product $\Pi(u_\epsilon)$ is given by the Weierstrass product expansion of $\log_{\mathcal{LT}}$ (see the end of Section 1.1.3). In fact we have the following period isomorphism.

Theorem 2.61. *The logarithm induces an isomorphism of E-Banach spaces*

$$\left(\mathfrak{m}_F, \underset{\mathcal{LT}}{+} \right) \xrightarrow{\sim} B^{\varphi=\pi}$$

$$\epsilon \longmapsto \log_{\mathcal{LT}}\left([\epsilon]_Q \right).$$

Remark 2.62. For $r, r' > 0$ the restrictions of v_r and $v_{r'}$ to $B^{\varphi=\pi}$ induce equivalent norms. This equivalence class of norms defines the Banach space topology of the preceding theorem. This is the same topology as the one induced by the embedding $B^{\varphi=\pi^d} \subset B$.

The Banach space topology on $(\mathfrak{m}_F, \underset{\mathcal{LT}}{+})$ is the one defined by the lattice $1 + \{x \in \mathfrak{m}_F \mid v(x) \geq r\}$ for any $r > 0$.

One has the formula

$$\log_{\mathcal{LT}}\left([\epsilon]_Q \right) = \lim_{n \to +\infty} \pi^n \log_{\mathcal{LT}}\left([\epsilon^{q^{-n}}] \right).$$

If $\mathcal{LT}$ is the Lubin–Tate group law whose logarithm is

$$\log_{\mathcal{LT}} = \sum_{n \geq 0} \frac{T^{q^n}}{\pi^n}$$

we then have the formula

$$\log_{\mathcal{LT}}\left([\epsilon]_Q \right) = \sum_{n \in \mathbb{Z}} \left[\epsilon^{q^{-n}} \right] \pi^n.$$

Remark 2.63. The preceding series $\sum_{n \in \mathbb{Z}} \left[\epsilon^{q^{-n}} \right] \pi^n$ is a Witt bivector, an element of $BW_{\mathcal{O}_E}(\mathcal{O}_F)$ (see the end of Section 1.2.1). The fact that such a series makes sense in the Witt bivectors is an essential ingredient in the proof of Theorem 2.61. For $d > 1$ we don't have such a description of the Banach space $B^{\varphi=\pi^d}$.

Suppose now $E = \mathbb{Q}_p$ and let $\mathcal{LT}$ be the formal group law with logarithm $\sum_{n \geq 0} \frac{T^{p^n}}{p^n}$. Let

$$E(T) = \exp\left(\sum_{n \geq 0} \frac{T^{p^n}}{p^n} \right) \in \mathbb{Z}_p[\![T]\!]$$

be the Artin–Hasse exponential:

$$E : \mathcal{LT} \xrightarrow{\sim} \mathbb{G}_m.$$

There is then a commutative diagram of isomorphisms

$$\left(\mathfrak{m}_F, + \atop \mathcal{LT} \right) \xrightarrow{\sim} B^{\varphi=p}$$
$$E \Big\downarrow \simeq \qquad \nearrow \atop \log \circ [\cdot]$$
$$\left(1 + \mathfrak{m}_F, \times \right)$$

where the horizontal map is $\epsilon \mapsto \sum_{n \in \mathbb{Z}} [\epsilon^{p^{-n}}] p^n$. We thus find back the usual formula: $t = \log[\epsilon]$ for $\epsilon \in 1 + \mathfrak{m}_F$. More precisely, for $\epsilon \in 1 + \mathfrak{m}_F$ and the group law $\mathbb{G}_m$ one has

$$u_{\epsilon-1} = 1 + \left[\epsilon^{\frac{1}{p}} \right] + \cdots + \left[\epsilon^{\frac{p-1}{p}} \right].$$

If $\mathfrak{m} = (u_{\epsilon-1})$ then $\epsilon \in \mathcal{R}(C_{\mathfrak{m}})$ is a generator of $\mathbb{Z}_p(1)$. Moreover if $\rho = |\epsilon - 1|^{1-1/p}$ we have

$$B^+_{cris}(C_{\mathfrak{m}}) = B^+_{cris,\rho}$$

where $B^+_{cris}(C_{\mathfrak{m}})$ is the crystalline ring of periods attached to $C_{\mathfrak{m}}$ ([16]) and $B^+_{cris,\rho}$ is the ring defined at the end of Section 1.2.1. Then $t = \log[\epsilon]$ is the usual period of μ_{p^∞} over $C_{\mathfrak{m}}$.

5. The curve

5.1. The fundamental exact sequence

Using the results from the preceding section we give a new proof of the fundamental exact sequence. In fact this fundamental exact sequence is a little bit more general than the usual one. If $t \in P_1 \setminus \{0\}$ we will say t is associated to $\mathfrak{m} \in |Y|$ if $\operatorname{div}(t) = \sum_{n \in \mathbb{Z}} [\varphi^n(\mathfrak{m})]$.

Theorem 2.64. *Suppose F is algebraically closed. Let $t_1, \ldots, t_n \in P_1$ be associated to $\mathfrak{m}_1, \ldots, \mathfrak{m}_n \in |Y|$ and such that for $i \neq j$, $t_i \notin Et_j$. Let $a_1, \ldots, a_n \in \mathbb{N}_{\geq 1}$ and set $d = \sum_i a_i$. Then for $r \geq 0$ there is an exact sequence*

$$0 \longrightarrow P_r . \prod_{i=1}^n t_i^{a_i} \longrightarrow P_{d+r} \xrightarrow{u} \prod_{i=1}^n B^+_{dR,\mathfrak{m}_i} / B^+_{dR,\mathfrak{m}_i} \mathfrak{m}_i^{a_i} \longrightarrow 0.$$

Proof. For $x \in P_{d+r}$, $u(x) = 0$ if and only if

$$\operatorname{div}(x) \geq \sum_{i=1}^{n} a_i[\mathfrak{m}_i].$$

But since $\operatorname{div}(x)$ is invariant under φ this is equivalent to

$$\operatorname{div}(x) \geq \sum_{i=1}^{n} a_i \sum_{n\in\mathbb{Z}}[\varphi^n(\mathfrak{m}_i)] = \operatorname{div}\left(\prod_{i=1}^{n} t_i^{a_i}\right).$$

According to Theorem 2.51 this is equivalent to

$$x = y \cdot \prod_{i=1}^{n} t_i^{a_i}$$

for some $y \in B$ (see Example 2.53). But such a y satisfies automatically $\varphi(y) = \pi^r y$.

By induction, the surjectivity of u reduces to the case $n = 1$ and $a_1 = 1$. Let us note $\mathfrak{m} = \mathfrak{m}_1$. We have to prove that the morphism $B^{\varphi=\pi} \xrightarrow{\theta_\mathfrak{m}} C_\mathfrak{m}$ is surjective. Note $\mathcal{G}$ the formal group associated to the Lubin–Tate group law $\mathcal{LT}$. We use the isomorphism

$$\mathcal{G}(\mathcal{O}_F) \xrightarrow{\sim} B^{\varphi-\pi}$$

of Theorem 2.61 together with the isomorphism

$$X(\mathcal{G})(\mathcal{O}_{C_\mathfrak{m}}) \xrightarrow{\sim} \mathcal{G}(\mathcal{O}_F)$$

of Section 2.4. One verifies the composite

$$X(\mathcal{G})(\mathcal{O}_{C_\mathfrak{m}}) \longrightarrow \mathcal{G}(\mathcal{O}_F) \longrightarrow B^{\varphi=\pi} \xrightarrow{\theta_\mathfrak{m}} C_\mathfrak{m}$$

is given by

$$\left(x^{(n)}\right)_{n\geq 0} \longmapsto \log_{\mathcal{LT}}\left(x^{(0)}\right).$$

We conclude since $C_\mathfrak{m}$ is algebraically closed. $\qquad\square$

We will use the following corollary.

Corollary 2.65. *Suppose F is algebraically closed. For $t \in P_1\setminus\{0\}$ associated to $\mathfrak{m} \in |Y|$ there is an isomorphism of graded E-algebras*

$$P/tP \xrightarrow{\sim} \{f \in C_\mathfrak{m}[T] \mid f(0) \in E\}$$

$$\sum_{d\geq 0} x_d \bmod tP \longmapsto \sum_{d\geq 0} \theta_\mathfrak{m}(x_d)T^d.$$

5.2. The curve when F is algebraically closed

Theorem 2.66. *Suppose F is algebraically closed. The scheme $X = Proj(P)$ is an integral noetherian regular scheme of dimension 1. Moreover:*

(1) *For $t \in P_1 \setminus \{0\}$, $D^+(t) = \mathrm{Spec}\left(P\left[\frac{1}{t}\right]_0\right)$ where $P\left[\frac{1}{t}\right]_0 = B\left[\frac{1}{t}\right]^{\varphi=\mathrm{Id}}$ is a principal ideal domain.*

(2) *For $t \in P_1 \setminus \{0\}$, $V^+(t) = \{\infty_t\}$ with ∞_t a closed point of X and if t is associated to $\mathfrak{m} \in |Y|$ there is a canonical identification of D.V.R.'s*

$$\mathcal{O}_{X,\infty_t} = B^+_{dR,\mathfrak{m}}.$$

(3) *If $|X|$ stands for the set of closed points of X, the application*

$$(P_1 \setminus \{0\})/E^\times \longrightarrow |X|$$
$$E^\times t \longmapsto \infty_t$$

is a bijection.

(4) *Let us note $E(X)$ the field of rational functions on X, that is to say the stalk of $\mathcal{O}_X$ at the generic point. Then, for all $f \in E(X)^\times$ one has*

$$\deg(\mathrm{div}(f)) = 0$$

where for $x \in |X|$ we set $\deg(x) = 1$.

Sketch of proof. As a consequence of Corollary 2.59 the ring $B_e := P\left[\frac{1}{t}\right]_0$ is factorial with irreducible elements the $\frac{t'}{t}$ where $t' \notin Et$. To prove it is a P.I.D. it thus suffices to verify those irreducible elements generate a maximal ideal. But for such a $t' \notin Et$, if t' is associated to $\mathfrak{m}' \in |Y|$ since $\theta_{\mathfrak{m}'}(t) \neq 0$, $\theta_{\mathfrak{m}'}$ induces a morphism $B_e \to C_{\mathfrak{m}'}$. Using the fundamental exact sequence one verifies it is surjective with kernel the principal ideal generated by $\frac{t'}{t}$.

Now, if $A = \{f \in C_{\mathfrak{m}}[T] \mid f(0) \in E\}$, one verifies $Proj(A)$ has only one element, the homogeneous prime ideal (0). Using Corollary 2.65 one deduces that $V^+(t) \simeq Proj(A)$ is one closed point of X.

We have the following description

$$\mathcal{O}_{X,\infty_t} = \left\{ \frac{x}{y} \in \mathrm{Frac}(P) \mid x \in P_d, \ y \in P_d \setminus t P_{d-1} \text{ for some } d \geq 0 \right\}$$

with uniformizing element $\frac{t}{t'}$ for some $t' \in P_1 \setminus Et$. Now, if $y \in P_d \setminus t P_{d-1}$, $y \in (B^+_{dR,\mathfrak{m}})^\times$ since according to the fundamental exact sequence $\theta_{\mathfrak{m}}(y) \neq 0$. We thus have

$$\mathcal{O}_{X,\infty_t} \subset B^+_{dR,\mathfrak{m}}.$$

Using the fundamental exact sequence one verifies this embedding of D.V.R.'s induces an isomorphism at the level of the residue fields. Moreover a uniformizing element of $\mathcal{O}_{X,\infty_t}$ is a uniformizing element of $B^+_{dR,\mathrm{m}}$. It thus induces $\mathcal{O}_{X,\infty} \xrightarrow{\sim} B^+_{dR,\mathrm{m}}$.

Other assertions of the theorem are easily verified. □

The following proposition makes clear the difference between X and $\mathbb{P}^1$ and will have important consequences on the classification of vector bundles on X. It is deduced from Corollary 2.65.

Proposition 2.67. *For a closed point $\infty \in |X|$ let* $B_e = \Gamma(X \setminus \{\infty\}, \mathcal{O}_X) \subset E(X)$, v_∞ *the valuation on $E(X)$ associated to ∞ and*

$$\deg = -v_{\infty|B_e} : B_e \longrightarrow \mathbb{N} \cup \{-\infty\}.$$

Then the couple $(B_e, \deg)$ is almost euclidean in the sense that

$$\forall x, y \in B_e, \ y \neq 0, \ \exists a, b \in B_e \ x = ay + b \text{ and } \deg(b) \leq \deg(y).$$

Moreover $(B_e, \deg)$ is not euclidean.

5.3. The curve in general

Note $\overline{F}$ an algebraic closure of F and $G_F = \mathrm{Gal}(\overline{F}|F)$. We put subscripts to indicate the dependence on the field F of the preceding constructions. The curve $X_{\widehat{\overline{F}}}$ of the preceding section is equipped with an action of G_F.

Theorem 2.68. *The scheme $X_F = Proj(P_F)$ is an integral noetherian regular scheme of dimension 1. It satisfies the following properties.*

(1) *The morphism of graded algebras $P_F \to P_{\widehat{\overline{F}}}$ induces a morphism*

$$\alpha : X_{\widehat{\overline{F}}} \longrightarrow X_F$$

satisfying:

- *for $x \in |X_F|$, $\alpha^{-1}(x)$ is a finite set of closed points of $|X_{\widehat{\overline{F}}}|$.*
- *for $x \in |X_{\widehat{\overline{F}}}|$:*
  - *if $G_F.x$ is infinite then $\alpha(x)$ is the generic point of X_F,*
  - *if $G_F.x$ is finite then $\alpha(x)$ is a closed point of X_F.*
- *it induces a bijection*

$$|X_{\widehat{\overline{F}}}|^{G_F-\mathrm{fin}}/G_F \xrightarrow{\sim} |X_F|$$

where $|X_{\widehat{\overline{F}}}|^{G_F-\mathrm{fin}}$ is the set of closed points with finite G_F-orbit.

(2) *For $x \in |X_F|$ set $\deg(x) = \#\alpha^{-1}(x)$. Then for $f \in E(X_F)^{\times}$*

$$\deg(\operatorname{div}(f)) = 0.$$

(3) *For $\mathfrak{m} \in |Y_F|$ define*

$$\mathfrak{p}_{\mathfrak{m}} = \Big\{ \sum_{d \geq \deg \mathfrak{m}} x_d \in P_F \mid x_d \in P_{F,d}, \ \operatorname{div}(x_d) \geq \sum_{n \in \mathbb{Z}} [\varphi^n(\mathfrak{m})] \Big\},$$

a prime homogeneous ideal of P. Then

$$|Y_F|/\varphi^{\mathbb{Z}} \xrightarrow{\sim} |X_F|$$
$$\varphi^{\mathbb{Z}}(\mathfrak{m}) \longmapsto \mathfrak{p}_{\mathfrak{m}}$$

and there is an identification $\mathcal{O}_{X_F, \mathfrak{p}_{\mathfrak{m}}} = B^+_{F,dR,\mathfrak{m}}$.

Sketch of proof. Let us give a few indications on the tools used in the proof.

Proposition 2.69. *One has $P^{G_F}_{\widehat{\overline{F}}} = P_F$.*

Proof. The divisor of $f \in P^{G_F}_{\widehat{\overline{F}},d}$ being G_F-invariant, there exists a primitive degree d element $x \in W_{\mathcal{O}_E}(\mathcal{O}_{\widehat{\overline{F}}})$ such that $\operatorname{div}(f) = \sum_{n \in \mathbb{Z}} \varphi^n(\operatorname{div}(x))$ and for all $\sigma \in G_F$, $(\sigma(x)) = (x)$. According to Proposition 2.33, one can choose $x \in W_{\mathcal{O}_E}(\mathcal{O}_F)$ and even $x \in \pi^d + W_{\mathcal{O}_E}(\mathfrak{m}_F)$. The Weierstrass product

$$\Pi^+(x) = \prod_{n \geq 0} \left(\frac{\varphi^n(x)}{\pi^d} \right)$$

is convergent in B_F. Applying Theorem 2.51 (see Example 2.53) one finds there exists $g \in B_{\widehat{\overline{F}},[0,1[}$ (see 3.2) such that

$$f = \Pi^+(x).g.$$

Of course, g is G_F-invariant and one concludes since

$$B^{G_F}_{\widehat{\overline{F}},[0,1[} = B_{F,[0,1[}.$$

$\square$

Let now $t \in P_{F,1} \setminus \{0\}$ and look at

$$B_{F,e} = B_F\Big[\tfrac{1}{t}\Big]^{\varphi = \operatorname{Id}} = \Gamma(X_F \setminus V^+(t), \mathcal{O}_{X_F})$$
$$B_{\widehat{\overline{F}},e} = B_{\widehat{\overline{F}}}\Big[\tfrac{1}{t}\Big]^{\varphi = \operatorname{Id}} = \Gamma(X_{\widehat{\overline{F}}} \setminus V^+(t), \mathcal{O}_{X_{\widehat{\overline{F}}}}).$$

According to the preceding proposition

$$B_{F,e} = (B_{\widehat{\overline{F}},e})^{G_F}.$$

We want to prove $B_{F,e}$ is a Dedekind ring such that the maps $I \mapsto B_{\widehat{F},e}I$ and $J \mapsto J \cap B_{F,e}$ are inverse bijections between non-zero ideals of $B_{F,e}$ and non-zero G_F-invariant ideals of $B_{\widehat{F},e}$. The key tool is the following cohomological computation.

Theorem 2.70. *For* $\chi : G_F \to E^\times$ *a continuous character one has*

$$H^1\big(G_F, B_{\widehat{F},e}(\chi)\big) = 0$$

$$H^0\big(G_F, B_{\widehat{F},e}(\chi)\big) \neq 0$$

where

$$H^1\big(G_F, B_{\widehat{F},e}(\chi)\big) := \varinjlim_{d \geq 0} H^1\big(G_F, t^{-d} P_{\widehat{F},d}(\chi)\big)$$

and $t^{-k}P_{\widehat{F},d}$ *is naturally an E-Banach space.*

Proof. We prove that for all $d \geq 1$, $H^1(G_F, P_{\widehat{F},d}(\chi)) = 0$. Let $\mathfrak{m} \in |Y_F|$ be associated to t. Note $\mathfrak{m}' = B_{\widehat{F}}\mathfrak{m} \in |Y_{\widehat{F}}|$ the unique element such that $\beta(\mathfrak{m}') = \mathfrak{m}$. For $d > 1$, using the fundamental exact sequence

$$0 \longrightarrow P_{\widehat{F},d-1}(\chi) \xrightarrow{\times t} P_{\widehat{F},d}(\chi) \xrightarrow{\theta_{\mathfrak{m}'}} C_{\mathfrak{m}'}(\chi) \longrightarrow 0$$

of G_F-modules together with the vanishing

$$H^1(G_F, C_{\mathfrak{m}'}(\chi)) = 0$$

(Theorem 2.42) one is reduced by induction to prove the case $d = 1$. Let $\mathcal{LT}$ be a Lubin–Tate group law. We have an isomorphism

$$P_{\widehat{F},1} \simeq \Big(\mathfrak{m}_{\widehat{F}}, \underset{\mathcal{LT}}{+}\Big).$$

For $r \in v(F^\times)_{>0}$ set

$$\mathfrak{m}_{\widehat{F},r} = \{x \in \mathfrak{m}_{\widehat{F}} \mid v(x) > r\}.$$

It defines a decreasing filtration of the Banach space $\Big(\mathfrak{m}_{\widehat{F}}, \underset{\mathcal{LT}}{+}\Big)$ by sub $\mathcal{O}_E$-modules. Moreover

$$\Big(\mathfrak{m}_{\widehat{F},r}/\mathfrak{m}_{\widehat{F},2r}, \underset{\mathcal{LT}}{+}\Big) = (\mathfrak{m}_{\widehat{F},r}/\mathfrak{m}_{\widehat{F},2r}, +).$$

It thus suffices to prove that for all $r \in \mathbb{Q}_{>0}$,

$$H^1\big(G_F, \mathfrak{m}_{\widehat{F},r}/\mathfrak{m}_{\widehat{F},2r}(\chi)\big) = 0$$

(discrete Galois cohomology). This is deduced from Theorem 2.31 which implies that all $r \in v(F^\times)_{>0}$ and $i > 0$,

$$H^i\left(G_F, \mathfrak{m}_{\widehat{\overline{F}},r}(\chi)\right) = 0.$$

To prove that $H^0(G_F, B_{\widehat{\overline{F}},e}(\chi)) \neq 0$ it suffices to prove $H^0(G_F, P_{\widehat{\overline{F}},1}(\chi)) \neq 0$. One checks easily that $H^0(G_F, \overline{F}(\chi)) \neq 0$ and thus for all $r \in v(F^\times)_{>0}$,

$$H^0\left(G_F, \mathfrak{m}_{\widehat{\overline{F}},r}/\mathfrak{m}_{\widehat{\overline{F}},2r}(\chi)\right) \neq 0.$$

Using the vanishing

$$H^1\left(G_F, \left(\mathfrak{m}_{\widehat{\overline{F}},2r}, \underset{LT}{+}\right)(\chi)\right) = 0$$

one deduces the morphism

$$H^0\left(G_F, \left(\mathfrak{m}_{\widehat{\overline{F}},r}, \underset{LT}{+}\right)(\chi)\right) \longrightarrow H^0\left(G_F, \mathfrak{m}_{\widehat{\overline{F}},r}/\mathfrak{m}_{\widehat{\overline{F}},2r}(\chi)\right)$$

is surjective and concludes.   □

Let now $f \in B_{F,e}$. Since $H^1(G_F, B_{\widehat{\overline{F}},e}) = 0$ one has

$$B_{F,e}/B_{F,e}f = \left(B_{\widehat{\overline{F}},e}/B_{\widehat{\overline{F}},e}f\right)^{G_F}.$$

Let

$$f = u.\prod_{i=1}^{r} f_i^{a_i}$$

be the decomposition of f in prime factors where $u \in B_{\widehat{\overline{F}},e}^\times = E^\times$. If f_i is associated to $\mathfrak{m}_i \in |Y_{\widehat{\overline{F}}}|$ then

$$B_{\widehat{\overline{F}},e}/B_{\widehat{\overline{F}},e}f \simeq \prod_{i=1}^{r} B_{\widehat{\overline{F}},dR,\mathfrak{m}_i}^+/B_{\widehat{\overline{F}},dR,\mathfrak{m}_i}^+ \mathfrak{m}_i^{a_i}.$$

Now, the finite subset $A = \{\mathfrak{m}_i\}_{1 \leq i \leq r} \subset |Y_{\widehat{\overline{F}}}|$ is stable under G_F and defines a subset $B = A/G_F \subset |Y_{\widehat{\overline{F}}}|^{G_F-\text{fin}} = |Y_F|$ (see Theorem 2.34). The multiplicity function $\mathfrak{m}_i \mapsto a_i$ on A is invariant under G_F and defines a function $\mathfrak{m} \mapsto a_\mathfrak{m}$ on B. Then, according to Theorem 2.43

$$\left(B_{\widehat{\overline{F}},e}/B_{\widehat{\overline{F}},e}f\right)^{G_F} \simeq \prod_{\mathfrak{m} \in A} B_{F,dR,\mathfrak{m}}^+/B_{F,dR,\mathfrak{m}}^+ \mathfrak{m}^{a_\mathfrak{m}}$$

and the functors $I \mapsto I^{G_F}$ and $J \mapsto \left(B_{\widehat{\overline{F}},e}/B_{\widehat{\overline{F}},e}f\right)J$ induce inverse bijections between G_F-invariant ideals of $B_{\widehat{\overline{F}},e}/B_{\widehat{\overline{F}},e}f$ and ideals of $B_{F,e}/B_{F,e}f$.

A ring A is a Dedekind ring if and only if for all $f \in A \setminus \{0\}$ the f-adic completion of A is isomorphic to a finite product of complete D.V.R.'s. From the preceding one deduces that $\mathrm{B}_{F,e}$ is a Dedekind ring such that the applications $I \mapsto I^{G_F}$ and $J \mapsto \mathrm{B}_{\widehat{F},e} J$ induce inverse bijections between

- non-zero ideals I of $\mathrm{B}_{\widehat{F},e}$ that are G_F-invariant and satisfy $I^{G_F} \neq 0$
- non-zero ideals J of $\mathrm{B}_{F,e}$.

But if I is a non-zero ideal of $\mathrm{B}_{\widehat{F},e}$ that is G_F-invariant, $I = (f)$, since $\mathrm{B}_{\widehat{F},e}^{\times} = E^{\times}$ there exists a continuous character

$$\chi : G_F \longrightarrow E^{\times}$$

such that for $\sigma \in G_F$, $\sigma(f) = \chi(\sigma)f$. According to Theorem 2.70, $H^0(G_F, \mathrm{B}_{\widehat{F},e}(\chi)) \neq 0$ and thus $I^{G_F} \neq 0$. Theorem 2.68 is easily deduced from those considerations. $\qquad\square$

With the notations from the preceding proof, if J is a fractional ideal of $\mathrm{B}_{F,e}$ there exists $f \in \mathrm{Frac}\left(\mathrm{B}_{\widehat{F},e}\right)$ well defined up to multiplication by $\mathrm{B}_{\widehat{F},e}^{\times} = E^{\times}$ such that $\mathrm{B}_{\widehat{F},e} J = \mathrm{B}_{\widehat{F},e} f$. This ideal being stable under G_F, there exists a continuous character

$$\chi_J : G_F \longrightarrow E^{\times}$$

such that for all $\sigma \in G_F$, $\sigma(f) = \chi_J(\sigma)f$. The arguments used in the proof of Theorem 2.68 give the following.

Theorem 2.71. *The morphism $J \mapsto \chi_J$ induces an isomorphism*

$$\mathcal{C}\ell(\mathrm{B}_{F,e}) \xrightarrow{\sim} \mathrm{Hom}(G_F, E^{\times}).$$

Let us remark the preceding theorem implies the following.

Theorem 2.72. *If $F'|F$ is a finite degree extension the morphism $P_F \to P_{F'}$ induces a finite étale cover $X_{F'} \to X_F$ of degree $[F' : F]$. If moreover $F'|F$ is Galois then $X_{F'} \to X_F$ is Galois with Galois group $\mathrm{Gal}(F'|F)$.*

5.4. Change of the base field E

By definition, the graded algebra P_F depends on the choice of the uniformizing element π of E. If the residue field of F is algebraically closed, the choice of another uniformizing element gives a graded algebra that is isomorphic to the preceding, but such an isomorphism is not canonical. In any case, taking the Proj, the curve X_F does not depend anymore on the choice of π. We now put a second subscript in our notations to indicate the dependence on E.

Proposition 2.73. *If $E'|E$ is a finite extension with residue field contained in F there is a canonical isomorphism*

$$X_{F,E'} \xrightarrow{\sim} X_{F,E} \otimes_E E'.$$

When $E' = E_h$ the degree h unramified extension of E with residue field $\mathbb{F}_{q^h} = F^{\varphi^h = \mathrm{Id}}$ the preceding isomorphism is described in the following way. One has $\mathrm{B}_{F,E_h} = \mathrm{B}_{F,E}$ with $\varphi_{E_h} = \varphi_E^h$. Thus, taking as a uniformizing element of E_h the uniformizing element π of E, one has

$$P_{F,E_h} = \bigoplus_{d \geq 0} \mathrm{B}_{F,E}^{\varphi_E^h = \pi^d}.$$

There is thus a morphism of graded algebras

$$P_{F,E,\bullet} \longrightarrow P_{F,E_h,h\bullet}$$

where the bullet "$\bullet$" indicates the grading. It induces an isomorphism

$$P_{F,E,\bullet} \otimes_E E_h \xrightarrow{\sim} P_{F,E_h,h\bullet}$$

and thus

$$X_{F,E} \otimes_E E_h = \mathrm{Proj}(P_{F,E,\bullet} \otimes_E E_h) \xrightarrow{\sim} \mathrm{Proj}(P_{F,E_h,h\bullet}) = \mathrm{Proj}(P_{F,E_h,\bullet}) = X_{F,E_h}.$$

Suppose we have fixed algebraic closures $\overline{F}$ and $\overline{E}$. We thus have a tower of finite étale coverings of $X_{F,E}$ with Galois group $\mathrm{Gal}(\overline{F}|F) \times \mathrm{Gal}(\overline{E}|E)$

$$\left(X_{F',E'}\right)_{F',E'} \longrightarrow X_{F,E}$$

where F' goes through the set of finite extensions of F in $\overline{F}$ and E' the set of finite extensions of E in $\overline{E}$. We can prove the following.

Theorem 2.74. *The tower of coverings $(X_{F',E'})_{F',E'} \to X_{F,E}$ is a universal covering and thus if $\bar{x}$ is a geometric point of $X_{F,E}$ then*

$$\pi_1(X_{F,E}, \bar{x}) \simeq \mathrm{Gal}(\overline{F}|F) \times \mathrm{Gal}(\overline{E}|E).$$

6. Vector bundles

6.1. Generalities

Definition 2.75. We note Bun_{X_F} the category of vector bundles on X_F.

Let ∞ be a closed point of $|X_F|$, $\mathrm{B}_{dR}^+ = \mathcal{O}_{X,\infty}$ and

$$\mathrm{B}_e = \Gamma(X \setminus \{\infty\}, \mathcal{O}_X).$$

We note t a uniformizing element of B_{dR}^+ and $B_{dR} = B_{dR}^+[\frac{1}{t}]$. Let $\mathscr{C}$ be the category of couples (M, W) where W is a free B_{dR}^+-module of finite type and $M \subset W[\frac{1}{t}]$ is a sub B_e-module of finite type (that is automatically projective since torsion free) such that

$$M \otimes_{B_e} B_{dR} \xrightarrow{\sim} W[\tfrac{1}{t}].$$

If F is algebraically closed, the ring B_e is a P.I.D. and such an M is a free module. There is an equivalence of categories

$$\mathrm{Bun}_{X_F} \xrightarrow{\sim} \mathscr{C}$$
$$\mathscr{E} \longmapsto \left(\Gamma(X \setminus \{\infty\}, \mathscr{E}), \widehat{\mathscr{E}}_\infty \right).$$

In particular if F is algebraically closed, isomorphism classes of rank n vector bundles are in bijection with the set

$$GL_n(B_e) \backslash GL_n(B_{dR}) / GL_n(B_{dR}^+).$$

If $\mathscr{E}$ corresponds to the pair (M, W) then Cech cohomology gives an isomorphism

$$R\Gamma(X, \mathscr{E}) \simeq \left[\overset{0}{M \oplus W} \xrightarrow{\partial} \overset{+1}{W[\tfrac{1}{t}]} \right]$$

where $\partial(x, y) = x - y$. In particular

$$H^0(X, \mathscr{E}) \simeq M \cap W$$
$$H^1(X, \mathscr{E}) \simeq W[\tfrac{1}{t}]/W + M.$$

6.2. Line bundles

6.2.1. Computation of the Picard group

One has the usual exact sequence

$$0 \longrightarrow E^\times \longrightarrow E(X)^\times \xrightarrow{\mathrm{div}} \mathrm{Div}(X) \longrightarrow \mathrm{Pic}(X) \longrightarrow 0$$
$$D \longmapsto [\mathcal{O}_X(D)]$$

where $\mathcal{O}_X(D)$ is the line bundle whose sections on the open subset U are

$$\Gamma(U, \mathcal{O}_X(D)) = \{f \in E(X) \mid \mathrm{div}(f)_{|U} + D_{|U} \geq 0\}.$$

Since the degree of a principal divisor is zero there is thus a degree function

$$\deg : \mathrm{Pic}(X) \longrightarrow \mathbb{Z}.$$

Definition 2.76. For $d \in \mathbb{Z}$ define

$$\mathcal{O}_X(d) = P[d],$$

a line bundle on X.

One has

$$H^0(X, \mathcal{O}_X(d)) = \begin{array}{ll} P_d & \text{if } d \geq 0 \\ 0 & \text{if } d < 0. \end{array}$$

If $d > 0$ and $t \in P_d \setminus \{0\}$, $V^+(t) = D$, a degree d Weil divisor on X, then

$$\times t : \mathcal{O}_X(D) \xrightarrow{\sim} \mathcal{O}_X(d).$$

In particular for all $d \in \mathbb{Z}$,

$$\deg(\mathcal{O}_X(d)) = d.$$

If F is algebraically closed, with the notation of Section 6.1, $B_{F,e}$ is a P.I.D. and since $B_{F,e}^\times = E^\times$

$$\mathrm{Pic}(X_F) \simeq E^\times \backslash B_{dR}^\times / (B_{dR}^+)^\times = B_{dR}^\times / (B_{dR}^+)^\times \xrightarrow[\sim]{\mathrm{ord}_\infty} \mathbb{Z}.$$

We thus obtain the following.

Proposition 2.77. *If F is algebraically closed then*

$$\deg : \mathrm{Pic}(X_F) \xrightarrow{\sim} \mathbb{Z}$$

with inverse $d \mapsto [\mathcal{O}_X(d)]$.

Suppose now F is general. There is thus an exact sequence of G_F-modules

$$0 \longrightarrow E^\times \longrightarrow E\left(X_{\widehat{\overline{F}}}\right)^\times \xrightarrow{\mathrm{div}} \mathrm{Div}^0\left(X_{\widehat{\overline{F}}}\right) \longrightarrow 0.$$

But according to Theorem 2.68

$$\mathrm{Div}^0(X_F) = \mathrm{Div}^0\left(X_{\widehat{\overline{F}}}\right)^{G_F}.$$

Applying $H^\bullet(G_F, -)$ to the preceding exact sequence one obtains a morphism

$$\mathrm{Div}^0(X_F) \longrightarrow H^1(G_F, E^\times) = \mathrm{Hom}(G_F, E^\times)$$
$$D \longmapsto \chi_D.$$

Theorem 2.71 translates in the following way.

Theorem 2.78. *The morphism $D \mapsto \chi_D$ induces an isomorphism*

$$\mathrm{Pic}^0(X_F) \xrightarrow{\sim} \mathrm{Hom}(G_F, E^\times).$$

6.2.2. Cohomology of line bundles

Suppose F is algebraically closed. With the notation of Section 6.1 the line bundle $\mathcal{O}_X(d[\infty])$ corresponds to the pair $(B_e, t^{-d}B_{dR}^+)$. The fact that $(B_e, \deg)$ is almost euclidean (2.67) is equivalent to saying that $B_{dR} = B_{dR}^+ + B_e$, that is to say $H^1(X, \mathcal{O}_X) = 0$. From this one obtains the following proposition.

Proposition 2.79. *If F is algebraically closed,*

$$H^1(X_F, \mathcal{O}_{X_F}(d)) = \begin{cases} 0 & \text{if } d \geq 0 \\ B_{dR}^+ / \text{Fil}^{-d} B_{dR}^+ + E & \text{if } d < 0. \end{cases}$$

Thus, like $\mathbb{P}^1$

$$H^1(X, \mathcal{O}_X) = 0.$$

But contrary to $\mathbb{P}^1$, $H^1(X, \mathcal{O}_X(-1))$ is non-zero and even infinite dimensional isomorphic to C/E where C is the residue field at a closed point of X.

Example 2.80. Let $t \in P_d = H^0(X, \mathcal{O}_X(d))$ be non-zero. It defines an exact sequence

$$0 \longrightarrow \mathcal{O}_X \xrightarrow{\times t} \mathcal{O}_X(d) \longrightarrow \mathcal{F} \longrightarrow 0$$

where $\mathcal{F}$ is a torsion coherent sheaf. If F is algebraically closed $H^1(X, \mathcal{O}_X) = 0$ and taking the global sections of the preceding exact sequence gives back the fundamental exact sequence (2.64).

Remark 2.81. If F is not algebraically closed then $H^1(X_F, \mathcal{O}_{X_F}) \neq 0$ in general (see 2.105).

6.3. The classification theorem when F is algebraically closed

6.3.1. Definition of some vector bundles

Suppose $\overline{\mathbb{F}}_q^F$ is algebraically closed and note for all $h \geq 1$, E_h the unramified extension of E with residue field $\mathbb{F}_{q^h} = F^{\varphi_E^h = Id}$. We thus have a pro-Galois cover

$$\left(X_{F, E_h}\right)_{h \geq 1} \longrightarrow X_{F, E}$$

with Galois group $\widehat{\mathbb{Z}}$. We note $X := X_{F, E}$, $X_h := X_{F, E_h}$ and $\pi_h : X_h \to X$. If F is algebraically closed the morphism π_h is totally decomposed at each point of X:

$$\forall x \in X, \ \#\pi_h^{-1}(x) = h.$$

For $\mathscr{E} \in \mathrm{Bun}_X$ one has

$$\deg(\pi_h^* \mathscr{E}) = h \deg(\mathscr{E})$$
$$\mathrm{rk}(\pi_h^* \mathscr{E}) = \mathrm{rk}(\mathscr{E}).$$

For example, $\pi_h^* \mathcal{O}_{X_h}(d) = \mathcal{O}_{X_h}(hd)$. If $\mathscr{E} \in \mathrm{Bun}_{X_h}$ one has

$$\deg(\pi_{h*} \mathscr{E}) = \deg(\mathscr{E})$$
$$\mathrm{rk}(\pi_{h*} \mathscr{E}) = h \, \mathrm{rk}(\mathscr{E}).$$

Definition 2.82. For $\lambda \in \mathbb{Q}$, $\lambda = \frac{d}{h}$ with $d \in \mathbb{Z}$, $h \in \mathbb{N}_{\geq 1}$ and $(d, h) = 1$ define

$$\mathcal{O}_X(\lambda) = \pi_{h*} \mathcal{O}_{X_h}(d).$$

We have

$$\mu(\mathcal{O}_X(\lambda)) = \lambda$$

where $\mu = \dfrac{\deg}{\mathrm{rk}}$ is the Harder–Narasimhan slope. The following properties are satisfied

$$\mathcal{O}_X(\lambda) \otimes \mathcal{O}_X(\mu) \simeq \bigoplus_{\text{finite}} \mathcal{O}_X(\lambda + \mu)$$
$$\mathcal{O}_X(\lambda)^\vee = \mathcal{O}_X(-\lambda)$$
$$H^0(X, \mathcal{O}_X(\lambda)) = 0 \text{ if } \lambda < 0$$
$$\mathrm{Hom}(\mathcal{O}_X(\lambda), \mathcal{O}_X(\mu)) = \bigoplus_{\text{finite}} H^0(X, \mathcal{O}_X(\mu - \lambda)) = 0 \text{ if } \lambda > \mu.$$

If F is algebraically closed then if $\lambda = \frac{d}{h}$ with $(d, h) = 1$

$$H^1(X, \mathcal{O}_X(\lambda)) = H^1(X_h, \mathcal{O}_{X_h}(d))$$
$$= 0 \text{ if } \lambda \geq 0$$

and thus

$$\mathrm{Ext}^1(\mathcal{O}_X(\lambda), \mathcal{O}_X(\mu)) = \bigoplus_{\text{finite}} H^1(X, \mathcal{O}_X(\mu - \lambda))$$
$$= 0 \text{ if } \lambda \leq \mu.$$

6.3.2. Statement of the theorem

Here is the main theorem about vector bundles. It is an analogue of Kedlaya ([21],[22]) or Hartl–Pink ([20]) classification theorems.

Theorem 2.83. *Suppose F is algebraically closed.*

(1) *The semi-stable vector bundles of slope λ on X are the direct sums of $\mathcal{O}_X(\lambda)$.*

(2) *The Harder–Narasimhan filtration of a vector bundle on X is split.*

(3) *There is a bijection*

$$\{\lambda_1 \geq \cdots \geq \lambda_n \mid n \in \mathbb{N}, \ \lambda_i \in \mathbb{Q}\} \xrightarrow{\sim} \mathrm{Bun}_X / \sim$$

$$(\lambda_1, \ldots, \lambda_n) \longmapsto \Big[\bigoplus_{i=1}^{n} \mathcal{O}_X(\lambda_i) \Big].$$

In this theorem, point (3) is equivalent to points (1) and (2) together. Moreover, since for $\lambda \geq \mu$ one has $\mathrm{Ext}^1(\mathcal{O}_X(\lambda), \mathcal{O}_X(\mu)) = 0$, point (2) is a consequence of point (1).

Remark 2.84. In any category with Harder–Narasimhan filtrations ([1]), the category of semi-stable objects of slope λ is abelian with simple objects the stable objects of slope λ. The preceding theorem tells more in our particular case: this category is semi-simple with one simple object $\mathcal{O}_X(\lambda)$. One computes easily that $\mathrm{End}(\mathcal{O}_X(\lambda)) = D_\lambda$ the division algebra with invariant λ over E. From this one deduces that the functor $\mathcal{E} \mapsto \mathrm{Hom}(\mathcal{O}_X(\lambda), \mathcal{E})$ induces an equivalence between the abelian category of semi-simple vector bundles of slope λ and the category of finite dimensional D_λ^{opp}-vector spaces. An inverse is given by the functor $V \mapsto V \otimes_{D_\lambda} \mathcal{O}_X(\lambda)$.

Example 2.85. The functors $V \mapsto V \otimes_E \mathcal{O}_X$ and $\mathcal{E} \mapsto H^0(X, \mathcal{E})$ are inverse equivalences between the category of finite dimensional E-vector spaces and the category of semi-stable vector bundles of slope 0 on X.

6.3.3. Proof of the classification theorem: a dévissage

We will now sketch a proof of Theorem 2.83. We mainly stick to the case of rank 2 vector bundles which is less technical but contains already all the ideas of the classification theorem. Before beginning let us remark that F algebraically closed being fixed we won't prove the classification theorem for the curve X_E with fixed E but simultaneously for all curves $X_{E_h}, h \geq 1$. As before we note $X = X_E$, $X_h = X_{E_h}$ and $\pi_h : X_h \to X$. We will use the following dévissage.

Proposition 2.86. *Let $\mathcal{E}$ be a vector bundle on X and $h \geq 1$ an integer.*

(1) *$\mathcal{E}$ is semi-stable of slope λ if and only if $\pi_h^* \mathcal{E}$ is semi-stable of slope $h\lambda$.*

(2) *$\mathcal{E} \simeq \mathcal{O}_X(\lambda)^r$ for some integer r if and only if $\pi_h^* \mathcal{E} \simeq \mathcal{O}_{X_h}(h\lambda)^{r'}$ for some integer r'.*

Proof. Since the morphism $\pi_h : X_h \to X$ is Galois with Galois group $\mathrm{Gal}(E_h|E)$, π_h^* induces an equivalence between Bun_X and $\mathrm{Gal}(E_h|E)$-equivariant vector bundles on X_h. Moreover, if $\mathscr{F}$ is a $\mathrm{Gal}(E_h|E)$-equivariant vector bundle on X_h then its Harder–Narasimhan filtration is $\mathrm{Gal}(E_h|E)$-invariant. This is a consequence of the uniqueness property of the Harder–Narasimhan filtration and the fact that for $\mathcal{G}$ a non-zero vector bundle on X_h and $\tau \in \mathrm{Gal}(E_h|E)$ one has $\mu(\tau^*\mathcal{G}) = \mu(\mathcal{G})$. From those considerations one deduces point (1). We skip point (2) that is, at the end, an easy application of Hilbert 90. $\qquad\square$

The following dévissage is an analogue of a dévissage contained in [20] (see prop. 9.1) and [22] (see prop. 2.1.7) which is itself a generalization of Grothendieck's method for classifying vector bundles on $\mathbb{P}^1$ ([18]).

Proposition 2.87. *Theorem 2.83 is equivalent to the following statement: for any $n \geq 1$ and any vector bundle $\mathscr{E}$ that is an extension*

$$0 \longrightarrow \mathcal{O}_X(-\tfrac{1}{n}) \longrightarrow \mathscr{E} \longrightarrow \mathcal{O}_X(1) \longrightarrow 0$$

one has $H^0(X, \mathscr{E}) \neq 0$.

Proof. Let $\mathscr{E}$ be a vector bundle that is an extension as in the statement. If Theorem 2.83 is true then $\mathscr{E} \simeq \bigoplus_{i \in I} \mathcal{O}_X(\lambda_i)$ but since $\deg(\mathscr{E}) = 0$, for an index $i \in I$, $\lambda_i \geq 0$. We thus have $H^0(X, \mathscr{E}) \neq 0$ since for $\lambda \geq 0$, $H^0(X, \mathcal{O}_X(\lambda)) \neq 0$.

In the other direction, let $\mathscr{E}$ be a semi-stable vector bundle on X. Up to replacing X by X_h and $\mathscr{E}$ by $\pi_h^*\mathscr{E}$ for $h \gg 1$, one can suppose $\mu(\mathscr{E}) \in \mathbb{Z}$ (here we use Proposition 2.86). Up to replacing $\mathscr{E}$ by a twist $\mathscr{E} \otimes \mathcal{O}_X(d)$ for some $d \in \mathbb{Z}$ one can moreover suppose that

$$\mu(\mathscr{E}) = 0.$$

Suppose now that $\mathrm{rk}\,\mathscr{E} = 2$ (the general case works the same but is more technical). Let $\mathscr{L} \subset \mathscr{E}$ be a sub line bundle of maximal degree (here sub line bundle means locally direct factor). Since $\mathscr{E}$ is semi-stable of slope 0, $\deg\mathscr{L} \leq 0$. Writing $\mathscr{L} \simeq \mathcal{O}_X(-d)$ with $d \geq 0$, we see that $\mathscr{E}$ is an extension

$$0 \longrightarrow \mathcal{O}_X(-d) \longrightarrow \mathscr{E} \longrightarrow \mathcal{O}_X(d) \longrightarrow 0. \qquad (2.5)$$

If $d = 0$, since $\mathrm{Ext}^1(\mathcal{O}_X, \mathcal{O}_X) = H^1(X, \mathcal{O}_X) = 0$, $\mathscr{E} \simeq \mathcal{O}_X^2$ and we are finished. Suppose thus that $d \geq 1$. Since $-d + 2 \leq d$ there exists a non-zero morphism

$$u : \mathcal{O}_X(-d + 2) \xrightarrow{\neq 0} \mathcal{O}_X(d).$$

Pulling back the exact sequence (2.5) via u one obtains an exact sequence

$$0 \longrightarrow \mathcal{O}_X(-d) \longrightarrow \mathcal{E}' \longrightarrow \mathcal{O}_X(-d+2) \longrightarrow 0$$

with a morphism $\mathcal{E}' \to \mathcal{E}$ that is generically an isomorphism. Twisting this exact sequence by $\mathcal{O}_X(d-1)$ one obtains

$$0 \longrightarrow \mathcal{O}_X(-1) \longrightarrow \mathcal{E}'(d-1) \longrightarrow \mathcal{O}_X(1) \longrightarrow 0.$$

By hypothesis,

$$H^0(X, \mathcal{E}'(d-1)) \neq 0$$

and thus there exists a non-zero morphism $\mathcal{O}_X(1-d) \to \mathcal{E}'$. Composed with $\mathcal{E}' \to \mathcal{E}$ this gives a non-zero morphism $\mathcal{O}_X(1-d) \to \mathcal{E}$. This contradicts the maximality of $\deg \mathcal{L}$ (the schematical closure of the image of $\mathcal{O}_X(1-d) \to \mathcal{E}$ has degree $\geq 1 - d$). $\qquad\square$

6.3.4. Modifications of vector bundles associated to p-divisible groups: Hodge–de Rham periods

We still suppose F is algebraically closed. Let $L|E$ be the completion of the maximal unramified extension of E with residue field $\overline{\mathbb{F}}_q := \overline{\mathbb{F}_q}^F$. We thus have

$$L = W_{\mathcal{O}_E}(\overline{\mathbb{F}}_q)[\tfrac{1}{\pi}]$$

equipped with a Frobenius σ that lifts $x \mapsto x^q$. Let

$$\varphi\text{-Mod}_L$$

be the associated category of isocrystals, that is to say couples (D, φ) where D is a finite dimensional L-vector space and φ a σ-linear automorphism of D. There is a functor

$$\varphi\text{-Mod}_L \longrightarrow \text{Bun}_X$$
$$(D, \varphi) \longmapsto \mathcal{E}(D, \varphi) := M(D, \varphi)$$

where $M(D, \varphi)$ is the P-graded module

$$M(D, \varphi) = \bigoplus_{d \geq 0} (D \otimes_L \mathrm{B})^{\varphi = \pi^d}.$$

In fact one checks that

$$\mathcal{E}(D, \varphi) \simeq \bigoplus_{\lambda \in \mathbb{Q}} \mathcal{O}_X(-\lambda)^{m_\lambda}$$

where m_λ is the multiplicity of the slope λ in the Dieudonné–Manin decomposition of (D, φ). One has the following concrete description: if $U \subset X$ is a non-empty open subset, $U = D^+(t)$ with $t \in P_d$ for some $d \geq 0$, then

$$\Gamma(U, \mathscr{E}(D, \varphi)) = \left(D \otimes_L B[\tfrac{1}{t}]\right)^{\varphi = \mathrm{Id}}.$$

Remark 2.88. With respect to the motivation given in Section 4.1 for the introduction of the curve, one sees that the vector bundle $\mathscr{E}(D, \varphi)$ should be understood as being the vector bundle on "$X = Y/\varphi^{\mathbb{Z}}$" associated to the φ-equivariant vector bundle on "Y" whose global sections are $D \otimes_L B$.

Let $\infty \in |X|$ be a closed point and $C|E$ the associated residue field. Note $B_{dR}^+ = \mathcal{O}_{X,\infty}$ with uniformizing element t. One checks there is a canonical identification

$$\mathscr{E}(D, \varphi)_\infty = D \otimes_L B_{dR}^+.$$

To any lattice $\Lambda \subset D \otimes_L B_{dR}^+$ there is associated an effective modification $\mathscr{E}(D, \varphi, \Lambda)$ of $\mathscr{E}(D, \varphi)$,

$$0 \longrightarrow \mathscr{E}(D, \varphi, \Lambda) \longrightarrow \mathscr{E}(D, \varphi) \longrightarrow i_{\infty*}\left(D \otimes B_{dR}^+/\Lambda\right) \longrightarrow 0.$$

Such lattices Λ that satisfy $t.D \otimes B_{dR}^+ \subset \Lambda \subset D \otimes B_{dR}^+$ are in bijection with sub-C-vector spaces

$$\mathrm{Fil}\, D_C \subset D_C := D \otimes_L C.$$

Thus, to any sub vector space $\mathrm{Fil}\, D_C \subset D_C$ there is associated a "minuscule" modification

$$0 \longrightarrow \mathscr{E}(D, \varphi, \mathrm{Fil}\, D_C) \longrightarrow \mathscr{E}(D, \varphi) \longrightarrow i_{\infty*}\left(D_C/\mathrm{Fil}\, D_C\right) \longrightarrow 0.$$

One has

$$H^0(X, \mathscr{E}(D, \varphi, \mathrm{Fil}\, D_C)) = \mathrm{Fil}\left(D \otimes_L B[\tfrac{1}{t}]\right)^{\varphi = Id}.$$

By definition, a π-divisible $\mathcal{O}_E$-module over an $\mathcal{O}_E$-scheme (or formal scheme) S is a p-divisible group H over S equipped with an action of $\mathcal{O}_E$ such that the induced action on $\mathrm{Lie}\, H$ is the canonical one deduced from the structural morphism $S \to \mathrm{Spec}(\mathcal{O}_E)$.

If H is a π-divisible $\mathcal{O}_E$-module over $\overline{\mathbb{F}}_q$ one can define its covariant $\mathcal{O}$-Dieudonné module $\mathbb{D}_{\mathcal{O}}(H)$. This is a free $\mathcal{O}_L = W_{\mathcal{O}_E}(\overline{\mathbb{F}}_q)$-module of rank

$$\mathrm{ht}_{\mathcal{O}}(H) := \frac{\mathrm{ht}(H)}{[E : \mathbb{Q}_p]}$$

equipped with a σ-linear morphism F and a σ^{-1}-linear one V_π satisfying $F V_\pi = \pi$ and $V_\pi F = \pi$. If $\mathbb{D}(H)$ is the covariant Dieudonné module of the underlying p-divisible group one has a decomposition given by the action of the maximal unramified extension of $\mathbb{Q}_p$ in E

$$\mathbb{D}(H) = \bigoplus_{\tau : \mathbb{F}_q \hookrightarrow \overline{\mathbb{F}}_q} \mathbb{D}(H)_\tau.$$

If τ_0 is the canonical embedding then by definition

$$\mathbb{D}_{\mathcal{O}}(H) = \mathbb{D}(H)_{\tau_0}.$$

Moreover if $F : \mathbb{D}(H) \to \mathbb{D}(H)$ is the usual Frobenius and $q = p^r$ then $F : \mathbb{D}_{\mathcal{O}}(H) \to \mathbb{D}_{\mathcal{O}}(H)$ is given by $(F^r)_{|\mathbb{D}(H)_{\tau_0}}$. From now we note φ for F acting on $\mathbb{D}_{\mathcal{O}}(H_0)$.

For H a π-divisible $\mathcal{O}_E$-module over $\mathcal{O}_C$ there is associated a universal $\mathcal{O}_E$-vector extension (see appendix B of [9])

$$0 \longrightarrow V_{\mathcal{O}}(H) \longrightarrow E_{\mathcal{O}}(H) \longrightarrow H \longrightarrow 0.$$

One has $V_{\mathcal{O}}(H) = \omega_{H^\vee}$ where $H^\vee$ is the strict dual of H as defined by Faltings ([7]), the usual Cartier dual when $E = \mathbb{Q}_p$. The Lie algebra of the preceding gives an exact sequence

$$0 \longrightarrow \omega_{H^\vee} \longrightarrow \mathrm{Lie}\, E_{\mathcal{O}}(H) \longrightarrow \mathrm{Lie}\, H \longrightarrow 0.$$

Suppose now given a π-divisible $\mathcal{O}_E$-module H_0 over $\overline{\mathbb{F}}_q$ and a quasi-isogeny

$$\rho : H_0 \otimes_{\overline{\mathbb{F}}_q} \mathcal{O}_C / p\mathcal{O}_C \longrightarrow H \otimes_{\mathcal{O}_C} \mathcal{O}_C / p\mathcal{O}_C.$$

Thanks to the crystalline nature of the universal $\mathcal{O}_E$-vector extension ρ induces an isomorphism

$$\mathbb{D}_{\mathcal{O}}(H_0) \otimes_{\mathcal{O}_L} C \xrightarrow{\sim} \mathrm{Lie}\, E_{\mathcal{O}}(H)\left[\tfrac{1}{\pi}\right].$$

Via this isomorphism we thus get a Hodge Filtration

$$\omega_{H^\vee}\left[\tfrac{1}{\pi}\right] \simeq \mathrm{Fil}\, \mathbb{D}_{\mathcal{O}}(H_0)_C \subset \mathbb{D}_{\mathcal{O}}(H)_C.$$

There is then a period morphism

$$V_\pi(H) \longrightarrow \mathbb{D}_{\mathcal{O}}(H_0) \otimes_{\mathcal{O}_L} B$$

inducing an isomorphism

$$V_\pi(H) \xrightarrow{\sim} \mathrm{Fil}\, (\mathbb{D}_{\mathcal{O}}(H_0) \otimes_{\mathcal{O}_L} B)^{\varphi = \pi}.$$

Here $V_\pi(H) = V_p(H)$ but we prefer to use the notation $V_\pi(H)$ since most of what we say can be adapted in equal characteristic when $E = \mathbb{F}_q((\pi))$, for example in the context of Drinfeld modules. This period morphism is such that the induced morphism

$$V_\pi(H) \otimes_E B \longrightarrow \mathbb{D}_{\mathcal{O}}(H_0) \otimes_{\mathcal{O}_L} B$$

is injective with cokernel killed by t where here $t \in B^{\varphi=\pi}$ is a non-zero period of a Lubin–Tate group attached to E over $\mathcal{O}_C$. It induces a morphism

$$V_p(H) \otimes_E \mathcal{O}_X \longrightarrow \mathcal{E}\big(\mathbb{D}_{\mathcal{O}}(H_0)[\tfrac{1}{\pi}], \pi^{-1}\varphi, \text{Fil } \mathbb{D}_{\mathcal{O}}(H_0)_C\big).$$

Since

$$V_\pi(H) \otimes_E B_e \xrightarrow{\sim} \big(D \otimes_E B[\tfrac{1}{t}]\big)^{\varphi=Id}$$

where $B_e = H^0(X \setminus \{\infty\}, \mathcal{O}_X)$, the preceding morphism is an isomorphism outside ∞. Since both vector bundles are of degree 0 this is an isomorphism. We thus obtain the following theorem.

Theorem 2.89. *If H is a π-divisible $\mathcal{O}_E$-module over $\mathcal{O}_C$, H_0 a π-divisible $\mathcal{O}_E$-module over $\overline{\mathbb{F}}_q$ and $\rho : H_0 \otimes \mathcal{O}_C/p\mathcal{O}_C \to H \otimes \mathcal{O}_C/p\mathcal{O}_C$ a quasi-isogeny then*

$$\mathcal{E}\big(\mathbb{D}_{\mathcal{O}}(H_0)[\tfrac{1}{\pi}], \pi^{-1}\varphi, \omega_{H^\vee}[\tfrac{1}{p}]\big) \simeq V_\pi(H) \otimes_E \mathcal{O}_X.$$

Let H_0 be fixed over $\overline{\mathbb{F}}_q$ and let $\widehat{\mathcal{M}}$ be its deformation space by quasi-isogenies as defined by Rapoport and Zink ([27]), a $\text{Spf}(\mathcal{O}_L)$-formal scheme. We note $\mathcal{M}$ for its generic fiber as a Berkovich analytic space over L. In fact, we won't use the analytic space structure on $\mathcal{M}$, but only the C-points $\mathcal{M}(C) = \widehat{\mathcal{M}}(\mathcal{O}_C)$. Let $\mathcal{F}^{dR}$ be the Grassmanian of subspaces of $\mathbb{D}_{\mathcal{O}}(H)[\tfrac{1}{\pi}]$ of codimension $\dim H_0$, seen as an L-analytic space. There is then a period morphism

$$\pi^{dR} : \mathcal{M} \longrightarrow \mathcal{F}^{dR}$$

associating to a deformation its associated Hodge filtration. This morphism is étale and its image $\mathcal{F}^{dR,a}$, the admissible locus, is thus open. To each point $z \in \mathscr{F}^{dR}(C)$ is associated a filtration $\text{Fil}_z \mathbb{D}_{\mathcal{O}}(H_0)_C$ and a vector bundle $\mathcal{E}(z)$ on X that is a modification of $\mathcal{E}\big(\mathbb{D}_{\mathcal{O}}(H_0)[\tfrac{1}{\pi}], \pi^{-1}\varphi\big)$. Now the preceding theorem says the following.

Theorem 2.90. *If $z \in \mathcal{F}^{dR,a}(C)$ then $\mathcal{E}(z)$ is a trivial vector bundle.*

Remark 2.91. In fact a theorem of Faltings ([8]), translated in the language of vector bundle on our curve, says that a point $z \in \mathcal{F}^{dR}(C)$ is in the admissible locus if and only if $\dim_E H^0(X, \mathcal{E}(z)) = \text{rk}(\mathcal{E}(z))$. Using the classification theorem 2.83 this amounts to saying that $\mathcal{E}(z)$ is trivial or equivalently semi-stable of slope 0. Thus, once Theorem 2.83 is proved, we have a characterization of $\mathcal{F}^{dR,a}$ in terms of semi-stability as this is the case for the weakly admissible locus $\mathcal{F}^{dR,wa}$ ([27] chap.I, [5]) (in the preceding, if we allow ourselves to vary the curve X, we can make a variation of the complete algebraically closed field $C|E$).

We will use the following theorem that gives us the image of the period morphism for Lubin–Tate spaces.

Theorem 2.92 (Laffaille [24], Gross–Hopkins [17]). *Let H_0 be a one dimensional formal π-divisible $\mathcal{O}_E$-module of $\mathcal{O}_E$-height n and $\mathcal{F}^{dR} = \mathbb{P}^{n-1}$ the associated Grassmanian. Then one has*

$$\mathcal{F}^{dR} = \mathcal{F}^{dR,a}.$$

Remark 2.93. The preceding theorem says that for Lubin–Tate spaces, the weakly admissible locus coincides with the admissible one (one has always $\mathcal{F}^{dR,a} \subset \mathcal{F}^{dR,wa}$). We will need this theorem for all points of the period domain $\mathbb{P}^{n-1}$, not only classical ones associated to finite extensions of L.

Translated in terms of vector bundles on the curve the preceding theorem gives the following.

Theorem 2.94. *Let*

$$0 \longrightarrow \mathcal{E} \longrightarrow \mathcal{O}_X\left(\tfrac{1}{n}\right) \longrightarrow \mathcal{F} \longrightarrow 0$$

be a degree one modification of $\mathcal{O}_X\left(\tfrac{1}{n}\right)$, that is to say $\mathcal{F}$ is a torsion coherent sheaf on X of length 1 (i.e. of the form $i_{x*}k(x)$ for some $x \in |X|$). Then $\mathcal{E}$ is a trivial vector bundle, $\mathcal{E} \simeq \mathcal{O}_X^n$.

Remark 2.95. We will use Theorem 2.94 to prove the classification theorem 2.83. Reciprocally, it is not difficult to see that the classification Theorem 2.83 implies Theorem 2.94. In fact, suppose

$$\mathcal{E} \simeq \oplus_{i \in I} \mathcal{O}_X(\lambda_i)$$

is a degree one modification of $\mathcal{O}_X\left(\tfrac{1}{n}\right)$. Write $\lambda_i = \frac{d_i}{h_i}$ with $(d_i, h_i) = 1$. Since rk $(\mathcal{E}) = n$ one has $h_i \leq n$. But

$$\mathrm{Hom}\left(\mathcal{O}_X(\lambda_i), \mathcal{O}_X\left(\tfrac{1}{n}\right)\right) \neq 0$$

implies $\lambda_i \leq \frac{1}{n}$. Thus, for $i \in I$, either $\lambda_i \leq 0$ or $\lambda_i = \frac{1}{n}$. Using deg $\mathcal{E} = 0$ one concludes that for all i, $\lambda_i = 0$.

6.3.5. Modifications of vector bundles associated to p-divisible groups: Hodge–Tate periods

Let V be a finite dimensional E-vector space and W a finite dimensional $C = k(\infty)$-vector space. Consider extensions of coherent sheaves on X

$$0 \longrightarrow V \otimes_E \mathcal{O}_X \longrightarrow \mathcal{H} \longrightarrow i_{\infty *} W \longrightarrow 0.$$

Those extensions are rigid since $\mathrm{Hom}(i_{\infty *}W, V \otimes_E \mathcal{O}_X) = 0$. Consider the category of triples (V, W, ξ) where V and W are vector spaces and ξ is an extension as before. Morphisms in this category are linear morphisms of vector spaces inducing morphisms of extensions. Fix a Lubin–Tate group over $\mathcal{O}_E$ and let $E\{1\}$ be its rational Tate module over $\mathcal{O}_C$, a one dimensional E-vector space. One has

$$E\{1\} \subset \mathrm{B}^{\varphi = \pi} = H^0(X, \mathcal{O}_X(1)).$$

There is a canonical extension

$$0 \longrightarrow \mathcal{O}_X\{1\} \longrightarrow \mathcal{O}_X(1) \longrightarrow i_{\infty *}C \longrightarrow 0,$$

where $\mathcal{O}_X\{1\} = \mathcal{O}_X \otimes_E E\{1\}$. If V and W are as before and

$$u : W \longrightarrow V\{-1\} \otimes_E C =: V_C\{-1\}$$

is C-linear there is an induced extension

$$
\begin{array}{ccccccccc}
0 & \longrightarrow & V \otimes_E \mathcal{O}_X & \longrightarrow & \mathcal{H}(V, W, u) & \longrightarrow & i_{\infty *}W & \longrightarrow & 0 \\
 & & \| & & \downarrow & & \downarrow {\scriptstyle i_{\infty *}u} & & \\
0 & \longrightarrow & V \otimes_E \mathcal{O}_X & \longrightarrow & V\{-1\} \otimes_E \mathcal{O}_X(1) & \longrightarrow & i_{\infty *}V_C\{-1\} & \longrightarrow & 0
\end{array}
$$

where the upper extension is obtained by pullback from the lower one via $i_{\infty *}u$ and we used the formula $V\{-1\} \otimes_E i_{\infty *}C = i_{\infty *}V_C\{-1\}$. It is easily seen that this induces a category equivalence between triplets (V, W, u) and the preceding category of extensions:

$$\mathrm{Hom}_C(W, V_C\{-1\}) \xrightarrow{\sim} \mathrm{Ext}^1(i_{\infty *}W, V \otimes_E \mathcal{O}_X)$$

canonically in W and V. The coherent sheaf $\mathcal{H}(V, W, u)$ is a vector bundle if and only if u is injective. In this case, $V \otimes_E \mathcal{O}_X$ is a "minuscule" modification of the vector bundle $\mathcal{H}(V, W, u)$.

There is another period morphism associated to p-divisible groups: Hodge–Tate periods. Let H be a π-divisible $\mathcal{O}_E$-module over $\mathcal{O}_C$. There is then a Hodge–Tate morphism, an E-linear morphism

$$\alpha_H : V_\pi(H) \longrightarrow \omega_{H^\vee}\left[\tfrac{1}{\pi}\right].$$

It is defined in the following way. An element of $T_\pi(H)$ can be interpreted as a morphism of π-divisible $\mathcal{O}_E$-modules

$$f : E/\mathcal{O}_E \longrightarrow H.$$

Using the duality of [7] it gives a morphism

$$f^\vee : H^\vee \longrightarrow \mathcal{LT}_{\mathcal{O}_C}$$

where $\mathcal{LT}$ is a fixed Lubin–Tate group over $\mathcal{O}_E$. Then, having fixed a generator α of $\omega_{\mathcal{LT}}$, one has

$$\alpha_H(f) = (f^\vee)^*\alpha.$$

Consider

$$\beta_H = {}^t(\alpha_{H^\vee} \otimes 1) : \mathrm{Lie}\, H\big[\tfrac{1}{\pi}\big] \longrightarrow V_\pi(H)_C\{-1\}$$

where

$$\alpha_{H^\vee} \otimes 1 : V_\pi(H)^*\{1\} \otimes_E C \longrightarrow \omega_H\big[\tfrac{1}{\pi}\big]$$

using the formula

$$V_\pi(H)^*\{1\} = V_\pi(H^\vee)$$

and β_H is the transpose of $\alpha_{H^\vee} \otimes 1$. All of this fits into an Hodge–Tate exact sequence of C-vector spaces ([11] chap.5 for $E = \mathbb{Q}_p$)

$$0 \longrightarrow \mathrm{Lie}\, H\big[\tfrac{1}{\pi}\big]\{1\} \xrightarrow{\beta_H\{1\}} V_\pi(H) \otimes_E C \xrightarrow{\alpha_H} \omega_{H^\vee}\big[\tfrac{1}{\pi}\big] \longrightarrow 0.$$

Suppose now (H_0, ρ) is as in the preceding section.

Theorem 2.96. *One has a canonical isomorphism*

$$\mathcal{H}\big(V_\pi(H), \mathrm{Lie}\, H\big[\tfrac{1}{\pi}\big], \beta_H\big) \simeq \mathcal{E}\big(\mathbb{D}_\mathcal{O}(H_0)\big[\tfrac{1}{\pi}\big], \pi^{-1}\varphi\big).$$

Sketch of proof. To prove the preceding theorem is suffices to construct a morphism f giving a commutative diagram

$$
\begin{array}{ccccccccc}
0 & \longrightarrow & V_\pi(H) \otimes_E \mathcal{O}_X & \longrightarrow & \mathcal{E}\big(\mathbb{D}_\mathcal{O}(H_0)\big[\tfrac{1}{\pi}\big], \pi^{-1}\varphi\big) & \longrightarrow & i_{\infty*} \mathrm{Lie}\, H\big[\tfrac{1}{\pi}\big] & \longrightarrow & 0 \\
& & \downarrow{\scriptstyle Id} & & \downarrow{\scriptstyle f} & & \downarrow{\scriptstyle i_{\infty*}\beta_H} & & \\
0 & \longrightarrow & V_\pi(H) \otimes_E \mathcal{O}_X & \longrightarrow & V_\pi(H)\{-1\} \otimes_E \mathcal{O}_X(1) & \longrightarrow & i_{\infty*} V_\pi(H)_C\{-1\} & \longrightarrow & 0.
\end{array}
$$

By duality and shifting, to give f is the same as to give its transpose twisted by $\mathcal{O}_X(1)$

$$f^\vee(1) : V_\pi(H^\vee) \otimes_E \mathcal{O}_X \longrightarrow \mathcal{E}\big(\mathbb{D}_O(H_0)\big[\tfrac{1}{\pi}\big], \pi^{-1}\varphi\big)^\vee(1)$$
$$= \mathcal{E}\big(\mathbb{D}_\mathcal{O}(H_0^\vee), \pi^{-1}\varphi\big).$$

One checks that taking $f^\vee(1)$ equal to the period morphism for $H^\vee$

$$V_\pi(H^\vee) \longrightarrow \mathbb{D}_\mathcal{O}(H_0^\vee) \otimes_{\mathcal{O}_L} B$$

makes the preceding diagram commutative.   □

Let's come back to Rapoport–Zink spaces. Let $\mathcal{M}$ be as in the preceding section. Let $n = \mathrm{ht}_\mathcal{O} H_0$ and for $K \subset \mathrm{GL}_n(\mathcal{O}_E)$ an open subgroup let $\mathcal{M}_K \to \mathcal{M}$ be the étale finite covering given by level K-structure on the universal deformation. Set

$$\text{``}\mathcal{M}_\infty = \varprojlim_K \mathcal{M}_K\text{''}.$$

There are different ways to give a meaning to this as a generalized rigid anaytic space (see [9] for the case of Lubin–Tate and Drinfeld spaces) but we don't need it for our purpose. The only thing we need is the points

$$\mathcal{M}_\infty(C) = \{(H, \rho, \eta\}/\sim$$

where $(H, \rho) \in \mathcal{M}(C) = \widehat{\mathcal{M}}(\mathcal{O}_C)$ is as before and

$$\eta : \mathcal{O}_E^n \xrightarrow{\sim} T_\pi(H).$$

Let $\mathcal{F}^{HT}$ be the Grassmanian of subspaces of E^n of dimension $\dim H_0$ as an analytic space over L. The Hodge–Tate map induces a morphism, at least at the level of the C-points,

$$\pi^{HT} : \mathcal{M}_\infty \longrightarrow \mathcal{F}^{HT}$$
$$(H, \rho, \eta) \longmapsto \left(\mathrm{Lie}\, H\!\left[\tfrac{1}{\pi}\right]\{1\} \xrightarrow{u} C^n\right)$$

where

$$u : \mathrm{Lie}\, H\!\left[\tfrac{1}{\pi}\right] \xrightarrow{\beta_H\{1\}} V_\pi(H)_C \xrightarrow{\eta_H^{-1} \otimes 1} C^n.$$

To each point $z \in \mathcal{F}^{HT}(C)$ there is associated a vector bundle $\mathcal{H}(z)$ on X. The preceding thus gives the following.

Theorem 2.97. *If $z \in \mathcal{F}^{HT}$ is in the image of the Hodge–Tate map $\pi^{HT} : \mathcal{M}_\infty(C) \to \mathcal{F}^{HT}(C)$ then $\mathcal{H}(z) \simeq \mathcal{E}\left(\mathbb{D}_\mathcal{O}(H_0)\!\left[\tfrac{1}{\pi}\right], \pi^{-1}\varphi\right).$*

Consider now the case when $\dim H_0^\vee = 1$, the dual Lubin–Tate case. Then, $\mathcal{F}^{HT} = \mathbb{P}^{n-1}$ as an analytic space over L. It is stratified in the following way. For $i \in \{0, \cdots, n-1\}$ let

$$(\mathbb{P}^{n-1})^{(i)} = \{x \in \mathbb{P}^{n-1} \mid \dim_E E^n \cap \mathrm{Fil}_x\, k(x)^n = i\}.$$

This is a locally closed subset of the Berkovich space $\mathbb{P}^{n-1}$ (but it has no analytic structure for $i > 0$). The open stratum

$$(\mathbb{P}^{n-1})^{(0)} = \Omega^{n-1}$$

is Drinfeld space. For each $i > 0$, $(\mathbb{P}^{n-1})^{(i)}$ is fibered over the Grassmanian Gr^i of i-dimensional subspaces of E^n (seen as a naïve analytic space)

$$(\mathbb{P}^{n-1})^{(i)} \longrightarrow \mathrm{Gr}^i$$

$$x \longmapsto E^n \cap \mathrm{Fil}_x k(x)^n$$

with fibers Drinfeld spaces Ω^{n-1-i}. The π-divisible $\mathcal{O}$-modules H_0 over $\overline{\mathbb{F}}_q$ of $\mathcal{O}$-height n and dimension $n-1$ are classified by the height of their étale part

$$\mathrm{ht}_{\mathcal{O}}(H_0^{\mathrm{et}}) \in \{0, \cdots, n-1\}.$$

Let $H_0^{(i)}$ over $\overline{\mathbb{F}}_q$ be such that $\mathrm{ht}_{\mathcal{O}}(H_0^{\mathrm{et}}) = i$. Let $\mathcal{M}^{(i)}$ be the corresponding Rapoport–Zink space of deformations by quasi-isogenies of $H_0^{(i)}$ and $\mathcal{M}_\infty^{(i)}$ the space "with infinite level". When $i = 0$, this is essentially the Rapoport–Zink space associated to Lubin–Tate space (the only difference is that the Hecke action is twisted by the automorphism $g \mapsto {}^t g^{-1}$ of $\mathrm{GL}_n(E)$) and for $i > 0$ this can be easily linked to a lower dimensional Lubin–Tate space of deformations of the dual of the connected component of $H_0^{(i)}$. One then has

$$\pi^{HT} : \mathcal{M}_\infty^{(i)} \longrightarrow (\mathbb{P}^{n-1})^{(i)}.$$

Theorem 2.98. *For all $i \in \{0, \cdots, n-1\}$, the Hodge–Tate period map*

$$\pi^{HT} : \mathcal{M}_\infty^{(i)} \longrightarrow (\mathbb{P}^{n-1})^{(i)}$$

is surjective, that is to say

$$\pi^{HT} : \coprod_{i=0,\cdots,n-1} \mathcal{M}_\infty^{(i)} \longrightarrow \mathbb{P}^{n-1}$$

is surjective.

The statement of this theorem is at the level of the points of the associated Berkovich topological spaces. It says that the associated maps are surjective at the level of the points with values in complete algebraically closed extensions of E. The proof of the preceding is easily reduced to the case when H_0 is formal, that is to say the Lubin–Tate case. One thus has to prove that

$$\pi^{HT} : \mathcal{M}_\infty^{(0)} \longrightarrow \Omega$$

is surjective: *up to varying the complete algebraically closed field $C|E$, any point in the Drinfeld space $\Omega(C)$ is the Hodge–Tate period of the dual of* a *Lubin–Tate group over* $\mathcal{O}_C$. The proof relies on elementary manipulations between Lubin–Tate and Drinfeld spaces and the following computation of the admissible locus for Drinfeld moduli spaces (see chapter II of [9] for more details).

Theorem 2.99 (Drinfeld [6]). *Let D be a division algebra central over E with invariant* $1/n$. *Let* $\mathcal{M}$ *be the analytic Rapoport–Zink space of deformations by quasi-isogenies of special formal* $\mathcal{O}_D$-*modules of* $\mathcal{O}_E$-*height* n^2. *Then, the image of*

$$\pi^{dR} : \mathcal{M} \longrightarrow \mathbb{P}^{n-1}$$

is Drinfeld's space Ω.

Remark 2.100. Of course, Drinfeld's theorem is more precise giving an explicit description of the formal scheme $\widehat{\mathcal{M}}$ and proving that if $\mathcal{M}^{[i]}$ is the open/closed subset where the $\mathcal{O}_E$-height of the universal quasi-isogeny is ni then for any $i \in \mathbb{Z}$

$$\pi^{dR} : \mathcal{M}^{[i]} \xrightarrow{\sim} \Omega.$$

In fact, to apply Drinfeld's result one need to compare its period morphism defined in terms of Cartier theory and the morphism π^{dR} defined in crystalline terms. This is done in [27] 5.19. Finally, one will notice that in [24] Laffaille gives another proof of Theorem 2.99.

As a consequence of Theorems 2.97 and 2.98 we obtain the following result that is the one we will use to prove Theorem 2.83.

Theorem 2.101. *Let* $\mathcal{E}$ *be a vector bundle on X having a degree one modification that is a trivial vector bundle of rank n,*

$$0 \longrightarrow \mathcal{O}_X^n \longrightarrow \mathcal{E} \longrightarrow i_{x*}k(x) \longrightarrow 0$$

for some $x \in |X|$. *Then, there exists an integer* $i \in \{0, \cdots, n-1\}$ *such that*

$$\mathcal{E} \simeq \mathcal{O}_X^i \oplus \mathcal{O}_X\left(\frac{1}{n-i}\right).$$

Remark 2.102. As in Remark 2.95, one checks that the classification theorem implies Theorem 2.101.

6.3.6. End of the proof of the classification theorem

We will now use Theorems 2.94 and 2.101 to prove the classification theorem 2.83. For this we use the dévissage given by Proposition 2.87. We treat the case of rank 2 vector bundles, the general case being analogous but longer and more technical. According to Proposition 2.87 we have to prove that if $\mathscr{E}$ is a rank 2 vector bundle that is an extension

$$0 \longrightarrow \mathcal{O}_X(-1) \longrightarrow \mathscr{E} \longrightarrow \mathcal{O}_X(1) \longrightarrow 0 \qquad (2.6)$$

then $H^0(X, \mathscr{E}) \neq 0$. Let us choose $t \in H^0(X, \mathcal{O}_X(2)) \setminus \{0\}$. It furnishes a degree 2 modification

$$0 \longrightarrow \mathcal{O}_X(-1) \xrightarrow{\times t} \mathcal{O}_X(1) \longrightarrow \mathscr{F} \longrightarrow 0$$

where $\mathscr{F}$ is a torsion coherent sheaf of length 2. We can push forward the exact sequence (2.6)

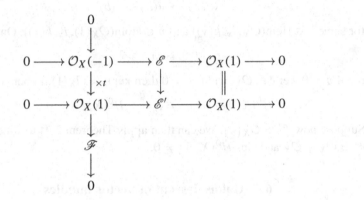

Since $\mathrm{Ext}^1(\mathcal{O}_X(1), \mathcal{O}_X(1)) = 0$ one has $\mathscr{E}' \simeq \mathcal{O}_X(1) \oplus \mathcal{O}_X(1)$. We thus obtain a degree 2 modification

$$0 \longrightarrow \mathscr{E} \longrightarrow \mathcal{O}_X(1) \oplus \mathcal{O}_X(1) \longrightarrow \mathscr{F} \longrightarrow 0.$$

Consider a dévissage of $\mathscr{F}$

$$0 \longrightarrow i_{y*}k(y) \longrightarrow \mathscr{F} \longrightarrow i_{x*}k(x) \longrightarrow 0$$

for some closed points $x, y \in |X|$. Consider the composite surjection

$$\mathcal{O}_X(1) \oplus \mathcal{O}_X(1) \longrightarrow \mathscr{F} \longrightarrow i_{x*}k(x),$$

write it

$$(a, b) \mapsto u(a) + v(b)$$

for two morphisms $u, v \in \mathrm{Hom}(\mathcal{O}_X(1), i_{x*}k(x))$ and let $\mathscr{E}' \subset \mathcal{O}_X(1) \oplus \mathcal{O}_X(1)$ be the kernel of this morphism. One has $(u, v) \neq (0, 0)$. Moreover, if $u = 0$ or $v = 0$ then

$$\mathscr{E}' \simeq \mathcal{O}_X \oplus \mathcal{O}_X(1).$$

Suppose now that $u \neq 0$ and $v \neq 0$. Then, $u_{|\mathscr{E}'}$ is surjective giving rise to an exact sequence

$$0 \longrightarrow \underbrace{\ker u \oplus \ker v}_{\simeq \mathcal{O}_X \oplus \mathcal{O}_X} \longrightarrow \mathscr{E}' \xrightarrow{\;u_{|\mathscr{E}'}\;} i_{x*}k(x) \longrightarrow 0.$$

Applying Theorem 2.101 we deduce that either $\mathscr{E}' \simeq \mathcal{O}_X \oplus \mathcal{O}_X(1)$ or $\mathscr{E}' \simeq \mathcal{O}_X\left(\frac{1}{2}\right)$. There is now a degree 1 modification

$$0 \longrightarrow \mathscr{E} \longrightarrow \mathscr{E}' \longrightarrow i_{y*}k(y) \longrightarrow 0.$$

If $\mathscr{E}' \simeq \mathcal{O}_X \oplus \mathcal{O}_X(1)$ the surjection $\mathscr{E}' \to i_{y*}k(y)$ is given by

$$(a, b) \mapsto u(a) + v(b)$$

for some $u \in \mathrm{Hom}(\mathcal{O}_X, i_{y*}k(y))$ and $v \in \mathrm{Hom}(\mathcal{O}_X(1), i_{y*}k(y))$. One has

$$0 \oplus \ker v \subset \mathscr{E}$$

but if $v \neq 0$, $\ker v \simeq \mathcal{O}_X$, and if $v = 0$ then $\ker v = \mathcal{O}_X(1)$. In both cases

$$H^0(X, 0 \oplus \ker v) \neq 0 \Longrightarrow H^0(X, \mathscr{E}) \neq 0.$$

Suppose now $\mathscr{E}' \simeq \mathcal{O}_X\left(\frac{1}{2}\right)$. We can then apply Theorem 2.94 to conclude that $\mathscr{E} \simeq \mathcal{O}_X \oplus \mathcal{O}_X$ and thus $H^0(X, \mathscr{E}) \neq 0$. $\qquad\square$

6.4. Galois descent of vector bundles

Now F is not necessarily algebraically closed. Let $\overline{F}$ be an algebraic closure of F. There is an action of G_F on $X_{\widehat{\overline{F}}}$. Define

$$\mathrm{Bun}_{X_{\widehat{\overline{F}}}}^{G_F}$$

to be the category of G_F-equivariant vector bundles on $X_{\widehat{\overline{F}}}$ together with a continuity condition on the action of G_F (we don't enter into the details).

Theorem 2.103. *If* $\alpha : X_{\widehat{\overline{F}}} \longrightarrow X_F$, *the functor*

$$\alpha^* : \mathrm{Bun}_{X_F} \longrightarrow \mathrm{Bun}_{X_{\widehat{\overline{F}}}}$$

is an equivalence.

For rank 1 vector bundles this theorem is nothing else than Theorem 2.78.

Example 2.104. Let $\mathrm{Rep}_E(G_F)$ be the category of continuous representations of G_F in finite dimensional E-vector spaces. The functor

$$\mathrm{Rep}_E(G_F) \longrightarrow \mathrm{Bun}_{X_{\widehat{\overline{F}}}}^{G_F}$$

$$V \longmapsto V \otimes_E \mathcal{O}_{X_{\widehat{\overline{F}}}}$$

induces an equivalence between $\text{Rep}_E(G_F)$ and the subcategory of $\text{Bun}_{X_{\widehat{\overline{F}}}}^{G_F}$ formed by equivariant vector bundles whose underlying vector bundle is trivial. An inverse is given by the global section functor $H^0(X_{\widehat{\overline{F}}}, -)$. Thus, via Theorem 2.103 the category $\text{Rep}_E(G_F)$ embeds in Bun_{X_F}. According to Theorem 2.83 this coincides with G_F-equivariant vector bundles semi-stable of slope 0.

Remark 2.105. With the notation of the preceding example if $\mathscr{E} \in \text{Bun}_{X_F}$ corresponds to $V \otimes_E \mathcal{O}_{X_{\widehat{\overline{F}}}}$ that is to say $\alpha^*\mathscr{E} \simeq V \otimes_E \mathcal{O}_{\widehat{\overline{F}}}$ then

$$H^1(X_F, \mathscr{E}) \simeq \text{Ext}^1_{\text{Fib}_{X_{\widehat{\overline{F}}}}^{G_F}}\left(\mathcal{O}_{\widehat{\overline{F}}}, V \otimes_E \mathcal{O}_{\widehat{\overline{F}}}\right)$$

$$\simeq H^1(G_F, V)$$

since $H^1(X_{\widehat{\overline{F}}}, \mathcal{O}_{\widehat{\overline{F}}}) = 0$. In particular, $H^1(X_F, \mathcal{O}_{X_F}) \simeq \text{Hom}(G_F, E)$ which is non-zero in general when F is not algebraically closed.

Sketch of proof of theorem 2.103. To prove the preceding theorem we use the following fundamental property of Harder–Narasimhan filtrations that is a consequence of their canonicity (see the proof of Proposition 2.86 for example).

Proposition 2.106. *Let $\Gamma \subset \text{Aut}(X)$ be a subgroup and $\mathscr{E}$ be a Γ-equivariant vector bundle on X. Then the Harder–Narasimhan filtration of $\mathscr{E}$ is Γ-invariant, that is to say is a filtration in the category of Γ-equivariant vector bundles.*

Let us fix $t \in B_F^{\varphi=\pi} \setminus \{0\}$ and note $\{\infty\} = V^+(t)$, $\infty \in |X_F|$

$$B_{F,e} = B_F[\tfrac{1}{t}]^{\varphi=Id} = \Gamma(X_F \setminus \{\infty\}, \mathcal{O}_{X_F}), \quad B_{F,dR}^+ = \mathcal{O}_{X_F,\infty}.$$

$$B_{\widehat{\overline{F}},e} = B_{\widehat{\overline{F}}}[\tfrac{1}{t}]^{\varphi=Id} = \Gamma(X_{\widehat{\overline{F}}} \setminus \{\infty\}, \mathcal{O}_{X_{\widehat{\overline{F}}}}), \quad B_{\widehat{\overline{F}},dR}^+ = \mathcal{O}_{X_{\widehat{\overline{F}}},\infty}.$$

The category $\text{Bun}_{X_{\widehat{\overline{F}}}}^{G_F}$ is equivalent to the following category of B-pairs ([3]) (M, W, u) where:

- $M \in \text{Rep}_{B_{\widehat{\overline{F}},e}}(G_F)$ is a semi-linear continuous representation of G_F in a free $B_{\widehat{\overline{F}},e}$-module,
- $M \in \text{Rep}_{B_{\widehat{\overline{F}},dR}^+}(G_F)$ is a semi-linear continuous representation of G_F in a free $B_{\widehat{\overline{F}},dR}^+$-module,
- $u : M \otimes_{B_{\widehat{\overline{F}},e}} B_{\widehat{\overline{F}},dR} \xrightarrow{\sim} W[\tfrac{1}{t}]$.

According to Theorem 2.43, to prove Theorem 2.103 it suffices to prove that if $\mathrm{Mod}_{\mathrm{B}_{F,e}}^{proj}$ is the category of projective $\mathrm{B}_{F,e}$-modules of finite type then

$$- \otimes_{\mathrm{B}_{F,e}} : \mathrm{Mod}_{\mathrm{B}_{F,e}}^{proj} \xrightarrow{\sim} \mathrm{Rep}_{\mathrm{B}_{\widehat{F},e}}(G_F).$$

Here in the definition of a $\mathrm{B}_{\widehat{F},e}$-representation M we impose that the G_F-action on $M \otimes \mathrm{B}_{\widehat{F},dR}$ stabilizes a $\mathrm{B}_{\widehat{F},dR}^{\pm}$-lattice (the continuity condition does not imply this) that is to say M comes from an equivariant vector bundle. Full faithfullness of the preceding functor is an easy consequence of the equality

$$\mathrm{B}_{F,e} = \left(\mathrm{B}_{\widehat{F},e}\right)^{G_F}.$$

We now treat the essential surjectivity. An easy Galois descent argument tells us we can replace the field E that was fixed by a finite extension of it (for this we may have to replace F by a finite extension of it so that the residue field of the finite extension of E is contained in F but this is harmless by Hilbert 90). Let M be a $\mathrm{B}_{\widehat{F},e}$-representation of G_F. Choose an equivariant vector bundle $\mathscr{E}$ on $X_{\widehat{F}}$ such that

$$M = \Gamma(X_{\widehat{F}} \setminus \{\infty\}, \mathscr{E}).$$

Applying the classification theorem 2.83 and Proposition 2.106 we see $\mathscr{E}$ is a successive extension of equivariant vector bundles whose underlying bundle is isomorphic to $\mathcal{O}_{X_{\widehat{F}}}(\lambda)^n$ for some $\lambda \in \mathbb{Q}$ and $n \in \mathbb{N}$. Up to making a finite extension of E one can suppose moreover that all such slopes λ are integers, $\lambda \in \mathbb{Z}$. Now, if $\lambda \in \mathbb{Z}$, an equivariant vector bundle whose underlying vector bundle is of the form $\mathcal{O}_{X_{\widehat{F}}}(\lambda)^n$ is isomorphic to

$$V \otimes_E \mathcal{O}_{X_{\widehat{F}}}(\lambda)$$

for a continuous representation V of G_F in a finite dimensional E-vector space $(\mathcal{O}_{X_{\widehat{F}}(\lambda)} = \alpha^* \mathcal{O}_{X_F}(\lambda)$ has a canonical G_F-equivariant structure). Let us remark that $\mathcal{O}_{X_F}(\lambda)$ become trivial on $X_F \setminus \{\infty\}$. From this one deduces that M has a filtration whose graded pieces are of the form

$$V \otimes_E \mathrm{B}_{\widehat{F},e}$$

for some finite dimensional E-representation of G_F. We now use the following result that generalizes Theorem 2.70. Its proof is identical to the proof of Theorem 2.70.

Theorem 2.107. *For V a continuous finite dimensional E-representation of G_F one has*

$$H^1\left(G_F, V \otimes_E \mathrm{B}_{\widehat{F},e}\right) = 0$$

$$H^0\left(G_F, V \otimes_E \mathrm{B}_{\widehat{F},e}\right) \neq 0.$$

The vanishing assertion in the preceding theorem tells us that in fact M is a direct sum of representations of the form $V \otimes_E B_{\widehat{\overline{F}},e}$. We can thus suppose $M = V \otimes_E B_{\widehat{\overline{F}},e}$. We now proceed by induction on the rank of M. Choose $x \in M^{G_F} \setminus \{0\}$ and let $N \subset M$ be the saturation of the submodule $B_{\widehat{\overline{F}},e}.x$ that is to say

$$N/B_{\widehat{\overline{F}},e}.x = (M/B_{\widehat{\overline{F}},e}.x)_{tor}.$$

According to Theorem 2.43 the $B_{F,e}$-module $(N/B_{\widehat{\overline{F}},e}.x)^{G_F}$ is of finite length and generates $N/B_{\widehat{\overline{F}},e}.x$. Using the vanishing $H^1(G_F, B_{\widehat{\overline{F}},e}) = 0$ one deduces that N^{G_F} is a torsion free finite type $B_{F,e}$-module satisfying

$$N^{G_F} \otimes_{B_{F,e}} B_{\widehat{\overline{F}},e} = N.$$

Now, by induction we know that $(M/N)^{G_F}$ is a finite rank projective $B_{F,e}$-module such that

$$(M/N)^{G_F} \otimes_{B_{F,e}} B_{\widehat{\overline{F}},e} = M/N.$$

Since $H^1(G_F, B_{\widehat{\overline{F}},e}) = 0$ one has $H^1(G_F, N) = 0$. From this one deduces that M^{G_F} is a finite rank projective module satisfying

$$M^{G_F} \otimes_{B_{F,e}} B_{\widehat{\overline{F}},e} = M.$$

$\square$

Here is an interesting corollary of Theorem 2.103.

Corollary 2.108. *Any G_F-equivariant vector bundle on $X_{\widehat{\overline{F}}}$ is a successive extension of G_F-equivariant line bundles.*

Example 2.109. Let V be a finite dimensional E-representation of G_F. Then, even if V is irreducible, $V \otimes_E \mathcal{O}_{X_{\widehat{\overline{F}}}}$ is a successive extension of line bundles of the form $\chi \otimes_E \mathcal{O}_{X_{\widehat{\overline{F}}}}(d)$ where $\chi : G_F \to E^\times$ and $d \in \mathbb{Z}$.

7. Vector bundles and φ-modules

7.1. The Robba ring and the bounded Robba ring

We define a new ring

$$\mathscr{R}_F = \varinjlim_{\rho \to 0} B_{]0,\rho]}$$

where $B_{]0,\rho]}$ is the completion of B^b with respect to $(|.|_{\rho'})_{0 < \rho' \le \rho}$. Since

$$\varphi : B_{]0,\rho]} \xrightarrow{\sim} B_{]0,\rho^q]}$$

the ring $\mathscr{R}_F$ is equipped with a bijective Frobenius φ. In equal characteristic, when $E = \mathbb{F}_q((\pi))$, $\mathscr{R}_F = \mathcal{O}_{\mathbb{D}^*,0}$ the germs of holomorphic functions at 0 on $\mathbb{D}^*$.

Theorem 2.110 (Kedlaya [21] theo.2.9.6). *For all $\rho \in]0, 1[$, the ring $B_{]0,\rho]}$ is Bezout. Any closed ideal of $B_{]0,\rho]}$ is principal.*

Proof. As in the proof of Theorem 2.51, using Theorem 2.51, the set of closed ideals of $B_{]0,\rho]}$ is in bijection with $\mathrm{Div}^+(Y_{]0,\rho]})$ where $|Y_{]0,\rho]}| = \{\mathfrak{m} \in |Y| \mid 0 < \|\mathfrak{m}\| \leq \rho\}$. If F is algebraically closed, one can then write any $D \in \mathrm{Div}^+(Y_{]0,\rho]})$ as

$$D = \sum_{i \geq 0}[\mathfrak{m}_i]$$

with $\mathfrak{m}_i = (\pi - [a_i]) \in |Y_{]0,\rho]}|$ and $\lim_{i \to +\infty} \|\mathfrak{m}_i\| = 0$. The Weierstrass product

$$\prod_{i \geq 0}\left(1 - \frac{[a_i]}{\pi}\right)$$

is convergent in $B_{]0,\rho]}$ and thus the divisor D is principal. For a general F this result remains true since one can prove that $H^1\left(G_F, B^\times_{\widehat{\bar{F}},]0,\rho]}\right) = 0$ (this uses Proposition 2.119). This proves that any closed ideal of $B_{]0,\rho]}$ is principal. It now remains to prove that for $f, g \in B_{]0,\rho]}$ non-zero satisfying $\mathrm{supp}(\mathrm{div}(f)) \cap \mathrm{supp}(\mathrm{div}(g)) = \emptyset$ the ideal generated by f and g is $B_{]0,\rho]}$. This is a consequence of the following more general fact:

$$B_{]0,\rho]}/(f) \xrightarrow{\sim} \prod_{\substack{\mathfrak{m} \in |Y| \\ \|\mathfrak{m}\| \leq \rho}} B^+_{dR,\mathfrak{m}}/\mathrm{Fil}^{\mathrm{ord}_\mathfrak{m}(f)}B^+_{dR,\mathfrak{m}}.$$

In fact, for $0 < \rho' \leq \rho$ one has

$$B_{[\rho',\rho]}/(f) \xrightarrow{\sim} \prod_{\substack{\mathfrak{m} \in |Y| \\ \rho' \leq \|\mathfrak{m}\| \leq \rho}} B^+_{dR,\mathfrak{m}}/\mathrm{Fil}^{\mathrm{ord}_\mathfrak{m}(f)}B^+_{dR,\mathfrak{m}}.$$

The preceding isomorphism inserts into a projective system of exact sequences

$$0 \longrightarrow B_{[\rho',\rho]} \xrightarrow{\times f} B_{[\rho',\rho]} \longrightarrow \prod_{\substack{\mathfrak{m} \in |Y| \\ \rho' \leq \|\mathfrak{m}\| \leq \rho}} B^+_{dR,\mathfrak{m}}/\mathrm{Fil}^{\mathrm{ord}_\mathfrak{m}(f)}B^+_{dR,\mathfrak{m}} \longrightarrow 0$$

when ρ' varies. Then, using remark 0.13.2.4 of [19] we have a Mittag Leffler type property and we can take the projective limit to obtain the result. □

Corollary 2.111. *The ring $\mathscr{R}_F$ is Bezout.*

Define now for $\rho \in {]}0, 1[$

$$B^b_{]0,\rho]} = \left\{ f \in B_{]0,\rho]} \mid \exists N \in \mathbb{Z}, \ \sup_{0<\rho'\leq\rho} |\pi^N f|_{\rho'} < +\infty \right\}.$$

One can define the Newton polygon of an element of $B_{]0,\rho]}$. This is defined only on an interval of $\mathbb{R}$ and has slopes inbetween $-\log_q \rho$ and $+\infty$. The part with slopes in ${]} -\log_q \rho, +\infty]$ is the Legendre transform of the function $r \mapsto v_r(x)$ as in Definition 2.11. As in the proof of Theorem 2.52 the definition of the $-\log_q \rho$ slope part is a little bit more tricky. Anyway, using those Newton polygons, we have the following proposition that is of the same type as Proposition 2.14.

Proposition 2.112. *Any element of* $B^b_{]0,\rho]}$ *is "meromorphic at* 0*", that is to say*

$$B^b_{]0,\rho]} = \left\{ \sum_{n \gg -\infty} [x_n]\pi^n \in W_{\mathcal{O}_E}(F)[\tfrac{1}{\pi}] \ \Big| \ \lim_{n \to +\infty} |x_n|\rho^n = 0 \right\}.$$

Define now

$$\mathscr{E}^\dagger_F = \varinjlim_{\rho \to 0} B^b_{]0,\rho]}.$$

One has $\mathscr{E}^\dagger_F \subset \mathscr{E}_F = W_{\mathcal{O}_E}(F)[\tfrac{1}{\pi}]$ (see the beginning of Section 1.2.1). The valuation v_π on $\mathscr{E}_F$ induces a valuation v_π on $\mathscr{E}^\dagger_F$. In equal characteristic we have $v_\pi = \mathrm{ord}_0$. One then verifies easily the following.

Proposition 2.113. *The ring* $\mathscr{E}^\dagger_F$ *is a Henselian valued field with completion the value field* $\mathscr{E}_F$.

7.2. Link with the "classical Robba rings"

Choose $\epsilon \in \mathfrak{m}_F \setminus \{0\}$ and consider $\pi_\epsilon := [\epsilon]_Q \in W_{\mathcal{O}_E}(\mathcal{O}_F)$ as in Section 2.4. Fix a perfect subfield $k \subset \mathcal{O}_F$ containing $\mathbb{F}_q$. Define the closed subfield

$$F_\epsilon = k((\epsilon)) \subset F.$$

The ring

$$\mathcal{O}_{\mathscr{E}_{F_\epsilon}} = W_{\mathcal{O}_E}(k)[\![u]\!][\tfrac{1}{u}]$$
$$= \left\{ \sum_{n \in \mathbb{Z}} a_n u^n \mid a_n \in W_{\mathcal{O}_E}(k), \ \lim_{n \to -\infty} a_n = 0 \right\}$$

is a Cohen ring for F_ϵ, that is to say a π-adic valuation ring with

$$\mathcal{O}_{\mathscr{E}_{F_\epsilon}} / \pi \mathcal{O}_{\mathscr{E}_{F_\epsilon}} = F_\epsilon.$$

Its fraction field is $\mathscr{E}_{F_\epsilon} = \mathcal{O}_{\mathscr{E}_{F_\epsilon}}\left[\frac{1}{\pi}\right]$. This complete valued field has a henselian approximation, the henselian valued field $\mathscr{E}_{F_\epsilon}^\dagger$ with ring of integers

$$\mathcal{O}_{\mathscr{E}_{F_\epsilon}^\dagger} = \left\{ \sum_{n \in \mathbb{Z}} a_n u^n \mid a_n \in W_{\mathcal{O}_E}(k), \exists \rho \in]0, 1[\lim_{n \to -\infty} |a_n| \rho^n = 0 \right\}.$$

Consider now the Robba ring

$$\mathscr{R}_{F_\epsilon} = \lim_{\substack{\rho \to 1 \\ <}} \mathcal{O}(\mathbb{D}_{[\rho,1[})$$

where $\mathcal{O}(\mathbb{D}_{[\rho,1[})$ is the ring of rigid analytic functions *of the variable u* on the annulus $\{\rho \le |u| < 1\}$. One then has

$$\mathscr{E}_{F_\epsilon}^\dagger = \mathscr{R}_{F_\epsilon}^b$$

the sub ring of analytic functions on some $\mathbb{D}_{[\rho,1[}$ that are bounded. Via this rigid analytic description of $\mathscr{E}_{F_\epsilon}^\dagger$, the valuation v_π on it is such that for $f \in \mathscr{E}_{F_\epsilon}^\dagger$ seen as an element of $\mathscr{R}_{F_\epsilon}$

$$q^{-v_\pi(f)} = \lim_{\rho \to 1} |f|_\rho = |f|_1$$

where $|.|_\rho$ is the Gauss supremum norm on the annulus $\{|u| = \rho\}$. We equip those rings with the Frobenius φ given by

$$\varphi(u) = Q(u).$$

Proposition 2.114. *The correspondence $u \mapsto \pi_\epsilon$ induces embeddings compatible with the Frobenius and the valuations*

$$\mathscr{E}_{F_\epsilon} \subset \mathscr{E}_F$$
$$\cup \qquad \cup$$
$$\mathscr{E}_{F_\epsilon}^\dagger \subset \mathscr{E}_F^\dagger$$
$$\cap \qquad \cap$$
$$\mathscr{R}_{F_\epsilon} \subset \mathscr{R}_F.$$

Proof. The injection $\mathcal{O}_{\mathscr{E}_{F_\epsilon}} \subset \mathcal{O}_{\mathscr{E}_F}$ is the natural injection between Cohen rings induced by the extension $F|F_\epsilon$. Since $\pi_\epsilon = \varphi(\Pi^-(u_\epsilon))$, the Newton polygon of π_ϵ is $+\infty$ on $]-\infty, 0[$, takes the value $v(\epsilon)$ at 0 and has slopes $\left(\frac{\lambda}{q^n}\right)_{n \ge 0}$ with multiplicities 1 on $[0, +\infty[$ where

$$\lambda = \frac{q-1}{q} v(\epsilon).$$

In particular, for $\rho > 0$ satisfying $\rho \le |\epsilon|^{\frac{q}{q-1}}$ one has

$$|\pi_\epsilon|_\rho = |\epsilon|.$$

Thus, if $a \in W_{\mathcal{O}_E}(k)_{\mathbb{Q}}$ and $n \in \mathbb{Z}$ then for $\rho = q^{-r} \in \,]0, |\epsilon|^{\frac{q}{q-1}}]$ one has

$$|a\pi_\epsilon|_\rho = |a|^r . |\epsilon|^n.$$

Thus, if $f(u) = \sum_{n \in \mathbb{Z}} a_n u^n \in \mathscr{R}_{F_\epsilon}$ then for $\rho = q^{-r} \in \,]0, |\epsilon|^{\frac{q}{q-1}}]$

$$|a_n \pi_\epsilon^n|_\rho \leq \left(|a_n| . (|\epsilon|^{1/r})^n \right)^r$$

where one has to be careful that on the left-hand side of this expression $|.|_\rho$ stands for the Gauss norm on B^b with respect to the "formal variable π" and the right-hand side the Gauss norm $|.|_{|\epsilon|^{1/r}}$ is taken with respect to the formal variable u. From this one deduces that if f is holomorphic on the annulus $\{|u| = |\epsilon|^{1/r}\}$ then the series $f(\pi_\epsilon) := \sum_{n \in \mathbb{Z}} a_n \pi_\epsilon^n$ converges in B_ρ. Since the condition $\rho \to 0$ is equivalent to $|\epsilon|^{1/r} \to 1$ one deduces a morphism

$$\mathscr{R}_{F_\epsilon} \longrightarrow \mathscr{R}_F$$
$$f \longmapsto f(\pi_\epsilon)$$

such that for all $\rho = q^{-r}$ sufficiently small,

$$|f(\pi_\epsilon)|_\rho \leq |f|^r_{|\epsilon|^{1/r}}.$$

A look at Newton polygons of elements of $\mathscr{R}_{F_\epsilon}$ tells us that that if $f \in \mathscr{E}^\dagger_{F_\epsilon}$ then for $r > 0$ sufficiently small, there exists $\alpha, \beta \in \mathbb{R}$ such that

$$v_r(f) = \alpha r + \beta.$$

This implies that for $r \gg 0$

$$|f|^r_{|\epsilon|^{1/r}} \leq A.B^r$$

for some constants $A, B \in \mathbb{R}_+$. From this one deduces easily that $f(\pi_\epsilon) \in \mathscr{E}^\dagger_F$. □

7.3. Harder–Narasimhan filtration of φ-modules over $\mathscr{E}^\dagger$

7.3.1. An analytic Dieudonné–Manin theorem

Let φ-Mod$_{\mathscr{E}^\dagger_F}$ be the category of finite dimensional $\mathscr{E}^\dagger_F$-vector spaces equipped with a semi-linear automorphism. For $(D, \varphi) \in \varphi$-Mod$_{\mathscr{E}^\dagger_F}$ set

$$\deg(D, \varphi) = -v_\pi(\det \varphi)$$

which is well defined independently of the choice of a base of D since v_π is φ-invariant. Since the category φ-Mod$_{\mathscr{E}^\dagger_F}$ is abelian, there are Harder–Narasimhan filtrations in φ-Mod$_{\mathscr{E}^\dagger_F}$ for the slope function $\mu = \frac{\deg}{rk}$. Let us remark that we also have such filtrations for the opposite slope function $-\mu$ that is to say up to replacing φ by φ^{-1}.

In the next theorem, if we replace $\mathscr{E}_F^\dagger$ by $\mathscr{E}_F = \widehat{\mathscr{E}_F^\dagger} = W_{O_E}(F)[\frac{1}{\pi}]$ we obtain the Dieudonné–Manin classification theorem. This theorem tells us that this classification extends to the Henselian case of $\mathscr{E}_F^\dagger$ that is to say the scalar extension induces an equivalence

$$\varphi\text{-Mod}_{\mathscr{E}_F^\dagger} \xrightarrow{\sim} \varphi\text{-Mod}_{\mathscr{E}_F}.$$

If F is algebraically closed, for each $\lambda \in \mathbb{Q}$ we note $\mathscr{E}_F^\dagger(\lambda)$ the standard isoclinic isocrystal with Dieudonné–Manin slope λ. One has $\mu(\mathscr{E}_F^\dagger(\lambda)) = -\lambda$.

Theorem 2.115. (1) *A φ-module $(D, \varphi) \in \varphi\text{-Mod}_{\mathscr{E}_F^\dagger}$ is semi-stable of slope $-\lambda = \frac{d}{h}$ if and only if there is a $\mathcal{O}_{\mathscr{E}_F^\dagger}$-lattice $\Lambda \subset M$ such that $\varphi^h(\Lambda) = \pi^d \Lambda$.*

(2) *If F is algebraically closed then semi-stable objects of slope λ in $\varphi\text{-Mod}_{\mathscr{E}_F^\dagger}$ are the ones isomorphic to a finite direct sum of $\mathscr{E}_F^\dagger(-\lambda)$.*

(3) *The category of semi-stable φ-modules of slope 0 is equivalent to the category of E-local systems on $\mathrm{Spec}(F)_{\acute{e}t}$. In concrete terms, after the choice of an algebraic closure $\overline{F}$ of F*

$$\varphi\text{-Mod}_{\mathscr{E}_F^\dagger}^{ss,0} \xrightarrow{\sim} \mathrm{Rep}_E(G_F)$$

$$(D, \varphi) \longmapsto \left(D \otimes_{\mathscr{E}_F^\dagger} \mathscr{E}_{\overline{F}}^\dagger \right)^{\varphi = Id}.$$

(4) *The Harder–Narasimhan filtrations of (D, φ) (associated to the slope function μ) and (D, φ^{-1}) (associated to the slope function $-\mu$) are opposite filtrations that define a canonical splitting of the Harder–Narasimhan filtration. There is a decomposition*

$$\varphi\text{-Mod}_{\mathscr{E}_F^\dagger} = \overset{\perp}{\underset{\lambda \in \mathbb{Q}}{\bigoplus}} \varphi\text{-Mod}_{\mathscr{E}_F^\dagger}^{ss,\lambda}$$

that is orthogonal in the sense that if A, resp. B, is semi-stable of slope λ, resp. μ, with $\lambda \neq \mu$ then $\mathrm{Hom}\,(A, B) = 0$. If F is algebraically closed the category $\varphi\text{-Mod}_{\mathscr{E}_F^\dagger}$ is semi-simple.

The main point we wanted to stress in this section is point (4) of the preceding theorem. In fact, if one replaces $\mathscr{E}_F^\dagger$ by $\mathscr{R}_F$ we will see in the following section that φ-modules of $\mathscr{R}_F$ have Harder–Narasimhan filtrations for the slope function μ. But, although φ is bijective on $\mathscr{R}_F$, there are no Harder–Narasimhan filtrations for the opposite slope function $-\mu$ that is to say for

φ^{-1}-modules (we have to use Proposition 2.120). In a sense, this is why there is no canonical splitting of the Harder–Narasimhan filtration for φ-modules over $\mathscr{R}_F$.

Sketch of proof of Theorem 2.115. The non-algebraically closed case is deduced from the algebraically closed one thanks to the following Galois descent result (see [4] III.3.1 that applies for any F thanks to the vanishing result 2.31).

Proposition 2.116 (Cherbonnier–Colmez). *The scalar extension functor is an equivalence between finite dimensional $\mathscr{E}_F^\dagger$-vector spaces and finite dimensional $\mathscr{E}_{\widehat{\overline{F}}}^\dagger$ vector spaces equipped with a continuous semi-linear action of* $\mathrm{Gal}(\overline{F}|F)$.

We now suppose F is algebraically closed. We note $\psi = \varphi^{-1}$ that is more suited than φ for what we want to do. One then checks the proof of the theorem is reduced to the following statements:

(1) If $a \in \mathbb{Z}$ and $b \in \mathbb{N}_{\geq 1}$, for any $(D, \varphi) \in \varphi\text{-Mod}_{\mathscr{E}_F^\dagger}$ admitting an $\mathcal{O}_{\mathscr{E}_F^\dagger}$-lattice Λ such that $\psi^b(\Lambda) = \pi^a \Lambda$ one has $D^{\psi^b = \pi^a} = (D \otimes \mathscr{E}_F)^{\psi^b = \pi^a}$.

(2) If $a \in \mathbb{Z}$ and $b \in \mathbb{N}_{\geq 1}$, $Id - \pi^a \psi^b : \mathscr{E}_F^\dagger \to \mathscr{E}_F^\dagger$ is surjective.

In fact, the first point implies that any (D, φ) has a decreasing filtration $(\mathrm{Fil}^\lambda D)_{\lambda \in \mathbb{Q}}$ satisfying $\mathrm{Gr}^\lambda D \simeq \mathscr{E}_F^\dagger(-\lambda)$. The second point shows that for $\mu < \lambda$ one has $\mathrm{Ext}^1(\mathscr{E}_F^\dagger(\lambda), \mathscr{E}_F^\dagger(\mu)) = 0$ and thus the preceding filtration is split.

For point (1), up to replacing ψ by a power, that is to say E by a finite unramified extension, and twisting we can suppose $a = 0$ and $b = 1$. For $\rho \in {]0, 1[}$ let

$$A_\rho = \left\{ x \in B^b_{]0,\rho]} \mid \forall \rho' \in {]0, \rho]},\ |x|_{\rho'} \leq 1 \right\}$$
$$= \left\{ \sum_{n \geq 0} [x_n] \pi^n \in B^b_{]0,\rho]} \cap \mathcal{O}_{\mathscr{E}_F^\dagger} \mid \forall n,\ |x_n| \rho^n \leq 1 \right\}.$$

Then A_ρ is stable under ψ,

$$A_\rho\left[\tfrac{1}{\pi}\right] = B^b_{]0,\rho]}$$

and

$$\varinjlim_{\rho \to 0} A_\rho = \left\{ x \in \mathcal{O}_{\mathscr{E}_F^\dagger} \mid x \bmod \pi \in \mathcal{O}_F \right\}.$$

Now, $\Lambda/\pi\Lambda$ is an F-vector space equipped with a Frob_q^{-1}-linear endomorphism $\overline{\psi}$ the reduction of ψ. One can find a basis of this vector space in which the matrix of $\overline{\psi}$ has coefficients in $\mathcal{O}_F$. Lifting such a basis we obtain a basis of Λ in which the matrix of ψ has coefficients in $\lim\limits_{\rho\to 0} A_\rho$. Let $C \in M_h(A_\rho)$, ρ sufficiently small, be the matrix of ψ in such a basis. For $k \geq 0$ and $x = \sum_{i\geq 0}[x_i]\pi^i \in \mathcal{O}_{\mathscr{E}_F}$ set

$$|x|_{k,\rho} = \sup_{0\leq i\leq k} |x_i|\rho^i.$$

One has for $x, y \in \mathcal{O}_{\mathscr{E}_F}$

$$|xy|_{k,\rho} \leq |x|_{k,\rho}|y|_{k,\rho}.$$

Now for $x = (x_1, \ldots, x_h) \in \mathcal{O}_{\mathscr{E}_F}^h$ set

$$\|x\|_{k,\rho} = \sup_{1\leq j\leq h} |x_j|_{k,\rho}.$$

If $x \in \mathcal{O}_{\mathscr{E}_F}^h$ satisfies

$$C\psi(x) = x$$

by iterating, we obtain for all n

$$C\psi(C)\cdots\psi^{n-1}(C).\psi^n(x) = x.$$

But since C has coefficients in A_ρ, for all $i \geq 0$, $\psi^i(C)$ has coefficients in A_ρ and one deduces that for all $k, n \geq 0$

$$\|x\|_{k,\rho} \leq \|\psi^n(x)\|_{k,\rho}.$$

But for $y \in \mathcal{O}_{\mathscr{E}_F}$,

$$\lim_{n\to+\infty} |\psi^n(y)|_{k,\rho} \leq 1.$$

From this one deduces that for all k, $\|x\|_{k,\rho} \leq 1$ and thus $x \in (B_{]0,\rho']}^b)^h$ as soon as $\rho' < \rho$. This proves point (1).

For point (2), $B_{]0,\rho]}^b$ is complete with respect to $(|.|_{\rho'})_{0\leq\rho'\leq\rho}$ where $|.|_0 = q^{-v_\pi}$. Moreover one checks that the operator $\pi^a\psi^b$ is topologically nilpotent with respect to those norms and thus $Id - \pi^a\psi^b$ is bijective on $B_{]0,\rho]}^b$. $\square$

7.3.2. The non-perfect case: Kedlaya's flat descent

Let $\epsilon \in \mathfrak{m}_F$ non-zero and $F_\epsilon = k((\epsilon))$ as in Section 7.2. The Frobenius φ of $\mathscr{E}_{F_\epsilon}^\dagger$ is not bijective like in the preceding sub-section. Let φ-Mod$_{\mathscr{E}_{F_\epsilon}^\dagger}$ be the category of couples (D, φ) where D is a finite dimensional $\mathscr{E}_{F_\epsilon}^\dagger$-vector space and φ a semi-linear automorphism, that is to say the linearization of φ is an isomorphism $\Phi : D^{(\varphi)} \xrightarrow{\sim} D$. We define in the same way φ-Mod$_{\mathscr{E}_{F_\epsilon}}$. As before, setting $\deg(D, \varphi) = -v_\pi(\det \varphi)$, there is a degree function on those abelian categories of φ-modules.

Theorem 2.117 (Kedlaya [22]). *A φ-module (D, φ) over $\mathscr{E}_{F_\epsilon}^\dagger$, resp. $\mathscr{E}_{F_\epsilon}$, is semi-stable of slope $-\lambda = \frac{d}{h}$ if and only if there exists an $\mathcal{O}_{\mathscr{E}_{F_\epsilon}^\dagger}$-lattice, resp. $\mathcal{O}_{\mathscr{E}_{F_\epsilon}}$-lattice, $\Lambda \subset D$ satisfying $\Phi^h(\Lambda) = \pi^d \Lambda$.*

Sketch of proof of Theorem 2.117. Let us recall how this theorem is deduced from Dieudonné–Manin by Kedlaya using a faithfully flat descent technique. We treat the case of φ-Mod$_{\mathscr{E}_{F_\epsilon}^\dagger}$, the other case being identical. We can suppose F is algebraically closed. We consider the scalar extension functor

$$- \otimes_{\mathscr{E}_{F_\epsilon}^\dagger} \mathscr{E}_F : \varphi\text{-Mod}_{\mathscr{E}_{F_\epsilon}^\dagger} \longrightarrow \varphi\text{-Mod}_{\mathscr{E}_F}.$$

Via this scalar extension, the Harder–Narasimhan slope functions correspond:

$$\mu\big(D \otimes_{\mathscr{E}_{F_\epsilon}^\dagger} \mathscr{E}_F, \varphi \otimes \varphi\big) = \mu(D, \varphi).$$

The first step is to prove that $(D, \varphi) \in \varphi\text{-Mod}_{\mathscr{E}_{F_\epsilon}^\dagger}$ is semi-stable of slope λ if and only its scalar extension to $\mathscr{E}_F$ is semi-stable of slope λ. One direction is clear: if $(D \otimes_{\mathscr{E}_{F_\epsilon}^\dagger} \mathscr{E}_F, \varphi \otimes \varphi)$ is semi-stable of slope λ then (D, φ) is semi-stable of slope λ. In the other direction, let us consider the diagram of rings equipped with Frobenius

$$\mathscr{E}_{F_\epsilon}^\dagger \longrightarrow \mathscr{E}_F \underset{i_2}{\overset{i_1}{\rightrightarrows}} \mathscr{E}_F \underset{\mathscr{E}_{F_\epsilon}^\dagger}{\otimes} \mathscr{E}_F$$

where the Frobenius on $\mathscr{E}_F \underset{\mathscr{E}_{F_\epsilon}^\dagger}{\otimes} \mathscr{E}_F$ is $\varphi \otimes \varphi$, $i_1(x) = x \otimes 1$ and $i_2(x) = 1 \otimes x$. This induces a diagram of categories of φ-modules

$$\varphi\text{-Mod}_{\mathscr{E}_{F_\epsilon}^\dagger} \longrightarrow \varphi\text{-Mod}_{\mathscr{E}_F} \underset{i_{2*}}{\overset{i_{1*}}{\rightrightarrows}} \varphi\text{-Mod}_{\mathscr{E}_F \underset{\mathscr{E}_{F_\epsilon}^\dagger}{\otimes} \mathscr{E}_F}.$$

Now, faithfully flat descent tells us that for $A \in \varphi\text{-Mod}_{\mathscr{E}_{F_\epsilon}^\dagger}$, the sub-objects of A are in bijection with the sub-objects B of $A \otimes_{\mathscr{E}_{F_\epsilon}^\dagger} \mathscr{E}_F$ satisfying $i_{1*}\text{B} = i_{2*}\text{B}$. Suppose now $A \in \varphi\text{-Mod}_{\mathscr{E}_{F_\epsilon}^\dagger}$ is semi-stable and $A' := A \otimes_{\mathscr{E}_{F_\epsilon}^\dagger} \mathscr{E}_F$ is not. Let

$$0 \subsetneq A_1' \subsetneq \cdots \subsetneq A_r' = A'$$

be the Harder–Narasimhan filtration of A'. The Dieudonné–Manin theorem gives us the complete structure of the graded pieces of this filtration in $\varphi\text{-Mod}_{\mathscr{E}_F}$. Let us prove by descending induction on $j \geq 1$ that

$$i_{1*}A_1' \subset i_{2*}A_j'.$$

In fact, if $j > 1$ and $i_{1*}A'_1 \subset i_{2*}A'_j$ then one can look at the composite morphism

$$i_{1*}A'_1 \hookrightarrow i_{2*}A'_j \longrightarrow i_{2*}A'_j/A'_{j-1}.$$

Dieudonné–Manin tells us that this morphism is given by a finite collection of elements in

$$\left(\mathscr{E}_F \otimes_{\mathscr{E}_{F_\epsilon}^\dagger} \mathscr{E}_F\right)^{\varphi^h = \pi^d}$$

where $h \in \mathbb{N}_{\geq 1}$, $d \in \mathbb{Z}$ and $\frac{d}{h}$ is the Dieudonné–Manin slope of A'_1 minus the one of A'_j/A'_{j-1} which is thus strictly negative (recall the Harder–Narasimhan slope is the opposite of the Dieudonné–Manin one). We thus have $d < 0$ and Lemma 2.118 tells us this is 0. We conclude $i_{1*}A'_1 \subset i_{2*}A'_{j-1}$ and obtain by induction that $i_{1*}A'_1 \subset i_{2*}A'_j$. By symmetry we thus have

$$i_{1*}A'_1 = i_{2*}A'_1$$

and A'_1 descends to a sub-object of A which contradicts the semi-stability of A. We thus have proved that $A \in \varphi\text{-Mod}_{\mathscr{E}_{F_\epsilon}^\dagger}$ is semi-stable if and only if $A \otimes_{\mathscr{E}_{F_\epsilon}^\dagger} \mathscr{E}_F$ is semi-stable.

Theorem 2.117 is easily reduced to the slope 0 case. Thus, let (D, φ) be semi-stable of slope 0. Let $\Lambda \subset D$ be a lattice. Then,

$$\Lambda' = \sum_{k \geq 0} \mathcal{O}_{\mathscr{E}_{F_\epsilon}^\dagger} \varphi^k(\Lambda) \subset D$$

is a lattice since after scalar extension to $\mathscr{E}_F$, $(D \otimes_{\mathscr{E}_{F_\epsilon}^\dagger} \mathscr{E}_F, \varphi \otimes \varphi)$ is isoclinic with slope 0. This lattice is stable under φ, but since (D, φ) has slope 0, automatically

$$\mathcal{O}_{\mathscr{E}_{F_\epsilon}^\dagger} \varphi(\Lambda') = \Lambda'.$$

$\square$

Lemma 2.118. *The ring* $\mathcal{O}_{\mathscr{E}_F} \otimes_{\mathcal{O}_{\mathscr{E}_{F_\epsilon}^\dagger}} \mathcal{O}_{\mathscr{E}_F}$ *is* π-*adically separated.*

7.4. The Harder–Narasimhan filtration of φ-modules over $\mathscr{R}_F$

Since the ring $\mathscr{R}_F$ is Bezout, for an $\mathscr{R}_F$-module M the following are equivalent:

- M is free of finite rank,
- M is torsion free of finite type,
- M is projective of finite type.

Moreover, if $\mathrm{Frac}(\mathscr{R}_F)$ is the fraction field of $\mathscr{R}_F$ and $\mathrm{Vect}_{\mathrm{Frac}(\mathscr{R}_F)}$ is the associated category of finite dimensional vector spaces, the functor $- \otimes_{\mathscr{R}_F} \mathrm{Frac}(\mathscr{R}_F)$ is a generic fiber functor in the sense that for a free $\mathscr{R}_F$-module of finite type M it induces a bijection

$$\{\text{direct factor sub modules of } M\} \xrightarrow{\sim} \{\text{sub-}\mathrm{Frac}(\mathscr{R}_F)\text{-vector spaces of}$$
$$M \otimes_{\mathscr{R}_F} \mathrm{Frac}(\mathscr{R}_F)$$

with inverse the map $W \mapsto W \cap M$, "the schematical closure of W in M". Let $\varphi\text{-Mod}_{\mathscr{R}_F}$ be the category of finite rank free $\mathscr{R}_F$-modules M equipped with a φ-linear isomorphism $\varphi : M \xrightarrow{\sim} M$. There are two additive functions on the exact category $\varphi\text{-Mod}_{\mathscr{R}_F}$

$$\deg, \mathrm{rk} : \varphi\text{-Mod}_{\mathscr{R}_F} \longrightarrow \mathbb{Z}$$

where the rk is the rank and the degree is defined using the following proposition that is deduced from Newton polygons considerations.

Proposition 2.119. *One has the equality* $\left(B_{]0,\rho]}\right)^{\times} = \left(B_{]0,\rho]}^{b}\right)^{\times}$ *and thus*

$$\mathscr{R}_F^{\times} = (\mathscr{E}_F^{\dagger})^{\times}.$$

Of course the valuation v_π on $\mathscr{E}_F^{\dagger}$ is invariant under φ. This allows us to define

$$\deg(M, \varphi) = -v_\pi(\det \varphi).$$

As in [10], to have Harder–Narasimhan filtrations in the exact category $\varphi\text{-Mod}_{\mathscr{R}_F}$ we now need to prove that any isomorphism that is "an isomorphism in generic fiber", that is to say after tensoring with $\mathrm{Frac}(\mathscr{R}_F)$,

$$f : (M, \varphi) \longrightarrow (M', \varphi')$$

induces the inequality

$$\deg(M, \varphi) \leq \deg(M', \varphi')$$

with equality if and only if f is an isomorphism. This is achieved by the following proposition.

Proposition 2.120. *Let* $x \in \mathscr{R}_F$ *non-zero such that* $\varphi(x)/x \in \mathscr{E}_F^{\dagger}$. *Then*

$$v_\pi(\varphi(x)/x) \leq 0$$

with equality if and only if $x \in \mathscr{R}_F^{\times}$.

Proof. For $x \in \mathscr{E}_F^{\dagger}$ one has

$$v_\pi(x) = \lim_{r \to +\infty} \frac{v_r(x)}{r}.$$

But, for $r \gg 0$,

$$v_r(\varphi(x)/x) = q v_{r/q}(x) - v_r(x).$$

But if $x \in B_{]0,\rho]}$, for $q^{-r} \in]0, \rho[$, the number $\frac{v_r(x)}{r}$ is the intersection with the x-axis of the line with slope r that is tangent to Newt(x). From this graphic interpretation, one deduces that as soon as r_0 is such that the intersection of the tangent line to Newt(x) of slope r_0 with $Newt(x)$ is in the upper half plane then for $r \geq r_0$, $r \mapsto \frac{v_r(x)}{r}$ is a decreasing function and it is bounded if and only if Newt(x)(t) $= +\infty$ for $t \ll 0$. $\qquad\square$

We thus have a good notion of Harder–Narasimhan filtrations in the exact category $\varphi\text{-Mod}_{\mathcal{R}_F}$. We note $\mu = \deg/\mathrm{rk}$ the associated slope function.

7.5. Classification of φ-modules over $\mathcal{R}_F$: Kedlaya's theorem

Suppose F is algebraically closed. For each slope $\lambda \in \mathbb{Q}$ there is associated an object

$$\mathcal{R}_F(\lambda) \in \varphi\text{-Mod}_{\mathcal{R}_F}$$

satisfying

$$\mu\big(\mathcal{R}_F(\lambda)\big) = -\lambda.$$

This is the image of the simple isocrystal with Dieudonné–Manin slope λ via the scalar extension functor

$$\varphi\text{-Mod}_{\mathcal{E}_F^\dagger} \longrightarrow \varphi\text{-Mod}_{\mathcal{R}_F}.$$

The next theorem tells us that this functor is essentially surjective (but not full).

Theorem 2.121 (Kedlaya [21]). *Suppose F is algebraically closed.*

(1) *The semi-stable objects of slope λ in $\varphi\text{-Mod}_{\mathcal{R}_F}$ are the direct sums of $\mathcal{R}_F(-\lambda)$.*
(2) *The Harder–Narasimhan filtration of a φ-module over $\mathcal{R}_F$ is split.*
(3) *There is a bijection*

$$\{\lambda_1 \geq \cdots \geq \lambda_n \mid n \in \mathbb{N}, \ \lambda_i \in \mathbb{Q}\} \xrightarrow{\sim} \varphi\text{-Mod}_{\mathcal{R}_F}/\sim$$

$$(\lambda_1, \ldots, \lambda_n) \longmapsto \Big[\bigoplus_{i=1}^{n} \mathcal{R}_F(-\lambda_i)\Big].$$

In particular for each slope $\lambda \in \mathbb{Q}$ scalar extension induces an equivalence

$$\varphi\text{-Mod}_{\mathcal{E}_F^\dagger}^{ss,\lambda} \xrightarrow{\sim} \varphi\text{-Mod}_{\mathcal{R}_F}^{ss,\lambda}$$

and $(M, \varphi) \in \varphi\text{-Mod}_{\mathscr{R}_F}$ is semi-stable of slope $\lambda = \frac{d}{h}$ if and only if there is a free of the same rank as M $\mathcal{O}_{\mathscr{E}_F^\dagger}$-sub-module $\Lambda \subset M$ generating M and satisfying $\varphi^h(\Lambda) = \pi^{-d}\Lambda$.

7.6. Application: classification of φ-modules over B

As a consequence of Theorem 2.52 one obtains the following.

Theorem 2.122. *The algebra* B *is a Frechet–Stein algebra in the sense of Schneider–Teitelbaum ([28]).*

Recall ([28]) there is a notion of coherent sheaf on the Frechet–Stein algebra B. A coherent sheaf on B is a collection of modules $(M_I)_I$ where I goes through the set of compact intervals in $]0, 1[$ and M_I is a B_I-module together with isomorphisms

$$M_I \otimes_{B_I} B_J \xrightarrow{\sim} M_J$$

for $J \subset I$, satisfying the evident compatibility relations for three intervals $K \subset J \subset I$. This is an abelian category. There is a global section functor

$$\Gamma : (M_I)_I \longmapsto \varprojlim_I M_I$$

from coherent sheaves to B-modules. It is fully faithful exact and identifies the category of coherent sheaves with an abelian subcategory of the category of B-modules. This functor has a left adjoint

$$M \longmapsto (M \otimes_B B_I)_I.$$

On the essential image of Γ this induces an equivalence with coherent sheaves. By definition, a coherent sheaf $(M_I)_I$ on B is a vector bundle if for all I, M_I is a free B_I-module of finite rank.

Proposition 2.123. *The global sections functor* Γ *induces an equivalence of categories between vector bundles on* B *and finite type projective* B-*modules.*

The proof is similar to the one of proposition 2.1.15 of [23]. More precisely, the main difficulty is to prove that the global sections M of a coherent sheaf $(M_I)_I$ such that for some integer r all M_I are generated by r elements is a finite type B-module. For this one writes $]0, 1[= F_1 \cup F_2$ where F_1 and F_2 are locally finite infinite *disjoint* unions of compact intervals. Then, one constructs for $i = 1, 2$ by approximation techniques global sections $f_{i,1}, \cdots, f_{i,r} \in M$ that generate each M_I for I a connected component of F_i. The sum of those sections furnishes a morphism $B^{2r} \rightarrow M$ that induces a surjection $B_I^{2r} \rightarrow M_I$

for all I a connected component of F_i, $i = 1, 2$. Thanks to Lemma 2.124 this induces surjections $B_I^{2r} \to M_I$ for any compact interval of $]0, 1[$.

The following lemma is an easy consequence of Theorem 2.52.

Lemma 2.124. *For a finite collection of compact intervals $I_1, \cdots, I_n \subset]0, 1[$ with union I the morphism $\coprod_{k=1,\cdots,n} \mathrm{Spec}(B_{I_k}) \to \mathrm{Spec}(B_I)$ is an fpqc covering.*

Let us come back to φ-modules. Let φ-Mod_B be the category of finite type projective B-modules M equipped with a semi-linear isomorphism $\varphi : M \xrightarrow{\sim} M$. For $\rho \in]0, 1[$ the ring $B_{]0,\rho]}$ is equipped with the endomorphism φ^{-1} satisfying $\varphi^{-1}(B_{]0,\rho]}) = B_{]0,\rho^{1/q}]}$ and thus

$$B = \bigcap_{n \geq 0} \varphi^{-n}(B_{]0,\rho]}).$$

Note φ^{-1}-$\mathrm{mod}_{B_{]0,\rho]}}$ the category of finite rank free $B_{]0,\rho]}$-modules M equipped with a semi-linear isomorphism $\varphi : M \to M$ (by a semi-linear isomorphism we mean a semi-linear morphism whose linearization is an isomorphism). Of course,

$$\varinjlim_{\rho \to 0} \varphi^{-1}\text{-}\mathrm{mod}_{B_{]0,\rho]}} \xrightarrow{\sim} \varphi^{-1}\text{-}\mathrm{mod}_{\mathscr{R}_F} = \varphi\text{-}\mathrm{Mod}_{\mathscr{R}_F}.$$

If $(M, \varphi^{-1}) \in \varphi^{-1}$-$\mathrm{mod}_{B_{]0,\rho]}}$ then the collection of modules $(\varphi^{-n}M)_{n \geq 0}$ defines a vector bundle on B whose global sections is

$$\bigcap_{n \geq 0} \varphi^{-n}(M).$$

Using Proposition 2.123 one obtains the following.

Proposition 2.125. *The scalar extension functor induces an equivalence*

$$\varphi\text{-}\mathrm{Mod}_B \xrightarrow{\sim} \varphi\text{-}\mathrm{Mod}_{\mathscr{R}_F}.$$

Applying Kedlaya's theorem [21] one thus obtains:

Theorem 2.126. *If F is algebraically closed there is a bijection*

$$\{\lambda_1 \geq \cdots \geq \lambda_n \mid n \in \mathbb{N}, \lambda_i \in \mathbb{Q}\} \xrightarrow{\sim} \varphi\text{-}\mathrm{Mod}_B/\sim$$

$$(\lambda_1, \ldots, \lambda_n) \longmapsto \Big[\bigoplus_{i=1}^{n} B(-\lambda_i)\Big].$$

For $(M, \varphi) \in \varphi$-Mod_B define

$$\mathscr{E}(M, \varphi) = \bigoplus_{d \geq 0} M^{\varphi = \pi^d},$$

a quasi-coherent sheaf on the curve X. Using Theorem 2.126 together with the classification of vector bundles theorem 2.83 one obtains the following theorem.

Theorem 2.127. *If F is algebraically closed there is an equivalence of exact categories*

$$\varphi\text{-Mod}_B \xrightarrow{\sim} \text{Bun}_X$$
$$(M, \varphi) \longmapsto \mathscr{E}(M, \varphi).$$

Via this equivalence one has

$$H^0(X, \mathscr{E}(M, \varphi)) = M^{\varphi = Id}$$
$$H^1(X, \mathscr{E}(M, \varphi)) = \text{coker}(M \xrightarrow{Id - \varphi} M).$$

Remark 2.128. In equal characteristic, when $E = \mathbb{F}_q((\pi))$, $Y = \mathbb{D}_F^*$ and the classification of φ-vector bundles on Y is due to Hartl and Pink ([20]). Via Theorem 2.127 this is the same as the classification of φ-modules over B. We explained the proof of the classification theorem 2.83 only when $E|\mathbb{Q}_p$. However, the same proof works when $E = \mathbb{F}_q((\pi))$ using periods of π-divisible $\mathcal{O}_E$-modules. In this case, Theorem 2.127 is thus still valid.

Sadly, there is no direct short proof of Theorem 2.127 that would allow us to recover the Kedlaya or Hartl Pink classification theorem from the classification of vector bundles on the curve.

However, one of the first steps in their proof is that if $(M, \varphi) \in \varphi\text{-Mod}_{\mathscr{R}_F}$ then $M^{\varphi = \pi^d} \neq 0$ for $d \gg 0$. As a consequence, any φ-module over $\mathscr{R}_F$ (and thus B) is an iterated extension of rank 1 modules. Those are easy to classify and thus any φ-module over B is a successive extension of B(λ) with $\lambda \in \mathbb{Z}$. Taking this granted plus the fact that for $\lambda \in \mathbb{Z}$, $H^1(B(\lambda + d))$ (the cokernel of $Id - \varphi$) is zero for $d \gg 0$, one deduces that for any $(M, \varphi) \in \varphi\text{-Mod}_B$, $\mathscr{E}(M, \varphi)$ is a vector bundle. Then, if one knows explicitly that for all $d \in \mathbb{Z}$ and $i = 0, 1$, $H^i(B(d)) \xrightarrow{\sim} H^i(X, \mathcal{O}_X(d))$ (this is easy for $i = 0$, and is deduced from the fundamental exact sequence plus computations found in the work of Kedlaya and Hartl Pink for $i = 1$) one can deduce a proof that the functor $(M, \varphi) \mapsto \mathscr{E}(M, \varphi)$ is fully faithful and thus the classification of vector bundles on the curve gives back the Kedlaya and Hartl Pink theorem.

7.7. Classification of φ-modules over B^+

Recall from Section 1.2.1 that for $\rho \in]0, 1[$, $B_\rho^+ = B_{[\rho, 1[}^+$ and that

$$B^+ = \bigcap_{\rho > 0} B_\rho^+ = \bigcap_{n \geq 0} \varphi^n(B_{\rho_0}^+)$$

for any ρ_0. Moreover, for any $x \in B_{[\rho,1[}$ there is defined a Newton polygon $\mathrm{Newt}(x)$ and

$$B_\rho^+ = \{x \in B_{[\rho,1[} \mid \mathrm{Newt}(x) \geq 0\}$$
$$B^+ = \{x \in B \mid \mathrm{Newt}(x) \geq 0\}.$$

In fact, by concavity of the Gauss valuation $r \mapsto v_r(x)$, for any $x \in B_{[\rho,1[}$ the limit

$$|x|_1 := \lim_{\rho \to 1} |x|_\rho$$

exists in $[0, +\infty]$ and equals $q^{-v_0(x)}$ for $x \in B^b$ and

$$B_\rho^+ = \{x \in B_{[0,\rho[} \mid |x|_1 \leq 1\}.$$

Let us note $|x|_1 := q^{-v_0(x)}$. One has

$$v_0(x) = \lim_{+\infty} \mathrm{Newt}(x)$$

and on B_ρ^+, v_0 is a valuation extending the valuation previously defined on B^b.

One has to be careful that, contrary to B, the Frechet algebra B^+ is not Frechet–Stein since the rings B_ρ^+ are not noetherian and it is not clear whether $\varphi : B_\rho^+ \to B_\rho^+$ (that is to say the inclusion $B_{\rho^q}^+ \subset B_\rho^+$) is flat or not.

Note $\varphi\text{-Mod}_{B^+}$ and $\varphi\text{-Mod}_{B_\rho^+}$ for the associated categories of finite rank *free* modules equipped with a semi-linear isomorphism. The category $\varphi\text{-Mod}_{B_\rho^+}$ does not depend on ρ. There is a scalar extension functor

$$\varphi\text{-Mod}_{B^+} \longrightarrow \varphi\text{-Mod}_B$$

and, using Proposition 2.123, a functor

$$\varphi\text{-Mod}_{B_\rho^+} \longrightarrow \varphi\text{-Mod}_B$$
$$(M, \varphi) \longmapsto \bigcap_{n \geq 0} \varphi^n \big(M \otimes_{B_\rho^+} B_{[\rho,1[}\big).$$

Proposition 2.129. *The functors* $\varphi\text{-Mod}_{B^+} \to \varphi\text{-Mod}_B$ *and* $\varphi\text{-Mod}_{B_\rho^+} \to \varphi\text{-Mod}_B$ *are fully faithful.*

Proof. Let's treat the case of $\varphi\text{-Mod}_{B^+}$, the case of $\varphi\text{-Mod}_{B_\rho^+}$ being identical. Using internal Hom's this is reduced to proving that for $(M, \varphi) \in \varphi\text{-Mod}_{B^+}$ one has

$$M^{\varphi=Id} \xrightarrow{\sim} \big(M \otimes_{B^+} B\big)^{\varphi=Id}.$$

Let us fix a basis of M and for $x = (x_1, \ldots, x_n) \in B^n \simeq M \otimes B$ and $r > 0$ set

$$W_r(x) = \inf_{1 \leq i \leq n} v_r(x_i).$$

An element $a \in B^+$ satisfies $v_r(a) \geq 0$ for $r \gg 0$. Let us fix $r_0 \gg 0$ such that all the coefficients $(a_{ij})_{i,j}$ of the matrix of φ in the fixed basis of M satisfy $v_{r_0}(a_{i,j}) \geq 0$. Then if $x \in M \otimes B$ satisfies $\varphi(x) = x$ one has

$$q W_{\frac{r_0}{q}}(x) = W_{r_0}(\varphi(x)) \geq W_{r_0}(x)$$

and thus for $k \geq 1$

$$W_{\frac{r_0}{q^n}}(x) \geq \frac{1}{q^n} W_{r_0}(x).$$

Taking the limit when $n \to +\infty$ one obtains $W_0(x) \geq 0$ that is to say $x \in (B^+)^n \simeq M$. $\qquad \square$

The preceding proposition together with Theorem 2.126 then gives the following.

Theorem 2.130. *Suppose F is algebraically closed. For $A \in \{B^+, B_\rho^+\}$ there is a bijection*

$$\{\lambda_1 \geq \cdots \geq \lambda_n \mid n \in \mathbb{N}, \; \lambda_i \in \mathbb{Q}\} \xrightarrow{\sim} \varphi\text{-Mod}_A / \sim$$

$$(\lambda_1, \ldots, \lambda_n) \longmapsto \left[\bigoplus_{i=1}^{n} A(-\lambda_i) \right].$$

One deduces there are equivalences of categories

$$\varphi\text{-Mod}_{B^+} \xrightarrow{\sim} \varphi\text{-Mod}_{B_\rho^+} \xrightarrow{\sim} \varphi\text{-Mod}_B \xrightarrow{\sim} \varphi\text{-Mod}_{\mathscr{R}_F}$$

where an inverse of the first equivalence is given by $M \mapsto \cap_{n \geq 0} \varphi^n(M)$.

7.8. Another proof of the classification of φ-modules over B^+ and B_ρ^+

We explain how to give a direct proof of Theorem 2.130 without using Kedlaya's Theorem 2.121. This proof is much simpler and in fact applies even if the field F is not algebraically closed:

Theorem 2.131. *Theorem 2.130 remains true for any F with algebraically closed residue field.*

This relies on the introduction of a new ring called $\overline{B}$. Set

$$\mathfrak{P} = \{x \in B^{b,+} \mid v_0(x) > 0\}$$

$$= \Big\{ \sum_{n \gg -\infty} [x_n] \pi^n \mid x_n \in \mathcal{O}_F, \; \exists C > 0, \; \forall n, \; v(x_n) \geq C \Big\}$$

and

$$\overline{B} = B^{b,+}/\mathfrak{P}.$$

If k stands for the residue field of $\mathcal{O}_F$, there is a reduction morphism

$$B^{b,+} \longrightarrow W_{\mathcal{O}_E}(k)_{\mathbb{Q}}.$$

Then, $\overline{B}$ is a local ring with residue field $W_{\mathcal{O}_E}(k)_{\mathbb{Q}}$. Let's begin by classifying φ-modules over $\overline{B}$.

Theorem 2.132. *Any φ-module over $\overline{B}$ is isomorphic to a direct sum of $\overline{B}(\lambda)$, $\lambda \in \mathbb{Q}$.*

Sketch of proof. One first proves that for $\lambda, \mu \in \mathbb{Q}$, $\text{Ext}^1(\overline{B}(\lambda), \overline{B}(\mu)) = 0$. For two φ-modules M and M' one has $\text{Ext}^1(M, M') = H^1(M^\vee \otimes M')$ where for a φ-module M'', $H^1(M'') = \text{coker}(Id - \varphi_{M''})$. Up to replacing φ by a power of itself, that is to say replacing E by an unramified extension, we are thus reduced to proving that for any $d \in \mathbb{Z}$,

$$Id - \pi^d\varphi : \overline{B} \longrightarrow \overline{B}$$

is surjective. For $d > 0$, this is a consequence of the fact that

$$Id - \pi^d\varphi : W_{\mathcal{O}_E}(\mathcal{O}_F) \longrightarrow W_{\mathcal{O}_E}(\mathcal{O}_F)$$

is surjective since $\pi^d\varphi$ is topologically nilpotent on $W_{\mathcal{O}_E}(\mathcal{O}_F)$ for the π-adic topology. For $d < 0$ this is deduced in the same way using $Id - \pi^{-d}\varphi^{-1}$. For $d = 0$, this is a consequence of the fact that $Id - \varphi$ is bijective on $W_{\mathcal{O}_E}(\mathfrak{m}_F)$ since φ is topologically nilpotent on $W_{\mathcal{O}_E}(\mathfrak{m}_F)$ for the $([a], \pi)$-adic topology for any $a \in \mathfrak{m}_F \setminus \{0\}$ (the topology induced by the Gauss norms $(|.|_\rho)_{\rho \in]0,1[}$) and since the residue field k of F is algebraically closed.

Let $(M, \varphi) \in \varphi\text{-Mod}_{\overline{B}}$ and $M_k = M \otimes W_{\mathcal{O}_E}(k)_{\mathbb{Q}}$ be the associated isocrystal. Let λ be the smallest slope of (M_k, φ). It suffices now to prove that M has a sub φ-module isomorphic to $\overline{B}(\lambda)$ whose underlying $\overline{B}$-module is a direct factor. Up to raising φ to a power and twisting one is reduced to the case $\lambda = 0$. Then M_k has a sub-lattice Λ stable under φ. Let us remark that any element of $B^{b,+}$ whose image in $W_{\mathcal{O}_E}(k)_{\mathbb{Q}}$ lies in $W_{\mathcal{O}_E}(k)$ is congruent modulo $\mathfrak{p}$ to an element of $W_{\mathcal{O}_E}(\mathcal{O}_F)$. Lifting the basis of Λ to a basis of M (recall $\overline{B}$ is a local ring), one then checks there is a free φ-module N over $W_{\mathcal{O}_E}(\mathcal{O}_F)$ together with a morphism $N \to M$ inducing an isomorphism $N \otimes_{W_{\mathcal{O}_E}(\mathcal{O}_F)} \overline{B} \xrightarrow{\sim} M$ and such that $N \otimes W_{\mathcal{O}_E}(k) \xrightarrow{\sim} \Lambda$. For such an N, there is an isomorphism

$$N^{\varphi=Id} \xrightarrow{\sim} \Lambda^{\varphi=Id}$$

such that $N^{\varphi=Id} \otimes_E W_{\mathcal{O}_E}(\mathcal{O}_F)$ is a direct factor in N. In fact, after fixing a basis of N, $N \simeq W_{\mathcal{O}_E}(\mathcal{O}_F)^n$ is complete with respect to the family of norms $(\|.\|_\rho)_{\rho \in]0,1[}$ where $\|(x_1, \ldots, x_n)\|_\rho = \sup_{1 \le i \le n} |x_i|_\rho$. Moreover the Frobenius of N is topologically nilpotent on $W_{\mathcal{O}_E}(\mathfrak{m}_F)^n$ for this set of norms. The result is deduced (for the direct factor assertion, one has to use that $W_{\mathcal{O}_E}(\mathfrak{m}_F)$ is contained in the Jacobson radical of $W_{\mathcal{O}_E}(\mathcal{O}_F)$ together with Nakayama lemma). $\qquad\square$

To make the link between B^+, B_ρ^+ and $\overline{B}$ we need the following.

Lemma 2.133. *For any* $a \in \mathfrak{m}_F \setminus \{0\}$, $B^+ = [a]B^+ + B^{b,+}$ *and* $B_\rho^+ = [a]B_\rho^+ + B^{b,+}$.

In fact, for any $x = \sum_{n \gg -\infty} [x_n]\pi^n$ let us note $x^+ = \sum_{n \ge 0} [x_n]\pi^n$ and $x^- = \sum_{n < 0} [x_n]\pi^n$. Then any $x \in B^+$, resp. B_ρ^+, can be written as $\sum_{n \ge 0} x_n$ with $x_n \in B^{b,+}$ going to zero when $n \to +\infty$. But one checks that if $x_n \underset{n \to +\infty}{\to} 0$ then for $n \gg 0$, $x_n^- \in [a]B^{b,+}$. This proves the lemma.

As a consequence of this lemma, if $r = v(a)$, $\{x \in B^+ \mid v_0(x) \ge r\} = [a]B^+$ and the same for B_ρ^+. Moreover, we deduce that the inclusion $B^{b,+} \to B^+$ induces an isomorphism

$$\overline{B} \xrightarrow{\sim} B^+/\{v_0 > 0\}$$

and the same for B_ρ^+. From this we deduce surjections $B^+ \to \overline{B}$ and $B_\rho^+ \to \overline{B}$. Now, Theorem 2.130 is a consequence of Theorem 2.132 and the following.

Proposition 2.134. *The reduction functor* $\varphi\text{-Mod}_{B^+} \to \varphi\text{-Mod}_{\overline{B}}$ *is fully faithful. The same holds for* B_ρ^+.

Sketch of proof. We treat the case of B^+, the case of B_ρ^+ being identical. This is reduced to proving that for $(M, \varphi) \in \varphi\text{-Mod}_{B^+}$, if $\overline{M}$ is the associated module over $\overline{B}$, then $M^{\varphi=Id} \xrightarrow{\sim} \overline{M}^{\varphi=Id}$. If $\Lambda = \mathfrak{p}M$ this is reduced to proving that

$$Id - \varphi : \Lambda \xrightarrow{\sim} \Lambda.$$

Fix a basis of $M \simeq (B^+)^n$ and note $A = (a_{ij})_{i,j} \in GL_n(B^+)$ the matrix of φ in this base. For each $r \ge 0$ and $m = (x_1, \ldots, x_n) \in M$ set $\|m\|_r = \inf_{1 \le i \le n} v_r(x_i)$. Let $r_0 > 0$ be fixed. Then $\Lambda = \mathfrak{p}^n$ is complete with respect to the set of additive norms $(\|.\|)_{r>0}$. Note $\|A\|_r = \inf_{i,j} v_r(a_{i,j})$.

Fix an $r > 0$. One first checks that for any $m \in M$ and $k \ge 1$

$$\|\varphi^k(m)\|_r \ge q^k \|m\|_{\frac{r}{q^k}} + \sum_{i=0}^{k-1} q^i \|A\|_{\frac{r}{q^i}}.$$

Now, according to inequality (2.3) of Section 1.2.1 (the inequality is stated for $B^{b,+}$ but extends by continuity to B^+), for any $r' \leq r$ one has

$$\|A\|_{r'} \geq \frac{r'}{r}\|A\|_r$$

and thus

$$\sum_{i=0}^{k-1} q^i \|A\|_{\frac{r}{q^k}} \geq \alpha k + \beta$$

for some constants $\alpha, \beta \in \mathbb{R}$. But now, if $m \in \Lambda$, $\lim_{r' \to 0} \|m\|_{r'} = \|m\|_0 > 0$ and thus

$$\lim_{k \to +\infty} \|\varphi^k(m)\|_r = +\infty.$$

We deduce that φ is topologically nilpotent on Λ and thus $Id - \varphi$ is bijective on it.   □

Remark 2.135. The preceding proof does not use the fact that F is algebraically closed and thus Theorem 2.130 remains true when F is any perfectoid field with algebraically closed residue field. On this point, there is a big difference between φ-modules over B^+ and the ones over B. In fact, one can prove that φ-modules over B satisfy Galois descent like vector bundles (Theorem 2.103) for any perfectoid field F and thus for a general F Theorem 2.126 is false.

Remark 2.136. As a consequence of the classification theorem and the first part of the proof of Theorem 2.132, for any $M, M' \in \varphi$-Mod_{B^+}, $\mathrm{Ext}^1(M, M') = 0$. This is not the case for φ-modules over B. The equivalence 2.127 is an equivalence of exact categories and for example $\mathrm{Ext}^1_{\varphi\text{-}\mathrm{Mod}_B}(B, B(1)) \neq 0$. In fact, although the scalar extension functor φ-$\mathrm{Mod}_{B^+} \xrightarrow{\sim} \varphi$-$\mathrm{Mod}_B$ is exact, its inverse is not.

Let us conclude with a geometric interpretation of the preceding result. Set $\Sigma = \mathrm{Spec}(\mathbb{Z}_p)$ and for an $\mathbb{F}_p$-scheme S note F-$\mathrm{Isocs}_{S/\Sigma}$ for the category of F-isocrystals. If $S \hookrightarrow S'$ is a thickening then F-$\mathrm{Isocs}_{S/\Sigma} \simeq F$-$\mathrm{Isocs}_{S'/\Sigma}$. Let now $a \in \mathfrak{m}_F \setminus \{0\}$, $\rho = |a|$ and $S_\rho = \mathrm{Spec}(\mathcal{O}_F/\mathcal{O}_F a)$. The category F-$\mathrm{Isocs}_{S_\rho/\Sigma}$ does not depend on the choice of $\rho \in]0, 1[$. The crystalline site $\mathrm{Cris}(S_\rho/\Sigma)$ has an initial object $A_{cris,\rho}$ such that $A_{cris,\rho}[\frac{1}{p}] = B_{cris,\rho}$ (see Section 1.2.1). We thus have an equivalence

$$F\text{-}\mathrm{Isocs}_{S_\rho/\Sigma} \simeq \varphi\text{-}\mathrm{Mod}_{B^+_{cris,\rho}}.$$

But since $B_{\rho^p}^+ \subset B_{cris,\rho}^+ \subset B_{\rho^{p-1}}^+$ we have an equivalence

$$\varphi\text{-Mod}_{B_{cris,\rho}^+} \simeq \varphi\text{-Mod}_{B_\rho^+}.$$

Moreover, one can think of φ-Mod$_{B^+}$ as being the category of "convergent F-isocrystals on S_ρ". We thus have proved the following.

Theorem 2.137. *Suppose F is a perfectoid field with algebraically closed residue field. Then any F-isocrystal, resp. convergent F-isocrystal, on S_ρ is isotrivial that is to say comes from the residue field of k after the choice of a splitting* $\mathcal{O}_F \xleftrightarrow{} k$

References

[1] Y. André. Slope filtrations. *Confluentes Mathematici*, 1:1–85, 2009.

[2] L. Berger. Représentations p-adiques et équations différentielles. *Invent. Math.*, 148(2):219–284, 2002.

[3] Laurent Berger. Construction de (ϕ, Γ)-modules: représentations p-adiques et B-paires. *Algebra Number Theory*, 2(1):91–120, 2008.

[4] F. Cherbonnier and P. Colmez. Représentations p-adiques surconvergentes. *Invent. Math.*, 133(3):581–611, 1998.

[5] J.-F. Dat, S. Orlik, and M. Rapoport. *Period domains over finite and p-adic fields*, volume 183 of Cambridge Tracts in Mathematics. Cambridge University Press, 2010.

[6] Vladimir G. Drinfeld. Coverings of p-adic symmetric domains. *Functional Analysis and its Applications*, 10(2):29–40, 1976.

[7] G. Faltings. Group schemes with strict $\mathcal{O}$-action. *Mosc. Math. J.*, 2(2):249–279, 2002.

[8] G. Faltings. Coverings of p-adic period domains. *J. Reine Angew. Math.*, 643:111–139, 2010.

[9] L. Fargues. L'isomorphisme entre les tours de Lubin-Tate et de Drinfeld et applications cohomologiques. In *L'isomorphisme entre les tours de Lubin-Tate et de Drinfeld*, Progress in math., 262, pages 1–325. Birkhäuser, 2008.

[10] L. Fargues. La filtration de Harder-Narasimhan des schémas en groupes finis et plats. *J. Reine Angew. Math.*, 645:1–39, 2010.

[11] L. Fargues. La filtration canonique des points de torsion des groupes p-divisibles. *Annales scientifiques de l'ENS*, 44(6):905–961, 2011.

[12] L. Fargues and J.-M. Fontaine. Courbes et fibrés vectoriels en théorie de Hodge p-adique. Prépublication.

[13] L. Fargues and J.-M. Fontaine. Factorization of analytic functions in mixed characteristic. To appear in the proceedings of a conference in Sanya.

[14] L. Fargues and J.-M. Fontaine. *Vector bundles and p-adic Galois representations*. Stud. Adv. Math. 51, 2011.

[15] J.-M. Fontaine. Groupes p-divisibles sur les corps locaux. Société Mathématique de France, Paris, 1977. *Astérisque*, No. 47-48.

[16] J.-M. Fontaine. Le corps des périodes p-adiques. *Astérisque*, (223):59–111, 1994. With an appendix by Pierre Colmez, Périodes p-adiques (Bures-sur-Yvette, 1988).

[17] B. Gross and M. Hopkins. Equivariant vector bundles on the Lubin-Tate moduli space. In *Topology and representation theory (Evanston, IL, 1992)*, volume 158 of Contemp. Math., pages 23–88. Amer. Math. Soc., 1994.

[18] A. Grothendieck. Sur la classification des fibrés holomorphes sur la sphère de Riemann. *Amer. J. Math.*, 79:121–138, 1957.

[19] A. Grothendieck. Éléments de géométrie algébrique. III. Étude cohomologique des faisceaux cohérents. I. *Inst. Hautes Études Sci. Publ. Math.*, (11):167, 1961.

[20] U. Hartl and R. Pink. Vector bundles with a Frobenius structure on the punctured unit disc. *Compos. Math.*, 140(3):689–716, 2004.

[21] K. Kedlaya. Slope filtrations revisited. *Doc. Math.*, 10:447–525, 2005.

[22] K. Kedlaya. Slope filtrations for relative Frobenius. *Astérisque*, (319):259–301, 2008. Représentations p-adiques de groupes p-adiques. I. Représentations galoisiennes et (ϕ, Γ)-modules.

[23] K. Kedlaya, J. Pottharst, and L. Xiao. Cohomology of arithmetic families of (phi,gamma)-modules. *arXiv:1203.5718v1*.

[24] G. Laffaille. Groupes p-divisibles et corps gauches. *Compositio Math.*, 56(2):221–232, 1985.

[25] M. Lazard. Les zéros des fonctions analytiques d'une variable sur un corps valué complet. *Inst. Hautes Études Sci. Publ. Math.*, (14):47–75, 1962.

[26] B. Poonen. Maximally complete fields. *Enseign. Math. (2)*, 39(1-2):87–106, 1993.

[27] M. Rapoport and T. Zink. *Period spaces for p-divisible groups*. Number 141 in Annals of Mathematics Studies. Princeton University Press, 1996.

[28] P. Schneider and J. Teitelbaum. Algebras of p-adic distributions and admissible representations. *Invent. Math.*, 153(1):145–196, 2003.

[29] P. Scholze. Perfectoid spaces. Preprint.

[30] S. Sen. Continuous cohomology and p-adic Galois representations. *Invent. Math.*, 62(1):89–116, 1980/81.

[31] J. T. Tate. p-divisible groups. In *Proc. Conf. Local Fields (Driebergen, 1966)*, pages 158–183. Springer, 1967.

[32] J.-P. Wintenberger. Le corps des normes de certaines extensions infinies de corps locaux; applications. *Ann. Sci. École Norm. Sup. (4)*, 16(1):59–89, 1983.

3
Around associators

Hidekazu Furusho

Abstract

This is a concise exposition of recent developments around the study of associators. It is based on the author's talk at the Mathematische Arbeitstagung in Bonn, June 2011 (cf. [F11b]) and at the Automorphic Forms and Galois Representations Symposium in Durham, July 2011. The first section is a review of Drinfeld's definition [Dr] of associators and the results [F10, F11a] concerning the definition. The second section explains the four pro-unipotent algebraic groups related to associators; the motivic Galois group, the Grothendieck–Teichmüller group, the double shuffle group and the Kashiwara–Vergne group. Relationships, actually inclusions, between them are also discussed.

1. Associators

We recall the definition of associators [Dr] and explain our main results in [F10, F11a] concerning the defining equations of associators.

The notion of associators was introduced by Drinfeld in [Dr]. They describe monodromies of the KZ (Knizhnik–Zamolodchikov) equations. They are essential for the construction of quasi-triangular quasi-Hopf quantized universal enveloping algebras ([Dr]), for the quantization of Lie-bialgebras (Etingof–Kazhdan quantization [EtK]), for the proof of formality of chain operad of little discs by Tamarkin [Ta] (see also Ševera and Willwacher [SW]) and also for the combinatorial reconstruction of the universal Vassiliev knot invariant (the Kontsevich invariant [Kon, Ba95]) by Bar-Natan [Ba97], Cartier [C], Kassel and Turaev [KssT], Le and Murakami [LM96a] and Piunikhin [P].

Automorphic Forms and Galois Representations, ed. Fred Diamond, Payman L. Kassaei and Minhyong Kim. Published by Cambridge University Press. © Cambridge University Press 2014.

Notation 3.1. Let k be a field of characteristic 0 and $\bar{k}$ be its algebraic closure. Denote by $U\mathfrak{F}_2 = k\langle\langle X_0, X_1 \rangle\rangle$ the non-commutative formal power series ring defined as the universal enveloping algebra of the completed free Lie algebra $\mathfrak{F}_2$ with two variables X_0 and X_1. An element $\varphi = \varphi(X_0, X_1)$ of $U\mathfrak{F}_2$ is called *group-like*[1] if it satisfies

$$\Delta(\varphi) = \varphi \otimes \varphi \text{ and } \varphi(0, 0) = 1 \qquad (3.1)$$

where $\Delta : U\mathfrak{F}_2 \to U\mathfrak{F}_2 \hat{\otimes} U\mathfrak{F}_2$ is given by $\Delta(X_0) = X_0 \otimes 1 + 1 \otimes X_0$ and $\Delta(X_1) = X_1 \otimes 1 + 1 \otimes X_1$. For any k-algebra homomorphism $\iota : U\mathfrak{F}_2 \to S$, the image $\iota(\varphi) \in S$ is denoted by $\varphi(\iota(X_0), \iota(X_1))$.

Denote by $U\mathfrak{a}_3$ (resp. $U\mathfrak{a}_4$) the universal enveloping algebra of the *completed pure braid Lie algebra* $\mathfrak{a}_3$ (resp. $\mathfrak{a}_4$) over k with 3 (resp. 4) strings, which is generated by t_{ij} ($1 \leqslant i, j \leqslant 3$ (resp. 4)) with defining relations

$$t_{ii} = 0, \ t_{ij} = t_{ji}, \ [t_{ij}, t_{ik} + t_{jk}] = 0 \ (i, j, k: \text{all distinct})$$

$$\text{and } [t_{ij}, t_{kl}] = 0 \ (i, j, k, l: \text{all distinct}).$$

Note that $X_0 \mapsto t_{12}$ and $X_1 \mapsto t_{23}$ give an isomorphism $U\mathfrak{F}_2 \simeq U\mathfrak{a}_3$.

Definition 3.2 ([Dr]). A pair (μ, φ) with a *non-zero* element μ in k and a group-like series $\varphi = \varphi(X_0, X_1) \in U\mathfrak{F}_2$ is called an *associator* if it satisfies *one pentagon equation*

$$\varphi(t_{12}, t_{23} + t_{24})\varphi(t_{13} + t_{23}, t_{34}) = \varphi(t_{23}, t_{34})\varphi(t_{12} + t_{13}, t_{24} + t_{34})\varphi(t_{12}, t_{23})$$
$$(3.2)$$

in $U\mathfrak{a}_4$ and *two hexagon equations*

$$\exp\{\frac{\mu(t_{13} + t_{23})}{2}\} = \varphi(t_{13}, t_{12}) \exp\{\frac{\mu t_{13}}{2}\}\varphi(t_{13}, t_{23})^{-1} \exp\{\frac{\mu t_{23}}{2}\}\varphi(t_{12}, t_{23}),$$
$$(3.3)$$

$$\exp\{\frac{\mu(t_{12} + t_{13})}{2}\} = \varphi(t_{23}, t_{13})^{-1} \exp\{\frac{\mu t_{13}}{2}\}\varphi(t_{12}, t_{13})\exp\{\frac{\mu t_{12}}{2}\}\varphi(t_{12}, t_{23})^{-1}$$
$$(3.4)$$

in $U\mathfrak{a}_3$.

Remark 3.3. (i) Drinfeld [Dr] proved that such a pair always exists for any field k of characteristic 0.

(ii) The equations (3.2)–(3.4) reflect the three axioms of braided monoidal categories [JS]. We note that for any k-linear *infinitesimal* tensor category $\mathcal{C}$, each associator gives a structure of a braided monoidal category on $\mathcal{C}[[h]]$ (cf. [C, Dr, KssT]). Here $\mathcal{C}[[h]]$ denotes the category whose set of objects is equal to that of $\mathcal{C}$ and whose set of morphisms $\text{Mor}_{\mathcal{C}[[h]]}(X, Y)$ is $\text{Mor}_{\mathcal{C}}(X, Y) \otimes k[[h]]$ (h: a formal parameter).

---

[1] It is equivalent to $\varphi \in \exp \mathfrak{F}_2$.

Actually, the two hexagon equations are a consequence of the one pentagon equation:

Theorem 3.4 ([F10]). *Let* $\varphi = \varphi(X_0, X_1)$ *be a group-like element of* $U\mathfrak{F}_2$. *Suppose that* φ *satisfies the pentagon equation* (3.2). *Then there always exists* $\mu \in \bar{k}$ *(unique up to signature) such that the pair* (μ, φ) *satisfies two hexagon equations* (3.3) *and* (3.4).

Recently several different proofs of the above theorem were obtained (see [AlT, BaD, Wi]).

One of the nicest examples of associators is the Drinfeld associator:

Example 3.5. The *Drinfeld associator* $\Phi_{KZ} = \Phi_{KZ}(X_0, X_1) \in \mathbf{C}\langle\langle X_0, X_1\rangle\rangle$ is defined to be the quotient $\Phi_{KZ} = G_1(z)^{-1}G_0(z)$ where G_0 and G_1 are the solutions of the *formal KZ-equation*, which is the following differential equation for multi-valued functions $G(z) : \mathbf{C}\backslash\{0, 1\} \to \mathbf{C}\langle\langle X_0, X_1\rangle\rangle$

$$\frac{d}{dz}G(z) = \left(\frac{X_0}{z} + \frac{X_1}{z-1}\right)G(z),$$

such that $G_0(z) \approx z^{X_0}$ when $z \to 0$ and $G_1(z) \approx (1-z)^{X_1}$ when $z \to 1$ (cf. [Dr]). It is shown in [Dr] (see also [Wo]) that the pair $(2\pi\sqrt{-1}, \Phi_{KZ})$ forms an associator for $k = \mathbf{C}$. Namely Φ_{KZ} satisfies (3.1)~(3.4) with $\mu = 2\pi\sqrt{-1}$.

Remark 3.6. (i) The Drinfeld associator is expressed as follows:

$$\Phi_{KZ}(X_0, X_1) = 1 + \sum_{\substack{m, k_1, \dots, k_m \in \mathbf{N} \\ k_m > 1}} (-1)^m \zeta(k_1, \cdots, k_m) X_0^{k_m - 1} X_1 \cdots X_0^{k_1 - 1} X_1$$

$$+ \text{(regularized terms)}.$$

Here $\zeta(k_1, \cdots, k_m)$ is the *multiple zeta value* (MZV in short), the real number defined by the following power series

$$\zeta(k_1, \cdots, k_m) := \sum_{0 < n_1 < \cdots < n_m} \frac{1}{n_1^{k_1} \cdots n_m^{k_m}} \tag{3.5}$$

for $m, k_1, \dots, k_m \in \mathbf{N}(= \mathbf{Z}_{>0})$ with $k_m > 1$ (its convergent condition). All of the coefficients of Φ_{KZ} (including its regularized terms) are explicitly calculated in terms of MZVs in [F03] Proposition 3.2.3 by Le–Murakami's method in [LM96b].

(ii) Since all of the coefficients of Φ_{KZ} are described by MZVs, the equations (3.1)–(3.4) for $(\mu, \varphi) = (2\pi\sqrt{-1}, \Phi_{KZ})$ yield algebraic relations among them, which are called *associator relations*. It is expected that the associator relations might produce all algebraic relations among MZVs.

The above MZVs were introduced by Euler in [Eu] and have recently under-gone a huge revival of interest due to their appearance in various different branches of mathematics and physics. In connection with motive theory, linear and algebraic relations among MZVs are particularly important. The regular-ized double shuffle relations which were initially introduced by Ecalle and Zagier in the early 1990s might be one of the most fascinating ones. To state them let us fix notation again:

Notation 3.7. Let $\pi_Y : k\langle\langle X_0, X_1\rangle\rangle \to k\langle\langle Y_1, Y_2, \dots\rangle\rangle$ be the k-linear map between non-commutative formal power series rings that sends all the words ending in X_0 to zero and the word $X_0^{n_m-1} X_1 \cdots X_0^{n_1-1} X_1$ $(n_1, \dots, n_m \in \mathbf{N})$ to $(-1)^m Y_{n_m} \cdots Y_{n_1}$. Define the coproduct Δ_* on $k\langle\langle Y_1, Y_2, \dots\rangle\rangle$ by

$$\Delta_*(Y_n) = \sum_{i=0}^{n} Y_i \otimes Y_{n-i}$$

for all $n \geqslant 0$ with $Y_0 := 1$. For $\varphi = \sum_{W:\text{word}} c_W(\varphi) W \in U\mathfrak{F}_2 = k\langle\langle X_0, X_1\rangle\rangle$ with $c_W(\varphi) \in k$ (a "word" is a monic monomial element or 1 in $U\mathfrak{F}_2$), put

$$\varphi_* = \exp\left(\sum_{n=1}^{\infty} \frac{(-1)^n}{n} c_{X_0^{n-1} X_1}(\varphi) Y_1^n\right) \cdot \pi_Y(\varphi).$$

The *regularized double shuffle relations* for a group-like series $\varphi \in U\mathfrak{F}_2$ is a relation of the form

$$\Delta_*(\varphi_*) = \varphi_* \otimes \varphi_*. \tag{3.6}$$

Remark 3.8. The *regularized double shuffle relations* for MZVs are the alge-braic relations among them obtained from (3.1) and (3.6) for $\varphi = \Phi_{KZ}$ (cf. [IkKZ, R]). It is also expected that the relations produce all algebraic relations among MZVs.

The following is the simplest example of the relations.

Example 3.9. For $a, b > 1$,

$$\zeta(a)\zeta(b) = \sum_{i=0}^{a-1} \binom{b-1+i}{i} \zeta(a-i, b+i)$$

$$+ \sum_{j=0}^{b-1} \binom{a-1+j}{j} \zeta(b-j, a+j),$$

$$\zeta(a)\zeta(b) = \zeta(a, b) + \zeta(a+b) + \zeta(b, a).$$

The former follows from (3.1) and the latter follows from (3.6).

The regularized double shuffle relations are also a consequence of the pentagon equation:

Theorem 3.10 ([F11a]). *Let $\varphi = \varphi(X_0, X_1)$ be a group-like element of $U\mathfrak{F}_2$. Suppose that φ satisfies the pentagon equation (3.2). Then it also satisfies the regularized double shuffle relations (3.6).*

This result attains the final goal of the project posed by Deligne–Terasoma [Te]. Their idea is to use some convolutions of perverse sheaves, whereas our proof is to use Chen's bar construction calculus. It would be our next project to complete their idea and to get another proof of Theorem 3.10.

Remark 3.11. Our Theorem 3.10 was extended cyclotomically in [F12].

The following Zagier's relation which is essential for Brown's proof of Theorem 3.17 might be also one of the most fascinating ones.

Theorem 3.12 ([Z]). *For $a, b \geqslant 0$*

$$\zeta(2^{\{a\}}, 3, 2^{\{b\}}) = 2 \sum_{r=1}^{a+b+1} (-1)^r (A_{a,b}^r - B_{a,b}^r)\zeta(2r + 1)\zeta(2^{\{a+b+1-r\}})$$

with $A_{a,b}^r = \binom{2r}{2a+2}$ and $B_{a,b}^r = (1 - 2^{-2r})\binom{2r}{2b+1}$.

2. Four groups

We explain recent developments on the four pro-unipotent algebraic groups related to associators; the motivic Galois group, the Grothendieck–Teichmüller group, the double shuffle group and the Kashiwara–Vergne group, all of which are regarded as subgroups of $Aut \exp \mathfrak{F}_2$. In the end of this section we discuss natural inclusions between them.

2.1. Motivic Galois group

We review the formulations of the motivic Galois groups (consult also [An] as a nice exposition).

Notation 3.13. We work in the triangulated category $DM(\mathbf{Q})_\mathbf{Q}$ of *mixed motives* over $\mathbf{Q}$ (a part of an idea of mixed motives is explained in [De] §1) constructed by Hanamura, Levine and Voevodsky. *Tate motives* $\mathbf{Q}(n)$ $(n \in \mathbf{Z})$ are (Tate) objects of the category. Let $DMT(\mathbf{Q})_\mathbf{Q}$ be the triangulated subcategory of $DM(\mathbf{Q})_\mathbf{Q}$ generated by Tate motives $\mathbf{Q}(n)$ $(n \in \mathbf{Z})$. By the work of Levine a neutral tannakian $\mathbf{Q}$-category $MT(\mathbf{Q}) = MT(\mathbf{Q})_\mathbf{Q}$ of *mixed Tate motives over* $\mathbf{Q}$ is extracted by taking the heart with respect to a t-structure of $DMT(\mathbf{Q})_\mathbf{Q}$. Deligne and Goncharov [DeG] introduced the full subcategory $MT(\mathbf{Z}) = MT(\mathbf{Z})_\mathbf{Q}$ of *unramified mixed Tate motives* inside of $MT(\mathbf{Q})_\mathbf{Q}$. All objects there are mixed Tate motives M (i.e. an object of $MT(\mathbf{Q})$) such that for each subquotient E of M which is an extension of $\mathbf{Q}(n)$ by $\mathbf{Q}(n+1)$ for $n \in \mathbf{Z}$, the extension class of E in

$$\mathrm{Ext}^1_{MT(\mathbf{Q})}(\mathbf{Q}(n), \mathbf{Q}(n+1)) = \mathrm{Ext}^1_{MT(\mathbf{Q})}(\mathbf{Q}(0), \mathbf{Q}(1)) = \mathbf{Q}^\times \otimes \mathbf{Q}$$

is equal to in $\mathbf{Z}^\times \otimes \mathbf{Q} = \{0\}$.

In the category $MT(\mathbf{Z})$ of unramified mixed Tate motives, the following holds:

$$\dim_\mathbf{Q} \mathrm{Ext}^1_{MT(\mathbf{Z})}(\mathbf{Q}(0), \mathbf{Q}(m)) = \begin{cases} 1 \ (m = 3, 5, 7, \dots), \\ 0 \ (m : \text{others}), \end{cases} \tag{3.7}$$

$$\dim_\mathbf{Q} \mathrm{Ext}^2_{MT(\mathbf{Z})}(\mathbf{Q}(0), \mathbf{Q}(m)) = 0. \tag{3.8}$$

The category $MT(\mathbf{Z})$ forms a neutral tannakian $\mathbf{Q}$-category (consult [DeM]) with the fiber functor

$$\omega_{\mathrm{can}} : MT(\mathbf{Z}) \to \mathrm{Vect}_\mathbf{Q}$$

($\mathrm{Vect}_\mathbf{Q}$: the category of $\mathbf{Q}$-vector spaces) sending each motive M to $\oplus_n \mathrm{Hom}(\mathbf{Q}(n), Gr^W_{-2n} M)$.

Definition 3.14. The *motivic Galois group* here is defined to be the Galois group of $MT(\mathbf{Z})$, which is the pro-$\mathbf{Q}$-algebraic group defined by $\mathrm{Gal}^\mathcal{M}(\mathbf{Z}) := \underline{\mathrm{Aut}}^\otimes(MT(\mathbf{Z}) : \omega_{\mathrm{can}})$.

By the fundamental theorem of tannakian category theory, ω_{can} induces an equivalence of categories

$$MT(\mathbf{Z}) \simeq \mathrm{Rep}\,\mathrm{Gal}^\mathcal{M}(\mathbf{Z}) \tag{3.9}$$

where the right-hand side of the isomorphism denotes the category of finite dimensional $\mathbf{Q}$-vector spaces with $\mathrm{Gal}^\mathcal{M}(\mathbf{Z})$-action.

Remark 3.15. The action of $\mathrm{Gal}^\mathcal{M}(\mathbf{Z})$ on $\omega_{\mathrm{can}}(\mathbf{Q}(1)) = \mathbf{Q}$ defines a surjection $\mathrm{Gal}^\mathcal{M}(\mathbf{Z}) \to \mathbf{G}_m$ and its kernel $\mathrm{Gal}^\mathcal{M}(\mathbf{Z})_1$ is the unipotent radical of $\mathrm{Gal}^\mathcal{M}(\mathbf{Z})$. There is a canonical splitting $\tau : \mathbf{G}_m \to \mathrm{Gal}^\mathcal{M}(\mathbf{Z})$ which gives a negative grading on its associated Lie algebra $\mathrm{LieGal}^\mathcal{M}(\mathbf{Z})_1$. From (3.7) and (3.8) it follows that the Lie algebra is the graded *free* Lie algebra generated by one element in each degree $-3, -5, -7, \dots$. (consult [De] §8 for the full story).

The *motivic fundamental group* $\pi_1^{\mathcal{M}}(\mathbf{P}^1 \backslash \{0, 1, \infty\} : \vec{01})$ constructed in [DeG] §4 is a (pro-) object of $MT(\mathbf{Z})$. The Drinfeld associator (cf. Example 3.5) is essential in describing the Hodge realization of the motive (cf. [An, DeG, F07]). By our tannakian equivalence (3.9), it gives a (pro-) object of the right-hand side of (3.9), which induces a (graded) action

$$\Psi : \mathrm{Gal}^{\mathcal{M}}(\mathbf{Z})_1 \to \mathrm{Aut} \exp \mathfrak{F}_2. \tag{3.10}$$

Remark 3.16. For each $\sigma \in \mathrm{Gal}^{\mathcal{M}}(\mathbf{Z})_1(k)$, its action on $\exp \mathfrak{F}_2$ is described by $e^{X_0} \mapsto e^{X_0}$ and $e^{X_1} \mapsto \varphi_\sigma^{-1} e^{X_1} \varphi_\sigma$ for some $\varphi_\sigma \in \exp \mathfrak{F}_2$.

The following has been conjectured (Deligne–Ihara conjecture) for a long time and finally proved by Brown by using Zagier's relation (Theorem 3.12).

Theorem 3.17 ([Br]). *The map Ψ is injective.*

It is a pro-unipotent analogue of the so-called Belyĭ's theorem [Bel] in the pro-finite group setting. The theorem says that all unramified mixed Tate motives are associated with MZVs.

2.2. Grothendieck–Teichmüller group

The Grothendieck–Teichmüller group was introduced by Drinfeld [Dr] in his study of deformations of quasi-triangular quasi-Hopf quantized universal enveloping algebras. It was defined to be the set of "degenerated" associators. The construction of the group was also stimulated by the previous idea of Grothendieck, *un jeu de Lego–Teichmüller*, which was posed in his article *Esquisse d'un programme* [G].

Definition 3.18 ([Dr]). The *Grothendieck–Teichmüller group GRT_1* is defined to be the pro-algebraic variety whose set of k-valued points consists of *degenerated associators*, which are group-like series $\varphi \in U\mathfrak{F}_2$ satisfying the defining equations (3.2)–(3.4) of associators with $\mu = 0$.

Remark 3.19. (i) By Theorem 3.4, GRT_1 is reformulated to be the set of group-like series satisfying (3.2) without quadratic terms.
(ii) It forms a group [Dr] by the multiplication below

$$\varphi_2 \circ \varphi_1 := \varphi_1(\varphi_2 X_0 \varphi_2^{-1}, X_1) \cdot \varphi_2 = \varphi_2 \cdot \varphi_1(X_0, \varphi_2^{-1} X_1 \varphi_2). \tag{3.11}$$

By the map $X_0 \mapsto X_0$ and $X_1 \mapsto \varphi^{-1} X_1 \varphi$, the group GRT_1 is regarded as a subgroup of $\mathrm{Aut} \exp \mathfrak{F}_2$.
(iii) Ihara came to the Lie algebra of GRT_1 independently of Drinfeld's work in his arithmetic study of Galois action on fundamental groups (cf. [Iy90]).

(iv) The cyclotomic analogues of associators and that of the Grothendieck–
Teichmüller group were introduced by Enriquez [En]. Some elimination
results on their defining equations in special case were obtained in [EnF].

Geometric interpretation (cf. [Dr, Iy90, Iy94]) of equations (3.2)–(3.4)
implies the following (for a proof, see also [An, F07])

Theorem 3.20. *Im$\Psi \subset GRT_1$.*

Related to the questions posed in [De, Dr, Iy90], it is expected that they are
isomorphic.

Remark 3.21. (i) The Drinfeld associator Φ_{KZ} is an associator (cf. Example 3.5) but is not a degenerated associator, i.e. $\Phi_{KZ} \notin GRT_1(\mathbf{C})$.
(ii) The *p*-adic Drinfeld associator Φ_{KZ}^p introduced in [F04] is not an
associator but a degenerated associator, i.e. $\Phi_{KZ}^p \in GRT_1(\mathbf{Q}_p)$ (cf. [F07]).

2.3. Double shuffle group

The double shuffle group was introduced by Racinet as the set of solutions
of the regularized double shuffle relations with "degeneration" condition (no
quadratic terms condition).

Definition 3.22 ([R]). The *double shuffle group* DMR_0 is the pro-algebraic
variety whose set of *k*-valued points consists of the group-like series $\varphi \in U\mathfrak{F}_2$
satisfying the regularized double shuffle relations (3.6) without linear terms
and quadratic terms.

Remark 3.23. (i) We note that DMR stands for *double mélange regularisé*
([R]).
(ii) It was shown in [R] that it forms a group by the operation (3.11).
(iii) In the same way as in Remark 3.19 (ii), the group DMR_0 is regarded as
a subgroup of Aut exp $\mathfrak{F}_2$.

It is also shown that ImΨ is contained in DMR_0 (cf. [F07])). Actually it is
expected that they are isomorphic. Theorem 3.10 follows the inclusion between
GRT_1 and DMR_0:

Theorem 3.24 ([F11a]). $GRT_1 \subset DMR_0$.

It is also expected that they are isomorphic.

Remark 3.25. (i) The Drinfeld associator Φ_{KZ} satisfies the regularized double shuffle relations (cf. Remark 3.8) but it is not an element of the double shuffle group, i.e. $\Phi_{KZ} \notin DMR_0(\mathbf{C})$, because its quadratic term is non-zero, actually is equal to $\zeta(2)X_1X_0 - \zeta(2)X_0X_1$.

(ii) The p-adic Drinfeld associator Φ_{KZ}^p satisfies the regularized double shuffle relations (cf. [BeF, FJ]) and it is an element of the double shuffle group, i.e. $\Phi_{KZ}^p \in DMR_0(\mathbf{Q}_p)$, which also follows from Remark 3.21.(ii) and Theorem 3.24.

2.4. Kashiwara–Vergne group

In [KswV] Kashiwara and Vergne proposed a conjecture related to the Campbell–Baker–Hausdorff series which generalizes Duflo's theorem (Duflo isomorphism) to some extent. The conjecture was settled generally by Alekseev and Meinrenken [AlM]. The Kashiwara–Vergne group was introduced as a "degeneration" of the set of solutions of the conjecture by Alekseev and Torossian in [AlT], where they gave another proof of the conjecture by using Drinfeld's [Dr] theory of associators.

The following is one of the formulations of the conjecture stated in [AlET].

Generalized Kashiwara–Vergne problem: Find a group automorphism $P : \exp \mathfrak{F}_2 \to \exp \mathfrak{F}_2$ such that P belongs to $T\mathrm{Aut}\exp \mathfrak{F}_2$ (that is, $P \in \mathrm{Aut}\exp \mathfrak{F}_2$ such that

$$P(e^{X_0}) = p_1 e^{X_0} p_1^{-1} \text{ and } P(e^{X_1}) = p_2 e^{X_1} p_2^{-1}$$

for some $p_1, p_2 \in \exp \mathfrak{F}_2$) and P satisfies

$$P(e^{X_0}e^{X_1}) = e^{(X_0+X_1)}$$

and the coboundary Jacobian condition

$$\delta \circ J(P) = 0.$$

Here J stands for the Jacobian cocycle $J : T\mathrm{Aut}\exp \mathfrak{F}_2 \to \mathfrak{tr}_2$ and δ denotes the differential map $\delta : \mathfrak{tr}_n \to \mathfrak{tr}_{n+1}$ for $n = 1, 2, \ldots$ (for their precise definitions see [AlT]). We note that P is uniquely determined by the pair (p_1, p_2).

The following is essential for the proof of the conjecture.

Theorem 3.26 ([AlT, AlET]). *Let (μ, φ) be an associator. Then the pair*

$$(p_1, p_2) = \left(\varphi(X_0/\mu, X_\infty/\mu),\ e^{X_\infty/2}\varphi(X_1/\mu, X_\infty/\mu)\right)$$

with $X_\infty = -X_0 - X_1$ gives a solution to the above problem.

The Kashiwara–Vergne group is defined to be the set of solutions of the problem with "degeneration condition" ("the condition $\mu = 0$"):

Definition 3.27 ([AlT, AlET]). The *Kashiwara–Vergne group KRV* is defined to be the set of $P \in Aut \exp \mathfrak{F}_2$ which satisfies $P \in T Aut \exp \mathfrak{F}_2$,

$$P(e^{(X_0+X_1)}) = e^{(X_0+X_1)}$$

and the coboundary Jacobian condition $\delta \circ J(P) = 0$.

The above KRV forms a subgroup of $Aut \exp \mathfrak{F}_2$. We denote by KRV_0 the subgroup of KRV consisting of P without linear terms in both p_1 and p_2. Theorem 3.26 yields the inclusion below.

Theorem 3.28 ([AlT, AlET]). $GRT_1 \subset KRV_0$.

Actually it is expected that they are isomorphic (cf. [AlT]). A recent result of Schneps in [S] also leads to

Theorem 3.29 ([S]). $DMR_0 \subset KRV_0$.

2.5. Comparison

By Theorem 3.17, 3.20, 3.24, 3.28 and 3.29, we obtain

Theorem 3.30. $Gal^{\mathcal{M}}(\mathbf{Z})_1 \subseteq GRT_1 \subseteq DMR_0 \subseteq KRV_0$.

We finish our exposition by posing the following question:

Question 3.31. Are they all equal? Namely,

$$Gal^{\mathcal{M}}(\mathbf{Z})_1 = GRT_1 = DMR_0 = KRV_0 \ ?$$

These four groups were constructed independently and there are no philosophical reasonings why we expect that they are all equal. Though it might be not so good mathematically to believe such equalities without any strong conceptual support, the author believes that it might be good at least spiritually to imagine a hidden theory to relate them.

References

[An]   André, Y.; *Une introduction aux motifs (motifs purs, motifs mixtes, périodes)*, Panoramas et Synthèses, 17, Société Mathématique de France, Paris, 2004.

[AlET] Alekseev, A., Enriquez, B. and Torossian, C.; Drinfeld associators, braid groups and explicit solutions of the Kashiwara-Vergne equations, *Publ. Math. Inst. Hautes Études Sci.* 112 (2010), 143–189.

[AlM] _____ and Meinrenken, E.; On the Kashiwara-Vergne conjecture, *Invent. Math.* 164 (2006), no. 3, 615–634.

[AlT] _____ and Torossian, C.; The Kashiwara-Vergne conjecture and Drinfeld's associators, *Ann. of Math.* (2) 175 (2012), no. 2, 415–463.

[Ba95] Bar-Natan, D.; On the Vassiliev knot invariants, *Topology* 34 (1995), no. 2, 423–472.

[Ba97] _____ ; Non-associative tangles, *Geometric topology* (Athens, GA, 1993), 139–183, AMAPIP Stud. Adv. Math., 2.1, Amer. Math. Soc., Providence, RI, 1997.

[BaD] _____ and Dancso, Z.; Pentagon and hexagon equations following Furusho, *Proc. Amer. Math. Soc.* 140 (2012), no. 4, 1243–1250.

[Bel] Belyĭ, G. V., Galois extensions of a maximal cyclotomic field, (Russian) *Izv. Akad. Nauk SSSR Ser. Mat.* 43 (1979), no. 2, 267–276, 479.

[BeF] Besser, A. and Furusho, H.; The double shuffle relations for p-adic multiple zeta values, *Primes and Knots*, AMS Contemporary Math, Vol 416, 2006, 9–29.

[Br] Brown, F.; Mixed Tate Motives over Spec(Z), *Annals of Math.*, 175, (2012) no. 2, 949–976.

[C] Cartier, P.; Construction combinatoire des invariants de Vassiliev-Kontsevich des nœuds, *C. R. Acad. Sci. Paris Ser. I Math.* 316 (1993), no. 11, 1205–1210.

[De] Deligne, P.; Le groupe fondamental de la droite projective moins trois points, *Galois groups over Q* (Berkeley, CA, 1987), 79–297, Math. S. Res. Inst. Publ., 16, Springer, New York-Berlin, 1989.

[DeG] _____ and Goncharov, A.; Groupes fondamentaux motiviques de Tate mixte, *Ann. Sci. Ecole Norm. Sup. (4)* 38 (2005), no. 1, 1–56.

[DeM] _____ and Milne, J.; Tannakian categories, in *Hodge cycles, motives, and Shimura varieties* (P.Deligne, J.Milne, A.Ogus, K.-Y.Shih editors), Lecture Notes in Mathematics 900, Springer-Verlag, 1982.

[Dr] Drinfel'd, V. G.; On quasitriangular quasi-Hopf algebras and a group closely connected with Gal($\overline{Q}/Q$), *Leningrad Math. J.* 2 (1991), no. 4, 829–860.

[En] Enriquez, B.; Quasi-reflection algebras and cyclotomic associators, *Selecta Math.* (N.S.) 13 (2007), no. 3, 391–463.

[EnF] _____ and Furusho, H.; Mixed Pentagon, octagon and Broadhurst duality equation, *J. Pure Appl. Algebra*, 216, (2012), no. 4, 982–995. no. 4,

[EtK] Etingof, P. and Kazhdan, D.; Quantization of Lie bialgebras. II, *Selecta Math.* 4 (1998), no. 2, 213–231.

[Eu] Euler, L., Meditationes circa singulare serierum genus, *Novi Commentarii academiae scientiarum Petropolitanae* 20, 1776, pp. 140–186 and Opera Omnia: Series 1, Volume 15, pp. 217–267 (also available from www.math.dartmouth.edu/euler/).

[F03] Furusho, H.; The multiple zeta value algebra and the stable derivation algebra, *Publ. Res. Inst. Math. Sci.* 39, (2003), no. 4, 695–720.

[F04] _____ ; p-adic multiple zeta values I – p-adic multiple polylogarithms and the p-adic KZ equation, *Invent. Math.* 155, (2004), no. 2, 253–286.

[F07] _____ ; p-adic multiple zeta values II – tannakian interpretations, *Amer. J. Math.* 129, (2007), no 4, 1105–1144.

[F10]   _____; Pentagon and hexagon equations, *Annals of Math.* 171 (2010), no. 1, 545–556.

[F11a]   _____; Double shuffle relation for associators, *Annals of Math.* 174 (2011), no. 1, 341–360.

[F11b]   _____; Four groups related to associators, report of the *Mathematische Arbeitstagung* in Bonn, June 2011, arXiv:1108.3389.

[F12]   _____; Geometric interpretation of double shuffle relation for multiple *L*-values, *Galois-Teichmüller theory and Arithmetic Geometry*, Advanced Studies in Pure Math 63 (2012), 163–187.

[FJ]   _____ and Jafari, A.; Regularization and generalized double shuffle relations for *p*-adic multiple zeta values, *Compositio Math.* 143, (2007), 1089–1107.

[G]   Grothendieck, A.; Esquisse d'un programme, 1983, available on pp. 243–283. London Math. Soc. LNS 242, *Geometric Galois actions*, 1, 5–48, Cambridge University Press.

[IkKZ]   Ihara, K., Kaneko, M. and Zagier, D.; Derivation and double shuffle relations for multiple zeta values, *Compos. Math.* 142 (2006), no. 2, 307–338.

[Iy90]   Ihara, Y.; Braids, Galois groups, and some arithmetic functions, *Proceedings of the International Congress of Mathematicians, Vol. I, II* (Kyoto, 1990), 99–120, Math. Soc. Japan, Tokyo, 1991.

[Iy94]   _____; On the embedding of $\mathrm{Gal}(\overline{Q}/Q)$ into $\widehat{\mathrm{GT}}$, London Math. Soc. LNS 200, *The Grothendieck theory of dessins d'enfants* (Luminy, 1993), 289–321, Cambridge University Press, Cambridge, 1994.

[JS]   Joyal, A. and Street, R.; Braided tensor categories, *Adv. Math.* 102 (1993), no. 1, 20–78.

[KswV]   Kashiwara, M. and Vergne, M.; The Campbell-Hausdorff formula and invariant hyperfunctions, *Invent. Math.* 47 (1978), no. 3, 249–272.

[KssT]   Kassel, C. and Turaev, V.; Chord diagram invariants of tangles and graphs, *Duke Math. J.* 92 (1998), no. 3, 497–552.

[Kon]   Kontsevich, M.; Vassiliev's knot invariants, *I. M. Gelfand Seminar*, 137–150, Adv. Soviet Math., 16, Part 2, Amer. Math. Soc., Providence, RI, 1993.

[LM96a]   Le, T.Q.T. and Murakami, J.; The universal Vassiliev-Kontsevich invariant for framed oriented links, *Compositio Math.* 102 (1996), no. 1, 41–64.

[LM96b]   _____; Kontsevich's integral for the Kauffman polynomial, *Nagoya Math. J.* 142 (1996), 39–65.

[P]   Piunikhin, S.; Combinatorial expression for universal Vassiliev link invariant, *Comm. Math. Phys.* 168 (1995), no. 1, 1–22.

[R]   Racinet, G.; Doubles melangés des polylogarithmes multiples aux racines de l'unité, *Publ. Math. Inst. Hautes Etudes Sci.* 95 (2002), 185–231.

[S]   Schneps, L.; Double shuffle and Kashiwara-Vergne Lie algebra, *J. Algebra* 367 (2012), 54–74.

[SW]   Ševera, P and Willwacher, T.; Equivalence of formalities of the little discs operad, *Duke Math. J.* 160 (2011), no. 1, 175–206.

[Ta]   Tamarkin, D. E.; Formality of chain operad of little discs, *Lett. Math. Phys.* 66 (2003), no. 1–2, 65–72.

[Te]   Terasoma, T.; Geometry of multiple zeta values, *International Congress of Mathematicians*. Vol. II, 627–635, Eur. Math. Soc., Zürich, 2006.

[Wi] Willwacher, T.; M. Kontsevich's graph complex and the Grothendieck-Teichmueller Lie algebra, arXiv:1009.1654, preprint (2010).

[Wo] Wojtkowiak, Z.; Monodromy of iterated integrals and non-abelian unipotent periods, *Geometric Galois actions*, 2, 219–289, London Math. Soc. LNS 243, Cambridge University Press, Cambridge, 1997.

[Z] Zagier, D.; Evaluation of the multiple zeta values $\zeta(2, ..., 2, 3, 2, ..., 2)$, *Annals of Math.*, 175, (2012), no. 2, 977–1000.

4

The stable Bernstein center and test functions for Shimura varieties

Thomas J. Haines

Abstract

We elaborate the theory of the stable Bernstein center of a p-adic group G, and apply it to state a general conjecture on test functions for Shimura varieties due to the author and R. Kottwitz. This conjecture provides a framework by which one might pursue the Langlands–Kottwitz method in a very general situation: not necessarily PEL Shimura varieties with arbitrary level structure at p. We give a concrete reinterpretation of the test function conjecture in the context of parahoric level structure. We also use the stable Bernstein center to formulate some of the transfer conjectures (the "fundamental lemmas") that would be needed if one attempts to use the test function conjecture to express the local Hasse–Weil zeta function of a Shimura variety in terms of automorphic L-functions.

Contents

Research partially supported by NSF DMS-0901723.

Automorphic Forms and Galois Representations, ed. Fred Diamond, Payman L. Kassaei and Minhyong Kim. Published by Cambridge University Press. © Cambridge University Press 2014.

1. Introduction

The main purpose of this chapter is to give precise statements of some conjectures on test functions for Shimura varieties with bad reduction.

In the Langlands–Kottwitz approach to studying the cohomology of a Shimura variety, one of the main steps is to identify a suitable test function that is "plugged into" the counting points formula that resembles the geometric side of the Arthur–Selberg trace formula. To be more precise, let $(\mathbf{G}, h^{-1}, K^p K_p)$ denote Shimura data where p is a fixed rational prime such that the level-structure group factorizes as $K^p K_p \subset \mathbf{G}(\mathbb{A}_f^p)\mathbf{G}(\mathbb{Q}_p)$. This data gives rise to a quasi-projective variety $Sh_{K_p} := Sh(\mathbf{G}, h^{-1}, K^p K_p)$ over a number field $\mathbf{E} \subset \mathbb{C}$. Let $\Phi_p \in \mathrm{Gal}(\overline{\mathbb{Q}}_p/\mathbb{Q}_p)$ denote a geometric Frobenius element. Then one seeks to prove a formula for the semi-simple Lefschetz number $\mathrm{Lef}^{\mathrm{ss}}(\Phi_p^r, Sh_{K_p})$

$$\mathrm{tr}^{\mathrm{ss}}(\Phi_p^r, \mathrm{H}_c^\bullet(Sh_{K_p} \otimes_{\mathbf{E}} \overline{\mathbb{Q}}_p, \overline{\mathbb{Q}}_\ell)) = \sum_{(\gamma_0; \gamma, \delta)} c(\gamma_0; \gamma, \delta)\, \mathrm{O}_\gamma(1_{K^p})\, \mathrm{TO}_{\delta\theta}(\phi_r),$$

$$(4.1)$$

(see §6.1 for more details).

The *test function* ϕ_r appearing here is the most interesting part of the formula. Experience has shown that we may often find a test function belonging to the center $\mathcal{Z}(\mathbf{G}(\mathbb{Q}_{p^r}), K_{p^r})$ of the Hecke algebra $\mathcal{H}(\mathbf{G}(\mathbb{Q}_{p^r}), K_{p^r})$, in a way that is explicitly determined by the **E**-rational conjugacy class $\{\mu\}$ of 1-parameter subgroups of **G** associated to the Shimura data. In most PEL cases with good reduction, where $K_p \subset \mathbf{G}(\mathbb{Q}_p)$ is a hyperspecial maximal compact subgroup, this was done by Kottwitz (cf. e.g. [Ko92a]). When K_p is a parahoric subgroup of $\mathbf{G}(\mathbb{Q}_p)$ and when $\mathbf{G}_{\mathbb{Q}_p}$ is unramified, the *Kottwitz conjecture* predicts that we can take ϕ_r to be a power of p times the *Bernstein function* $z_{-\mu,j}^{K_p}$ arising from the Bernstein isomorphism for the center $\mathcal{Z}(\mathbf{G}(\mathbb{Q}_{p^r}), K_{p^r})$ of the parahoric Hecke algebra $\mathcal{H}(\mathbf{G}(\mathbb{Q}_{p^r}), K_{p^r})$ (see Conjecture 4.43 and §11).

In fact Kottwitz formulated (again, for unramified groups coming from Shimura data) a closely related conjecture concerning nearby cycles on Rapoport–Zink local models of Shimura varieties, which subsequently played an important role in the study of local models (Conjecture 4.44). It also inspired important developments in the geometric Langlands program, e.g. [Ga]. Both versions of Kottwitz' conjectures were later proved in several parahoric cases attached to linear or symplectic groups (see [HN02a, H05]). In a recent breakthrough, Pappas and Zhu [PZ] defined group-theoretic versions of Rapoport–Zink local models for quite general groups, and proved in the unramified situations the analogue of Kottwitz' nearby cycles conjecture for them. These matters are discussed in more detail in §7 and §8.

Until around 2009 it was still not clear how one could describe the test functions ϕ_r in all deeper level situations. In the spring of 2009 the author and Kottwitz formulated a conjecture predicting test functions ϕ_r for general level structure K_p. This is the **test function conjecture**, Conjecture 4.30. It postulates that we may express ϕ_r in terms of a distribution $Z_{V^{E_{j0}}_{-\mu,j}}$ in the Bernstein center $\mathfrak{Z}(\mathbf{G}(\mathbb{Q}_{p^r}))$ associated to a certain representation $V^{E_{j0}}_{-\mu,j}$ (defined in (4.16)) of the Langlands L-group ${}^L(G_{\mathbb{Q}_{p^r}})$. Let $d = \dim(Sh_{K_p})$. Then Conjecture 4.30 asserts that we may take

$$\phi_r = p^{rd/2}\big(Z_{V^{E_{j0}}_{-\mu,j}} * 1_{K_{p^r}}\big) \in \mathcal{Z}(\mathbf{G}(\mathbb{Q}_{p^r}), K_{p^r})$$

the convolution of the distribution $Z_{V^{E_{j0}}_{-\mu,j}}$ with the characteristic function $1_{K_{p^r}}$ of the subgroup K_{p^r}. As shown in §7, this specializes to the Kottwitz conjecture for parahoric subgroups in unramified groups. Conjecture 4.30 was subsequently proved for Drinfeld case Shimura varieties with $\Gamma_1(p)$-level structure by the author and Rapoport [HRa1], and for modular curves and for Drinfeld case Shimura varieties with arbitrary level structure by Scholze [Sch1, Sch2].

The distributions in Conjecture 4.30 are best seen as examples of a construction $V \rightsquigarrow Z_V$ which attaches to any algebraic representation V of the Langlands dual group ${}^L G$ (for G any connected reductive group over any p-adic field F), an element Z_V in the **stable Bernstein center** of G/F. This chapter elaborates the theory of the stable Bernstein center, following the lead of Vogan [Vo]. The set of all *infinitesimal characters*, i.e. the set of all $\widehat{G}$-conjugacy classes of admissible homomorphisms $\lambda : W_F \to {}^L G$ (where W_F is the Weil group of the local field F), is given the structure of an affine algebraic variety over $\mathbb{C}$, and the stable Bernstein center $\mathfrak{Z}^{st}(G/F)$ is defined to be the ring of regular functions on this variety.[1]

In order to describe the precise conjectural relation between the Bernstein and stable Bernstein centers of a p-adic group, it was necessary to formulate an

---

[1] The difference between our treatment and Vogan's is in the definition of the variety structure on the set of infinitesimal characters.

enhancement LLC+ of the usual conjectural local Langlands correspondence LLC for that group. Having this relation in hand, the construction $V \rightsquigarrow Z_V$ provides a supply of elements in the usual Bernstein center of G/F, which we call the *geometric Bernstein center*.

It is for such distributions that one can formulate natural candidates for (Frobenius-twisted) endoscopic transfer, which we illustrate for standard endoscopy in Conjecture 4.35 and for stable base change in Conjecture 4.36. These form part of the cadre of "fundamental lemmas" that one would need to pursue the "pseudostabilization" of (4.1) and thereby express the cohomology of Sh_{K_p} in terms of automorphic representations along the lines envisioned by Kottwitz [Ko90] but for places with arbitrary bad reduction. In the compact and non-endoscopic situations, we prove in Theorem 4.40 that the various Conjectures we have made yield an expression of the semi-simple local Hasse–Weil zeta function in terms of semi-simple automorphic L-functions. Earlier unconditional results in this direction, for nice PEL situations, were established in [H05], [HRa1], [Sch1, Sch2]. We stress that the framework here should not be limited to PEL Shimura varieties, but should work more generally.

In recent work of Scholze and Shin [SS], the connection of the stable Bernstein center with Shimura varieties helped them to give nearly complete descriptions of the cohomology of many compact unitary Shimura varieties with bad reduction at p; they consider the "EL cases" where $\mathbf{G}_{\mathbb{Q}_p}$ is a product of Weil restrictions of general linear groups. It would be interesting to extend the connection to further examples.

Returning to the original Kottwitz conjecture for parahoric level structure, Conjecture 4.30 in some sense subsumes it, since it makes sense for arbitrary level structure and without the hypothesis that $\mathbf{G}_{\mathbb{Q}_{p^r}}$ be unramified. However, Conjecture 4.30 has the drawback that it assumes LLC+ for $\mathbf{G}_{\mathbb{Q}_{p^r}}$. Further, it is still of interest to formulate the Kottwitz conjecture in the parahoric cases for *arbitrary groups* in a concrete way that can be checked (for example) by explicit comparison of test functions with nearby cycles. In §7 we formulate the Kottwitz conjecture for general groups, making use of the transfer homomorphisms of the Appendix §11 to determine test functions on arbitrary groups from test functions on their quasi-split inner forms. The definition of transfer homomorphisms requires a theory of Bernstein isomorphisms more general than what was heretofore available. Therefore, in the Appendix we establish these isomorphisms in complete generality in a nearly self-contained way, and also provide some related structure theory results that should be of independent interest.

Here is an outline of the contents of this chapter. In §3 we review the
Bernstein center of a p-adic group, including the algebraic structure on the
Bernstein variety of all supercuspidal supports. In §4 we recall the conjectural
local Langlands correspondence (LLC), and discuss additional desiderata we
need in our elaboration of the stable Bernstein center in §5. In particular in §5.2
we describe the enhancement (LLC+) which plays a significant role throughout the chapter, and explain why it holds for general linear groups in Remark
4.8 and Corollary 4.11. The distributions Z_V are defined in §5.7, and are used
to formulate the test function conjecture, Conjecture 4.30, in §6.1. In the rest
of §6, we describe the nearby cycles variant Conjecture 4.31 along with some
of the endoscopic transfer conjectures needed for the "pseudostabilization",
and assuming these conjectures we prove in Theorem 4.40 the expected form
of the semi-simple local Hasse–Weil zeta functions, in the compact and non-
endoscopic cases. In §7 we give a concrete reformulation of the test function
conjecture in parahoric cases, recovering the Kottwitz conjecture and generalizing it to all groups using the material from the Appendix. The purpose
of §8 and §9 is to list some of the available evidence for Conjectures 4.30
and 4.36. In §10 certain test functions are described very explicitly. Finally,
the Appendix gives the treatment of Bernstein isomorphisms and the transfer
homomorphisms, alluded to above.

Acknowledgments. I am very grateful to Guy Henniart for supplying the proof
of Proposition 4.10 and for allowing me to include his proof in this chapter. I
warmly thank Timo Richarz for sending me his unpublished article [Ri] and for
letting me quote a few of his results in Lemma 4.56. I am indebted to Brooks
Roberts for proving Conjecture 4.7 for GSp(4) (see Remark 4.8). I thank my
colleagues Jeffrey Adams and Niranjan Ramachandran for useful conversations. I also thank Robert Kottwitz for his influence on the ideas in this chapter
and for his comments on a preliminary version. I thank Michael Rapoport
for many stimulating conversations about test functions over the years. I am
grateful to the referee for helpful suggestions and remarks.

2. Notation

If G is a connected reductive group over a p-adic field F, then $\mathfrak{R}(G)$ will
denote the category of smooth representations of $G(F)$ on $\mathbb{C}$-vector spaces. We
will write $\pi \in \mathfrak{R}(G)_{\text{irred}}$ or $\pi \in \Pi(G/F)$ if π is an irreducible object in $\mathfrak{R}(G)$.

If G as above contains an F-rational parabolic subgroup P with F-Levi
factor M and unipotent radical N, define the modulus function $\delta_P : M(F) \rightarrow$

$\mathbb{R}_{>0}$ by

$$\delta_P(m) = |\det(\mathrm{Ad}(m)\,;\,\mathrm{Lie}(N(F)))|_F$$

where $|\cdot|_F$ is the normalized absolute value on F. By $\delta_P^{1/2}(m)$ we mean the positive square-root of the positive real number $\delta_P(m)$. For $\sigma \in \mathfrak{R}(M)$, we frequently consider the normalized induced representation

$$i_P^G(\sigma) = \mathrm{Ind}_{P(F)}^{G(F)}(\delta_P^{1/2}\sigma).$$

We let 1_S denote the characteristic function of a subset S of some ambient space. If $S \subset G$, let ${}^g S = gSg^{-1}$. If f is a function on S, define the function ${}^g f$ on ${}^g S$ by ${}^g f(\cdot) = f(g^{-1} \cdot g)$.

Throughout the chapter we use the Weil form of the local or global Langlands L-group ${}^L G$.

3. Review of the Bernstein center

We shall give a brief synopsis of [BD] that is suitable for our purposes. Other useful references are [Be92], [Ren], and [Roc].

The Bernstein center $\mathfrak{Z}(G)$ of a p-adic group G is defined as the ring of endomorphisms of the identity functor on the category of smooth representations $\mathfrak{R}(G)$. It can also be realized as an algebra of certain distributions, as the projective limit of the centers of the finite-level Hecke algebras, and as the ring of regular functions on a certain algebraic variety. We describe these in turn.

3.1. Distributions

We start by defining the convolution algebra of distributions.

We write G for the rational points of a connected reductive group over a p-adic field. Thus G is a totally disconnected locally compact Hausdorff topological group. Further G is unimodular; fix a Haar measure dx. Let $C_c^\infty(G)$ denote the set of $\mathbb{C}$-valued compactly supported and locally constant functions on G. Let $\mathcal{H}(G, dx) = (C_c^\infty(G), *_{dx})$, the convolution product $*_{dx}$ being defined using the Haar measure dx.

A *distribution* is a $\mathbb{C}$-linear map $D : C_c^\infty(G) \to \mathbb{C}$. For each $f \in C^\infty(G)$ we define $\check{f} \in C^\infty(G)$ by $\check{f}(x) = f(x^{-1})$ for $x \in G$. We set

$$\check{D}(f) := D(\check{f}).$$

We can convolve a distribution D with a function $f \in C_c^\infty(G)$ and get a new function $D * f \in C^\infty(G)$, by setting

$$(D * f)(g) = \check{D}(g \cdot f),$$

where $(g \cdot f)(x) := f(xg)$. The function $D * f$ does not automatically have compact support. We say D is *essentially compact* provided that $D * f \in C_c^\infty(G)$ for every $f \in C_c^\infty(G)$.

We define ${}^g f$ by ${}^g f(x) := f(g^{-1}xg)$ for $x, g \in G$. We say that D is *G-invariant* if $D({}^g f) = D(f)$ for all g, f. The set $\mathcal{D}(G)_{ec}^G$ of G-invariant essentially compact distributions on $C_c^\infty(G)$ turns out to have the structure of a commutative $\mathbb{C}$-algebra. We describe next the convolution product and its properties.

Given distributions D_1, D_2 with D_2 essentially compact, we define another distribution $D_1 * D_2$ by

$$(D_1 * D_2)(f) = \check{D}_1(D_2 * \check{f}).$$

Lemma 4.1. *The convolution products $D * f$ and $D_1 * D_2$ have the following properties:*

(a) *For $\phi \in C_c^\infty(G)$ let $D_{\phi\,dx}$ (sometimes abbreviated $\phi\,dx$) denote the essentially compact distribution given by $f \mapsto \int_G f(x)\phi(x)\,dx$. Then $D_{\phi\,dx} * f = \phi *_{dx} f$.*

(b) *If $f \in C_c^\infty(G)$, then $D * (f\,dx) = (D * f)\,dx$. In particular, $D_{\phi_1\,dx} * D_{\phi_2\,dx} = D_{\phi_1 *_{dx} \phi_2\,dx}$.*

(c) *If D_2 is essentially compact, then $(D_1 * D_2) * f = D_1 * (D_2 * f)$. If D_1 and D_2 are each essentially compact, so is $D_1 * D_2$.*

(d) *If D_2 and D_3 are essentially compact, then $(D_1*D_2)*D_3 = D_1*(D_2*D_3)$.*

(e) *An essentially compact distribution D is G-invariant if and only if $D * (1_{Ug}\,dx) = (1_{Ug}\,dx) * D$ for all compact open subgroups $U \subset G$ and $g \in G$. Here 1_{Ug} is the characteristic function of the set Ug.*

(f) *If D is essentially compact and $f_1, f_2 \in C_c^\infty(G)$, then $D * (f_1 *_{dx} f_2) = (D * f_1) *_{dx} f_2$.*

Corollary 4.2. *The pair $(\mathcal{D}(G)_{ec}^G, *)$ is a commutative and associative $\mathbb{C}$-algebra.*

3.2. The projective limit

Let $J \subset G$ range over the set of all compact open subgroups of G. Let $\mathcal{H}(G)$ denote the convolution algebra of compactly-supported measures on G, and let $\mathcal{H}_J(G) \subset \mathcal{H}(G)$ denote the ring of J-bi-invariant compactly-supported measures, with center $\mathcal{Z}_J(G)$. The ring $\mathcal{H}_J(G)$ has as unit $e_J = 1_J\,dx_J$, where 1_J is the characteristic function of J and dx_J is the Haar measure with $\text{vol}_{dx_J}(J) = 1$. Note that if $J' \subset J$, then $dx_{J'} = [J : J']\,dx_J$.

Let $\mathcal{Z}(G, J)$ denote the center of the algebra $\mathcal{H}(G, J)$ consisting of compactly-supported J-bi-invariant functions on G with product $*_{dx_J}$. There is an isomorphism $\mathcal{Z}(G, J) \overset{\sim}{\to} \mathcal{Z}_J(G)$ by $z_J \mapsto z_J dx_J$.

For $J' \subset J$ there is an algebra map $\mathcal{Z}(G, J') \to \mathcal{Z}(G, J)$, given by $z_{J'} \mapsto z_{J'} *_{dx_{J'}} 1_J$. Equivalently, we have $\mathcal{Z}_{J'}(G) \to \mathcal{Z}_J(G)$ given by $z_{J'} dx_{J'} \mapsto z_{J'} dx_{J'} * (1_J dx_J)$.

We can view $\mathfrak{R}(G)$ as the category of non-degenerate smooth $\mathcal{H}(G)$-modules, and any element of $\varprojlim \mathcal{Z}(G, J)$ acts on objects in $\mathfrak{R}(G)$ in a way that commutes with the action of $\mathcal{H}(G)$. Hence there is a canonical homomorphism $\varprojlim \mathcal{Z}(G, J) \to \mathfrak{Z}(G)$.

There is also a canonical homomorphism

$$\varprojlim Z(G, J) \to \mathcal{D}(G)^G_{\mathrm{ec}}$$

since $Z = (z_J)_J \in \varprojlim \mathcal{Z}(G, J)$ gives a distribution on $f \in C_c^\infty(G)$ as follows: choose $J \subset G$ sufficiently small that f is right-J-invariant, and set

$$Z(f) = \int_G z_J(x) \, f(x) \, dx_J. \tag{4.2}$$

This is independent of the choice of J. Note that for $f \in \mathcal{H}(G, J)$ we have $Z * f = z_J *_{dx_J} f$, and in particular $Z * 1_J = z_J$, for all J. To see $Z = (z_J)_J$ as a distribution is really G-invariant, note that for $f \in \mathcal{H}(G, J)$, the identities $Z * f = z_J *_{dx_J} f = f *_{dx_J} z_J$ imply that $Z * (f dx) = (f dx) * Z$. This in turn shows that Z is G-invariant by Lemma 4.1(e).

Now §1.4–1.7 of [BD] show that the above maps yield isomorphisms

$$\mathfrak{Z}(G) \overset{\sim}{\leftarrow} \varprojlim \mathcal{Z}(G, J) \overset{\sim}{\to} \mathcal{D}(G)^G_{\mathrm{ec}}. \tag{4.3}$$

Corollary 4.3. *Let $Z \in \mathfrak{Z}(G)$, and suppose a finite-length representation $\pi \in \mathfrak{R}(G)$ has the property that Z acts on π by a scalar $Z(\pi)$.*

(a) *For every compact open subgroup $J \subset G$, $Z * 1_J$ acts on the left on π^J by the scalar $Z(\pi)$.*

(b) *For every $f \in \mathcal{H}(G)$, $\mathrm{tr}(Z * f \mid \pi) = Z(\pi) \, \mathrm{tr}(f \mid \pi)$.*

3.3. Regular functions on the variety of supercuspidal supports

3.3.1. Variety structure on set of supercuspidal supports

We describe the variety of supercuspidal supports in some detail. Also we will describe it in a slightly unconventional way, in that we use the Kottwitz homomorphism to parametrize the (weakly) unramified characters on $G(F)$. This will be useful later on, when we compare the Bernstein center with the stable Bernstein center.

Let us recall the basic facts on the Kottwitz homomorphism [Ko97]. Let L be the completion $\widehat{F^{\mathrm{un}}}$ of the maximal unramified extension F^{un} in some algebraic closure of F, and let $\bar{L} \supset \bar{F}$ denote an algebraic closure of L. Let $I = \mathrm{Gal}(\bar{L}/L) \cong \mathrm{Gal}(\bar{F}/F^{\mathrm{un}})$ denote the inertia group. Let $\Phi \in \mathrm{Aut}(L/F)$ be the inverse of the Frobenius automorphism σ. In [Ko97] is defined a functorial surjective homomorphism for any connected reductive F-group H

$$\kappa_H : H(L) \twoheadrightarrow X^*(Z(\widehat{H}))_I, \tag{4.4}$$

where $\widehat{H} = \widehat{H}(\mathbb{C})$ denotes the Langlands dual group of H. By [Ko97, §7], it remains surjective on taking Φ-fixed points:

$$\kappa_H : H(F) \twoheadrightarrow X^*(Z(\widehat{H}))_I^{\Phi}.$$

We define

$$H(L)_1 := \ker(\kappa_H)$$
$$H(F)_1 := \ker(\kappa_H) \cap H(F).$$

We also define $H(F)^1 \supseteq H(F)_1$ to be the kernel of the map $H(F) \to X^*(Z(\widehat{H}))_I^{\Phi}/tors$ derived from κ_H.

If H is anisotropic modulo center, then $H(F)^1$ is the unique maximal compact subgroup of $H(F)$ and $H(F)_1$ is the unique parahoric subgroup of $H(F)$ (see e.g. [HRo]). Sometimes the two subgroups coincide: for example if H is any *unramified* F-torus, then $H(F)_1 = H(F)^1$.

We define

$$X(H) := \mathrm{Hom}_{\mathrm{grp}}(H(F)/H(F)^1, \mathbb{C}^{\times}),$$

the group of *unramified characters* on $H(F)$. This definition of $X(H)$ agrees with the usual one as in [BD]. We define

$$X^{\mathrm{w}}(H) := \mathrm{Hom}_{\mathrm{grp}}(H(F)/H(F)_1, \mathbb{C}^{\times})$$

and call it the group of *weakly unramified characters* on $H(F)$.

We follow the notation of [BK] in discussing supercuspidal supports and inertial equivalence classes. As indicated earlier in §3.1, for convenience we will sometimes write G when we mean the group $G(F)$ of F-points of an F-group G.

A *cuspidal pair* (M, σ) consists of an F-Levi subgroup $M \subseteq G$ and a supercuspidal representation σ on M. The G-conjugacy class of the cuspidal pair (M, σ) will be denoted $(M, \sigma)_G$. We define the inertial equivalence classes: we write $(M, \sigma) \sim (L, \tau)$ if there exists $g \in G$ such that $gMg^{-1} = L$ and ${}^g\sigma = \tau \otimes \chi$ for some $\chi \in X(L)$. Let $[M, \sigma]_G$ denote the equivalence class of $(M, \sigma)_G$.

If $\pi \in \mathfrak{R}(G)_{\text{irred}}$, then the supercuspidal support of π is the unique element $(M, \sigma)_G$ such that π is a subquotient of the induced representation $i_P^G(\sigma)$, where P is any F-parabolic subgroup having M as a Levi subgroup. Let $\mathfrak{X}_G$ denote the set of all supercuspidal supports $(M, \sigma)_G$. Denote by the symbol $\mathfrak{s} = [M, \sigma]_G$ a typical inertial class.

For an inertial class $\mathfrak{s} = [M, \sigma]_G$, define the set $\mathfrak{X}_{\mathfrak{s}} = \{(L, \tau)_G \mid (L, \tau) \sim (M, \sigma)\}$. We have

$$\mathfrak{X}_G = \coprod_{\mathfrak{s}} \mathfrak{X}_{\mathfrak{s}}.$$

We shall see below that $\mathfrak{X}_G$ has a natural structure of an algebraic variety, and the Bernstein components $\mathfrak{X}_{\mathfrak{s}}$ form the connected components of that variety.

First we need to recall the variety structure on $X(M)$. As is well-known, $X(M)$ has the structure of a complex torus. To describe this, we first consider the weakly unramified character group $X^w(M)$. This is a diagonalizable group over $\mathbb{C}$. In fact, by Kottwitz we have an isomorphism

$$M(F)/M(F)_1 \cong X^*(Z(\widehat{M})^I)^\Phi = X^*((Z(\widehat{M})^I)_\Phi).$$

This means that

$$X^w(M)(\mathbb{C}) = \text{Hom}_{\text{grp}}(M(F)/M(F)_1, \mathbb{C}^\times) = \text{Hom}_{\text{alg}}(\mathbb{C}[X^*((Z(\widehat{M})^I)_\Phi)], \mathbb{C}),$$

in other words,

$$X^w(M) = (Z(\widehat{M})^I)_\Phi. \tag{4.5}$$

Another way to see (4.5) is to use Langlands' duality for quasicharacters, which is an isomorphism

$$\text{Hom}_{\text{cont}}(M(F), \mathbb{C}^\times) \cong H^1(W_F, Z(\widehat{M})).$$

(Here W_F is the Weil group of F; see §4.) Under this isomorphism, $X^w(M)$ is identified with the image of the inflation map $H^1(W_F/I, Z(\widehat{M})^I) \to H^1(W_F, Z(\widehat{M}))$, that is, with $H^1(\langle \Phi \rangle, Z(\widehat{M})^I) = (Z(\widehat{M})^I)_\Phi$. This last identification is given by the map sending a cocycle $\varphi^{cocyc} \in Z^1(\langle \Phi \rangle, Z(\widehat{M})^I)$ to $\varphi^{cocyc}(\Phi) \in (Z(\widehat{M})^I)_\Phi$. The two ways of identifying $X^w(M)$ with $(Z(\widehat{M})^I)_\Phi$, that via the Kottwitz isomorphism and that via Langlands duality, agree.[2] For a more general result which implies this agreement, see [Kal], Prop. 4.5.2.

Now we can apply the same argument to identify the torus $X(M)$. We first get

$$X(M)(\mathbb{C}) = \text{Hom}_{\text{grp}}(M(F)/M(F)^1, \mathbb{C}^\times).$$

---

[2] We normalize the Kottwitz homomorphism as in [Ko97], so that $\kappa_{\mathbb{G}_m} : L^\times \to \mathbb{Z}$ is the valuation map sending a uniformizer ϖ to 1. Then the claimed agreement holds provided we normalize the Langlands duality for tori as in (4.7).

Since $M(F)/M(F)^1$ is the quotient of $M(F)/M(F)_1$ by its torsion, it follows that

$$X(M) = ((Z(\widehat{M})^I)_\Phi)^\circ =: (Z(\widehat{M})^I)_\Phi^\circ,$$

the neutral component of $X^w(M)$.

Now we turn to the variety structure on $\mathfrak{X}_\mathfrak{s}$. We fix a cuspidal pair (M, σ) representing $\mathfrak{s}_M = [M, \sigma]_M$ and $\mathfrak{s} = [M, \sigma]_G$. Let the corresponding Bernstein components be denoted $\mathfrak{X}_\mathfrak{s}$ and $\mathfrak{X}_{\mathfrak{s}_M}$. As sets, we have $\mathfrak{X}_\mathfrak{s} = \{(M, \sigma\chi)_G\}$ and $\mathfrak{X}_{\mathfrak{s}_M} = \{(M, \sigma\chi)_M\}$, where $\chi \in X(M)$. The torus $X(M)$ acts on $\mathfrak{X}_{\mathfrak{s}_M}$ by $\chi \mapsto (M, \sigma\chi)_M$. The isotropy group is $\mathrm{stab}_\sigma := \{\chi \mid \sigma \cong \sigma\chi\}$. Let $Z(M)^\circ$ denote the neutral component of the center of M. Then stab_σ belongs to the kernel of the map $X(M) \to X(Z(M)^\circ)$, $\chi \mapsto \chi|_{Z(M)^\circ(F)}$, hence stab_σ is a *finite* subgroup of $X(M)$. Thus $\mathfrak{X}_{\mathfrak{s}_M}$ is a torsor under the torus $X(M)/\mathrm{stab}_\sigma$, and thus has the structure of an affine variety over $\mathbb{C}$.

There is a surjective map

$$\mathfrak{X}_{\mathfrak{s}_M} \to \mathfrak{X}_\mathfrak{s}$$
$$(M, \sigma\chi)_M \mapsto (M, \sigma\chi)_G, \quad \chi \in X(M)/\mathrm{stab}_\sigma.$$

Let $N_G([M, \sigma]_M) := \{n \in N_G(M) \mid {}^n\sigma \cong \sigma\chi \text{ for some } \chi \in X(M)\}$. Then the fibers of $\mathfrak{X}_{\mathfrak{s}_M} \to \mathfrak{X}_\mathfrak{s}$ are precisely the orbits the finite group $W_{[M,\sigma]_M}^G := N_G([M, \sigma]_G)/M$ on $\mathfrak{X}_{\mathfrak{s}_M}$. Via

$$\mathfrak{X}_\mathfrak{s} = W_{[M,\sigma]_M}^G \backslash \mathfrak{X}_{\mathfrak{s}_M}$$

the set $\mathfrak{X}_\mathfrak{s}$ acquires the structure of an irreducible affine variety over $\mathbb{C}$. Up to isomorphism, this structure does not depend on the choice of the cuspidal pair (M, σ).

3.3.2. The center as regular functions on $\mathfrak{X}_G$

An element $z \in \mathfrak{z}(G)$ determines a regular function $\mathfrak{X}$: for a point $(M, \sigma)_G \in \mathfrak{X}_\mathfrak{s}$, z acts on $i_P^G(\sigma)$ by a scalar $z(\sigma)$ and the function $(M, \sigma)_G \mapsto z(\sigma)$ is a regular function on $\mathfrak{X}_G$. This is the content of [BD, Prop. 2.11]. In fact we have by [BD, Thm. 2.13] an isomorphism

$$\mathfrak{z}(G) \xrightarrow{\sim} \mathbb{C}[\mathfrak{X}_G]. \tag{4.6}$$

Together with (4.3) this gives all the equivalent ways of realizing the Bernstein center of G.

4. The local Langlands correspondence

We need to recall the general form of the conjectural local Langlands correspondence (LLC) for a connected reductive group G over a p-adic field F. Let $\bar{F}$ denote an algebraic closure of F. Let $W_F \subset \mathrm{Gal}(\bar{F}/F) =: \Gamma_F$ be the *Weil group of F*. It fits into an exact sequence of topological groups

$$1 \longrightarrow I_F \longrightarrow W_F \xrightarrow{\mathrm{val}} \mathbb{Z} \longrightarrow 1,$$

where I_F is the inertia subgroup of Γ_F and where, if $\Phi \in W_F$ is a geometric Frobenius element (the inverse of an arithmetic Frobenius element), then $\mathrm{val}(\Phi) = -1$. Here I_F has its profinite topology and $\mathbb{Z}$ has the discrete topology. Sometimes we write I for I_F in what follows.

Recall the *Weil–Deligne* group is $W_F' := W_F \ltimes \mathbb{C}$, where $wzw^{-1} = |w|z$ for $w \in W_F$ and $z \in \mathbb{C}$, with $|w| := q_F^{\mathrm{val}(w)}$ for $q_F = \#(\mathcal{O}_F/(\varpi_F))$, the cardinality of the residue field of F.

A Langlands parameter is an *admissible* homomorphism $\varphi : W_F' \to {}^L G$, where ${}^L G := \widehat{G} \rtimes W_F$. This means:

- φ is compatible with the projections $W_F' \to W_F$ and $\nu : {}^L G \to W_F$;
- φ is continuous and respects Jordan decompositions of elements in W_F' and ${}^L G$ (cf. [Bo79, §8], for the definition of Jordan decomposition in the group $W_F \ltimes \mathbb{C}$ and what it means to respect Jordan decompositions here);
- if $\varphi(W_F')$ is contained in a Levi subgroup of a parabolic subgroup of ${}^L G$, then that parabolic subgroup is *relevant* in the sense of [Bo79, §3.3]. (This condition is automatic if G/F is quasi-split.)

Let $\Phi(G/F)$ denote the set of $\widehat{G}$-conjugacy classes of admissible homomorphisms $\varphi : W_F' \to {}^L G$ and let $\Pi(G/F) = \mathfrak{R}(G(F))_{\mathrm{irred}}$ the set of irreducible smooth (or admissible) representations of $G(F)$ up to isomorphism.

Conjecture 4.4 (LLC). *There is a finite-to-one surjective map $\Pi(G/F) \to \Phi(G/F)$, which satisfies the desiderata of* [Bo79, §10].

The fiber Π_φ over $\varphi \in \Phi(G/F)$ is called the *L-packet for φ*.

We mention a few desiderata of the LLC that will come up in what follows. First, LLC for $\mathbb{G}_m$ is nothing other than Langlands duality for $\mathbb{G}_m$, which we normalize as follows: for T any split torus torus over F, with dual torus $\widehat{T}$,

$$\mathrm{Hom}_{\mathrm{conts}}(T(F), \mathbb{C}^\times) = \mathrm{Hom}_{\mathrm{conts}}(W_F, \widehat{T})$$
$$\xi \leftrightarrow \varphi_\xi$$

satisfies, for every $\nu \in X_*(T) = X^*(\widehat{T})$ and $w \in W_F$,

$$\nu(\varphi_\xi(w)) = \xi(\nu(\mathrm{Art}_F^{-1}(w))). \qquad (4.7)$$

Here $\mathrm{Art}_F^{-1} : W_F^{ab} \to F^\times$ is the reciprocity map of local class field theory which sends any geometric Frobenius element $\Phi \in W_F$ to a uniformizer in F.

Next, we think of Langlands parameters in two ways, either as continuous *L-homomorphisms*

$$\varphi : W_F' \to {}^L G$$

modulo $\widehat{G}$-conjugation, or as continuous *1-cocycles*

$$\varphi^{cocyc} : W_F' \to \widehat{G}$$

modulo 1-coboundaries (where W_F' acts on $\widehat{G}$ via the projection $W_F' \to W_F$). The dictionary between these is

$$\varphi(w) = (\varphi^{cocyc}(w), \bar{w}) \in \widehat{G} \rtimes W_F$$

for $w \in W_F'$ and $\bar{w}$ the image of w under $W_F' \to W_F$.

The desideratum we will use explicitly is the following (a special case of [Bo79, 10.3(2)]): given any Levi pair (M, σ) (where $\sigma \in \Pi(M/F)$) with representing 1-cocycle $\varphi_\sigma^{cocyc} : W_F' \to \widehat{M}$, and any unramified 1-cocycle $z_\chi^{cocyc} : W_F \to Z(\widehat{M})^I$ representing $\chi \in X(M)$ via the Langlands correspondence for quasi-characters on $M(F)$, we have

$$\varphi_{\sigma\chi}^{cocyc} = \varphi_\sigma^{cocyc} \cdot z_\chi^{cocyc}$$

modulo 1-coboundaries with values in $\widehat{M}$. We may view z_χ^{cocyc} as a 1-cocycle on W_F' which is trivial on $\mathbb{C}$; since it takes values in the center of $\widehat{M}$, the right-hand side is a 1-cocycle whose cohomology class is independent of the choices of 1-cocycles φ_σ^{cocyc} and z_χ^{cocyc} in their respective cohomology classes. Hence the condition just stated makes sense. Concretely, if $\chi \in X(M)$ lifts to an element $z \in Z(\widehat{M})^I$, then up to $\widehat{M}$-conjugacy we have

$$\varphi_{\sigma\chi}(\Phi) = (z, 1)\varphi_\sigma(\Phi) \in \widehat{M} \rtimes W_F. \qquad (4.8)$$

Remark 4.5. There is a well-known dictionary between equivalence classes of admissible homomorphisms $\varphi : W_F \ltimes \mathbb{C} \to {}^L G$ and equivalence classes of admissible homomorphisms $W_F \times \mathrm{SL}_2(\mathbb{C}) \to {}^L G$. For a complete explanation, see [GR, Prop. 2.2]. Because of this equivalence, it is common in the literature for the Weil–Deligne group W_F' to sometimes be defined as $W_F \ltimes \mathbb{C}$, and sometimes as $W_F \times \mathrm{SL}_2(\mathbb{C})$.

5. The stable Bernstein center

5.1. Infinitesimal characters

Following Vogan [Vo], we term a $\widehat{G}$-conjugacy class of an admissible[3] homomorphism

$$\lambda : W_F \to {}^L G$$

an *infinitesimal character*. Denote the $\widehat{G}$-conjugacy class of λ by $(\lambda)_{\widehat{G}}$. In this section we give a geometric structure to the set of all infinitesimal characters for a group G. It should be noted that the variety structure we define here differs from that put forth by Vogan in [Vo, §7].

If $\varphi : W_F' \to {}^L G$ is an admissible homomorphism, then its restriction $\varphi|_{W_F}$ represents an infinitesimal character. Here it is essential to consider restriction along the proper embedding $W_F \hookrightarrow W_F'$: if W_F' is thought of as $W_F \ltimes \mathbb{C}$, then this inclusion is $w \mapsto (w, 0)$; if W_F' is thought of as $W_F \times \mathrm{SL}_2(\mathbb{C})$, then the inclusion is $w \mapsto (w, \mathrm{diag}(|w|^{1/2}, |w|^{-1/2}))$. If $\varphi_\pi \in \Phi(G/F)$ is attached by LLC to $\pi \in \Pi(G/F)$, then following Vogan [Vo] we shall call the $\widehat{G}$-conjugacy class $(\varphi_\pi|_{W_F})_{\widehat{G}}$ the *infinitesimal character of π*.

If G is quasi-split over F, then conjecturally *every* infinitesimal character λ is represented by a restriction $\varphi_\pi|_{W_F} : W_F \to {}^L G$ for *some* $\pi \in \Pi(G/F)$.

Assume LLC holds for G/F. Let λ be an infinitesimal character for G. Define the **infinitesimal class** to be the following finite union of L-packets

$$\Pi_\lambda := \coprod_{\varphi \rightsquigarrow \lambda} \Pi_\varphi.$$

Here φ ranges over $\widehat{G}$-conjugacy classes of admissible homomorphisms $W_F' \to {}^L G$ such that $(\varphi|_{W_F})_{\widehat{G}} = (\lambda)_{\widehat{G}}$, and Π_φ is the corresponding L-packet of smooth irreducible representations of $G(F)$.

5.2. LLC+

In order to relate the Bernstein variety $\mathfrak{X}$ with the variety $\mathfrak{Y}$ of infinitesimal characters, we will assume the Local Langlands Correspondence (LLC) for G and all of its F-Levi subgroups. We assume all the desiderata listed by Borel in [Bo79].

There are two additional desiderata of LLC we need.

---

[3] "Admissible" is defined as for the parameters $\varphi : W_F' \to {}^L G$ (e.g. $\lambda(W_F)$ consists of semisimple elements of ${}^L G$) **except** that we omit the "relevance condition". This is because the restriction $\varphi|_{W_F}$ of a Langlands parameter could conceivably factor through a non-relevant Levi subgroup of ${}^L G$ (even though φ does not) and we want to include such restrictions in what we call infinitesimal characters.

Definition 4.6. We will declare that G satisfies LLC+ if the LLC holds for G and its F-Levi subgroups, and these correspondences are compatible with normalized parabolic induction in the sense of the Conjecture 4.7 below, and invariant under certain isomorphisms in the sense of Conjecture 4.12 below.

Let $M \subset G$ denote an F-Levi subgroup. Then the inclusion $M \hookrightarrow G$ induces an embedding $^L M \hookrightarrow {}^L G$ which is well-defined up to $\widehat{G}$-conjugacy (cf. [Bo79, §3]).

Conjecture 4.7. (Compatibility of LLC with parabolic induction) *Let $\sigma \in \Pi(M/F)$ and $\pi \in \Pi(G/F)$ and assume π is an irreducible subquotient of $i_P^G(\sigma)$, where $P = MN$ is any F-parabolic subgroup of G with F-Levi factor M. Then the infinitesimal characters*

$$\varphi_\pi|_{W_F} : W_F \to {}^L G$$

and

$$\varphi_\sigma|_{W_F} : W_F \to {}^L M \hookrightarrow {}^L G$$

are $\widehat{G}$-conjugate.

Remark 4.8. (1) The conjecture implies that the restriction $\varphi_\pi|_{W_F}$ depends only on the supercuspidal support of π. This latter statement is a formal consequence of Vogan's Conjecture 7.18 in [Vo], but the Conjecture 4.7 is slightly more precise. In Proposition 4.23 we will give a construction of the map f in Vogan's Conjecture 7.18, by sending a supercuspidal support $(M, \sigma)_G$ (a "classical infinitesimal character" in [Vo]) to the infinitesimal character $(\varphi_\sigma|_{W_F})_{\widehat{G}}$. With this formulation, the condition on f imposed in Vogan's Conjecture 7.18 is exactly the compatibility in the conjecture above.
(2) The conjecture holds for GL_n, and is implicit in the way the local Langlands correspondence for GL_n is extended from supercuspidals to all representations (see Remark 13.1.1 of [HRa1]). It was a point of departure in Scholze's new characterization of LLC for GL_n [Sch3], and that paper also provides another proof of the conjecture in that case.
(3) I was informed by Brooks Roberts (private communication), that the conjecture holds for $GSp(4)$.
(4) Given a parameter $\varphi : W_F' \to {}^L G$, there exists a certain $P = MN$ and a certain tempered parameter $\varphi_M : W_F' \to {}^L M$ and a certain real-valued unramified character χ_M on $M(F)$ whose parameter is in the interior of the Weyl chamber determined by P, such that the L-packet Π_ϕ consists of Langlands quotients $J(\pi_M \otimes \chi_M)$, for π_M ranging over the packet

Π_{φ_M}. The parameter φ is the twist of φ_M by the parameter associated to the character χ_M. This reduces the conjecture to the case of tempered representations. One can further reduce to the case of discrete series representations.

The following is a very natural kind of functoriality which should be satisfied for all groups.

Conjecture 4.9. (Invariance of LLC under isomorphisms) *Suppose ϕ : $(G, \pi) \overset{\sim}{\to} (G', \pi')$ is an isomorphism of connected reductive F-groups together with irreducible smooth representations on them. Then the induced isomorphism ${}^L\phi : {}^L G' \overset{\sim}{\to} {}^L G$ (well-defined up to an inner automorphism of $\widehat{G}$), takes the $\widehat{G}'$-conjugacy class of $\varphi_{\pi'} : W'_F \to {}^L G'$ to the $\widehat{G}$-conjugacy class of $\varphi_\pi : W'_F \to {}^L G$.*

Proposition 4.10. *Conjecture 4.9 holds when $G = \mathrm{GL}_n$.*

Proof. (Guy Henniart). It is enough to consider the case where $G' = \mathrm{GL}_n$ and ϕ is an F-automorphism of GL_n.

The functorial properties in the Langlands correspondence for GL_n are:

(i) Compatibility with class field theory, that is, with the case where $n = 1$.

(ii) The determinant of the Weil–Deligne group representation corresponds to the central character: this is Langlands functoriality for the homomorphism $\det : \mathrm{GL}_n(\mathbb{C}) \to \mathrm{GL}_1(\mathbb{C})$.

(iii) Compatibility with twists by characters, i.e., Langlands functoriality for the obvious homomorphism of dual groups $\mathrm{GL}_1(\mathbb{C}) \times \mathrm{GL}_n(\mathbb{C}) \to \mathrm{GL}_n(\mathbb{C})$.

(iv) Compatibility with taking contragredients: this is Langlands functoriality with respect to the automorphism $g \mapsto {}^t g^{-1}$ (transpose inverse), since it is known that for $\mathrm{GL}_n(F)$ this sends an irreducible representation to a representation isomorphic to its contragredient.

These properties are enough to imply the desired functoriality for F-automorphisms of GL_n. When $n = 1$, the functoriality is obvious for any F-endomorphism of GL_1. When n is at least 2, an F-automorphism of GL_n induces an automorphism of SL_n hence an automorphism of the Dynkin diagram which must be the identity or, (when $n \geq 3$) the opposition automorphism. Hence up to conjugation by $\mathrm{GL}_n(F)$, the F-automorphism is the identity on SL_n, or possibly (when $n \geq 3$) transpose inverse. Consequently the F-automorphism can be reduced (by composing with an inner automorphism or possibly with transpose inverse) to one which is the identity on SL_n, hence is of the form $g \mapsto g \cdot c(\det(g))$ where $c \in X_*(Z(\mathrm{GL}_n))$. But this implies that it is the identity unless $n = 2$, in which case it could also be $g \mapsto g \cdot \det(g)^{-1}$. In that exceptional case, the map induced on the dual group $GL_2(\mathbb{C})$ is also $g \mapsto g \cdot \det(g)^{-1}$, and the desired result holds by invoking (ii) and (iii) above. $\qquad\square$

Corollary 4.11. *Let $M = GL_{n_1} \times \cdots \times GL_{n_r} \subset GL_n$ be a standard Levi subgroup. Let $g \in GL_n(F)$. Then Conjecture 4.9 holds for the isomorphism $c_g : M \overset{\sim}{\to} {}^g M$ given by conjugation by g.*

Proof. It is enough to consider the case where g belongs to the normalizer of M in GL_n. Let $T \subset M$ be the standard diagonal torus in GL_n. Then $g \in N_G(T)M$. Thus composing g with a permutation matrix which normalizes M we may assume that c_g preserves each diagonal factor GL_{n_i}. The desired functoriality follows by applying Proposition 4.10 to each GL_{n_i}. $\qquad\square$

For the purposes of comparing the Bernstein center and the stable Bernstein center as in Proposition 4.23, we need only this weaker variant of Conjecture 4.9.

Conjecture 4.12. (Weak invariance of LLC) *Let $M \subseteq G$ be any F-Levi subgroup and let $g \in G(F)$. Then Conjecture 4.9 holds for the isomorphism $c_g : M \overset{\sim}{\to} {}^g M$.*

5.3. Variety structure on the set of infinitesimal characters

It is helpful to rigidify things on the dual side by choosing the data $\widehat{G} \supset \widehat{B} \supset \widehat{T}$ of a Borel subgroup and maximal torus which are stable under the action of Γ_F on $\widehat{G}$ and which form part of the data of a Γ_F-invariant splitting for $\widehat{G}$ (cf. [Ko84a, §1]). The variety structure we will define will be independent of this choice, up to isomorphism, since different choices such that $\widehat{B} \supset \widehat{T}$ are conjugate under $\widehat{G}^{\Gamma_F}$ ([Ko84a, Cor. 1.7]). Let ${}^L B := \widehat{B} \rtimes W_F$ and ${}^L T = \widehat{T} \rtimes W_F$.

Following [Bo79, §3.3], we say a parabolic subgroup $\mathcal{P} \subseteq {}^L G$ is *standard* if $\mathcal{P} \supseteq {}^L B$. Then its neutral component $\mathcal{P}^\circ := \mathcal{P} \cap \widehat{G}$ is a W_F-stable standard parabolic subgroup of $\widehat{G}$ (containing $\widehat{B}$), and $\mathcal{P} = \mathcal{P}^\circ \rtimes W_F$. Every parabolic subgroup in ${}^L G$ is $\widehat{G}$-conjugate to a unique standard parabolic subgroup.

Assume $\mathcal{P}$ is standard and let $\mathcal{M}^\circ \subset \mathcal{P}^\circ$ be the unique Levi factor with $\mathcal{M}^\circ \supseteq \widehat{T}$; it is W_F-stable. Then $\mathcal{M} := N_{\mathcal{P}}(\mathcal{M}^\circ)$ is a Levi subgroup of $\mathcal{P}$ in the sense of [Bo79, §3.3], and $\mathcal{M} = \mathcal{M}^\circ \rtimes W_F$. The Levi subgroups $\mathcal{M} \subset {}^L G$ which arise this way are called *standard*. Every Levi subgroup in ${}^L G$ is $\widehat{G}$-conjugate to at least one standard Levi subgroup; two different standard Levi subgroups may be conjugate under $\widehat{G}$. Denote by $\{\mathcal{M}\}$ the set of standard Levi subgroups in $^L G$ which are $\widehat{G}$-conjugate to a fixed standard Levi subgroup $\mathcal{M}$.

Now suppose $\lambda : W_F \to {}^L G$ is an admissible homomorphism. Then there exists a minimal Levi subgroup of $^L G$ containing $\lambda(W_F)$. Any two such are conjugate by an element of C_λ°, where C_λ is the subgroup of $\widehat{G}$ commuting with $\lambda(W_F)$, by (the proof of) [Bo79, Prop. 3.6].

Suppose $\lambda_1, \lambda_2 : W_F \to {}^L G$ are $\widehat{G}$-conjugate. Then there exists a $\widehat{G}$-conjugate λ_1^+ (resp. λ_2^+) of λ_1 (resp. λ_2) and a standard Levi subgroup $\mathcal{M}_1$ (resp. $\mathcal{M}_2$) containing $\lambda_1^+(W_F)$ (resp. $\lambda_2^+(W_F)$) minimally. Write $g\lambda_1^+ g^{-1} = \lambda_2^+$ for some $g \in \widehat{G}$. Then the Levi subgroups $g\mathcal{M}_1 g^{-1}$ and $\mathcal{M}_2$ contain $\lambda_2^+(W_F)$ minimally, hence by [Bo79, Prop. 3.6] are conjugate by an element $s \in C_{\lambda_2^+}^\circ$. Then $sg(\mathcal{M}_1)(sg)^{-1} = \mathcal{M}_2$, and thus $\{\mathcal{M}_1\} = \{\mathcal{M}_2\}$.

Hence any $\widehat{G}$-conjugacy class $(\lambda)_{\widehat{G}}$ gives rise to a *unique* class of standard Levi subgroups $\{\mathcal{M}_\lambda\}$, with the property that the image of some element $\lambda^+ \in (\lambda)_{\widehat{G}}$ is contained minimally by $\mathcal{M}_\lambda$ for some $\mathcal{M}_\lambda$ in this class.

A similar argument shows the following lemma.

Lemma 4.13. *Let λ_1^+ and λ_2^+ be admissible homomorphisms with $(\lambda_1^+)_{\widehat{G}} = (\lambda_2^+)_{\widehat{G}}$, and suppose $\lambda_1^+(W_F)$ and $\lambda_2^+(W_F)$ are contained minimally by a standard Levi subgroup $\mathcal{M}$. Then there exists $n \in N_{\widehat{G}}(\mathcal{M})$ such that $^n\lambda_1^+ = \lambda_2^+$.*

The following lemma is left to the reader.

Lemma 4.14. *If $\mathcal{M} \subseteq {}^L G$ is a standard Levi subgroup, then*

$$N_{\widehat{G}}(\mathcal{M}) = \{n \in N_{\widehat{G}}(\mathcal{M}^\circ) \mid n\mathcal{M}^\circ \text{ is } W_F\text{-stable}\}.$$

Consequently, conjugation by $n \in N_{\widehat{G}}(\mathcal{M})$ preserves the set $(Z(\mathcal{M}^\circ)^I)_\Phi^\circ$. More generally, if $\mathcal{M}_1$ and $\mathcal{M}_2$ are standard Levi subgroups of $^L G$ and if we define the transporter subset by

$$\mathrm{Trans}_{\widehat{G}}(\mathcal{M}_1, \mathcal{M}_2) := \{g \in \widehat{G} \mid g\mathcal{M}_1 g^{-1} = \mathcal{M}_2\},$$

then

$$\mathrm{Trans}_{\widehat{G}}(\mathcal{M}_1, \mathcal{M}_2) = \{g \in \mathrm{Trans}_{\widehat{G}}(\mathcal{M}_1^\circ, \mathcal{M}_2^\circ) \mid g\mathcal{M}_1^\circ = \mathcal{M}_2^\circ g \text{ is } W_F\text{-stable}\}.$$

Consequently, conjugation by $g \in \mathrm{Trans}_{\widehat{G}}(\mathcal{M}_1, \mathcal{M}_2)$ sends $(Z(\mathcal{M}_1^\circ)^I)_\Phi^\circ$ into $(Z(\mathcal{M}_2^\circ)^I)_\Phi^\circ$.

We can now define the notion of *inertial equivalence* $(\lambda_1)_{\widehat{G}} \sim (\lambda_2)_{\widehat{G}}$ of infinitesimal characters.

Definition 4.15. We say $(\lambda_1)_{\widehat{G}}$ and $(\lambda_2)_{\widehat{G}}$ are *inertially equivalent* if

- $\{\mathcal{M}_{\lambda_1}\} = \{\mathcal{M}_{\lambda_2}\}$;
- there exists $\mathcal{M} \in \{\mathcal{M}_{\lambda_1}\}$, and $\lambda_1^+ \in (\lambda_1)_{\widehat{G}}$ and $\lambda_2^+ \in (\lambda_2)_{\widehat{G}}$ whose images are minimally contained by $\mathcal{M}$, and an element $z \in (Z(\mathcal{M}^\circ)^I)_\Phi^\circ$, such that

$$(z\lambda_1^+)_{\mathcal{M}^\circ} = (\lambda_2^+)_{\mathcal{M}^\circ}.$$

We write $[\lambda]_{\widehat{G}}$ for the inertial equivalence class of $(\lambda)_{\widehat{G}}$.

Note that $\mathcal{M}$ automatically contains $(z\lambda_1^+)(W_F)$ minimally if it contains $\lambda_1^+(W_F)$ minimally.

Lemma 4.16. *The relation $\sim$ is an equivalence relation on the set of infinitesimal characters.*

Proof. Use Lemmas 4.13 and 4.14.   □

Remark 4.17. To define $(\lambda_1)_{\widehat{G}} \sim (\lambda_2)_{\widehat{G}}$ we used the choice of $\widehat{G} \supset \widehat{B} \supset \widehat{T}$ (which was assumed to form part of a Γ_F-invariant splitting for $\widehat{G}$) in order to define the notion of standard Levi subgroup of ${}^L G$. However, the equivalence relation $\sim$ is independent of this choice, since as remarked above, any two Γ_F-invariant splittings for $\widehat{G}$ are conjugate under $\widehat{G}^{\Gamma_F}$, by [Ko84a, Cor. 1.7].

Remark 4.18. The property we need of standard Levi subgroups $\mathcal{M} \subseteq {}^L G$ is that they are *decomposable*, that is, $\mathcal{M}^\circ := \mathcal{M} \cap \widehat{G}$ is W_F-stable, and $\mathcal{M} = \mathcal{M}^\circ \rtimes W_F$. Any standard Levi subgroup is decomposable. In our discussion, we could have avoided choosing a notion of standard Levi, by associating to each $(\lambda)_{\widehat{G}}$ a unique class of decomposable Levi subgroups $\{\mathcal{M}\}$, all of which are $\widehat{G}$-conjugate, such that λ factors minimally through some $\mathcal{M} \in \{\mathcal{M}\}$.

Now fix a standard Levi subgroup $\mathcal{M} \subseteq {}^L G$. We write $\mathfrak{t}_{\mathcal{M}^\circ}$ for an inertial equivalence class of admissible homomorphisms $W_F \to \mathcal{M}$. We write $\mathfrak{Y}_{\mathfrak{t}_{\mathcal{M}^\circ}}$ for the set of $\mathcal{M}^\circ$-conjugacy classes contained in this inertial class. We want to give this set the structure of an affine algebraic variety over $\mathbb{C}$. Define the torus

$$Y(\mathcal{M}^\circ) := (Z(\mathcal{M}^\circ)^I)_\Phi^\circ. \qquad (4.9)$$

Then $Y(\mathcal{M}^\circ)$ acts transitively on $\mathfrak{Y}_{\mathfrak{t}_{\mathcal{M}^\circ}}$. Fix a representative

$$\lambda : W_F \to \mathcal{M}$$

for this inertial class, so that $\mathfrak{t}_{\mathcal{M}^\circ} = [\lambda]_{\mathcal{M}^\circ}$.

Lemma 4.19. *The $Y(\mathcal{M}^\circ)$-stabilizer*

$$\mathrm{stab}_\lambda := \{z \in Y(\mathcal{M}^\circ) \mid (z\lambda)_{\mathcal{M}^\circ} = (\lambda)_{\mathcal{M}^\circ}\}$$

is finite.

Proof. There exists an integer $r \geq 1$ such that Φ^r acts trivially on $\mathcal{M}^\circ$. The group stab_λ is contained in the preimage of the finite group $Z(\mathcal{M}^\circ)^{\Gamma_F} \cap (\mathcal{M}^\circ)_{\mathrm{der}}$ under the norm homomorphism

$$N_r : (Z(\mathcal{M}^\circ)^I)_\Phi \to Z(\mathcal{M}^\circ)^{\Gamma_F},$$

$$z \mapsto z\Phi(z) \cdots \Phi^{r-1}(z)$$

and the kernel of this homomorphism is finite. $\qquad\square$

Then $\mathfrak{Y}_{t_{\mathcal{M}^\circ}}$ is a torsor under the quotient torus

$$\mathfrak{Y}_{t_{\mathcal{M}^\circ}} \cong Y(\mathcal{M}^\circ)/\mathrm{stab}_\lambda.$$

In this way the left-hand side acquires the structure of an affine algebraic variety. Up to isomorphism, this structure is independent of the choice of λ representing $t_{\mathcal{M}^\circ}$.

Now let t denote an inertial class of infinitesimal characters for G, and let $\mathfrak{Y}_t$ denote the set of infinitesimal characters in t. Recall t gives rise to a unique class of standard Levi subgroups $\{\mathcal{M}\}$, having the property that some representative λ for t factors minimally through some $\mathcal{M} \in \{\mathcal{M}\}$. Fix such a representative $\lambda : W_F \to \mathcal{M} \hookrightarrow {}^L G$ for t, so that $t = [\lambda]_{\widehat{G}}$ and $t_{\mathcal{M}^\circ} = [\lambda]_{\mathcal{M}^\circ}$. By our previous work, there is a *surjective* map

$$\mathfrak{Y}_{t_{\mathcal{M}^\circ}} \to \mathfrak{Y}_t$$

$$(z\lambda)_{\mathcal{M}^\circ} \mapsto (z\lambda)_{\widehat{G}}.$$

where $z \in Y(\mathcal{M}^\circ)/\mathrm{stab}_\lambda$. Let

$$N_{\widehat{G}}(\mathcal{M}, [\lambda]_{\mathcal{M}^\circ}) = \{n \in N_{\widehat{G}}(\mathcal{M}) \mid ({}^n\lambda)_{\mathcal{M}^\circ} = (z\lambda)_{\mathcal{M}^\circ}, \text{ for some } z \in Y(\mathcal{M}^\circ)\}.$$

From the above discussion we see the following.

Lemma 4.20. *The fibers of $\mathfrak{Y}_{t_{\mathcal{M}^\circ}} \to \mathfrak{Y}_t$ are precisely the orbits of the finite group $W^{\widehat{G}}_{[\lambda]_{\mathcal{M}^\circ}} := N_{\widehat{G}}(\mathcal{M}, [\lambda]_{\mathcal{M}^\circ})/\mathcal{M}^\circ$ on $\mathfrak{Y}_{t_{\mathcal{M}^\circ}}$.*

Hence $\mathfrak{Y}_t = W^{\widehat{G}}_{t_{\mathcal{M}^\circ}} \backslash \mathfrak{Y}_{t_{\mathcal{M}^\circ}}$ acquires the structure of an affine variety over $\mathbb{C}$. Thus $\mathfrak{Y} = \coprod_t \mathfrak{Y}_t$ is an affine variety over $\mathbb{C}$ and each $\mathfrak{Y}_t$ is a connected component.

Let $\mathfrak{Z}^{\mathrm{st}}(G)$ denote the ring of regular functions on the affine variety $\mathfrak{Y}$. We call this ring the **stable Bernstein center** of G/F.

5.4. Base change homomorphism of the stable Bernstein center

Let E/F be a finite extension in $\overline{F}/F$ with ramification index e and residue field extension k_E/k_F of degree f. Then $W_E \subset W_F$ and $I_E \subseteq I_F$. Further, we can take $\Phi_E := \Phi_F^f$ as a geometric Frobenius element in W_E. Let $\mathfrak{Y}^{G/F}$ resp. $\mathfrak{Y}^{G/E}$ denote the variety of infinitesimal characters associated to G resp. G_E.

Proposition 4.21. *The map* $(\lambda)_{\widehat{G}} \mapsto (\lambda|_{W_E})_{\widehat{G}}$ *determines a morphism of algebraic varieties* $\mathfrak{Y}^{G/F} \to \mathfrak{Y}^{G/E}$.

Definition 4.22. We call the corresponding map $b_{E/F} : 3^{\mathrm{st}}(G_E) \to 3^{\mathrm{st}}(G)$ the *base change homomorphism* for the stable Bernstein center.

Proof. Suppose $\lambda : W_F \to \widehat{G} \rtimes W_F$ factors minimally through the standard Levi subgroup $\mathcal{M} \subset \widehat{G} \rtimes W_F$ and that its restriction $\lambda|_{W_E} : W_E \to \widehat{G} \rtimes W_E$ factors minimally through the standard Levi subgroup $\mathcal{M}_E \subset \widehat{G} \rtimes W_E$. We may assume $\mathcal{M}_E^\circ \subseteq \mathcal{M}^\circ$ and thus $Z(\mathcal{M}^\circ) \subset Z(\mathcal{M}_E^\circ)$.

There is a homomorphism of tori

$$Y(\mathcal{M}^\circ) = (Z(\mathcal{M}^\circ)^{I_F})_{\Phi_F}^\circ \longrightarrow (Z(\mathcal{M}_E^\circ)^{I_E})_{\Phi_F^f}^\circ = Y(\mathcal{M}_E^\circ) \qquad (4.10)$$

$$z \longmapsto z_f := N_f(z) := z \cdot \Phi_F(z) \cdots \Phi_F^{f-1}(z).$$

Recall that $z \in (Z(\mathcal{M}^\circ)^{I_F})_{\Phi_F}$ is identified with the image of the element $z(\Phi_F) \in Z(\mathcal{M}^\circ)^{I_F}$, where z is viewed as a cohomology class $z \in H^1(\langle \Phi_F \rangle, Z(\mathcal{M}^\circ)^{I_F})$. Using the same fact for E in place of F, it follows that $(z\lambda)|_{W_E} = z_f \lambda|_{W_E}$, where z_f is defined as above. Thus the map $(z\lambda)_{\widehat{G}} \mapsto ((z\lambda)|_{W_E})_{\widehat{G}}$ lifts to the map

$$(z\lambda)_{\mathcal{M}^\circ} \mapsto (z_f \lambda|_{W_E})_{\mathcal{M}^\circ} \mapsto (z_f \lambda|_{W_E})_{\widehat{G}},$$

and being induced by (4.10), the latter is an algebraic morphism. $\qquad \square$

5.5. Relation between the Bernstein center and the stable Bernstein center

The varieties $\mathfrak{X}$ and $\mathfrak{Y}$ are defined unconditionally. In order to relate them, we need to assume LLC+ holds.

Proposition 4.23. *Assume* LLC+ *holds for the group G. Then the map* $(M, \sigma)_G \mapsto (\varphi_\sigma|_{W_F})_{\widehat{G}}$ *defines a quasi-finite morphism of affine algebraic varieties*

$$f : \mathfrak{X} \to \mathfrak{Y}.$$

It is surjective if G/F is quasi-split.

The reader should compare this with Conjecture 7.18 in [Vo]. Our variety structure on the set $\mathfrak{Y}$ is different from that put forth by Vogan, and our f is given by a simple and explicit rule. In view of LLC+ our f automatically satisfies the condition which Vogan imposed on the map in his Conjecture 7.18: if π has supercuspidal support $(M, \sigma)_G$, then the infinitesimal character of π is $f((M, \sigma)_G)$.

Proof. It is easy to see that the map $(M, \sigma)_G \mapsto (\varphi_\sigma|_{W_F})_{\widehat{G}}$ is well-defined. We need to show that an isomorphism $c_g : (M, \sigma) \overset{\sim}{\to} (^g M, {}^g\sigma)$ given by conjugation by $g \in G(F)$ gives rise to parameters $\varphi_\sigma : W'_F \to {}^L M \hookrightarrow {}^L G$ and $\varphi_{^g\sigma} : W'_F \to {}^L(^g M) \hookrightarrow {}^L G$ which differ by an inner automorphism of $\widehat{G}$. In view of Conjecture 4.12 applied to M, the isomorphism ${}^L(^g M) \overset{\sim}{\to} {}^L M$ takes $\varphi_{^g\sigma}$ to an $\widehat{M}$-conjugate of φ_σ. On the other hand the embeddings ${}^L M \hookrightarrow {}^L G$ and ${}^L(^g M) \hookrightarrow {}^L G$ are defined using based root systems in such a way that it is obvious that they are $\widehat{G}$-conjugate.

To examine the local structure of this map, we first fix a λ and a standard $\mathcal{M}_\lambda$ through which λ factors minimally. Let $\mathfrak{t} = [\lambda]_{\widehat{G}}$. Then over $\mathfrak{Y}_\mathfrak{t}$ the map f takes the form

$$\coprod_{\mathfrak{s}_M \rightsquigarrow \mathfrak{t}} \mathfrak{X}_{\mathfrak{s}_M} \to \mathfrak{Y}_\mathfrak{t}. \qquad (4.11)$$

Here $\mathfrak{s}_M$ ranges over the inertial classes $[M, \sigma]_G$ such that $(\varphi_\sigma|_{W_F})_{\widehat{G}}$ is inertially equivalent to $(\lambda)_{\widehat{G}}$. We now fix a representative (M, σ) for $\mathfrak{s}_M$. Given such a φ_σ, its restriction $\varphi_\sigma|_{W_F}$ factors through a $\widehat{G}$-conjugate of ${}^L M$. But $(\varphi_\sigma|_{W_F})_{\widehat{G}} \sim (\lambda)_{\widehat{G}}$ implies that (up to conjugation by $\widehat{G}$) $\varphi_\sigma|_{W_F}$ factors minimally through $\mathcal{M}_\lambda$. Thus we may assume that $\mathcal{M}_\lambda \subseteq {}^L M$. The corresponding inclusion $Z(\widehat{M}) \hookrightarrow Z(\mathcal{M}_\lambda^\circ)$ induces a morphism of algebraic tori

$$Y(\widehat{M}) = (Z(\widehat{M})^I)^\circ_\Phi \to (Z(\mathcal{M}_\lambda^\circ)^I)^\circ_\Phi = Y(\mathcal{M}_\lambda^\circ).$$

Further, recall $X(M) \cong Y(\widehat{M})$ by the Kottwitz isomorphism (or the Langlands duality for quasi-characters), by the rule $\chi \mapsto z_\chi^{cocyc}(\Phi)$.

Taking (4.8) into account, we see that (4.11) on $\mathfrak{X}_{\mathfrak{s}_M}$, given by

$$(M, \sigma\chi)_G \mapsto (\varphi_{\sigma\chi}|_{W_F})_{\widehat{G}}$$

for $\chi \in X(M)/\mathrm{stab}_\sigma$, lifts to the map

$$X(M)/\mathrm{stab}_\sigma \to Y(\mathcal{M}_\lambda^\circ)/\mathrm{stab}_\lambda, \qquad (4.12)$$

which is the obvious map induced by $X(M) \to Y(\widehat{M}) \to Y(\mathcal{M}_\lambda^\circ)$, up to translation by an element in $Y(\mathcal{M}_\lambda^\circ)$ measuring the difference between $(\varphi_\sigma|_{W_F})_{\mathcal{M}_\lambda^\circ}$ and $(\lambda)_{\mathcal{M}_\lambda^\circ}$. The map (4.12) is clearly a morphism of algebraic varieties. Hence the map f is a morphism of algebraic varieties.

The fibers of f are finite by a property of LLC. Finally, if G/F is quasi-split, the morphism f is surjective by another property of LLC.   □

Corollary 4.24. *Assume G/F satisfies LLC+, so that the map f in Proposition 4.23 exists. Then f induces a $\mathbb{C}$-algebra homomorphism $\mathfrak{Z}^{st}(G) \to \mathfrak{Z}(G)$. It is injective if G/F is quasi-split.*

Remark 4.25. For the group GL_n the constructions above are unconditional because the local Langlands correspondence and its enhancement LLC+ are known (cf. Remark 4.8(2) and Corollary 4.11). One can see that $\mathfrak{X}_{GL_n} \to \mathfrak{Y}_{GL_n}$ is an isomorphism and hence $\mathfrak{Z}^{st}(GL_n) = \mathfrak{Z}(GL_n)$.

Remark 4.26. As remarked by Scholze and Shin [SS, §6], one may conjecturally characterize the image of $\mathfrak{Z}^{st}(G) \to \mathfrak{Z}(G)$ in a way that avoids direct mention of L-parameters. According to them it should consist of the distributions $D \in \mathcal{D}(G)_{ec}^G$ such that, for any function $f \in C_c^\infty(G(F)$ whose stable orbital integrals vanish at semi-simple elements, the function $D * f$ also has this property. See [SS, §6] for further discussion of this. From conjectured relations between stable characters and stable orbital integrals, one can conjecturally rephrase the condition on D in terms of stable characters, as

$$SO_\varphi(D * f) = 0, \; \forall \varphi, \; \text{if } SO_\varphi(f) = 0, \; \forall \varphi. \tag{4.13}$$

An element of $\mathfrak{Z}^{st}(G)$ acts by the same scalar on all $\pi \in \Pi_\varphi$, and so the above condition holds if $D \in f(\mathfrak{Z}^{st}(G))$. The converse direction is much less clear, and implies non-trivial statements about the relation between supercuspidal supports, L-packets, and infinitesimal classes. Indeed, suppose we are given $D \in \mathfrak{Z}(G)$ that satisfies (4.13). This should mean that it acts by the same scalar on all $\pi \in \Pi_\varphi$. On the other hand, saying D comes from $\mathfrak{Z}^{st}(G)$ would mean that D acts by the same scalar on all $\pi \in \Pi_\lambda$ and those scalars vary algebraically as λ ranges over $\mathfrak{Y}$. So if for some λ the infinitesimal class

$$\Pi_\lambda = \coprod_{\varphi \rightsquigarrow \lambda} \Pi_\varphi$$

contains an L-packet $\Pi_{\varphi_0} \subsetneq \Pi_\lambda$ such that the set of supercuspidal supports coming from Π_{φ_0} does not meet the set of those coming from any $\Pi_{\varphi'}$ with φ' not conjugate to φ_0, then one could construct a regular function $D \in \mathfrak{Z}(G)$ which is constant on the L-packets Π_φ but not constant on Π_λ, and thus not in $f(\mathfrak{Z}^{st}(G))$. In that case (4.13) would not be sufficient to force $D \in f(\mathfrak{Z}^{st}(G))$, and the conjecture of Scholze–Shin would be false. In that case, one could define the subring $\mathfrak{Z}^{st*}(G) \subseteq \mathfrak{Z}(G)$ of regular functions on the Bernstein variety which take the same value on all supercuspidal supports of representations in the same L-packet. This would then perhaps better deserve the title "stable

Bernstein center" and it would be strictly larger than $f(3^{st}(G))$ at least in some cases.

To illustrate this in a more specific setting, suppose G/F is quasi-split and λ does not factor through any proper Levi subgroup of $^L G$. Then by Proposition 4.27 below, we expect Π_λ to consist entirely of supercuspidal representations. If Π_λ contains at least two L-packets Π_φ, **then** there would exist a $D \in 3(G)$ which is constant on the Π_φ's yet not constant on Π_λ, and the Scholze–Shin conjecture should be false. Put another way, if the Scholze–Shin conjecture is true, we expect that whenever λ does not factor through a proper Levi in $^L G$, the infinitesimal class Π_λ consists of at most one L-packet.[4]

5.6. Aside: when does an infinitesimal class consist only of supercuspidal representations?

Proposition 4.27. *Assume G/F is quasi-split and* LLC+ *holds for G. Then Π_λ consists entirely of supercuspidal representations if and only if λ does not factor through any proper Levi subgroup $^L M \subsetneq {}^L G$.*

Proof. If Π_λ contains a nonsupercuspidal representation π with supercuspidal support $(M, \sigma)_G$ for $M \subsetneq G$, then by LLC+, we may assume $\varphi_\pi|_{W_F}$, and hence λ, factors through the proper Levi subgroup $^L M \subsetneq {}^L G$.

Conversely, if λ factors minimally through a standard Levi subgroup $\mathcal{M}_\lambda \subsetneq {}^L G$, then we must show that Π_λ contains a nonsupercuspidal representation of G. Since G/F is quasi-split, we may identify $\mathcal{M}_\lambda = {}^L M_\lambda$ for an F-Levi subgroup $M_\lambda \subsetneq G$.

Now for $\mathfrak{t} = [\lambda]_{\widehat{G}}$, the map (4.11) is surjective. For any F-Levi subgroup $M \supsetneq M_\lambda$, a component of the form $\mathfrak{X}_{[M,\sigma]_G}$ has dimension $\dim (Z(\widehat{M})^I)^\circ_\Phi < \dim (Z(\widehat{M_\lambda})^I)^\circ_\Phi = \dim \mathfrak{Y}_\mathfrak{t}$. Thus the union of the components of the form $\mathfrak{X}_{[M,\sigma]_G}$ with $M \supsetneq M_\lambda$ cannot surject onto $\mathfrak{Y}_\mathfrak{t}$. Thus there must be a component of the form $\mathfrak{X}_{[M_\lambda,\sigma_\lambda]}$ appearing in the left-hand side of (4.11). We may assume φ_{σ_λ} factors through $^L M_\lambda$ along with λ. Writing $(\lambda)_{\widehat{M}_\lambda} = (z_\chi^{cocyc} \varphi_{\sigma_\lambda})_{\widehat{M}_\lambda}$ for some $\chi \in X(M_\lambda)$, it follows that Π_λ contains the nonsupercuspidal representations with supercuspidal support $(M_\lambda, \sigma_\lambda \chi)_G$. $\qquad\square$

5.7. Construction of the distributions Z_V

Let (r, V) be a finite-dimensional algebraic representation of $^L G$ on a complex vector space. Given a geometric Frobenius element $\Phi \in W_F$ and an admissible homomorphism $\lambda : W_F \to {}^L G$, we may define the *semi-simple trace*

---

[4] In fact this statement holds: if λ does not factor through a proper Levi subgroup of $^L G$, then there is at most one way to extend it to an admissible homomorphism $\varphi \colon W'_F \to {}^L G$.

$$\mathrm{tr}^{\mathrm{ss}}(\lambda(\Phi), V)) := \mathrm{tr}(r\lambda(\Phi), V^{r\lambda(I_F)}).$$

Note this is independent of the choice of Φ.

This notion was introduced by Rapoport [Ra90] in order to define semi-simple local L-functions $L(s, \pi_p, r)$, and is parallel to the notion for ℓ-adic Galois representations used in [Ra90] to define semi-simple local zeta functions $\zeta_{\mathfrak{p}}^{\mathrm{ss}}(X, s)$; see also [HN02a], [H05].

The following result is an easy consequence of the material in §5.3.

Proposition 4.28. *The map* $\lambda \mapsto \mathrm{tr}^{\mathrm{ss}}(\lambda(\Phi), V)$ *defines a regular function on the variety* $\mathfrak{Y}$ *hence defines an element* $Z_V \in \mathfrak{Z}^{\mathrm{st}}(G)$ *by*

$$Z_V((\lambda)_{\widehat{G}}) = \mathrm{tr}^{\mathrm{ss}}(\lambda(\Phi), V).$$

We use the same symbol Z_V *to denote the corresponding element in* $\mathfrak{Z}(G)$ *given via* $\mathfrak{Z}^{\mathrm{st}}(G) \to \mathfrak{Z}(G)$. *The latter has the property*

$$Z_V(\pi) = \mathrm{tr}^{\mathrm{ss}}(\varphi_\pi(\Phi), V) \qquad (4.14)$$

for every $\pi \in \Pi(G/F)$, *where* $Z_V(\pi)$ *stands for* $Z_V((M, \sigma)_G)$ *if* $(M, \sigma)_G$ *is the supercuspidal support of* π.

Remark 4.29. One does not really need the full geometric structure on the set $\mathfrak{Y}$ in order to construct $Z_V \in \mathfrak{Z}(G)$: one may show directly, assuming that LLC and Conjecture 4.7 hold, that $\pi \mapsto \mathrm{tr}^{\mathrm{ss}}(\varphi_\pi(\Phi), V)$ descends to give a regular function on $\mathfrak{X}$ and hence (4.14) defines an element $Z_V \in \mathfrak{Z}(G)$. Using the map f simply makes the construction more transparent (but has the drawback that we also need to assume Conjecture 4.12).

6. The Langlands–Kottwitz approach for arbitrary level structure

6.1. The test functions

Let $(\mathbf{G}, X)$ be a Shimura datum, where X is the $\mathbf{G}(\mathbb{R})$-conjugacy class of an $\mathbb{R}$-group homomorphism $h : \mathrm{R}_{\mathbb{C}/\mathbb{R}}\mathbb{G}_m \to \mathbf{G}_\mathbb{R}$. This gives rise to the reflex field $\mathbf{E} \subset \mathbb{C}$ and a $\mathbf{G}(\mathbb{C})$-conjugacy class $\{\mu\} \subset X_*(\mathbf{G}_\mathbb{C})$ which is defined over $\mathbf{E}$.

Choose a quasi-split group $\mathbf{G}^*$ over $\mathbb{Q}$ and an inner twisting $\psi : \mathbf{G}^* \to \mathbf{G}$ of $\mathbb{Q}$-groups. In particular we get an inner twisting $\mathbf{G}_\mathbf{E}^* \to \mathbf{G}_\mathbf{E}$ as well as an isomorphism of L-groups ${}^L(\mathbf{G}_\mathbf{E}) \overset{\sim}{\to} {}^L(\mathbf{G}_\mathbf{E}^*)$.

Let $\overline{\mathbb{Q}} \subset \mathbb{C}$ denote the algebraic numbers, so that we have an inclusion $\mathbf{E} \subset \overline{\mathbb{Q}}$ and we can regard $\{\mu\}$ as a $\mathbf{G}(\overline{\mathbb{Q}})$-conjugacy class in $X_*(\mathbf{G}_{\overline{\mathbb{Q}}})$ which is defined over $\mathbf{E}$ (cf. [Ko84b, Lemma 1.1.3]). Using ψ regard $\{\mu\}$ as a $\mathbf{G}^*(\overline{\mathbb{Q}})$-conjugacy class in $X_*(\mathbf{G}^*_{\overline{\mathbb{Q}}})$, defined over $\mathbf{E}$. By Kottwitz' lemma ([Ko84b, 1.1.3]), $\{\mu\}$ is represented by an $\mathbf{E}$-rational cocharacter $\mu : \mathbb{G}_m \to \mathbf{G}^*_{\mathbf{E}}$. Following Kottwitz' argument in [Ko84b, 2.1.2], it is easy to show that there exists a unique representation $(\mathbf{r}_{-\mu}, V_{-\mu})$ of $^L(\mathbf{G}_{\mathbf{E}})$ such that as a representation of $\widehat{\mathbf{G}}^*$ it is an irreducible representation with extreme weight $-\mu$ and the Weil group $W_{\mathbf{E}}$ acts trivially on the highest-weight space corresponding to any $\Gamma_{\mathbf{E}}$-fixed splitting for $\widehat{\mathbf{G}}^*_{\mathbf{E}}$.

Using ψ we can regard $(\mathbf{r}_{-\mu}, V_{-\mu})$ as a representation of $^L(\mathbf{G}_{\mathbf{E}})$. The isomorphism class of this representation depends only on the equivalence class of the inner twisting ψ, thus only on $\mathbf{G}$ and $\{\mu\}$.

Now we fix a rational prime p and set $G := \mathbf{G}_{\mathbb{Q}_p}$. Choose a prime ideal $\mathfrak{p} \subset \mathbf{E}$ lying above p, and set $E := \mathbf{E}_{\mathfrak{p}}$. Choose an algebraic closure $\overline{\mathbb{Q}}_p$ of $\mathbb{Q}_p$ and fix henceforth an isomorphism of fields $\mathbb{C} \cong \overline{\mathbb{Q}}_p$ such that the embedding $\mathbf{E} \hookrightarrow \mathbb{C} \cong \overline{\mathbb{Q}}_p$ corresponds to the prime ideal $\mathfrak{p}$. This gives rise to an embedding $\overline{\mathbb{Q}} \hookrightarrow \overline{\mathbb{Q}}_p$ extending $\mathbf{E} \hookrightarrow \overline{\mathbb{Q}}_p$, and thus to an embedding $W_E \hookrightarrow W_{\mathbf{E}}$. We get from this an embedding $^L(G_E) \hookrightarrow {}^L(\mathbf{G}_{\mathbf{E}})$. Via this embedding we can regard $(\mathbf{r}_{-\mu}, V_{-\mu})$ as a representation $(r_{-\mu}, V_{-\mu})$ of $^L(G_E)$.

Associated to $(r_{-\mu}, V_{-\mu}) \in \text{Rep}(^L(G_E))$ we have an element $Z_{V_{-\mu}}$ in the Bernstein center $\mathfrak{Z}(G_E)$. Of course here and in what follows, we are assuming LLC+ holds for G_E.

Now we review briefly the Langlands–Kottwitz approach to studying the local Hasse–Weil zeta functions of Shimura varieties. Let $Sh_{K_p} = Sh(\mathbf{G}, h^{-1}, K^p K_p)$ denote the canonical model[5] over $\mathbf{E}$ for the Shimura variety attached to the data $(\mathbf{G}, h^{-1}, K^p K_p)$ for some sufficiently small compact open subgroup $K^p \subset \mathbf{G}(\mathbb{A}_f^p)$ and some compact open subgroup $K_p \subset \mathbf{G}(\mathbb{Q}_p)$. We limit ourselves to constant coefficients $\overline{\mathbb{Q}}_\ell$ in the generic fiber of Sh_{K_p} (here $\ell \neq p$ is a fixed rational prime). Let Φ_p denote any geometric Frobenius element in $\text{Gal}(\overline{\mathbb{Q}}_p/\mathbb{Q}_p)$. Then in the Langlands–Kottwitz approach to the semi-simple local zeta function $\zeta_{\mathfrak{p}}^{ss}(s, Sh_{K_p})$, one needs to prove an identity of the form

$$\text{tr}^{ss}(\Phi_p^r, H_c^\bullet(Sh_{K_p} \otimes_{\mathbf{E}} \overline{\mathbb{Q}}_p, \overline{\mathbb{Q}}_\ell)) = \sum_{(\gamma_0; \gamma, \delta)} c(\gamma_0; \gamma, \delta)\, O_\gamma(1_{K^p})\, TO_{\delta\theta}(\phi_r).$$

$$(4.15)$$

Here the semi-simple Lefschetz number $\text{Lef}^{ss}(\Phi_p^r, Sh_{K_p})$ on the left-hand side is the alternating semi-simple trace of Frobenius on the compactly-supported

[5] We use this term in the same sense as Kottwitz [Ko92a], comp. Milne [Mil, §1, esp. 1.10].

ℓ-adic cohomology groups[6] of Sh_{K_p} (see [Ra90] and [HN02a] for the notion of semi-simple trace). The expression on the right has precisely the same form as the counting points formula proved by Kottwitz in certain good reduction cases (PEL type A or C, K_p hyperspecial; cf. [Ko92a, (19.6)]). The integer $r \geq 1$ ranges over integers of the form $j \cdot [k_{E_0} : \mathbb{F}_p]$, $j \geq 1$, where $E_0/\mathbb{Q}_p$ is the maximal unramified subextension of $E/\mathbb{Q}_p$ and k_{E_0} is its residue field. Thus $\Phi_p^r = \Phi_{\mathfrak{p}}^j$ where $\Phi_{\mathfrak{p}}$ is a geometric Frobenius element in $\mathrm{Gal}(\overline{\mathbb{Q}}_p/E)$. Finally, ϕ_r is an element in the Hecke algebra $\mathcal{H}(G(\mathbb{Q}_{p^r}), K_{p^r})$ with values in $\overline{\mathbb{Q}}_\ell$, where $\mathbb{Q}_{p^r}$ is the unique degree r unramified extension in $\overline{\mathbb{Q}}_p/\mathbb{Q}_p$, and where $K_{p^r} \subset G(\mathbb{Q}_{p^r})$ is a suitable compact open subgroup which is assumed to be a natural analogue of $K_p \subset G(\mathbb{Q}_p)$. To be more precise about K_{p^r}, in practice there is a smooth connected $\mathbb{Z}_p$-model $\mathcal{G}$ for G, such that $K_p = \mathcal{G}(\mathbb{Z}_p)$. In that case, we always take $K_{p^r} = \mathcal{G}(\mathbb{Z}_{p^r})$, where $\mathbb{Z}_{p^r}$ is the ring of integers in $\mathbb{Q}_{p^r}$. In forming $\mathrm{TO}_{\delta\sigma}(\phi_r)$, the Haar measure on $G(\mathbb{Q}_{p^r})$ is normalized to give K_{p^r} measure 1.

Let E_j/E be the unique unramified extension of degree j in $\overline{\mathbb{Q}}_p/E$. Let $E_{j0}/\mathbb{Q}_p$ be the maximal unramified subextension of $E_j/\mathbb{Q}_p$. So E/E_0 and E_j/E_{j0} are totally ramified of the same degree, and $E_{j0} = \mathbb{Q}_{p^r}$.

We make the choice of $\sqrt{p} \in \overline{\mathbb{Q}}_\ell$, and use it to define $\delta_p^{1/2}$ as a function with values in $\mathbb{Q}(\sqrt{p}) \subset \overline{\mathbb{Q}}_\ell$. We can now specify the test function $\phi_r \in \mathcal{Z}(G(E_{j0}), K_{j0})$, which will take values in $\overline{\mathbb{Q}}_\ell$.

In the construction of the elements $Z_V \in \mathfrak{Z}^{\mathrm{st}}(G)$, everything works the same way for (r, V) a representation of $^L G := \widehat{G}(\overline{\mathbb{Q}}_\ell) \rtimes W_F$ on a $\overline{\mathbb{Q}}_\ell$-vector space. We henceforth take this point of view. Let $(r_{-\mu,j}, V_{-\mu,j}) \in \mathrm{Rep}(^L(G_{E_j}))$ denote the restriction of $(r_{-\mu}, V_{-\mu}) \in \mathrm{Rep}(^L(G_E))$ via $^L(G_{E_j}) \hookrightarrow {}^L(G_E)$. We can then induce to get a representation $(r_{-\mu,j}^{E_{j0}}, V_{-\mu,j}^{E_{j0}})$ of $^L(G_{E_{j0}})$. By Mackey theory, we get the same representation if we first induce to $^L(G_{E_0})$ and then restrict to $^L(G_{E_{j0}})$, that is, we have

$$(r_{-\mu,j}^{E_{j0}}, V_{-\mu,j}^{E_{j0}}) := \mathrm{Ind}_{\widehat{G} \rtimes W_{E_j}}^{\widehat{G} \rtimes W_{E_{j0}}} \mathrm{Res}_{\widehat{G} \rtimes W_{E_j}}^{\widehat{G} \rtimes W_E} r_{-\mu} = \mathrm{Res}_{\widehat{G} \rtimes W_{E_{j0}}}^{\widehat{G} \rtimes W_{E_0}} \mathrm{Ind}_{\widehat{G} \rtimes W_E}^{\widehat{G} \rtimes W_{E_0}} r_{-\mu}.$$

$$(4.16)$$

This gives rise to $Z_{V_{-\mu,j}^{E_{j0}}} \in \mathfrak{Z}^{\mathrm{st}}(G_{E_{j0}})$. By abuse of notation, we use the same symbol to denote the image of this in the Bernstein center: $Z_{V_{-\mu,j}^{E_{j0}}} \in \mathfrak{Z}(G_{E_{j0}})$.

---

[6] The Langlands–Kottwitz method really applies to the middle intersection cohomology groups of the Baily–Borel compactification and not just to the cohomology groups with compact supports; see [Ko90] and [Mor] for some general conjectures and results in this context, at primes of good reduction. The identity (4.15) corresponds to the contribution of the interior, at primes of arbitrary reduction, and is a first step toward understanding the intersection cohomology groups.

Of course here we are viewing $3(G_{E_{j0}})$ as $\overline{\mathbb{Q}}_\ell$-valued regular functions on the Bernstein variety, or equivalently as $\overline{\mathbb{Q}}_\ell$-valued invariant essentially compact distributions: the topology on $\mathbb{C}$ playing no role, it is harmless to identify it with $\overline{\mathbb{Q}}_\ell$.

The following is the conjecture formulated jointly with R. Kottwitz.

Conjecture 4.30. (Test function conjecture) *Let* $d = \dim(Sh_{K_p})$. *The test function* ϕ_r *in (4.15) may be taken to be* $p^{rd/2}\big(Z_{V_{-\mu,j}^{E_{j0}}} * 1_{K_{p^r}}\big)$. *In particular,* ϕ_r *may be taken in the center* $\mathcal{Z}(G(\mathbb{Q}_{p^r}), K_{p^r})$ *of* $\mathcal{H}(G(\mathbb{Q}_{p^r}), K_{p^r})$ *and these test functions vary compatibly with change in the level* K_p *in an obvious sense.*

The same test functions should be used when one incorporates arbitrary Hecke operators away from p into (4.15).

Following Rapoport's strategy (cf. [Ra90], [Ra05], [H05]), one seeks to find a natural integral model $\mathcal{M}_{K_p}$ over $\mathcal{O}_E$ for Sh_{K_p}, and then rephrase the above conjecture using the method of nearby cycles $R\Psi := R\Psi^{\mathcal{M}_{K_p}}(\overline{\mathbb{Q}}_\ell)$.

Conjecture 4.31. *There exists a natural integral model* $\mathcal{M}_{K_p}/\mathcal{O}_E$ *for* Sh_{K_p}, *such that*

$$\sum_{x \in \mathcal{M}_{K_p}(k_{E_{j0}})} \mathrm{tr}^{ss}(\Phi_p^r, R\Psi_x) = \sum_{(\gamma_0;\gamma,\delta)} c(\gamma_0; \gamma, \delta) \, \mathrm{O}(1_{K^p}) \, \mathrm{TO}_{\delta\theta}(\phi_r), \quad (4.17)$$

where $\phi_r = p^{rd/2}\big(Z_{V_{-\mu,j}^{E_{j0}}} * 1_{K_{p^r}}\big)$ *as in Conjecture 4.30.*

Remark 4.32. Implicit in this conjecture is that the method of nearby cycles can be used for compactly-supported cohomology. In fact we could conjecture there exists a suitably nice compactification of $\mathcal{M}_{K_p}/\mathcal{O}_E$ so that the natural map

$$\mathrm{H}_c^i(\mathcal{M}_{K_p} \otimes_{\mathcal{O}_E} \overline{\mathbb{F}}_p, \, R\Psi(\overline{\mathbb{Q}}_\ell)) \to \mathrm{H}_c^i(Sh_{K_p} \otimes_E \overline{\mathbb{Q}}_p, \, \overline{\mathbb{Q}}_\ell)$$

is a Galois-equivariant isomorphism. For $G = \mathrm{GSp}_{2g}$ and where $\mathcal{M}_{K_p}$ is the natural integral model for Sh_{K_p} for K_p an Iwahori subgroup, this was proved by Benoit Stroh. Of course, one is really interested in intersection cohomology groups of the Baily–Borel compactification (see footnote 5), and in fact Stroh [Str] computed the nearby cycles and verified the analogue of the Kottwitz conjecture on nearby cycles (see Conjecture 4.44 below) for these compactifications.

Remark 4.33. Some unconditional versions of Conjectures 4.30 and 4.31 have been proved. See §8.

6.2. Endoscopic transfer of the stable Bernstein center

Part of the Langlands–Kottwitz approach is to perform a "pseudostabilization" of (4.15), and in particular prove the "fundamental lemmas" that are required for this. The *stabilization* expresses (4.15) in the form $\sum_{\mathbf{H}} i(\mathbf{G}, \mathbf{H}) \, ST_e^*(\mathbf{h})$, the sum over global $\mathbb{Q}$-elliptic endoscopic groups $\mathbf{H}$ for $\mathbf{G}$ of the $(\mathbf{G}, \mathbf{H})$-regular $\mathbb{Q}$-elliptic part of the geometric side of the stable trace formula for $(\mathbf{H}, \mathbf{h})$ (cf. notation of [Ko90]), for a certain *transfer* function $\mathbf{h} \in C_c^\infty(\mathbf{H}(\mathbb{A}))$. (By contrast in "pseudostabilization" which is used in certain situations, one instead writes (4.15) in terms of the trace formula for $\mathbf{G}$ and not its quasi-split inner form, and this is sometimes enough, as in e.g. Theorem 4.40 below.) For stabilization one needs to produce elements $\mathbf{h}_p \in C_c^\infty(\mathbf{H}(\mathbb{Q}_p))$ which are Frobenius-twisted endoscopic transfers of ϕ_r. The existence of such transfers $\mathbf{h}_p$ is due mainly to the work of Ngô [Ngo] and Waldspurger [Wal97, Wal04, Wal08]. But we hope to have a priori spectral information about the transferred functions $\mathbf{h}_p$.

A guiding principle is that the nearby cycles on an appropriate "local model" for Sh_{K_p} should *naturally* produce a central element as a test function ϕ_r, which should coincide with that given by the test function conjecture (cf. Conjecture 4.31); then its spectral behavior is known by construction. In that case one can formulate a conjectural endoscopic transfer h_p of ϕ_r with known spectral behavior.

General Frobenius-twisted endoscopic transfer homomorphisms $\mathfrak{Z}^{st}(G_{\mathbb{Q}_{p^r}})$ $\rightarrow \mathfrak{Z}^{st}(H_{\mathbb{Q}_p})$ will be described elsewhere. Here for simplicity we content ourselves to describe two special cases: standard (untwisted) endoscopic transfer of the geometric Bernstein center, and the base change transfer for the stable Bernstein center.

6.2.1. Endoscopic transfer of the geometric Bernstein center

Let us fix an endoscopic triple (H, s, η_0) for G over a p-adic field F (cf. [Ko84a, §7]), and suppose we have fixed an extension $\eta : {}^L H \rightarrow {}^L G$ of $\eta_0 : \widehat{H} \hookrightarrow \widehat{G}$ (we suppose we are in a situation, e.g. $G_{\mathrm{der}} = G_{\mathrm{sc}}$, where such extensions always exist). We could hope the natural map

$$\mathfrak{Y}^{H/F} \longrightarrow \mathfrak{Y}^{G/F}$$
$$(\lambda)_{\widehat{H}} \longmapsto (\eta \circ \lambda)_{\widehat{G}}$$

would be algebraic and hence would induce an *endoscopic transfer homomorphism* $\mathfrak{Z}^{st}(G) \rightarrow \mathfrak{Z}^{st}(H)$. By invoking further expectations about endoscopic lifting, one would then formulate a map on the level of Bernstein centers, $\mathfrak{Z}(G) \rightarrow \mathfrak{Z}(H)$, which we could write as $Z \mapsto Z|_\eta$. But these assertions are not obvious. Fortunately, in practice we need this construction rather on the *geometric Bernstein center*.

Definition 4.34. Assume LLC+ holds for G/F. We define the **geometric Bernstein center** $\mathfrak{Z}^{\mathrm{geom}}(G)$ to be the subalgebra of $\mathfrak{Z}^{\mathrm{st}}(G)$ generated by the elements Z_V as V ranges over $\mathrm{Rep}(^L G)$.

The terminology *geometric Bernstein center* is motivated by § 6.4 below. Let $V|_\eta \in \mathrm{Rep}(^L H)$ denote the restriction of $V \in \mathrm{Rep}(^L G)$ along η. Further assume LLC+ also holds for H/F. Then $Z_V \mapsto Z_{V|_\eta}$ determines a map $\mathfrak{Z}^{\mathrm{geom}}(G) \to \mathfrak{Z}^{\mathrm{geom}}(H)$. Write Z_V^G (resp. $Z_{V|_\eta}^H$) for the image of Z_V (resp. $Z_{V|_\eta}$) in $\mathfrak{Z}(G)$ (resp. $\mathfrak{Z}(H)$).

Conjecture 4.35. *Assume* LLC+ *holds for both G and H. Then in the above situation the distribution $Z_{V|_\eta}^H \in \mathfrak{Z}(H)$ is the endoscopic transfer of $Z_V^G \in \mathfrak{Z}(G)$ in the following sense: whenever a function $\phi^H \in C_c^\infty(H(F))$ is a transfer of a function $\phi \in C_c^\infty(G(F))$, then $Z_{V|_\eta}^H * \phi^H$ is a transfer of $Z_V^G * \phi$.*

This conjecture and its Frobenius-twisted analogue were announced by the author in April 2011 at Princeton [H11]. A very similar statement subsequently appeared as Conjecture 7.2 in [SS]. Considering the untwisted case for simplicity, the difference is that in [SS], the authors take in place of Z_V an element in the stable Bernstein center essentially of the form

$$(\lambda)_{\widehat{G}} \mapsto \mathrm{tr}(\lambda(\Phi_F), V_{-\mu}),$$

where here the usual trace, not the semi-simple trace, is used. That conjecture is proved in [SS] in all EL or quasi-EL cases, by invoking special features of general linear groups such as the existence of base change representations.

Formally, Conjecture 4.35 contains as a special case the "fundamental lemma implies spherical transfer" result of Hales [Hal] (see also Waldspurger [Wal97]). Indeed if K, K_H are hyperspecial maximal compact subgroups in $G(F)$, $H(F)$, then 1_{K_H} is a transfer of 1_K by the fundamental lemma, and hence $Z_{V|_\eta}^H * 1_{K_H}$ is a transfer of $Z_V^G * 1_K$. But by the Satake isomorphism, every K-spherical function on $G(F)$ is a linear combination of functions of the form $Z_V^G * 1_K$ for some representation V (comp. §6.4).

Even in more general situations, Conjecture 4.35 is most useful when applied to a pair ϕ, ϕ^H of unit elements in appropriate Hecke algebras. At least when G splits over F^{un}, Kazhdan–Varshavsky proved in [KV] that for some explicit scalar c, the Iwahori unit $c 1_{I_H}$ is a transfer of the Iwahori unit 1_I. As another example, if $K_n^G \subset G(F)$ is the n-th principal congruence subgroup in $G(F)$, then for some explicit scalar c the function $c 1_{K_n^H}$ is a transfer of $1_{K_n^G}$ (proved by Ferrari [Fer] under some mild restrictions on the residue characteristic of F), and thus $c(Z_{V|_\eta}^H * 1_{K_n^H})$ should be an explicit transfer of $Z_V^G * 1_{K_n^G}$. A Frobenius-twisted analogue of Ferrari's theorem together with the Frobenius-twisted analogue of Conjecture 4.35 would give an explicit Frobenius-twisted transfer of the test function ϕ_r from Conjecture 4.30, if K_p is a principal congruence subgroup.

6.2.2. Base change of the stable Bernstein center

We return to the situation of Proposition 4.21, but we specialize it to cyclic Galois extensions of F and furthermore we assume G/F is quasi-split. Let E/F be any finite cyclic Galois subextension of $\overline{F}/F$ with Galois group $\langle \theta \rangle$, and with corresponding inclusion of Weil groups $W_E \hookrightarrow W_F$.

If $\phi \in \mathcal{H}(G(E))$ and $f \in \mathcal{H}(G(F))$ are functions in the corresponding Hecke algebras of locally constant compactly-supported functions, then we say ϕ, f are *associated* (or f *is a base-change transfer of* ϕ), if the following result holds for the stable (twisted) orbital integrals: for every semisimple element $\gamma \in G(F)$, we have

$$\mathrm{SO}_\gamma(f) = \sum_\delta \Delta(\gamma, \delta) \, \mathrm{SO}_{\delta\theta}(\phi) \qquad (4.18)$$

where the sum is over stable θ-conjugacy classes $\delta \in G(E)$ with semisimple norm $\mathcal{N}\delta$, and where $\Delta(\gamma, \delta) = 1$ if $\mathcal{N}\delta = \gamma$ and $\Delta(\gamma, \delta) = 0$ otherwise. See e.g. [Ko86], [Ko88], [Cl90], or [H09] for further discussion.

Conjecture 4.36. *In the above situation, consider $Z \in \mathfrak{Z}^{\mathrm{st}}(G_E)$, and consider its image, also denoted by Z, in $\mathfrak{Z}(G_E)$. Consider $b_{E/F}(Z) \in \mathfrak{Z}^{\mathrm{st}}(G)$ (cf. Definition 4.22) and also denote by $b_{E/F}(Z)$ its image in $\mathfrak{Z}(G)$. Then $b_{E/F}(Z)$ is the base-change transfer of $Z \in \mathfrak{Z}(G_E)$, in the following sense: whenever a function $f \in C_c^\infty(G(F))$ is a base-change transfer of $\phi \in C_c^\infty(G(E))$, then $b_{E/F}(Z) * f$ is a base-change transfer of $Z * \phi$.*

Proposition 4.37. *Conjecture 4.36 holds for* GL_n.

Proof. The most efficient proof follows Scholze's proof of Theorem C in [Sch2] which makes essential use of the existence of cyclic base change lifts for GL_n. Let $\pi \in \Pi(\mathrm{GL}_n/F)$ be a tempered irreducible representation with base change lift $\Pi \in \Pi(\mathrm{GL}_n/E)$, a tempered representation which is characterized by the character identity $\Theta_\Pi((g, \theta)) = \Theta_\pi(Ng)$ for all elements $g \in \mathrm{GL}_n(E)$ with regular semisimple norm Ng ([AC, Thm. 6.2, p. 51]). Here $(g, \theta) \in \mathrm{GL}_n(E) \rtimes \mathrm{Gal}(E/F)$ and θ acts on Π by the normalized intertwiner $I_\theta : \Pi \to \Pi$ of [AC, p. 11].

Suppose f is a base-change transfer of ϕ. Using the Weyl character formula and its twisted analogue (cf. [AC, p. 36]), we see that

$$\mathrm{tr}((\phi, \theta) \mid \Pi) = \mathrm{tr}(f \mid \pi).$$

Multiplying by the constant $Z(\Pi) = b_{E/F}(Z)(\pi)$, we get

$$\mathrm{tr}((Z * \phi, \theta) \mid \Pi) = \mathrm{tr}(b_{E/F}(Z) * f \mid \pi).$$

(Use Corollary 4.3 and its twisted analogue.) There exists a base-change trans-fer $h \in C_c^\infty(\mathrm{GL}_n(F))$ of $Z * \phi$ ([AC, Prop. 3.1]). Using the same argument as above for the pair $Z * \phi$ and h, we conclude that $\mathrm{tr}(b_{E/F}(Z) * f - h \mid \pi) = 0$ for every tempered irreducible $\pi \in \Pi(\mathrm{GL}_n/F)$. By Kazhdan's density theorem (Theorem 1 in [Kaz]) the regular semi-simple orbital integrals of $b_{E/F}(Z) * f$ and h agree. Thus the (twisted) orbitals integrals of $b_{E/F}(Z) * f$ and ϕ match at all regular semi-simple elements, and hence at all semi-simple elements by Clozel's Shalika germ argument [Cl90, Prop. 7.2]. $\qquad\square$

Remark 4.38. Unconditional versions of Conjecture 4.36 are available for parahoric and pro-p Iwahori–Hecke algebras, when G/F is unramified.[7] See §9.

6.3. Application: local Hasse–Weil zeta functions

By Kottwitz' base change fundamental lemma for units [Ko86], we know 1_{K_p} is a base-change transfer of $1_{K_{p^r}}$ whenever $K_p = \mathcal{G}(\mathbb{Z}_p)$ and $K_{p^r} = \mathcal{G}(\mathbb{Z}_{p^r})$ for a smooth connected $\mathbb{Z}_p$-model $\mathcal{G}$ for G. Then Conjectures 4.30 and 4.36 together say that

$$f_p^{(j)} := p^{rd/2} \, b_{E_{j0}/\mathbb{Q}_p}(Z_{V_{-\mu,j}^{E_{j0}}}) * 1_{K_p} \qquad (4.19)$$

is a base-change transfer of a test function ϕ_r that satisfies (4.15).

Setting

$$(r_{-\mu}^{E_0}, V_{-\mu}^{E_0}) := \mathrm{Ind}_{\widehat{G} \rtimes W_E}^{\widehat{G} \rtimes W_{E_0}} r_{-\mu} \qquad (4.20)$$

we have for any admissible parameter $\varphi : W_{\mathbb{Q}_p}' \to {}^L(G_{\mathbb{Q}_p})$ and any $\pi_p \in \Pi_\varphi(G/\mathbb{Q}_p)$ the identity

$$\mathrm{tr}(f_p^{(j)} \mid \pi_p) = p^{rd/2} \dim(\pi_p^{K_p}) \, \mathrm{tr}^{ss}(\varphi(\Phi_p^r), V_{-\mu}^{E_0}), \qquad (4.21)$$

---

[7] The pro-p Sylow subgroup of an Iwahori subgroup $I \subset G(F)$ coincides with its pro-unipotent radical I^+, and it has become conventional to term the Hecke algebra $C_c^\infty(I^+ \backslash G(F)/I^+)$ the *pro-p Iwahori–Hecke algebra*.

where $r = j[E_0 : \mathbb{Q}_p]$. In the compact and non-endoscopic cases, the above discussion allows us to express $\zeta_{\mathfrak{p}}^{ss}(s, Sh_{K_p})$ in terms of semi-simple automorphic L-functions. To explain this we need a detour on the point of view taken in [L1, L2] (comp. [Ko84b, §2.2]).

Recall $(r_{-\mu}, V_{-\mu}) \in \text{Rep}(^L(G_E))$. Consider the Langlands representation

$$\mathbf{r} := \text{Ind}_{\widehat{G} \rtimes W_{\mathbf{E}}}^{\widehat{G} \rtimes W_{\mathbb{Q}}} r_{-\mu},$$

and for each prime $\mathfrak{p}$ of $\mathbf{E}$ dividing p, consider

$$\mathbf{r}_{\mathfrak{p}} := \text{Ind}_{\widehat{G} \rtimes W_{\mathbf{E}_{\mathfrak{p}}}}^{\widehat{G} \rtimes W_{\mathbb{Q}_p}} \text{Res}_{\widehat{G} \rtimes W_{\mathbf{E}_{\mathfrak{p}}}}^{\widehat{G} \rtimes W_{\mathbf{E}}} r_{-\mu} = \text{Ind}_{\widehat{G} \rtimes W_{\mathbf{E}_{\mathfrak{p}}}}^{\widehat{G} \rtimes W_{\mathbb{Q}_p}} r_{-\mu}.$$

Mackey theory gives

$$\text{Res}_{\widehat{G} \rtimes W_{\mathbb{Q}_p}}^{\widehat{G} \rtimes W_{\mathbb{Q}}} \mathbf{r} = \bigoplus_{\mathfrak{p}|p} \mathbf{r}_{\mathfrak{p}}.$$

If $\mathfrak{p}$ is understood, let $\mathbf{E}_{\mathfrak{p}0}/\mathbb{Q}_p$ denote the maximal unramified subextension of $\mathbf{E}_{\mathfrak{p}}/\mathbb{Q}_p$, and set $E = \mathbf{E}_{\mathfrak{p}}$ and $E_0 := \mathbf{E}_{\mathfrak{p}0}$. Then we have

$$\mathbf{r}_{\mathfrak{p}} = \text{Ind}_{\widehat{G} \rtimes W_E}^{\widehat{G} \rtimes W_{\mathbb{Q}_p}} r_{-\mu} = \text{Ind}_{\widehat{G} \rtimes W_{E_0}}^{\widehat{G} \rtimes W_{\mathbb{Q}_p}} r_{-\mu}^{E_0}. \tag{4.22}$$

Lemma 4.39. *Suppose $\pi_p \in \Pi_{\varphi}(G/\mathbb{Q}_p)$. Then*

$$[E_0 : \mathbb{Q}_p]^{-1}\text{tr}^{ss}(\varphi(\Phi_p^r), \mathbf{r}_{\mathfrak{p}}) = \begin{cases} \text{tr}^{ss}(\varphi(\Phi_{\mathfrak{p}}^j), r_{-\mu}^{E_0}) & \text{if } r = j[E_0 : \mathbb{Q}_p] \\ 0, & \text{if } [E_0 : \mathbb{Q}_p] \nmid r. \end{cases}$$
$$\tag{4.23}$$

Proof. There is an isomorphism of $\widehat{G} \rtimes W_{\mathbb{Q}_p}$-modules

$$\mathbf{r}_{\mathfrak{p}} \cong \mathbb{C}[\widehat{G} \rtimes W_{\mathbb{Q}_p}] \otimes_{\mathbb{C}[\widehat{G} \rtimes W_{E_0}]} r_{-\mu}^{E_0},$$

and $\mathbf{r}_{\mathfrak{p}}^{\varphi(I_{\mathbb{Q}_p})}$ has a $\mathbb{C}$-basis of the form $\{\varphi(\Phi_p^i) \otimes w_k\}$ where $0 \leq i \leq [E_0 : \mathbb{Q}_p]-1$ and $\{w_k\}$ comprises a $\mathbb{C}$-basis for $(r_{-\mu}^{E_0})^{\varphi(I_{\mathbb{Q}_p})}$. The lemma follows. $\square$

The following result shows the potential utility of Conjectures 4.30 and 4.36. It applies not just to PEL Shimura varieties, but to any Shimura variety where these conjectures are known. Similar results will hold when incorporating Hecke operators away from p.

Theorem 4.40. *Suppose $\mathbf{G}_{\text{der}}$ is anisotropic over $\mathbb{Q}$, so that the associated Shimura variety $Sh_{K_p} = Sh(\mathbf{G}, h^{-1}, K^p K_p)$ is proper over $\mathbf{E}$. Suppose $\mathbf{G}$ has "no endoscopy", in the sense that the group $\mathfrak{K}(\mathbf{G}_{\gamma_0}/\mathbb{Q})$ is trivial for every semisimple element $\gamma_0 \in \mathbf{G}(\mathbb{Q})$, as in e.g. [Ko92b]. Let $\mathfrak{p}$ be a prime ideal of $\mathbf{E}$*

dividing p. Assume (LLC+) *(cf. §5.2), and Conjectures 4.30 and 4.36 hold for all groups* $\mathbf{G}_{\mathbb{Q}_{p^r}}$.

Then in the notation above, we have

$$\zeta_p^{ss}(s, Sh_{K_p}) = \prod_{\pi_f} L^{ss}\left(s - \frac{d}{2}, \pi_p, \mathbf{r}_p\right)^{a(\pi_f)\dim(\pi_f^K)}, \qquad (4.24)$$

where $\pi_f = \pi^p \otimes \pi_p$ *runs over irreducible admissible representations of* $\mathbf{G}(\mathbb{A}_f)$ *and the integer* $a(\pi_f)$ *is given by*

$$a(\pi_f) = \sum_{\pi_\infty \in \Pi_\infty} m(\pi_f \otimes \pi_\infty)\, \mathrm{tr}(f_\infty|\pi_\infty),$$

where $m(\pi_f \otimes \pi_\infty)$ *is the multiplicity of* $\pi_f \otimes \pi_\infty$ *in* $L^2(\mathbf{G}(\mathbb{Q})A_\mathbf{G}(\mathbb{R})^\circ\backslash\mathbf{G}(\mathbb{A}))$. *Here* $A_\mathbf{G}$ *is the* $\mathbb{Q}$-split *component of the center of* $\mathbf{G}$ *(which we assume is also its* $\mathbb{R}$-split *component). Further* Π_∞ *is the set of irreducible admissible representations of* $\mathbf{G}(\mathbb{R})$ *which have trivial infinitesimal and central characters, and* f_∞ *is defined as in* [Ko92b] *to be* $(-1)^{\dim(Sh_K)}$ *times a pseudo-coefficent of an essentially discrete series member* $\pi_\infty^0 \in \Pi_\infty$.

Proof. The method follows closely the argument of Kottwitz in [Ko92b] (comp. [HRa1, §13.4]), so we just give an outline. We will use freely the notation of Kottwitz and [HRa1]. Set $f = [\mathbf{E}_{p0} : \mathbb{Q}_p]$. By definition we have

$$\log \zeta_p^{ss}(s, Sh_{K_p}) = \sum_{j=1}^\infty \mathrm{Lef}^{ss}(\Phi_p^j, Sh_{K_p}) \frac{p^{-jfs}}{j}. \qquad (4.25)$$

By using (4.15) together with Conjectures 4.30 and 4.36, the arguments of Kottwitz [Ko92b] show that for each $j \geq 1$

$$\mathrm{Lef}^{ss}(\Phi_p^j, Sh_{K_p}) = \tau(\mathbf{G}) \sum_{\gamma_0} \mathrm{SO}_{\gamma_0}(f^p\, f_p^{(j)}\, f_\infty), \qquad (4.26)$$

where $f_p^{(j)}$ is defined as in (4.19) and f^p is the characteristic function of $K^p \subset \mathbf{G}(\mathbb{A}_f^p)$. Here γ_0 ranges over all stable conjugacy classes in $\mathbf{G}(\mathbb{Q})$.

Since $\mathbf{G}_{\mathrm{der}}$ is anisotropic over $\mathbb{Q}$, the trace formula for any $f \in C_c^\infty(A_\mathbf{G}(\mathbb{R})^\circ\backslash\mathbf{G}(\mathbb{A}))$ takes the simple form

$$\sum_\gamma \tau(\mathbf{G}_\gamma)O_\gamma(f) = \sum_\pi m(\pi)\,\mathrm{tr}(f|\pi), \qquad (4.27)$$

where γ ranges over conjugacy classes in $\mathbf{G}(\mathbb{Q})$ and π ranges over irreducible representations in $L^2(\mathbf{G}(\mathbb{Q})A_\mathbf{G}(\mathbb{R})^\circ\backslash\mathbf{G}(\mathbb{A}))$. By [Ko92b, Lemma 4.1], the vanishing of all $\mathfrak{K}(\mathbf{G}_{\gamma_0}/\mathbb{Q})$ means that

$$\sum_\gamma \tau(\mathbf{G}_\gamma)\, \mathrm{O}_\gamma(f) = \tau(\mathbf{G}) \sum_{\gamma_0} \mathrm{SO}_{\gamma_0}(f).$$

It follows that

$$
\begin{aligned}
\mathrm{Lef}^{ss}(\Phi_\mathfrak{p}^j, Sh_{K_p}) \\
&= \sum_\pi m(\pi)\, \mathrm{tr}(f^p f_p^{(j)} f_\infty | \pi) \\
&= \sum_{\pi_f} \sum_{\pi_\infty \in \Pi_\infty} m(\pi_f \otimes \pi_\infty) \cdot \mathrm{tr}(f^p | \pi_f^p) \cdot \mathrm{tr}(f_p^{(j)} | \pi_p) \cdot \mathrm{tr}(f_\infty | \pi_\infty) \\
&= \sum_{\pi_f} a(\pi_f)\, \dim(\pi_f^K)\, p^{jfd/2}\, \mathrm{tr}^{ss}(\varphi_{\pi_p}(\Phi_\mathfrak{p}^j), V_{-\mu}^{E_0}),
\end{aligned}
$$

the last equality by (4.21).

By definition we have

$$\log L^{ss}(s, \pi_p, \mathbf{r}_\mathfrak{p}) = \sum_{r=1}^\infty \mathrm{tr}^{ss}(\varphi_{\pi_p}(\Phi_p^r), \mathbf{r}_\mathfrak{p}) \frac{p^{-rs}}{r}.$$

Now (4.24) follows by invoking (4.23). □

Remark 4.41. Unconditional versions of Theorem 4.40 are available for some parahoric or pro-p-Iwahori level cases, or for certain compact "Drinfeld case" Shimura varieties with arbitrary level; these cases are alluded to in §8.

6.4. Relation with geometric Langlands

For simplicity, assume G is split over a p-adic or local function field F. Assume G satisfies LLC+. From the construction of Z_V in Proposition 4.28, we have a map

$$K_0 \mathrm{Rep}_\mathbb{C}(\widehat{G}) \to \mathcal{Z}(G, J) \qquad (4.28)$$

$$V \mapsto Z_V * 1_J$$

for any compact open subgroup $J \subset G(F)$, which gives rise to a commutative diagram

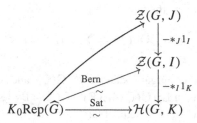

whenever $J \subseteq I \subset K$ where I resp. K is an Iwahori resp. special maximal compact subgroup, and where the bottom two arrows are the Bernstein resp. Satake isomorphisms. We warn the reader that the oblique arrow $K_0\mathrm{Rep}(\widehat{G}) \to \mathcal{Z}(G, J)$ is injective but not surjective in general, and also it is additive but not an algebra homomorphism in general.

Gaitsgory [Ga] constructed the two arrows Sat and Bern geometrically when F is a function field, using nearby cycles for a degeneration of the affine Grassmannian Gr_G to the affine flag variety Fl for G. One can hope that, as in the Iwahori case [Ga], one can construct the arrow $K_0\mathrm{Rep}(\widehat{G}) \to \mathcal{Z}(G, J)$ categorically using nearby cycles for a similar degeneration of Gr_G to a "partial affine flag variety", namely an *fpqc*-quotient $L\mathfrak{J}/L^+\mathfrak{J}$ where $\mathfrak{J}$ is a smooth connected group scheme over $\overline{\mathbb{F}}_p[[t]]$ with generic fiber $\mathfrak{J}_{\mathbb{F}_p((t))} = G_{\mathbb{F}_p((t))}$ and $\mathfrak{J}(\mathbb{F}_p[[t]]) = J$. Here $L\mathfrak{J}$ (resp. $L^+\mathfrak{J}$) is the ind-scheme (resp. scheme) over $\mathbb{F}_p$ representing the sheaf of groups for the *fpqc*-topology whose sections for an $\mathbb{F}_p$-algebra R are given by $L\mathfrak{J}(\mathrm{Spec}\, R) = \mathfrak{J}(R[[t]][\frac{1}{t}])$ (resp. $L^+\mathfrak{J}(\mathrm{Spec}\, R) = \mathfrak{J}(R[[t]])$). At least for $J = I^+$, the pro-p Iwahori subgroup, this will be realized in forthcoming joint work of the author and Benoit Stroh.

In a related vein, the geometric Satake equivalence of Mirkovic–Vilonen [MV] is a categorical version of the Satake isomorphism Sat, and this is usually stated when G is a split group over $F = \mathbb{F}_p((t))$. One can ask for a version of this when G is nonsplit, possibly not even quasi-split, over such a field F. The correct Satake isomorphism to "categorify" appears to be the one described in [HRo]. In many cases where G is quasi-split and split over a tamely ramified extension of F, this has been carried out in recent work of X. Zhu [Zhu].

7. Test functions in the parahoric case

We fix $r = j[E_0 : \mathbb{Q}_p]$ for some $j \in \mathbb{N}$. We assume K_p is a parahoric subgroup of $G(\mathbb{Q}_p)$, and we let K_{p^r} denote the corresponding parahoric subgroup of $G(\mathbb{Q}_{p^r})$. Assuming LLC+ holds for $G_{\mathbb{Q}_{p^r}}$, we can speak of the test function

$$\phi_r = p^{rd/2}\big(Z_{V^{E_{j0}}_{-\mu,j}} * 1_{K_{p^r}}\big) \in \mathcal{Z}(G(\mathbb{Q}_{p^r}), K_{p^r}). \tag{4.29}$$

We wish to give a more concrete description of this function, making use of Bernstein's isomorphism for $\mathcal{Z}(G(\mathbb{Q}_{p^r}), K_{p^r})$ which is detailed in the Appendix, §11.

In the next two subsections, we are concerned with the case where $G_{\mathbb{Q}_{p^r}}$ is quasi-split. We write $F := \mathbb{Q}_{p^r}$. Choose a maximal F-split torus A in G, and let T denote its centralizer in G. Fix an F-rational Borel subgroup B containing T. Let $K_F \subset G(F)$ denote the parahoric subgroup corresponding to K_p.

By Kottwitz [Ko84b, Lemma (1.1.3)], the $G(\overline{\mathbb{Q}}_p)$-conjugacy class $\{\mu\}$ is represented by an F-rational cocharacter $\mu \in X_*(T)^{\Phi_F} = X_*(A)$. It is clear that E, the field of definition of $\{\mu\}$, is contained in any subfield of $\overline{\mathbb{Q}}_p$ which splits G.

Given $\pi \in \Pi(G/F)$ with $\pi^{K_F} \neq 0$, to understand (4.29) we need to compute the scalar

$$\mathrm{tr}(\varphi_\pi(\Phi_F), (V_{-\mu}^{E_0})^{\varphi_\pi(I_F)}).   \tag{4.30}$$

There is an unramified character χ of $T(F)$ such that π is a subquotient of $i_B^G(\chi)$, and we may assume $\varphi_\pi|_{W_F} = \varphi_\chi|_{W_F}$. Since χ is unramified, $\varphi_\chi(I_F) = 1 \rtimes I_F \subset \widehat{T} \rtimes W_F$. Regarding χ as an element of $\widehat{T}$, (4.7) implies that we may write $\varphi_\chi(\Phi_F) = \chi \rtimes \Phi_F \in \widehat{T} \rtimes W_F$. Then we need to compute

$$\mathrm{tr}(\chi \rtimes \Phi_F, (V_{-\mu}^{E_0})^{1 \rtimes I_F}).   \tag{4.31}$$

7.1. Unramified groups and the Kottwitz conjecture

Let us consider the case where $G_{\mathbb{Q}_{p^r}}$ is unramified. Since we are assuming G splits over an unramified extension of $\mathbb{Q}_p$, it follows that $E/\mathbb{Q}_p$ is unramified, i.e. $E = E_0$ and $V_{-\mu}^{E_0} = V_{-\mu}$. Moreover $F = E_{j0}$ contains E with degree j.

Further, since G splits over F^{un}, we have $V_{-\mu}^{1 \rtimes I_F} = V_{-\mu}$. So we are reduced to computing $\mathrm{tr}(\chi \rtimes \Phi_F, V_{-\mu})$. Exactly as in Kottwitz' calculation of the Satake transform in [Ko84b, p. 295], we see that (4.30) is

$$\mathrm{tr}(\chi \rtimes \Phi_F, V_{-\mu}) = \sum_{\lambda \in W(F) \cdot \mu} (-\lambda)(\chi).   \tag{4.32}$$

Here $W(F) = W(G, A)$ is the relative Weyl group for G/F, and we view $\lambda \in X_*(A) = X_*(T)^{\Phi_F}$ as a character on $\widehat{T}$. This proves the following result.

Lemma 4.42. *In the above situation,*

$$Z_{V_{-\mu,j}^{E_{j0}}} * 1_{K_{p^r}} = z_{-\mu,j},   \tag{4.33}$$

where the Bernstein function $z_{-\mu,j}$ (cf. Definition 4.72) is the unique element of $\mathcal{Z}(G(F), K_F)$ which acts (on the left) on the normalized induced representation $i_B^G(\chi)^{K_F}$ by the scalar $\sum_{\lambda \in W(F) \cdot \mu} (-\lambda)(\chi)$, for any unramified character $\chi : T(F) \to \mathbb{C}^\times$.

Of course the advantage of $z_{-\mu,j}$ is that unlike the left-hand side of (4.33), it is defined unconditionally. A relatively self-contained, elementary, and efficient approach to Bernstein functions is given in §11.

Thus Conjecture 4.30 in this situation is equivalent to the *Kottwitz Conjecture*.

Conjecture 4.43. (Kottwitz conjecture) *In the situation where* $\mathbf{G}_{\mathbb{Q}_{p^r}}$ *is unramified and* K_p *is a parahoric subgroup, the function* ϕ_r *in (4.15) may be taken to be* $p^{rd/2}z_{-\mu,j}$.

Conjecture 4.43 was formulated by Kottwitz in 1998, about 11 years earlier than Conjecture 4.30. There is a closely related conjecture of Kottwitz concerning nearby cycles on Rapoport–Zink local models $\mathbf{M}_{K_p}^{\mathrm{loc}}$ for Sh_{K_p}. We refer to [RZ, Ra05] for definitions of local models, and to [H05, HN02a] for further details about the following conjecture in various special cases.

Conjecture 4.44. (Kottwitz Conjecture for Nearby Cycles) *Write* $\mathcal{G}$ *for the Bruhat–Tits parahoric group scheme over* $\mathbb{Z}_{p^r}$ *with generic fiber* $G_{\mathbb{Q}_{p^r}}$ *and with* $\mathcal{G}(\mathbb{Z}_{p^r}) = K_{p^r}$. *Let* $\mathcal{G}_t$ *denote the analogous parahoric group scheme over* $\mathbb{F}_{p^r}[[t]]$ *with the "same" special fiber as* $\mathcal{G}$. *Then there is an* $L^+\mathcal{G}_{t,\mathbb{F}_{p^r}}$-*equivariant embedding of* $\mathbf{M}_{K_p,\mathbb{F}_{p^r}}^{\mathrm{loc}}$ *into the affine flag variety* $L\mathcal{G}_{t,\mathbb{F}_{p^r}}/L^+\mathcal{G}_{t,\mathbb{F}_{p^r}}$, *via which we can identify the semisimple trace of Frobenius function* $x \mapsto \mathrm{tr}^{\mathrm{ss}}(\mathrm{Fr}_{p^r}, R\Psi_x^{\mathbf{M}_{K_p}^{\mathrm{loc}}})$ *on* $x \in \mathbf{M}_{K_p}^{\mathrm{loc}}(\mathbb{F}_{p^r})$ *with the function* $p^{dr/2}z_{-\mu,j} \in \mathcal{Z}(\mathcal{G}_t(\mathbb{F}_{p^r}((t))), \mathcal{G}_t(\mathbb{F}_{p^r}[[t]]))$.

7.2. The quasi-split case

The group $\widehat{G}^{I_F}$ is a possibly disconnected reductive group, with maximal torus $(\widehat{T}^{I_F})^\circ$ (see the proof of Theorem 8.2 of [St]). Now we may restrict the representation $V_{-\mu}^{E_0}$ to the subgroup $\widehat{G}^{I_F} \rtimes W_F \subset \widehat{G} \rtimes W_F$. Let χ be a weakly unramified character of $T(F)$; by (4.5) we can view $\chi \in (\widehat{T}^{I_F})_{\Phi_F}$. The only $\widehat{T}^{I_F}$-weight spaces of $(V_{-\mu}^{E_0})^{1 \rtimes I_F}$ which contribute to (4.31) are indexed by the Φ_F-fixed weights, i.e. by those in $X^*(\widehat{T}^{I_F})^{\Phi_F}$. (It is important to note that it is the weight spaces for the diagonalizable group $\widehat{T}^{I_F}$, and not for the maximal torus $(\widehat{T}^{I_F})^\circ$, which come in here.) This is consistent with Theorem 4.71 of the Appendix, and may be expressed as follows.

Proposition 4.45. *In the general quasi-split situation,* $Z_{V_{-\mu,j}^{E_{j0}}} * 1_{K_{p^r}}$ *is the unique function in* $\mathcal{Z}(G(\mathbb{Q}_{p^r}), K_{p^r})$ *which acts on the left on each weakly unramified principal series representation* $i_B^G(\chi)^{K_F}$ *by the scalar (4.31), and thus is a certain linear combination of Bernstein functions* $z_{-\lambda,j}$ *where* $-\lambda \in X^*(\widehat{T}^{I_F})^{\Phi_F}$ *ranges over the* $W(G, A)$-*orbits of* Φ_F-*fixed* $\widehat{T}^{I_F}$-*weights in* $V_{-\mu}^{E_0}$.

It is an interesting exercise to write out the linear combinations of Bernstein functions explicitly in each given case. Once this is done, the result can be used to find explicit descriptions of test functions for inner forms of quasi-split groups. This is the subject of the next subsection.

7.3. Passing from quasi-split to general cases via transfer homomorphisms

7.3.1. Test function conjecture via transfer homomorphisms

We use freely the notation and set-up explained in the Appendix §11.12. Let G^* be a quasi-split F-group with an inner twisting $\psi : G \to G^*$. Let $J^* \subset G^*(F)$ resp. $J \subset G(F)$ be parahoric subgroups and consider the *normalized transfer homomorphism* $\tilde{t} : \mathcal{Z}(G^*(F), J^*) \to \mathcal{Z}(G(F), J)$ from Definition 4.78.

The following conjecture indicates that test functions for the quasi-split group G^* should determine test functions for G. This is compatible with the global considerations which led to Theorem 4.40.

Conjecture 4.46. *Let K_{p^r} resp. $K_{p^r}^*$ be parahoric subgroups of $G(\mathbb{Q}_{p^r})$ resp. $G^*(\mathbb{Q}_{p^r})$, with corresponding normalized transfer homomorphism $\tilde{t}$: $\mathcal{Z}(G^*(\mathbb{Q}_{p^r}), K_{p^r}^*) \to \mathcal{Z}(G(\mathbb{Q}_{p^r}), K_{p^r})$. If $\phi_r^* \in \mathcal{Z}(G^*(\mathbb{Q}_{p^r}), K_{p^r}^*)$ is the function $p^{rd/2}(Z_{V_{-\mu,j}^{E j 0}}^{G^*} * 1_{K_{p^r}^*})$ described in Proposition 4.45 for the data $(G_{\mathbb{Q}_{p^r}}^*, \{-\mu\}, K_{p^r}^*)$, then $\phi_r := \tilde{t}(\phi_r^*) \in \mathcal{Z}(G(\mathbb{Q}_{p^r}), K_{p^r})$ is a test function satisfying (4.15) for the original data $(G_{\mathbb{Q}_{p^r}}, \{-\mu\}, K_{p^r})$.*

Assuming Conjecture 4.30 holds, another way to formulate this is that the normalized transfer homomorphism $\tilde{t}$ takes the function $Z_{V_{-\mu,j}^{E j 0}}^{G^*} * 1_{K_{p^r}^*} \in \mathcal{Z}(G^*(\mathbb{Q}_{p^r}), K_{p^r}^*)$ to the function $Z_{V_{-\mu,j}^{E j 0}}^{G} * 1_{K_{p^r}} \in \mathcal{Z}(G(\mathbb{Q}_{p^r}), K_{p^r})$. But the point of Conjecture 4.46 is to provide an explicit test function for the non-quasi-split data $(G_{\mathbb{Q}_{p^r}}, \{-\mu\}, K_{p^r})$ which can be compared with direct geometric calculations of the nearby cycles attached to this data, and thus to provide a method to prove an unconditional analogue of Conjecture 4.30 for such data. This is illustrated in §7.3.3 below.

The next two paragraphs show that Conjecture 4.46 is indeed reasonable.

7.3.2. A calculation for GL_2

Take $G^* = GL_{2,F}$ and $G = D^\times$, where D is the central simple division algebra over F of dimension 4.

Here we will explicitly calculate and compare the test functions associated to $(GL_{2,F}, \{-\mu\}, I_F)$ and $(D^\times, \{-\mu\}, \mathcal{O}_D^\times)$, where $\mu = (1, 0)$, and where $I_F \subset GL_2(F)$ and $\mathcal{O}_D^\times \subset D^\times$ are the standard Iwahori subgroups. This calculation will show that the normalized transfer homomorphism takes one test function to the other. This is required in order for both Conjectures 4.30 and 4.46 to hold true.

Write $z^*_{-\mu} = Z^{\mathrm{GL}_2}_{V-\mu} * 1_{I_F} \in \mathcal{Z}(\mathrm{GL}_2(F), I_F)$ and $z_{-\mu} = Z^{D^\times}_{V-\mu} * 1_{\mathcal{O}^\times_D} \in \mathcal{H}(D^\times, \mathcal{O}^\times_D) \cong \mathbb{C}[\mathbb{Z}]$. The last isomorphism is induced by the Kottwitz homomorphism, which in this case is the normalized valuation $\mathrm{val}_F \circ \mathrm{Nrd}_D$: $D^\times \twoheadrightarrow \mathbb{Z}$, where val_F is the normalized valuation for F and $\mathrm{Nrd}_D : D \to F$ is the reduced norm.

Write $\bar{\mu} = (0, 1)$ and let B^* denote the Borel subgroup of *lower triangular* matrices in GL_2. Then $z^*_{-\mu}$ acts on the left on $i^{\mathrm{GL}_2}_{B^*}(\chi)^{I_F}$ by the scalar

$$\mathrm{tr}(\chi \rtimes \Phi_F, V_{-\mu}) = (-\mu - \bar{\mu})(\chi),$$

for any unramified character $\chi \in \mathrm{Hom}(T^*(F)/T^*(F)_1, \mathbb{C}^\times)$. We may view χ as a diagonal 2×2 complex matrix $\chi = \mathrm{diag}(\chi_1, \chi_2)$.

To calculate $z_{-\mu}$ we need a few preliminary remarks. First we parametrize unramified characters $\eta \in \mathrm{Hom}(D^\times/\mathcal{O}^\times_D, \mathbb{C}^\times)$ by writing $\eta = \eta_0 \circ \mathrm{Nrd}_D$, where $\eta_0 \in \mathbb{C}^\times$ is viewed as the unramified character on $F^\times$ which sends $\varpi_F \mapsto \eta_0$. The map $\mathrm{Nrd}_D : D^\times \to F^\times$ is Langlands dual to the diagonal embedding $\mathbb{G}_m(\mathbb{C}) \to \mathrm{GL}_2(\mathbb{C})$, and it follows that the cocycles z_η and z_{η_0} attached by Langlands duality to the quasicharacters η and η_0 satisfy

$$z_\eta = \begin{bmatrix} z_{\eta_0} & 0 \\ 0 & z_{\eta_0} \end{bmatrix}, \quad \text{and thus,} \quad z_\eta(\Phi_F) = \begin{bmatrix} \eta_0 & 0 \\ 0 & \eta_0 \end{bmatrix}.$$

On the other hand, if $\mathbf{1}$ denotes the trivial 1-dimensional representation of $D^\times$, then its Langlands parameter $\varphi^{D^\times}_1$ satisfies $\varphi^{D^\times}_1(\Phi_F) = \mathrm{diag}(q^{-1/2}, q^{1/2})$ (see, e.g., [PrRa, Thm. 4.4]). So using (4.8), we obtain

$$\mathrm{tr}(\varphi^{D^\times}_\eta(\Phi_F), V_{-\mu})$$

$$= \mathrm{tr}\left(\begin{bmatrix} \eta_0 q^{-1/2} & 0 \\ 0 & \eta_0 q^{1/2} \end{bmatrix}, V_{-\mu} \right)$$

$$= (\delta^{-1/2}_{B^*}(\varpi^{-\mu}) \cdot {-\mu}|_{\hat{Z}} + \delta^{-1/2}_{B^*}(\varpi^{-\bar{\mu}}) \cdot {-\bar{\mu}}|_{\hat{Z}})\left(\begin{bmatrix} \eta_0 & 0 \\ 0 & \eta_0 \end{bmatrix} \right).$$

Here $\hat{Z}$ is the center of $\widehat{G^*}$. Using the definition of $\tilde{t}$ we deduce the following result.

Proposition 4.47. *The normalized transfer homomorphism* $\tilde{t}$: $\mathcal{Z}(\mathrm{GL}_2(F), I_F) \to \mathcal{H}(D^\times, \mathcal{O}^\times_D)$ *sends* $z^*_{-\mu}$ *to* $z_{-\mu}$.

7.3.3. Compatibility with nearby cycles in some anisotropic cases

Suppose we are in a situation where $E = \mathbb{Q}_p$. As before, write $F = \mathbb{Q}_{p^r}$. Suppose $G_F = (D \otimes F)^\times \times \mathbb{G}_m$, where D is a central division algebra over E of degree n^2, for $n > 2$. This situation arises in the setting of "fake unitary" simple Shimura varieties (see, e.g. [H01, §5]). Let $G^* = GL_n \times \mathbb{G}_m$, a split inner form of G over $\mathbb{Q}_p$.

Suppose that $V_{-\mu} = \wedge^m(\mathbb{C}^n)$ for $0 < m < n$, i.e. the representation of $\widehat{G^*} = GL_n(\mathbb{C}) \times \mathbb{C}^\times$ where the first factor acts via the irreducible representation with highest weight $(1^m, 0^{n-m})$ and the second factor acts via scalars.

Consider the local models $\mathbf{M}^{*\mathrm{loc}} = \mathbf{M}^{\mathrm{loc}}(G^*, \{-\mu\}, K_p^*)$ and $\mathbf{M}^{\mathrm{loc}} = \mathbf{M}^{\mathrm{loc}}(G, \{-\mu\}, K_p)$, where $K_p^* \subset G^*(F)$ and $K_p \subset G(F)$ are Iwahori subgroups. We can choose the inner twist $G \to G^*$ and the subgroups K_p^* and K_p so that

$$\mathbf{M}^{*\mathrm{loc}}(\overline{\mathbb{F}}_p) = \mathbf{M}^{\mathrm{loc}}(\overline{\mathbb{F}}_p)$$

and where the action of geometric Frobenius Φ_p on the right-hand side is given by $\Phi_p = \mathrm{Ad}(c_{\Phi_p}) \cdot \Phi_p^*$ where Φ_p^* is the usual Frobenius action (on the left-hand side) and where $\tau \mapsto \mathrm{Ad}(c_\tau)$ represents the class in $H^1(\mathbb{Q}_p, \mathrm{PGL}_n)$ corresponding to the inner twist $G \to G^*$.

Assume $(r, n) = 1$ and set $q = p^r$. Then $\mathbf{M}^{\mathrm{loc}}(G, \{-\mu\}, K_p)(\mathbb{F}_q)$ consists of a single point. To understand the corresponding test function we may ignore the $\mathbb{G}_m$-factor and pretend that $G = D^\times$ and $G^* = GL_n$. Then the Kottwitz homomorphism $\kappa_G : G(F) \to \mathbb{Z}$ induces an isomorphism

$$\mathcal{H}(G(\mathbb{Q}_{p^r}), K_{p^r}) \cong \mathbb{C}[\mathbb{Z}].$$

The test function for the Shimura variety giving rise to the local Shimura data $(G, \{-\mu\}, K_p)$ should be calculated by understanding the function trace of Frobenius on nearby cycles on $\mathbf{M}^{\mathrm{loc}}$, similarly to Conjecture 4.44 in the unramified case. The test function should be of the form $C_q \cdot 1_m \in \mathbb{C}[\mathbb{Z}] = \mathcal{H}(G(\mathbb{Q}_p), K_{p^r})$ for some scalar C_q.

Proposition 4.48. *In the above situation, Conjecture 4.46 predicts that the coefficient C_q is given by $C_q = \#\mathrm{Gr}(m, n)(\mathbb{F}_q)$, the number of $\mathbb{F}_q$-rational points on the Grassmannian variety $\mathrm{Gr}(m, n)$ parametrizing m-planes in n-space.*

This is compatible with calculations of Rapoport of the trace of Frobenius on nearby cycles of the local models for such situations, see [Ra90]. Thus the normalized transfer homomorphism gives a group-theoretic framework with which we could make further predictions about nearby cycles on the local models attached to non-quasisplit groups G, assuming we know explicitly the corresponding test function for a quasi-split inner form of G.

Proof. By the final sentence of Proposition 4.79, we simply need to integrate the function $p^{rd/2}z_{-\mu}^* \in \mathcal{Z}(GL_n(\mathbb{Q}_{p^r}), K_{p^r}^*)$ over the fiber of the

Kottwitz homomorphism val ∘ det over $1_m \in \mathbb{C}[\mathbb{Z}]$. This is a combinatorial problem which could be solved since we know $p^{rd/2}z^*_{-\mu}$ explicitly. However, it is easier to use geometry. Translating "integration over the fiber of the Kottwitz homomorphism" in terms of local models gives us the equality

$$C_q = \sum_{x \in \mathbf{M}^{*\text{loc}}(\mathbb{F}_q)} \text{Tr}(\Phi^r_p, R\Psi^{\mathbf{M}^{*\text{loc}}}(\overline{\mathbb{Q}}_\ell)_x).$$

(Here ℓ is a rational prime with $\ell \neq p$.) But the special fiber of $\mathbf{M}^{*\text{loc}}$ embeds into the affine flag variety Fl_{GL_n} for $\text{GL}_n/\mathbb{F}_p$, and under the projection $p :$ $\text{Fl}_{\text{GL}_n} \to \text{Gr}_{\text{GL}_n}$ to the affine Grassmannian, $\mathbf{M}^{*\text{loc}}$ maps onto $\text{Gr}(m,n)$ and $Rp_*(R\Psi^{\mathbf{M}^{*\text{loc}}}(\overline{\mathbb{Q}}_\ell)) = \overline{\mathbb{Q}}_\ell$, the constant ℓ-adic sheaf on $\text{Gr}(m,n)$ in degree 0. Thus we obtain

$$C_q = \sum_{x \in \text{Gr}(m,n)(\mathbb{F}_q)} \text{Tr}(\Phi^r_p, (\overline{\mathbb{Q}}_\ell)_x) = \#\text{Gr}(m,n)(\mathbb{F}_q)$$

as desired. (The reader should note the similarity with Prop. 3.17 in [Ra90], which is justified in a slightly different way.) $\qquad \square$

•

8. Overview of evidence for the test function conjecture

8.1. Good reduction cases

In case $\mathbf{G}_{\mathbb{Q}_p}$ is unramified and K_p is a hyperspecial maximal compact subgroup of $\mathbf{G}(\mathbb{Q}_p)$, we expect $Sh(\mathbf{G}, h^{-1}, K^p K_p)$ to have good reduction over $\mathcal{O}_{\mathbf{E}_p}$. In PEL cases this was proved by Kottwitz [Ko92a]. In the same paper for PEL cases of type A or C, it is proved that the function $\phi_r = 1_{K_{p^r}\mu(p^{-1})K_{p^r}}$ satisfies (4.15), which can be viewed as verifying Conjecture 4.30 for these cases.

8.2. Parahoric cases

Assume K_p is a parahoric subgroup. We will discuss only PEL Shimura varieties.

Here the approach is via the Rapoport–Zink local model $\mathbf{M}^{\text{loc}}_{K_p}$ for a suitable integral model $\mathcal{M}_{K_p}$ for Sh_{K_p} and the main ideas are due to Rapoport. We refer to the survey articles [Ra90], [Ra05], and [H05] for more about how local models fit in with the Langlands–Kottwitz approach. For much more about the geometry of local models, we refer the reader to the survey article of Pappas–Rapoport–Smithing [PRS] and the references therein.

Using local models, the first step to proving Conjecture 4.31 is to prove Conjecture 4.44. The first evidence was purely computational: in [H01], $z_{-\mu,j}$ was computed explicitly in the Drinfeld case and the result was compared with Rapoport's computation of the nearby cycles in that setting, proving Conjecture 7.1.3 directly. This result motivated Kottwitz' more general conjecture and also inspired Beilinson and Gaitsgory to construct the center of an affine Hecke algebra via a nearby cycles construction, a feat carried out in [Ga]. Then in [HN02a] Gaitsgory's method was adapted to prove Conjecture 4.44 for the split groups GL_n and GSp_{2n}. This in turn was used to demonstrate Conjecture 4.43 for certain special Shimura varieties in [H05], and then to prove the analogue of Theorem 4.40 for those special Shimura varieties with parahoric level structure at p. The harmonic analysis ingredient needed for the latter was provided by [H09].

In his 2011 PhD thesis, Sean Rostami proved Conjecture 4.44 when G is an unramified unitary group. In a recent breakthrough, Pappas and Zhu defined group-theoretic local models $\mathbf{M}^{\mathrm{loc}}_{K_p}$ whenever G splits over a tamely ramified extension, and for unramified groups G proved Conjecture 4.44, see [PZ], esp. Theorem 10.16.

8.3. Deeper level cases

We again limit our discussion to PEL situations, where progress to date has occurred.

It is again natural to study directly the nearby cycles relative to a suitable integral model for the Shimura variety and hope that it gives rise to a test distribution in the Bernstein center. For Shimura varieties in the "Drinfeld case" with K_p a pro-p Iwahori subgroup of $\mathbf{G}(\mathbb{Q}_p) = \mathrm{GL}_n(\mathbb{Q}_p) \times \mathbb{Q}_p^\times$ ("$\Gamma_1(p)$-level structure at p"), one may use Oort–Tate theory to define suitable integral models and prove Conjectures 4.31 and 4.30 for them. This was done by the author and Rapoport [HRa1] (and [H12] provided the harmonic analysis ingredient needed to go further and prove Theorem 4.40 in this case).

Around the same time as [HRa1], Scholze studied in [Sch1] nearby cycles on suitable integral models for the modular curves with arbitrarily deep full level structure at p. In this way he proved Conjectures 4.31 and 4.30 in these cases, taking the compactifications also into account, and thereby proved the analogue of Theorem 4.40 for the compactified modular curves at nearly all primes of bad reduction. The nearby cycles on his integral models naturally gave rise to some remarkable distributions in the Bernstein center, for which he gave explicit formulas (see § 10).

Then in [Sch2] Scholze generalized the approach of [Sch1] to compact Shimura varieties in the Drinfeld case, again finding an explicit description of nearby cycles. In this case, he was still able to produce a test function to plug into (4.15), or rather, simultaneously incorporating the base-change transfer results he needed in precisely this case, he found a test function that goes directly into the pseudostabilization of (4.15). This allowed him to prove Theorem 4.40. In contrast to the modular curve situation, in higher rank the nearby cycles on Scholze's integral models do not directly produce distributions in the Bernstein center, and an explicit description of his test functions seems hopeless. But nevertheless Scholze was able to prove by indirect means Conjecture 4.30 in this case.

The description of the nearby cycles in [Sch2] provided one ingredient for Scholze's subsequent paper [Sch3] which gave a new and streamlined proof of the local Langlands conjecture for general linear groups.

In later work Scholze [Sch4] formalized his method of producing test functions in many cases, using deformation spaces of p-divisible groups, and this is used to give a nearly complete description of the cohomology groups of many compact unitary Shimura varieties in his joint work [SS] with S. W. Shin; their main assumption at p is that $\mathbf{G}_{\mathbb{Q}_p}$ is a product of Weil restrictions of general linear groups. The advantage of what we could call the Langlands–Kottwitz–Scholze approach in this situation is that it yields in [SS] a new construction of the Galois representations constructed earlier by Shin [Sh], in a shorter way that avoids Igusa varieties.

In these later developments, Conjecture 4.30 does not play a central part, but the stable Bernstein center does nevertheless still play a clarifying role in the pseudostabilization process (e.g. in [SS]). It seems that only certain integral models, such as those we see in many parahoric or pro-p Iwahori level cases, have the favorable property that their nearby cycles naturally give rise to distributions in the Bernstein center. It remains an interesting problem to find such integral models in more cases, and to better understand the role of the Bernstein center in the study of Shimura varieties.

9. Evidence for conjectures on transfer of the Bernstein center

Here we present some evidence for the general principle that the (stable/geometric) Bernstein center is particularly well-behaved with respect to (twisted) endoscopic transfer. The primary evidence thus far consists of some unconditional analogues of Conjecture 4.36.

Let G/F be an unramified group, and let F_r/F be the degree r unramified extension of F in some algebraic closure of F. In [H09, H12], the author defined *base change homomorphisms*

$$b_r : \mathcal{Z}(G(F_r), J_r) \to \mathcal{Z}(G(F), J),$$

where $J \subset G(F)$ is either a parahoric subgroup or a pro-p Iwahori subgroup, and where J_r is the corresponding subgroup of $G(F_r)$. Then we have "base-change fundamental lemmas" of the following form.[8]

Theorem 4.49. *For any* $\phi_r \in \mathcal{Z}(G(F_r), J_r)$, *the function* $b_r(\phi_r)$ *is a base-change transfer of* ϕ_r *in the sense of (4.18).*

By Kottwitz [Ko86], the function 1_J is a base-change transfer of 1_{J_r}. Hence for any $V_r \in \mathrm{Rep}(\,^L(G_{F_r}))$, Conjecture 4.36 predicts that $b_{F_r/F}(Z_{V_r}) * 1_J$ is a base-change transfer of $Z_{V_r} * 1_{J_r}$. This is a consequence of Theorem 4.49, because of the following compatibility between the base-change operations in [H09, H12] and in the context of stable Bernstein centers (cf. Prop. 4.21).

Lemma 4.50. *In the above situations,* $b_r(Z_{V_r} * 1_{J_r}) = b_{F_r/F}(Z_{V_r}) * 1_J$.

Proof. First assume J is a parahoric subgroup. Let χ be any unramified character of $T(F)$. It is enough to show that the two functions act on the left by the same scalar on every unramified principal series representation $i_B^G(\chi)^J$.

Let $N_r : T(F_r) \to T(F)$ be the norm homomorphism. By the definition of b_r in [H09], $b_r(Z_{V_r} * 1_{J_r})$ acts by the scalar by which $Z_{V_r} * 1_{J_r}$ acts on $i_{B_r}^{G_r}(\chi \circ N_r)^{J_r}$. This is the scalar by which Z_{V_r} acts on $i_{B_r}^{G_r}(\chi \circ N_r)$, which in view of LLC+ is

$$\mathrm{tr}^{\mathrm{ss}}(\varphi_{\chi \circ N_r}^{T_r}(\Phi_F^r), V_r) = \mathrm{tr}^{\mathrm{ss}}(\varphi_\chi^T(\Phi_F^r), V_r). \tag{4.34}$$

But the right-hand side is the scalar by which $b_{F_r/F}(Z_{V_r}) * 1_J$ acts on $i_B^G(\chi)^J$.

The equality $\varphi_{\chi \circ N_r}^{T_r}(\Phi_F^r) = \varphi_\chi^T(\Phi_F^r)$ we used in (4.34) follows from the commutativity of the diagram of Langlands dualities for tori

$$
\begin{array}{ccc}
\mathrm{Hom}_{\mathrm{conts}}(T(F), \mathbb{C}^\times) & \longrightarrow & H^1_{\mathrm{conts}}(W_F, \,^L T) \\
\downarrow{\scriptstyle N_r} & & {\scriptstyle \mathrm{Res}}\downarrow \\
\mathrm{Hom}_{\mathrm{conts}}(T(F_r), \mathbb{C}^\times) & \longrightarrow & H^1_{\mathrm{conts}}(W_{F_r}, \,^L T_r)
\end{array}
$$

which was proved in [KV, Lemma 8.1.3].

---

[8] Relating to pro-p Iwahori level, a much stronger result is proved in [H12] concerning the base-change transfer of Bernstein centers of Bernstein blocks for depth-zero principal series representations.

Now suppose $J = I^+$ is a pro-p Iwahori subgroup. Then the same argument works given the following fact: for any depth-zero character $\chi : T(F)_1 \to \mathbb{C}^\times$ and any extension of it to a character $\tilde{\chi}$ on $T(F)$, and any $z_r \in \mathcal{Z}(G(F_r), I_r^+)$, the function $b_r(z_r)$ acts on $i_B^G(\tilde{\chi})^{I^+}$ by the same scalar by which z_r acts on $i_{B_r}^{G_r}(\tilde{\chi} \circ N_r)^{I_r^+}$. This follows from the definition of b_r given in Definition 10.0.3 of [H12], using [H12, Lemma 4.2.1]. $\qquad\square$

Let us also mention again Scholze's Theorem C in [Sch2], which essentially proves Conjecture 4.36 for GL_n (see Proposition 4.37).

10. Explicit computation of the test functions

10.1. Parahoric cases

Conjecture 4.30 implies that test functions are compatible with change of level. Therefore for the purposes of computing them for parahoric level, the key case is where K_p is an Iwahori subgroup. Thus, for the rest of this subsection we consider only Iwahori level structure. Since test functions attached to quasi-split groups should determine, in an computable way, those for inner forms (by Conjecture 4.46 and Proposition 4.79), it is enough to understand quasi-split groups. Via Proposition 4.45 this boils down to giving explicit descriptions of the Bernstein functions $z_{-\lambda, j}$, assuming we have already expressed the test function explicitly in term of these – this is automatic for unramified groups using the Kottwitz Conjecture (Conjecture 4.43).

Let us therefore consider the problem of explicitly computing Bernstein functions z_μ attached to any group G/F and an Iwahori subgroup $I \subset G(F)$ (F being any local nonarchimedean field). For simplicity consider the case where G/F is unramified, and regard μ as a dominant coweight in $X_*(A)$. The μ which arise in Conjecture 4.43 are *minuscule*; however, we consider μ which are **not necessarily** minuscule here. Let $\tilde{W}$ denote the extended affine Weyl group of G over F (cf. §11). Attached to μ is an the μ-*admissible set* $\mathrm{Adm}(\mu) \subset \tilde{W}$, defined by

$$\mathrm{Adm}(\mu) = \{x \in \tilde{W} \mid x \leq t_\lambda, \text{ for some } \lambda \in W(G, A) \cdot \mu\},$$

where $\leq$ denotes the Bruhat order on $\tilde{W}$ determined by the Iwahori subgroup I and where t_λ denotes the translation element in $\tilde{W}$ corresponding to $\lambda \in X_*(A)$. The μ-admissible set has been studied for its relation to the stratification by Iwahori-orbits in the local model $\mathbf{M}_{K_p}^{\mathrm{loc}}$; for much information see [KR], [HN02b], [Ra05]. The strongest combinatorial results relating local models and $\mathrm{Adm}(\mu)$ are due to Brian Smithling, see e.g. [Sm1, Sm2, Sm3].

For our purposes, the set $\text{Adm}(\mu)$ enters because it is the set indexing the double cosets in the support of z_μ.

Proposition 4.51. *The support of z_μ is the union $\bigcup_{x \in \text{Adm}(\mu)} IxI$.*

Proof. This was proved using the theory of alcove walks as elaborated by Görtz [G], in the Appendix to [HRa1]. It applies to affine Hecke algebras with arbitrary parameters, hence the corresponding result holds for arbitrary groups, not just unramified groups.   □

The following explicit formula was proved in [H01] and in [HP]. Let $T_x = 1_{IxI}$ for $x \in \widetilde{W}$. In the formulas here and below, $q = p^r$ is the cardinality of the residue field of F.

Proposition 4.52. *Assume μ is minuscule. Assume the parameters for the Iwahori Hecke algebra are all equal. Then*

$$q^{\ell(t_\mu)/2} z_\mu = (-1)^{\ell(t_\mu)} \sum_{x \in \text{Adm}(\mu)} (-1)^{\ell(x)} R_{x, t_{\lambda(x)}}(q)\, T_x,$$

where x decomposes as $x = t_{\lambda(x)} w \in X_*(A) \rtimes W = \widetilde{W}$ and where $R_{x,y}(q)$ denotes the R-polynomial of Kazhdan–Lusztig [KL], and ℓ the length function, for the quasi-Coxeter group $\widetilde{W}$. A similar formula holds in the context of affine Hecke algebras with arbitrary parameters. In the Drinfeld case $G = \text{GL}_n$ and $\mu = (1, 0^{n-1})$, the coefficient of T_x is $(1 - q)^{\ell(t_\mu) - \ell(x)}$.

There are also explicit formulas for Bernstein functions z_μ when μ is not minuscule, but they tend to be much more complicated. For related computations see [HP] and [GH].

10.2. A Pro-p Iwahori level case

In the Drinfeld case where $\mathbf{G}_{\mathbb{Q}_p} = \text{GL}_n \times \mathbb{G}_m$ and $\mu = (1, 0^{n-1}) \times 1$, and where K_p is a pro-p Iwahori subgroup, an explicit formula for the test function ϕ_r for $Sh(\mathbf{G}, h^{-1}, K^p K_p)$ was found by the author and Rapoport. We shall rephrase this slightly by ignoring the $\mathbb{G}_m$ factor and giving the formula for $G = \text{GL}_n$.

Proposition 4.53 ([HRa1], Prop. 12.2.1). *Let $q = p^r$. Let I_r^+ denote the standard pro-p Iwahori subgroup of $G_r := \text{GL}_n(\mathbb{Q}_{p^r})$. Let T denote the standard diagonal torus in GL_n. In terms of natural embeddings $T(\mathbb{F}_q) \hookrightarrow G_r$, and $w \in \widetilde{W} \hookrightarrow G_r$ giving elements $tw^{-1} \in G_r$ representing $I_r^+ \backslash G_r / I_r^+$, we have*

$$\phi_r(I_r^+ \, tw^{-1} \, I_r^+) = \begin{cases} 0, & \text{if } w \notin \mathrm{Adm}(\mu) \\ 0, & \text{if } w \in \mathrm{Adm}(\mu) \text{ but } N_r(t) \notin T_{S(w)}(\mathbb{F}_p) \\ (-1)^n \, (p-1)^{n-|S(w)|} \, (1-q)^{|S(w)|-n-1}, & \text{otherwise.} \end{cases}$$

(4.35)

Here $S(w)$ is the set of critical indices for w, equivalently $S(w)$ is the set of standard basis vectors $e_j \in \mathbb{Z}^n$ such that $w \le t_{e_j}$ in the Bruhat order on $\widetilde{W}$ determined by the standard Iwahori subgroup of GL_n.

10.3. Deeper level structures

Here the known explicit descriptions pertain only to $G = \mathrm{GL}_2$ and first were proved by Scholze [Sch1]. It remains an interesting question whether one can find explicit descriptions of test functions in higher rank groups with arbitrary level structure: even the Drinfeld case $G = \mathrm{GL}_n$, $\mu = (1, 0^{n-1})$ looks difficult, cf. [Sch2].

To state Scholze's result, we need some notation. As usual let F be a nonarchimedean local field with ring of integers $\mathcal{O}$, uniformizer ϖ, and residue field cardinality q. Let B denote the $\mathcal{O}$-subalgebra $\mathrm{M}_2(\mathcal{O})$ of $\mathrm{M}_2(F)$. For any $j \in \mathbb{Z}$ set $B_j = \varpi^j B$. Let $K = B^\times$, the standard maximal compact subgroup of $G = \mathrm{GL}_2(F)$. For $n \ge 1$, let $K_n = 1 + B_n$; so K_n is a principal congruence subgroup and is a normal subgroup of K.

Scholze defines a (compatible) family of functions $\phi_n \in \mathcal{H}(G, K_n)$ for $n \ge 1$. His definition uses two functions, $\ell : G \to \mathbb{Z} \cup \{\infty\}$ and $k : G \to \mathbb{Z}$. Let $\ell(g) = \mathrm{val} \circ \det(1 - g)$. Let $k(g)$ be the unique integer k such that $g \in B_k$ and $g \notin B_{k+1}$. By definition ϕ_n is 0 unless $\mathrm{val} \circ \det(g) = 1$, $\mathrm{tr}(g) \in \mathcal{O}$, and $g \in B_{1-n}$. Assume these conditions, in which case one can check that $1 - n \le k(g) \le 0$ and $\ell(g) \ge 0$. Now define

$$\phi_n(g) := \begin{cases} -1 - q, & \text{if } \mathrm{tr}(g) \in \varpi\mathcal{O}, \\ 1 - q^{2\ell(g)}, & \text{if } \mathrm{tr}(g) \in \mathcal{O}^\times \text{ and } \ell(g) < n + k(g), \\ 1 + q^{2(n+k(g))-1}, & \text{if } \mathrm{tr}(g) \in \mathcal{O}^\times \text{ and } \ell(g) \ge n + k(g). \end{cases}$$

Proposition 4.54. (Scholze [Sch1]) *For each $n \ge 1$, the function $z_n := \frac{q-1}{[K:K_n]} \cdot \phi_n$ belongs to the center $\mathcal{Z}(\mathrm{GL}_2(F), K_n)$, and the family $(z_n)_n$ is compatible with change of level and thus defines a distribution in the sense of (4.2). This distribution is $q^{1/2} Z_V$ where V is the standard representation $\mathbb{C}^2$ of the Langlands dual group $\mathrm{GL}_2(\mathbb{C})$.*

In an unpublished work, Kottwitz gave another proof of this proposition and also described the same distribution in terms of a family $(\Phi_n)_n$ of functions

$\Phi_n \in \mathcal{Z}(\mathrm{GL}_2(F), I_n)$ where I_n ranges over the "barycentric" Moy–Prasad filtration in the standard Iwahori subgroup $I \subset \mathrm{GL}_2(F)$.

By a completely different technique, in [Var] Sandeep Varma extended both the results of Scholze and Kottwitz stated above, by describing the distributions attached to $V = \mathrm{Sym}^r(\mathbb{C}^2)$ where r is any odd natural number less than p, the residual characteristic of F.

11. Appendix: Bernstein isomorphisms via types

11.1. Statement of Purpose

Alan Roche proved the following beautiful result in [Roc, Theorem 1.10.3.1].

Theorem 4.55 (Roche). *Let e be an idempotent in the Hecke algebra $\mathcal{H} = \mathcal{H}(G(F))$. View $\mathcal{H}$ as a smooth $G(F)$-module via the left regular representation, and write $e = \sum_{\mathfrak{s} \in \mathfrak{S}} e_{\mathfrak{s}}$ according to the Bernstein decomposition $\mathcal{H} = \bigoplus_{\mathfrak{s} \in \mathfrak{S}} \mathcal{H}_{\mathfrak{s}}$. Let $\mathfrak{S}_e = \{\mathfrak{s} \in \mathfrak{S} \mid e_{\mathfrak{s}} \neq 0\}$, and consider the category $\mathfrak{R}^{\mathfrak{S}_e}(G(F)) = \prod_{\mathfrak{s} \in \mathfrak{S}_e} \mathfrak{R}_{\mathfrak{s}}(G(F))$ and its categorical center $\mathfrak{Z}^{\mathfrak{S}_e} = \prod_{\mathfrak{s} \in \mathfrak{S}_e} \mathfrak{Z}_{\mathfrak{s}}$. Let $\mathcal{Z}(e\mathcal{H}e)$ denote the center of the algebra $e\mathcal{H}e$. Then the map $z \mapsto z(e)$ defines an algebra isomorphism $\mathfrak{Z}^{\mathfrak{S}_e} \xrightarrow{\sim} \mathcal{Z}(e\mathcal{H}e)$.*

Roche's proof is decidedly non-elementary: besides the material developed in [Roc], it relies on some deep results of Bernstein cited there, most importantly Bernstein's Second Adjointness Theorem and the construction of an explicit progenerator for each Bernstein block $\mathfrak{R}_{\mathfrak{s}}(G(F))$.

In this chapter we use only the very special case of Roche's theorem where $e = e_J$ for a parahoric subgroup $J \subset G(F)$. We will explain a more elementary approach to this special case. It will rely only on the part of Bernstein's theory embodied in Proposition 4.68 below. Formally, the inputs needed are, first, the existence of Bernstein's categorical decomposition $\mathfrak{R}(G) = \prod_{\mathfrak{s}} \mathfrak{R}_{\mathfrak{s}}(G)$, which is proved for instance in [Roc, Theorem 1.7.3.1], in an elementary way, and, second, the internal structure of the Bernstein block $\mathfrak{R}_{\mathfrak{s}}(G)$ associated to a cuspidal pair $\mathfrak{s} = [(M(F), \tilde{\chi}]_G$ where M is a minimal F-Levi subgroup of G and $\tilde{\chi}$ is a character on $M(F)$ which is trivial on its unique parahoric subgroup. For such components, progenerators can be constructed in an elementary way, without using Bernstein's Second Adjointness Theorem. In fact in what follows we describe this internal structure using a few straightforward elements of the theory of Bushnell–Kutzko types, all of which are contained in [BK].

For $e = e_J$ Roche's theorem gives the identification of the center of the parahoric Hecke algebra, in other words a Bernstein isomorphism for the most general case, where G/F is arbitrary and $J \subset G(F)$ is an arbitrary parahoric subgroup. However we will provide a proof only for the crucial case of $J = I$, an Iwahori subgroup of $G(F)$. The general parahoric case should follow formally from the Iwahori case, following the method of Theorem 3.1.1 of [H09], provided one is willing to rely on some basic properties of intertwiners for principal series representations (a purely algebraic theory of such intertwiners was detailed for split resp. unramified groups in [HKP] resp. [H07], and the extension to arbitrary groups should be similar to [H07]).

The Iwahori case is approached in a different way by S. Rostami [Ro]. Rostami's proof yields more information, describing the Iwahori–Matsumoto and Bernstein presentations for the Iwahori–Hecke algebra and deducing the description of its center from its Bernstein presentation.

11.2. Some notation

The notation will largely come from [HRo]. Recall $L = \widehat{F^{\mathrm{un}}}$ and let $\sigma \in$ $\mathrm{Aut}(L/F)$ the Frobenius automorphism, which has fixed field F. We use the symbol Λ_G as an abbreviation for $X^*(Z(\widehat{G}))^\sigma_{I_F}$. Moreover, if S denotes a maximal L-split torus in G which is defined over F, with centralizer $T = \mathrm{Cent}_G(S)$, then Ω_G will denote the subgroup of $\widetilde{W}^{\mathrm{un}} = N_G(S)(L)/T(L)_1$, the extended affine Weyl group for G/L, which preserves the alcove $\mathbf{a}$ in the apartment $\mathbf{S}$ corresponding to S, in the building $\mathcal{B}(G, L)$ of G over L.

As always, we let I be the Iwahori subgroup $I = \mathcal{G}_{\mathbf{a}}^{\mathrm{o\natural}}(\mathcal{O}_F)$, which we recall we have chosen to be in good position relative to A: the corresponding alcove $\mathbf{a}^\sigma \subset \mathcal{B}(G, F)$ is required to belong in the apartment $\mathbf{A}$ corresponding to A.

Let $v_F \in \overline{\mathbf{a}^\sigma}$ be a special vertex with corresponding special maximal parahoric subgroup $K = \mathcal{G}_{v_F}^{\mathrm{o\natural}}(\mathcal{O}_F)$. Thus $K \supset I$.

Recall M is a minimal F-Levi subgroup of G. Further, if I is an Iwahori subgroup of $G(F)$, then $I_M := M(F) \cap I = M(F)_1$ is the corresponding Iwahori subgroup of $M(F)$ (cf. [HRo, Lemma 4.1.1]).

Use the symbol $\mathbf{1}$ to denote the trivial 1-dimensional representation of any group.

11.3. Preliminary structure theory results

Several of the results discussed here were proved independently by S. Rostami and will appear with somewhat different proofs in [Ro].

11.3.1. Iwahori–Weyl group over F

The following lemma concerns variations on well-known results, and were first proved by Timo Richarz [Ri].

Let G_1 denote the subgroup of $G(F)$ generated by all parahoric subgroups of $G(F)$. By [HRa1, Lemma 17] and [Ri], we have $G_1 = G(F)_1$. Let $N_1 = N_G(A)(F) \cap G_1$, and let **S** denote the set of reflections through the walls of **a**. Then by [BT2, Prop. 5.2.12], the quadruple $(G_1, I, N_1, \mathbf{S})$ is a (double) Tits system with affine Weyl group $W_{\mathrm{aff}} = N_1/I \cap N_1$, and the inclusion $G_1 \to G(F)$ is BN-adapted[9] of connected type.

Lemma 4.56 (T. Richarz [Ri]). (a) *Let $M_1 = M(F)_1$. Define the Iwahori–Weyl group as $\widetilde{W} := N_G(A)(F)/M_1$. Then there is an isomorphism $\widetilde{W} = W_{\mathrm{aff}} \rtimes \Lambda_G$, which is canonical given the choice of base alcove* **a**. *This gives $\widetilde{W}$ the structure of a quasi-Coxeter group.*

(b) *If $S \subset G$ is a maximal L-split torus which is F-rational and contains A, and if we set $T := \mathrm{Cent}_G(S)$ and $\widetilde{W}^{\mathrm{un}} := N_G(S)(L)/T(L)_1$, then the natural map $N_G(S)(L)^\sigma \to N_G(A)(F)$ induces an isomorphism $(\widetilde{W}^{\mathrm{un}})^\sigma \overset{\sim}{\to} \widetilde{W}$.*

Thus, in light of (b) we may reformulate the Bruhat–Tits decomposition of [HRa1, Prop. 8 and Rem. 9], as follows.

Lemma 4.57. *The map $N_G(A)(F) \to G(F)$ induces a bijection*

$$\widetilde{W} \cong I \backslash G(F) / I. \tag{4.36}$$

Further, the Bruhat order $\le$ and length function ℓ on W_{aff} extend in the usual way to $\widetilde{W}$, and we have for $w \in \widetilde{W}$ and $s \in W_{\mathrm{aff}}$ representing a simple affine reflection, the usual BN-pair relations

$$IsIwI = \begin{cases} IswI, & \text{if } w < sw \\ IwI \cup IswI, & \text{if } sw < w. \end{cases} \tag{4.37}$$

11.3.2. Iwahori factorization

Let $P = MN$ be an F-rational parabolic subgroup with Levi factor M, unipotent radical N and opposite unipotent radical $\overline{N}$. Let $I_H = I \cap H$ for $H = N$, $\overline{N}$, or M.

Lemma 4.58. *In the above situation, we have the Iwahori factorization*

$$I = I_N \cdot I_M \cdot I_{\overline{N}}. \tag{4.38}$$

Proof. We use the notation of [BT2]. By [BT2, 5.2.4] with $\Omega := \mathbf{a}$, we have

$$\mathfrak{G}_{\mathbf{a}}^\circ(\mathcal{O}^\natural) = U_{\mathbf{a}}^{+\natural} U_{\mathbf{a}}^{-\natural} N_{\mathbf{a}}^{\circ\natural},$$

---

[9] In [BT2] the symbol B is used in place of I.

where $N_{\mathbf{a}}^{o\natural} := N^\natural \cap 3^\circ(\mathcal{O}^\natural) U_{\mathbf{a}}^\natural$. Since $3^\circ(\mathcal{O}^\natural) U_{\mathbf{a}}^\natural \subset \mathfrak{G}_{\mathbf{a}}^\circ(\mathcal{O}^\natural)$, we have

$$N_{\mathbf{a}}^{o\natural} = N^\natural \cap \mathfrak{G}_{\mathbf{a}}^\circ(\mathcal{O}^\natural) = 3^\circ(\mathcal{O}^\natural).$$

The key inclusion here, $N^\natural \cap \mathfrak{G}_{\mathbf{a}}^\circ(\mathcal{O}^\natural) \subseteq 3^\circ(\mathcal{O}^\natural)$, translates in our notation to $N_G(A)(F) \cap I \subseteq M(F)_1$, which can be deduced from Lemma 4.56(a).

Translating again back to our notation we get $I = I_N \cdot I_{\overline{N}} \cdot I_M$ which is the desired equality since I_M normalizes $I_{\overline{N}}$. $\qquad\square$

11.3.3. On $M(F)^1/M(F)_1$

Lemma 4.59. *The following basic structure theory results hold:*

(a) *In the notation of* [HRo], *we have* $M(F)^1/M(F)_1 = \tilde{K}/K$, *which injects into* $G(F)^1/G(F)_1$. *Thus* $M(F)_1 = M(F)^1 \cap G(F)_1$.
(b) *The Weyl group* $W(G, A)$ *acts trivially on* $M(F)^1/M(F)_1$.
(c) *Let* $\mathbf{a} \subset \mathcal{B}(G, L)$ *denote the alcove invariant under the group* $\mathrm{Aut}(L/F) \supset \langle \sigma \rangle$ *which corresponds to the Iwahori* $I \subset G(F)$. *We assume* $I \subset K$. *Then the naive Iwahori* $\tilde{I} := G(F)^1 \cap \mathrm{Fix}(\mathbf{a}^\sigma)$ *has the following properties*

- $M(F)^1/M(F)_1 = \tilde{I}/I = \tilde{K}/K$.
- $\tilde{I} = I \cdot M(F)^1$.

Proof. Part (a): in the notation of [HRo], we know that $\Lambda_{M,\mathrm{tors}} = \tilde{K}/K$ ([HRo, Prop. 11.1.4]). Applying this to $G = M$ we get $\Lambda_{M,\mathrm{tors}} = M(F)^1/M(F)_1$. So $M(F)^1/M(F)_1 = \tilde{K}/K$. By (8.0.1) and Lemma 8.0.1 in [HRo], the latter injects into $G(F)^1/G(F)_1$. The final statement follows.

Part (b): By [HRo, Lemma 5.0.1], $W(G, A)$ has representatives in $K \cap N_G(A)(F)$. Thus it is enough to show that if $n \in K \cap N_G(A)(F)$ and $m \in M(F)^1$, then $nmn^{-1}m^{-1} \in M(F)_1$. This follows from (a), since we clearly have $nmn^{-1}m^{-1} \in M(F)^1 \cap G(F)_1$.

Part (c): First note that $M(F)^1 \subset \tilde{I}$ and $M(F)_1 \subset I$. Thus there is a commutative diagram

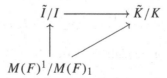

The oblique arrow is bijective by (a). We claim the horizontal arrow is injective, that is, $\tilde{I} \cap K = I$. But $\tilde{I} \cap K = G(F)^1 \cap \mathrm{Fix}(\mathbf{a}^\sigma) \cap G(F)_1 \cap \mathrm{Fix}(v_F)$, where v_F is the special vertex in $\mathcal{B}(G_{\mathrm{ad}}, F)$ corresponding to K (cf. [HRo, Lem. 8.0.1]). Thus $\tilde{I} \cap K = G(F)_1 \cap \mathrm{Fix}(\mathbf{a}^\sigma) = I$ by Remark 8.0.2 of [HRo].

It now follows that all arrows in the diagram are bijective. This implies both statements in (c).   □

Remark 4.60. Let $P = MN$ be as above. We deduce from (c) and (4.38) the Iwahori factorization for $\tilde{I}$

$$\tilde{I} = I_N \cdot M(F)^1 \cdot I_{\overline{N}}, \tag{4.39}$$

using the fact that $M(F)^1$ normalizes I_N and $I_{\overline{N}}$.

11.3.4. Iwasawa decomposition

Next we need to establish a suitable form of the Iwasawa decomposition. Let $P = MN$ be as above.

Lemma 4.61. *The inclusion $N_G(A)(F) \hookrightarrow G(F)$ induces bijections*

$$\tilde{W} := N_G(A)(F)/M(F)_1 \stackrel{\sim}{\to} M(F)_1 N(F)\backslash G(F)/I \tag{4.40}$$

$$W(G, A) = N_G(A)(F)/M(F) \stackrel{\sim}{\to} P(F)\backslash G(F)/I. \tag{4.41}$$

Proof. In view of the decomposition $\tilde{W} = \Omega_M^\sigma \rtimes W(G, A)$ (cf. Lemma 3.0.1(III) of [HRo] plus Lemma 4.56(b)) and the Kottwitz isomorphism $\Omega_M^\sigma \stackrel{\sim}{\to} M(F)/M(F)_1$ (cf. Lemma 3.0.1 [HRo]), it suffices to prove (4.40).

For $x \in \mathcal{B}(G, F)$, let $P_x \subset G(F)$ denote the subgroup fixing x. By [Land, Prop. 12.9], we have

$$G(F) = N(F) \cdot N_G(A)(F) \cdot P_x.$$

For sufficiently generic points $x \in \mathbf{a}^\sigma$, we have $P_x = \tilde{I}$, which is $M(F)^1 I$ by Lemma 4.59(c). Since $M(F)^1 \subset N_G(A)(F)$, we have $G(F) = N(F) \cdot N_G(A)(F) \cdot I$ and the map (4.40) is surjective.

To prove injectivity, assume $n_1 = um_0 \cdot n_2 \cdot j$ for $u \in N(F)$, $m_0 \in M(F)_1$, $n_1, n_2 \in N_G(A)(F)$, and $j \in I$. There exists $z \in Z(M)(F)$ such that $zuz^{-1} \in I_N$ (cf. e.g. [BK, Lem. 6.14]). Then

$$zn_2 = (zuz^{-1})m_0 \cdot zn_2 \cdot j \in Izn_2 I,$$

and so by (4.36), $zn_2 \equiv zn_2$ modulo $M(F)_1$.   □

Lemma 4.62. *If $x, y \in \tilde{W}$ and $M(F)_1 N(F)xI \cap IyI \neq \emptyset$, then $x \leq y$ in the Bruhat order on $\tilde{W}$ determined by I.*

Proof. This follows from the BN-pair relations (4.37) as in the proof of the Claim in Lemma 1.6.1 of [HKP].   □

11.3.5. The universal unramified principal series module M

Define

$$\mathbf{M} = C_c(M(F)_1 N(F) \backslash G(F)/I).$$

The subscript "c" means we are considering functions supported on finitely many double cosets. Some basic facts about $\mathbf{M}$ were given in [HKP] for the special case where G is split, and here we need to state those facts in the current general situation.

Abbreviate by setting $H = \mathcal{H}(G(F), I)$ and $R = \mathbb{C}[\Lambda_M]$. Then $f \in H$ acts on the left on $\mathbf{M}$ by right convolutions by $\check{f}$, which is defined by $\check{f}(g) = f(g^{-1})$. The same goes for the normalized induced representation $i_P^G(\tilde{\chi})^I = \mathrm{Ind}_P^G(\delta_P^{1/2} \tilde{\chi})^I$, where $\tilde{\chi}$ is a character on $M(F)/M(F)_1$. Moreover, R acts on the left on $\mathbf{M}$ by normalized left convolutions: for $r \in R$ and $\phi \in \mathbf{M}, m \in M(F)$,

$$(r \cdot \phi)(m) = \int_{M(F)} r(y) \delta_P^{1/2}(y) \phi(y^{-1}m) \, dy$$

where $\mathrm{vol}_{dy}(M(F)_1) = 1$. The actions of R and H commute, so $\mathbf{M}$ is an (R, H)-bimodule.

Lemma 4.63. *The following statements hold.*

(a) *Any character $\tilde{\chi}^{-1} : M(F)/M(F)_1 \to \mathbb{C}^\times$ extends to an algebra homomorphism $\tilde{\chi}^{-1} : R \to \mathbb{C}$, and there is an isomorphism of left H-modules*

$$\mathbb{C} \otimes_{R, \tilde{\chi}^{-1}} \mathbf{M} = i_P^G(\tilde{\chi})^I.$$

(b) *For $w \in W(G, A) =: W$, set $v_w = 1_{M(F)_1 N(F)wI} \in \mathbf{M}$. Then $\mathbf{M}$ is free of rank 1 over H with canonical generator v_1.*

(c) $\mathbf{M}$ *is free as an R-module, with basis $\{v_w\}_{w \in W}$.*

Proof. The proofs for (a–b) are nearly identical to their analogues for split groups in [HKP]. Part (a) is formal. Part (b) relies on the Bruhat–Tits decomposition (4.36), the Iwasawa decomposition (4.40), and Lemma 4.62.

Part (c) was not explicitly mentioned in [HKP]. But it can be proved using (4.40) along with the relations analogous to [HKP, (1.6.1–1.6.2)], for which the Iwahori factorization (4.38) is the main ingredient.   □

11.4. Why $(M(F)_1, 1)$ is an $\mathfrak{S}_M$-type

We let χ range over the characters of $M(F)^1/M(F)_1$. Let $\tilde{\chi}$ denote any extension to a character of the finitely generated abelian group $M(F)/M(F)_1$. Fix one such extension $\tilde{\chi}_0$. Note that the inertial class $[M(F), \tilde{\chi}_0]_M$ consists of all pairs $(M(F), \tilde{\chi})$, since $M(F)$-conjugation does not introduce any new characters on $M(F)$. Therefore we may abuse notation and denote this inertial class by $[M(F), \tilde{\chi}]_M =: \mathfrak{s}_{\chi}^M$. Let $\mathfrak{S}_M := \{[M(F), \tilde{\chi}]_M \mid \chi \text{ ranges as above}\}$. This is a finite set of inertial classes, in bijective correspondence with $M(F)^1/M(F)_1$.

Proposition 4.64. *The pair* $(M(F)_1, 1)$ *is a Bushnell–Kutzko type for* $\mathfrak{S}_M$.

Note: This proposition simply makes precise the last paragraph of [BK, 9.2].

Proof. Let σ be an irreducible smooth representation of $M(F)$. We must show that $\sigma = \tilde{\chi}$ for some $\tilde{\chi}$ iff $\sigma|_{M(F)_1} \supset 1$.

($\Rightarrow$): Obvious.

($\Leftarrow$): We see that $\sigma \ni v \neq 0$ on which $M(F)_1$ acts trivially. Since σ is irreducible, it coincides with the smallest $M(F)$-subrepresentation containing v, and then since $M(F)_1 \triangleleft M(F)$, we see that $M(F)_1$ acts trivially on all of σ; further, σ is necessarily finite-dimensional over $\mathbb{C}$. Since $M(F)/M(F)_1$ is abelian, σ contains an $M(F)$-stable line, since a commuting set of matrices can be simultaneously triangularized. This line is all of σ since σ is irreducible. Thus σ is 1-dimensional, and so $\sigma = \tilde{\chi}$ for some $\tilde{\chi}$. $\qquad\square$

11.5. Why $(I, 1)$ is an $\mathfrak{S}_G$-type

We define $\mathfrak{S}_G = \{[t]_G \mid [t]_M \in \mathfrak{S}_M\}$. The map $[M, \tilde{\chi}]_M \mapsto [M, \tilde{\chi}]_G$ is injective: if $[M, \tilde{\chi}_1]_G = [M, \tilde{\chi}_2]_G$, then there exists $n \in N_G(A)(F)$ such that $^n(\tilde{\chi}_1) = \tilde{\chi}_2 \eta$ for some character η on $M(F)/M(F)^1$. Restricting to $M(F)^1$ and using $^n(\chi_1) = \chi_1$ (Lemma 4.59(b)), we see $\chi_1 = \chi_2$. So $\mathfrak{S}_M \cong \mathfrak{S}_G$ via $[t]_M \mapsto [t]_G$.

We saw above that $(M(F)_1, 1)$ is an $\mathfrak{S}_M$-type. The fact that $(I, 1)$ is an $\mathfrak{S}_G$-type follows from [BK], Theorem 8.3, once we check the following proposition.

Proposition 4.65. *The pair* $(I, 1)$ *is a G-cover for* $(M(F)_1, 1)$ *in the sense of* [BK, Definition 8.1].

Proof. We need to check the three conditions (i–iii) of Definition 8.1. First (i), the fact that $(I, 1)$ is decomposed with respect to (M, P) in the sense of [BK, (6.1)], follows from the Iwahori factorization $I = I_N \cdot I_M \cdot I_{\overline{N}}$ discussed in Remark 4.60. The equality $I \cap M(F) = M(F)_1$ gives condition (ii).

Finally we must prove (iii): for every F-parabolic P with Levi factor M, there exists an invertible element of $\mathcal{H}(G(F), I)$ supported on $Iz_P I$, where z_P belongs to $Z(M)(F)$ and is strongly (P, I)-positive. The existence of elements $z_P \in Z(M)(F)$ which are strongly (P, I)-positive is proved in [BK, Lemma 6.14]. Any corresponding characteristic function $1_{Iz_P I}$ is invertible in $\mathcal{H}(G(F), I)$, as follows from the Iwahori–Matsumoto presentation of $\mathcal{H}(G(F), I)$. (This presentation itself is easy to prove using (4.37).)   □

11.6. Structure of the Bernstein varieties

Let $\mathfrak{R}(G)$ denote the category of smooth representations of $G(F)$, and let $\mathfrak{R}_\chi(G)$ denote the full subcategory corresponding to the inertial class $[M, \tilde{\chi}]_G$. That is, a representation $(\pi, V) \in \mathfrak{R}(G)$ is an object of $\mathfrak{R}_\chi(G)$ if and only if for each irreducible subquotient π' of π, there exists an extension $\tilde{\chi}$ of χ such that π' is a subquotient of $\mathrm{Ind}_P^G(\delta_P^{1/2}\tilde{\chi})$.

We review the structure of the Bernstein varieties $\mathfrak{X}_\chi^G$ and $\mathfrak{X}_\chi^M$. In this discussion, for each χ we fix an extension $\tilde{\chi}$ of χ once and for all – the structures we define will be independent of the choice of $\tilde{\chi}$, i.e. uniquely determined by (M, χ) up to a unique isomorphism.

As a set $\mathfrak{X}_\chi^G$ (resp. $\mathfrak{X}_\chi^M$) consists of the elements $(M, \tilde{\chi}\eta)_G$ (resp. $(M, \tilde{\chi}\eta)_M$) belonging to the inertial equivalence class $[M, \tilde{\chi}]_G$ (resp. $[M, \tilde{\chi}]_M$) as η ranges over the set $X(M)$ of unramified $\mathbb{C}^\times$-valued characters on $M(F)$ (unramified means it factors through $M(F)/M(F)^1$).

The map $X(M) \to \mathfrak{X}_\chi^M$, $\eta \mapsto (M, \tilde{\chi}\eta)_M$, is a bijection. Since $X(M)$ is a complex torus, this gives $\mathfrak{X}_\chi^M$ the structure of a complex torus. More canonically, $\mathfrak{X}_\chi^M$ is just the variety of *all* extensions $\tilde{\chi}$ of χ, and it is a torsor under the torus $X(M)$.

Now fix $\tilde{\chi}$ again. There is a surjective map

$$\mathfrak{X}_\chi^M \to \mathfrak{X}_\chi^G$$

$$(M, \tilde{\chi}\eta)_M \mapsto (M, \tilde{\chi}\eta)_G.$$

Since $W := W(G, A)$ acts trivially on $M(F)^1/M(F)_1$ (Lemma 4.59), one can prove that the fibers of this map are precisely the W-orbits on $\mathfrak{X}_\chi^M$. Thus as sets

$$W\backslash\mathfrak{X}_\chi^M = \mathfrak{X}_\chi^G,$$

and this gives $\mathfrak{X}_\chi^G$ the structure of an affine variety over $\mathbb{C}$. Having chosen the isomorphism $X(M) \cong \mathfrak{X}_\chi^M$ as above, we can transport the W-action on $\mathfrak{X}_\chi^M$ over to an action on $X(M)$. *This action depends on the choice of $\tilde{\chi}$ and is not the usual action unless $\tilde{\chi}$ is W-invariant.* We obtain $\mathfrak{X}_\chi^G = W\backslash_{\tilde{\chi}} X(M)$, where the latter denotes the quotient with respect to this *unusual* action on $X(M)$.

Let $\mathbb{C}[\mathfrak{X}_\chi^G]$ denote the ring of regular functions on the variety $\mathfrak{X}_\chi^G$. The algebraic morphism $\mathfrak{X}_\chi^M \to \mathcal{X}_\chi^G$ induces an isomorphism of algebras

$$\mathbb{C}[\mathfrak{X}_\chi^G] \overset{\sim}{\to} \mathbb{C}[\mathfrak{X}_\chi^M]^W. \qquad (4.42)$$

11.7. Consequences of the theory of types

Let us define convolution in $\mathcal{H}(G(F), I)$ using the Haar measure dx on $G(F)$ which gives I volume 1. Let $\mathcal{Z}(G(F), I)$ denote the center of $\mathcal{H}(G(F), I)$.

We define for each $\chi \in (M(F)^1/M(F)_1)^\vee$ a function $e_\chi \in \mathcal{H}(G(F), I)$ by requiring it to be supported on $\tilde{I}$, and by setting $e_\chi(y) = \mathrm{vol}_{dx}(\tilde{I})^{-1} \chi(\bar{y})^{-1}$ if $y \in \tilde{I}$. Here we regard χ as a character on $\tilde{I}/I$ (cf. Lemma 4.59) and let $\bar{y} \in \tilde{I}/I$ denote the image of y. If $y = n_+ \cdot m^1 \cdot n_- \in I_N \cdot M(F)^1 \cdot I_{\overline{N}}$, then $e_\chi(y) = \mathrm{vol}_{dx}(\tilde{I})^{-1} \chi(m^1)^{-1}$.

Lemma 4.66. *The functions $\{e_\chi\}_\chi$ give a complete set of central orthogonal idempotents for $\mathcal{H}(G(F), I)$:*

(a) $e_\chi \in \mathcal{Z}(G(F), I)$;
(b) $e_\chi e_{\chi'} = \delta_{\chi,\chi'} e_\chi$, *there $\delta_{\chi,\chi'}$ is the Kronecker delta function;*
(c) $1_I = \sum_\chi e_\chi$.

Proof. The proof is a straightforward exercise for the reader. For parts (a–b), use the fact that $M(F)^1$ normalizes I, that $G(F) = I \cdot N_G(A)(F) \cdot I$, and that $W(G, A)$ acts trivially on $M(F)^1/M(F)_1$ (cf. Lemma 4.59). $\qquad\square$

Proposition 4.67. *The idempotents e_χ are the elements in the Bernstein center which project the category $\mathfrak{R}(G)$ onto the various Bernstein components $\mathfrak{R}_\chi(G)$. That is, there is a canonical isomorphism of algebras*

$$\mathcal{H}(G(F), I) = \prod_\chi e_\chi \mathcal{H}(G(F), I) e_\chi$$

and, for any smooth representation $(\pi, V) \in \mathfrak{R}(G)$, the $G(F)$-module spanned by the χ-isotypical vectors $V^\chi = \pi(e_\chi)V$ is the component of V lying in the subcategory $\mathfrak{R}_\chi(G)$. Finally,

$$e_\chi \mathcal{H}(G(F), I) e_\chi = \mathcal{H}(G(F), \tilde{I}, \chi),$$

the right-hand side being the algebra of I-bi-invariant $\mathbb{C}$-valued functions $f \in C_c(G(F))$ such that $f(\tilde{i}_1 g \tilde{i}_2) = \chi(\tilde{i}_1)^{-1} f(g) \chi(\tilde{i}_2)^{-1}$ for all $g \in G(F)$ and $\tilde{i}_1, \tilde{i}_2 \in \tilde{I}$.

The following records the standard consequences of the fact that $(I, 1)$ is an $\mathfrak{S}_G$-type (see [BK, Theorem 4.3]). Let $\mathfrak{R}_I(G)$ denote the full subcategory of $\mathfrak{R}(G)$ whose objects are generated as G-modules by their I-invariant vectors.

Proposition 4.68. *As subcategories of $\mathfrak{R}(G)$, we have the equality $\mathfrak{R}_I(G) = \prod_\chi \mathfrak{R}_\chi(G)$. In particular, an irreducible representation $(\pi, V) \in \mathfrak{R}(G)$*

belongs to $\mathfrak{R}_I(G)$ if and only if $\pi \in \mathfrak{R}_\chi(G)$ for some χ. Furthermore, there is an equivalence of categories

$$\mathfrak{R}_I(G) \xrightarrow{\sim} \mathcal{H}(G(F), I)\text{-Mod}$$

$$(\pi, V) \mapsto V^I.$$

Finally, $\mathcal{Z}(G(F), I)$ is isomorphic with the center of the category $\prod_\chi \mathfrak{R}_\chi(G)$, which according to Bernstein's theory is the ring $\prod_\chi \mathbb{C}[\mathfrak{X}_\chi^G]$.

Concretely, the map $\mathcal{Z}(G(F), I) \to \prod_\lambda \mathbb{C}[\mathfrak{X}_\chi^G]$, $z \mapsto \hat{z}$, is characterized as follows: for every χ and every $(M, \bar{\chi})_G \in \mathfrak{X}_\chi^G$, $z \in \mathcal{Z}(G(F), I)$ acts on $\mathrm{Ind}_P^G(\delta_P^{1/2}\bar{\chi})^I$ by the scalar $\hat{z}(\bar{\chi})$.

Let us single out what happens in the special case of $G = M$. We can identify $\mathcal{H}(M(F), M(F)_1) = \mathbb{C}[\Lambda_M]$. Let e_χ^M denote the idempotent in $\mathcal{H}(M(F), M(F)_1)$ analogous to e_χ, for the case $G = M$. By Propositions 4.67 and 4.68 for $G = M$, we have

$$\mathcal{H}(M(F), M(F)_1) = \prod_\chi e_\chi^M \mathcal{H}(M(F), M(F)_1) e_\chi^M = \prod_\chi \mathbb{C}[\mathfrak{X}_\chi^M], \quad (4.43)$$

the last equality holding since $\mathcal{H}(M(F), M(F)_1)$ is already commutative. Thus, the ring

$$e_\chi^M \mathcal{H}(MF), M(F)_1) e_\chi^M$$

can be regarded as the ring of regular functions on the variety $\mathfrak{X}_\chi^M$ of all extensions $\bar{\chi}$ of χ.

11.8. The embedding of $\mathbb{C}[\Lambda_M]^W$ into $\mathcal{Z}(G(F), I)$

We make use of the following special case of a general construction of Bushnell–Kutzko [BK]: for any F-parabolic P with Levi factor M, there is an injective algebra homomorphism

$$t_P : \mathcal{H}(M(F), M(F)_1) \to \mathcal{H}(G(F), I)$$

which is uniquely characterized by the property that for each $(\pi, V) \in \mathfrak{R}_I(G)$, $v \in V^I$, and $h \in \mathcal{H}(M(F), M(F)_1)$, we have the identity

$$q_\pi(t_P(h) \cdot v) = h \cdot q_\pi(v). \quad (4.44)$$

Here $q_\pi : V^I \xrightarrow{\sim} V_N^{M(F)_1}$ is an isomorphism, which is induced by the canonical projection $V \to V_N$ to the (unnormalized) Jacquet module. See [BK, Thm. 7.9].

It turns out that it is better to work with a different normalization. We define another injective algebra homomorphism

$$\theta_P : \mathcal{H}(M(F), M(F)_1) \;\to\; \mathcal{H}(G(F), I)$$

$$h \;\mapsto\; t_P(\delta_P^{-1/2} h).$$

Then using (4.44) θ_P satisfies

$$q_\pi(\theta_P(h) \cdot v) = (\delta_P^{-1/2} h) \cdot q_\pi(v). \tag{4.45}$$

We view $\tilde{\chi}$ as a varying element of the Zariski-dense subset S of the variety of all characters on the finitely generated abelian group $M(F)/M(F)_1$, consisting of those regular characters $\tilde{\chi}$ such that $V(\tilde{\chi}) := i_P^G(\tilde{\chi}) := \mathrm{Ind}_P^G(\delta_P^{1/2}\tilde{\chi})$ is irreducible as an object of $\mathfrak{R}(G)$. We apply the above discussion to the representations $V := V(\tilde{\chi})$ with $\tilde{\chi} \in S$. By a result of Casselman [Cas], we know that as $M(F)$-modules

$$V_N = \bigoplus_{w \in W} \delta_P^{1/2}(^w\tilde{\chi})$$

and that $M(F)_1$ acts trivially on this module. Now suppose $h \in \mathbb{C}[\Lambda_M]^W$. Then $\delta_P^{-1/2} h$ acts on $V_N = V_N^{M(F)_1}$ by the scalar $h(\tilde{\chi})$ (viewing h as a regular function on $\mathfrak{X}_\chi^M$). It follows from (4.45) that

$$\theta_P(h) \text{ acts by the scalar } h(\tilde{\chi}) \text{ on } i_P^G(\tilde{\chi})^I, \text{ for } \tilde{\chi} \in S.$$

Now let $f \in H$ be arbitrary, and set $\epsilon := f * \theta_P(h) - \theta_P(h) * f \in H$. We see that

$$\epsilon \text{ acts by zero on } i_P^G(\tilde{\chi})^I \text{ for every } \tilde{\chi} \in S. \tag{4.46}$$

We claim that $\epsilon = 0$. Recall that $\epsilon \in H$ gives an R-linear endomorphism of $\mathbf{M}$, hence by Lemma 4.63(c) may be represented by an $|W| \times |W|$ matrix E with entries in R. Now $\mathrm{Spec}(R) = \mathrm{Spec}(\mathbb{C}[\Lambda_M])$ is a diagonalizable group scheme over $\mathbb{C}$ with character group Λ_M. Hence R is a reduced finite-type $\mathbb{C}$-algebra and its maximal ideals are precisely the kernels of the $\mathbb{C}$-algebra homomorphisms $\tilde{\chi}^{-1} : R \to \mathbb{C}$ coming into Lemma 4.63(a). By that Lemma and (4.46), we see that $E \equiv 0 \pmod{\mathfrak{m}}$ for a Zariski-dense set of maximal ideals $\mathfrak{m}$ in $\mathrm{Spec}(R)$. Since R is reduced and finite-type over $\mathbb{C}$, this implies that the entries of E are identically zero. This proves the claim because $\mathbf{M}$ is free of rank 1 over H (Lemma 4.63(b)).

Since f was arbitrary, we get $\theta_P(h) \in \mathcal{Z}(G(F), I)$, as desired. We have proved the following result.

Lemma 4.69. *The map* $\theta_P : \mathbb{C}[\Lambda_M] \to \mathcal{H}(G(F), I)$ *restricts to give an embedding* $\mathbb{C}[\Lambda_M]^W \to \mathcal{Z}(G(F), I)$.

11.9. The center of the Iwahori–Hecke algebra

Theorem 4.70. *The map* θ_P *gives an algebra isomorphism* $\mathbb{C}[\Lambda_M]^W \xrightarrow{\sim} \mathcal{Z}(G(F), I)$. *Further, this isomorphism is independent of the choice of parabolic* P *containing* M *as a Levi factor.*

Proof. The description of θ_P above, and the preceding discussion, show that we have a commutative diagram

$$
\begin{array}{ccc}
\mathbb{C}[\Lambda_M]^W & \xrightarrow{\ \sim\ } & \prod_\chi \mathbb{C}[\mathfrak{X}_\chi^M]^W \\
\Big\downarrow{\scriptstyle \theta_P} & & \Big\uparrow \\
\mathcal{Z}(G(F), I) & \xrightarrow{\ \sim\ } & \prod_\chi \mathbb{C}[\mathfrak{X}_\chi^G].
\end{array}
$$

The left vertical arrow is bijective because the right vertical arrow is, by (4.42). $\qquad\square$

11.10. Bernstein isomorphisms and functions

Putting together Roche's theorem 4.55 with Theorem 4.70, we deduce a more general result that holds for any parahoric subgroup $J \supseteq I$.

Theorem 4.71. *The composition*

$$
\mathbb{C}[\Lambda_M]^W \xrightarrow{\ \theta_P\ } \mathcal{Z}(G(F), I) \xrightarrow{\ -*_I 1_J\ } \mathcal{Z}(G(F), J) \tag{4.47}
$$

is an isomorphism. We call this map the Bernstein isomorphism.

Definition 4.72. Given $\mu \in \Lambda_M$, we define the *Bernstein function* $z_\mu \in \mathcal{Z}(G(F), J)$ to be the image of the symmetric monomial function $\sum_{\lambda \in W \cdot \mu} \lambda \in \mathbb{C}[\Lambda_M]^W$ under the Bernstein isomorphism (4.47).

11.11. Compatibility with constant terms

Recall $M = \mathrm{Cent}_G(A)$ is a minimal F-Levi subgroup of G and $P = MN$ is a minimal F-parabolic subgroup with Levi factor M and unipotent radical N. Let $Q = LR$ be another F-parabolic subgroup with F-Levi factor L and unipotent radical R. Assume $Q \supseteq P$; then $L \supseteq M$ and $R \subseteq N$. Further L contains a minimal F-parabolic subgroup $L \cap P = M \cdot (L \cap N)$, and $N = L \cap N \cdot R$.

If $J \subset G(F)$ is a parahoric subgroup corresponding to a facet in the apartment of the Bruhat–Tits building of $G(F)$ corresponding to A, then $J_L := J \cap L$ is a parahoric subgroup of $L(F)$ (by [HRo, Lem. 4.1.1]).

Given $f \in \mathcal{H}(G(F), J)$, define $f^{(Q)} \in \mathcal{H}(L(F), J_L)$ by

$$f^{(Q)}(l) = \delta_Q^{1/2}(l) \int_R f(lr)\, dr = \delta_Q^{-1/2}(l) \int_R f(rl)\, dr,$$

where $\mathrm{vol}_{dr}(J \cap R) = 1$. An argument similar to Lemma 4.7.2 of [H09] shows that $f \mapsto f^{(Q)}$ sends $\mathcal{Z}(G(F), J)$ into $\mathcal{Z}(L(F), J_L)$, and determines a map c_L^G making the following diagram commute:

$$
\begin{array}{ccc}
\mathbb{C}[\Lambda_M]^{W(G,A)} & \xrightarrow{\ \sim\ } & \mathcal{Z}(G(F), J) \\
\Big\uparrow & & \Big\downarrow{\scriptstyle c_L^G} \\
\mathbb{C}[\Lambda_M]^{W(L,A)} & \xrightarrow{\ \sim\ } & \mathcal{Z}(L(F), J_L).
\end{array}
\qquad (4.48)
$$

The diagram shows that c_L^G is indeed an (injective) algebra homomorphism and, as the notation suggests, is independent of the choice of parabolic subgroup Q having L as a Levi factor. We call c_L^G the *constant term homomorphism*.

By its very construction, the map $\theta_M : \mathbb{C}[\Lambda_M] \xrightarrow{\sim} \mathcal{H}(M(F), M(F)_1)$ has its inverse induced by the Kottwitz isomorphism $\kappa_M(F) : M(F)/M(F)_1 \xrightarrow{\sim} \Lambda_M$. By taking $L = M$ in (4.48), this remark allows us to write down the inverse of θ_P in general.

Corollary 4.73. *The composition $\kappa_M(F) \circ c_M^G$ takes values in $\mathbb{C}[\Lambda_M]^{W(G,A)}$ and is the inverse of the Bernstein isomorphism θ_P.*

11.12. Transfer homomorphisms

11.12.1. Construction

Transfer homomorphisms were defined for special maximal parahoric Hecke algebras in [HRo]. By virtue of the Bernstein isomorphisms (4.47), we can now define these homomorphisms on the level of centers of arbitrary parahoric Hecke algebras.

Let us recall the general set-up from [HRo, §11.2]. Let G^* be a quasi-split group over F. Let F^s denote a separable closure of F, and set $\Gamma = \mathrm{Gal}(F^s/F)$. Recall that an inner form of G^* is a pair (G, Ψ) consisting of a connected reductive F-group G and a Γ-stable $G_{\mathrm{ad}}^*(F^s)$-orbit Ψ of F^s-isomorphisms $\psi : G \to G^*$. The set of isomorphism classes of pairs (G, Ψ) corresponds bijectively to $H^1(F, G_{\mathrm{ad}}^*)$.

Fix once and for all parahoric subgroups $J \subset G(F)$ and $J^* \subset G^*(F)$. Choose any maximal F-split tori $A \subset G$ and $A^* \subset G^*$ such that the facet fixed by J (resp. J^*) is contained in the apartment of the building $\mathcal{B}(G, F)$

(resp. $\mathcal{B}(G^*, F)$) corresponding to the torus A (resp. A^*). Let $M = \mathrm{Cent}_G(A)$ and $T^* = \mathrm{Cent}_{G^*}(A^*)$, a maximal F-torus in G^*.

Now *choose* an F-parabolic subgroup $P \subset G$ having M as Levi factor, and an F-rational Borel subgroup $B^* \subset G^*$ having T^* as Levi factor. Then there exists a unique parabolic subgroup $P^* \subset G^*$ such that $P^* \supseteq B^*$ and P^* is $G^*(F^s)$-conjugate to $\psi(P)$ for every $\psi \in \Psi$. Let M^* be the unique Levi factor of P^* containing T^*. Then define

$$\Psi_M = \{\psi \in \Psi \mid \psi(P) = P^*, \ \psi(M) = M^*\}.$$

The set Ψ_M is a nonempty Γ-stable $M_{\mathrm{ad}}^*(F^s)$-orbit of F^s-isomorphisms $M \to M^*$, and so (M, Ψ_M) is an inner form of M^*. Choose any $\psi_0 \in \Psi_M$. Then since $\psi_0|A$ is F-rational, $\psi_0(A)$ is an F-split torus in $Z(M^*)$ and hence $\psi_0(A) \subseteq A^*$.

Now ψ_0 induces a Γ-equivariant map $Z(\widehat{M}) \overset{\sim}{\to} Z(\widehat{M^*}) \hookrightarrow \widehat{T^*}$ and hence a homomorphism

$$t_{A^*,A} : \mathbb{C}[X^*(\widehat{T^*})_{I_F^*}^{\Phi_F^*}]^{W(G^*,A^*)} \longrightarrow \mathbb{C}[X^*(Z(\widehat{M}))_{I_F}^{\Phi_F}]^{W(G,A)},$$

where $(\cdot)^*$ designates the Galois action on G^* (for Weyl-group equivariance see [HRo, §12.2]). Since Ψ_M is a torsor for M_{ad}^* this homomorphism does not depend on the choice of $\psi_0 \in \Psi_M$. Further, it depends only on the choice of A^* and A, and not on the choice of the parabolic subgroups $P \supset M$ and $B^* \supset T^*$ we made in constructing it.

Definition 4.74. Let $J \subset G(F)$ and $J^* \subset G^*(F)$ be any parahoric subgroups and choose compatible maximal F-split tori A resp. A^* as above. Then we define the *transfer homomorphism* $t : \mathcal{Z}(G^*(F), J^*) \to \mathcal{Z}(G(F), J)$ to be the unique homomorphism making the following diagram commute

$$
\begin{array}{ccc}
\mathcal{Z}(G^*(F), J^*) & \overset{t}{\longrightarrow} & \mathcal{Z}(G(F), J) \\
\wr \downarrow & & \downarrow \wr \\
\mathbb{C}[X^*(\widehat{T^*})_{I_F^*}^{\Phi_F^*}]^{W(G^*,A^*)} & \overset{t_{A^*,A}}{\longrightarrow} & \mathbb{C}[X^*(Z(\widehat{M}))_{I_F}^{\Phi_F}]^{W(G,A)},
\end{array}
$$

where the vertical arrows are the Bernstein isomorphisms.

By [BT2, 4.6.28], any two choices for A (resp. A^*) are J-(resp. J^*-) conjugate. Using Corollary 4.73 it follows that t is independent of the choice of A and A^* and is a completely canonical homomorphism.

Remark 4.75. The map

$$t_{A^*,A} : X^*(\widehat{T^*})_{I_F^*}^{\Phi_F^*} \to X^*(Z(\widehat{M}))_{I_F}^{\Phi_F}$$

is surjective. Via the Kottwitz homomorphism we may view this as the composition

$$T^*(F)/T^*(F)_1 \longrightarrow M^*(F)/M^*(F)_1 \xrightarrow[\sim]{\psi_0^{-1}} M(F)/M(F)_1 \qquad (4.49)$$

where the first arrow is induced by the inclusion $T^* \hookrightarrow M^*$. It is enough to observe that $M^*(F) = T^*(F) \cdot M^*(F)_1$, which in turn follows from the Iwasawa decomposition (cf. (4.41)) for $M^*(F)$, which states that $M^*(F) = T^*(F) \cdot U_{M^*}^*(F) \cdot K_{M^*}$ for an F-rational Borel subgroup $B_{M^*}^* = T^* \cdot U_{M^*}^*$ and a special maximal parahoric subgroup K_{M^*} in M^*, and from the vanishing of the Kottwitz homomorphism on $U_{M^*}^*(F) \cdot K_{M^*}$.

11.12.2. Normalized transfer homomorphism

The transfer homomorphism is slightly too naive, and it is necessary to normalize it in order to get a homomorphism which has the required properties. We need to define normalized homomorphisms $\widetilde{t}_{A^*,A}$ on Weyl-group invariants, for which the following lemma is needed.

Lemma 4.76. *Recall that $T^* = \mathrm{Cent}_{G^*}(A^*)$ is a maximal torus in G^* defined over F; let S^* be the F^{un}-split component of T^*, a maximal F^{un}-split torus in G^* defined over F and containing A^*. We have $T^* = \mathrm{Cent}_{G^*}(A^*) = \mathrm{Cent}_{G^*}(S^*)$. Choose a maximal F^{un}-split torus $S \subset G$ which is defined over F and which contains A, and set $T = \mathrm{Cent}_G(S)$. Choose $\psi_0 \in \Psi_M$ such that ψ_0 is defined over F^{un} and satisfies $\psi_0(S) = S^*$ and hence $\psi_0(T) = T^*$. Then the diagram*

$$
\begin{array}{ccc}
W(G, A) & \xdashrightarrow[\sim]{\psi_0^\natural} & W(G^*, A^*)/W(M^*, A^*) \\
\Big\downarrow{\wr} & & \Big\downarrow{\wr} \\
\big[W(G, S)/W(M, S)\big]^{\Phi_F} & \xrightarrow[\sim]{\psi_0} & \big[W(G^*, S^*)/W(M^*, S^*)\big]^{\Phi_F^*}
\end{array}
$$

defines a bijective map $\psi_0^\natural$. It depends on the choice of the data P, B^* used to define Ψ_M and M^*, but it is independent of the choices of S and $\psi_0 \in \Psi_M$ with the stated properties.

Proof. The left vertical arrow is [HRo, Lem. 6.1.2]. The right vertical arrow is described in [HRo, Prop. 12.1.1]. The proof of the latter justifies the lower horizontal arrow. Indeed, given $w \in W(G, A)$ we may choose a representative $n \in N_G(S)(L)^{\Phi_F}$ (cf. [HRo]). We have $\psi_0^{-1} \circ \Phi_F^* \circ \psi_0 \circ \Phi_F^{-1} = \mathrm{Int}(m_\Phi)$ for some $m_\Phi \in N_M(S)(F^s)$. Since n is Φ_F-fixed, we get

$$\Phi_F^*(\psi_0(n)) = \psi_0(n) \cdot \big[\psi_0(n)^{-1}\psi_0(m_\Phi)\psi_0(n)\psi_0(m_\Phi)^{-1}\big].$$

As n normalizes M and hence $\psi_0(n)$ normalizes M^*, this shows that $\psi_0(n)W(M^*, S^*)$ is Φ_F^*-fixed.

There exists $m_n^* \in N_{M^*}(S^*)(L)$ such that $\psi_0(n)m_n^* \in N_{G^*}(A^*)(F)$. Then $\psi_0^\flat(w)$ is the image of $\psi_0(n)m_n^*$ in $W(G^*, A^*)/W(M^*, A^*)$. The independence statement is proved using this description. $\qquad\square$

Via the Kottwitz homomorphism we can view $t_{A^*, A}$ as induced by the composition (4.49). We now alter this slightly.

Lemma 4.77. *Given the choices of $P \supset M$ and $B^* \supset T^*$ needed to define Ψ_M and given any F^{un}-rational $\psi_0 \in \Psi_M$, we define an algebra homomorphism*

$$\mathbb{C}[T^*(F)/T^*(F)_1] \longrightarrow \mathbb{C}[M(F)/M(F)_1] \qquad (4.50)$$

$$\sum_{t^*} a_{t^*} t^* \longmapsto \sum_m \left(\sum_{t^* \mapsto m} a_{t^*} \delta_{B^*}^{-1/2}(t^*) \delta_P^{1/2}(m) \right) \cdot m,$$

where t^* ranges over $T^*(F)/T^*(F)_1$ and m ranges over $M(F)/M(F)_1$ and $t^* \mapsto m$ means that $\psi_0^{-1}(t^*) \in mM(F)_1$, (cf. (4.49)).

Then (4.50) takes $W(G^*, A^*)$-invariants to $W(G, A)$-invariants, and the resulting homomorphism

$$\tilde{t}_{A^*, A} : \mathbb{C}[T^*(F)/T^*(F)_1]^{W(G^*, A^*)} \to \mathbb{C}[M(F)/M(F)_1]^{W(G, A)}$$

is independent of the choices of P, B^*, and F^{un}-rational $\psi_0 \in \Psi_M$.

Proof. To check the Weyl-group invariance, we may fix P and B^*, and choose S and ψ_0 as in Lemma 4.76. Suppose $\sum_{t^*} a_{t^*} t^*$ is $W(G^*, A^*)$-invariant. We need to show that the function on $M(F)/M(F)_1$

$$m \mapsto \sum_{t^* \mapsto m} a_{t^*} \delta_{B^*}^{-1/2}(t^*) \delta_P^{1/2}(m) \qquad (4.51)$$

is $W(G, A)$-invariant, and independent of the choice of P and B^*.

For $w \in W(G, A)$ choose n and m_n^* as in the proof of Lemma 4.76, and define n' by $\psi_0(n') = \psi_0(n)m_n^*$. Thus $\psi_0(n') \in N_{G^*}(A^*)(F)$ and hence it normalizes $T^*(F)$.

We claim that (4.51) takes the same values on $mM(F)_1$ and on ${}^n mM(F)_1$. First we observe that ${}^n mM(F)_1 = {}^{n'} mM(F)_1$. Setting $m_n := \psi_0^{-1}(m_n^*)$, it is enough to prove ${}^{m_n} mM(F)_1 = mM(F)_1$. But as ψ_0 is L-rational we have $m_n \in M(L)$ and so conjugation by m_n induces the identity map on $M(L)/M(L)_1$ and hence on its subset $M(F)/M(F)_1$ as well.

Now we write the value of (4.51) on ${}^{n'} mM(F)_1$ as

$$\sum_{t^{**} \mapsto n' mM(F)_1} a_{t^{**}} \delta_{B^*}^{-1/2}(t^{**}) \delta_P^{1/2}({}^{n'} m).$$

Setting $t^* = {}^{\psi_0(n')^{-1}} t^{**}$ and using $a_{t^*} = a_{t^{**}}$ (which follows from $W(G^*, A^*)$-invariance), we write the above as

$$\sum_{t^* \mapsto m} a_{t^*} \delta_{B^*}^{-1/2} ({}^{\psi_0(n')} t^*) \delta_P^{1/2} (n'm).$$

The index set is stable under the $W(M^*, A^*)$-action on $T^*(F)/T^*(F)_1$. If we look at the sum over the $W(M^*, A^*)$-orbit of a single element t_0^*, with stabilizer group $\mathrm{Stab}(t_0^*)$, we get

$$\frac{1}{|\mathrm{Stab}(t_0^*)|} \cdot a_{t_0^*} \cdot \sum_y \delta_{B_{M^*}^*}^{-1/2} ({}^{\psi_0(n')y} t_0^*), \qquad (4.52)$$

where y ranges over $W(M^*, A^*)$. Now $n \in N_G(S)(L)^{\Phi_F} \subseteq N_G(A)(F) = N_G(M)(F)$. Hence $\psi_0(n) m_n^* = \psi_0(n')$ normalizes M^* as well as T^*, and thus conjugation by $\psi_0(n')$ takes $B_{M^*}^*$ to another F-rational Borel subgroup of M^* containing T^*. Using this it is clear that (4.52) is unchanged if the superscript $\psi_0(n')$ is omitted, and this proves our claim. For the same reason (4.51) is independent of the choice of P and B^*, and similarly $\tilde{t}_{A^*,A}$ is independent of the choice of P and B^*, and of the choice of F^{un}-rational $\psi_0 \in \Psi_M$. $\qquad \square$

Now we give a normalized version of Definition 4.74.

Definition 4.78. We define the *normalized transfer homomorphism* $\tilde{t}$: $\mathcal{Z}(G^*(F), J^*) \to \mathcal{Z}(G(F), J)$ to be the unique homomorphism making the following diagram commute

$$\begin{array}{ccc}
\mathcal{Z}(G^*(F), J^*) & \xrightarrow{\tilde{t}} & \mathcal{Z}(G(F), J) \\
\downarrow{\scriptstyle\wr} & & \downarrow{\scriptstyle\wr} \\
\mathbb{C}[X^*(\widehat{T^*})_{I_F^*}^{\Phi_F^*}]^{W(G^*,A^*)} & \xrightarrow{\tilde{t}_{A^*,A}} & \mathbb{C}[X^*(Z(\widehat{M}))_{I_F}^{\Phi_F}]^{W(G,A)},
\end{array}$$

where the vertical arrows are the Bernstein isomorphisms.

As was the case for t, the homomorphism $\tilde{t}$ is independent of the choice of A and A^*, and it is a completely canonical homomorphism.

The following shows it is compatible with constant term homomorphisms.

11.12.3. Normalized transfer homomorphisms and constant terms

We use the notation of §11.11. Write $L = \mathrm{Cent}_G(A_L)$ for some torus $A_L \subseteq A$. Let $L^* = \mathrm{Cent}_{G^*}(A_{L^*}^*)$ for a subtorus $A_{L^*}^* \subseteq A^*$. Without loss of generality, we may assume that our inner twist $G \to G^*$ restricts to give an inner twist $L \to L^*$.

Proposition 4.79. *In the above situation, the following diagram commutes:*

$$\mathcal{Z}(G^*(F), J^*) \xrightarrow{\tilde{t}} \mathcal{Z}(G(F), J) \qquad (4.53)$$

$$c_{L^*}^{G^*} \downarrow \qquad\qquad c_L^G \downarrow$$

$$\mathcal{Z}(L^*(F), J_{L^*}^*) \xrightarrow{\tilde{t}} \mathcal{Z}(L(F), J_L).$$

Taking $L = M$, the diagram shows in order to compute $\tilde{t}$ it is enough to compute it in the case where G_{ad} is anisotropic. In that case if $z \in \mathcal{Z}(G^*(F), J^*)$, the function $\tilde{t}(z)$ is given by integrating z over the fibers of the Kottwitz homomorphism $\kappa_{G^*}(F)$.

Proof. The commutativity boils down to the fact that the quantities (4.52) do not depend on the ambient group G. $\qquad\qquad\qquad\qquad\qquad\qquad\qquad\square$

Remark 4.80. The final sentence in Proposition 4.79 is the key to explicit computation of $\tilde{t}(z)$ given z, and is illustrated in §7.3.3. This final sentence was incorrectly asserted to hold for the *unnormalized* transfer homomorphisms (for special maximal parahoric Hecke algebras) in Prop. 12.3.1 of [HRo].

References

[AC] J. Arthur, L. Clozel, *Simple algebras, base change, and the advanced theory of the trace formula*, Annals of Math. Studies 120, Princeton University Press, 1989.

[BD] J.-N. Bernstein, rédigé par P. Deligne, Le "centre" de Bernstein, In: *Représentations des groupes réductifs sur un corps local*, Hermann (1984).

[Be92] J. Bernstein, Representations of p-adic groups, Notes taken by K. Rumelhart of lectures by J. Bernstein at Harvard in the Fall of 1992.

[Bo79] A. Borel, Automorphic L-functions, In: *Automorphic forms, representations and L-functions*, Proc. Sympos. Pure Math., vol. 33, part 2, Amer. Math. Soc., 1979, pp. 27–61.

[BT2] F. Bruhat, J. Tits, Groupes réductifs sur un corps local. II, *Inst. Hautes Études Sci. Publ. Math.* **60** (1984), 5–184.

[BK] C. J. Bushnell, P. C. Kutzko, Smooth representations of reductive p-adic groups: structure theory via types, *Proc. London Math. Soc.* (3) **77** (1998), 582–634.

[Cas] W. Casselman, Characters and Jacquet modules, *Math. Ann.* **230** (1977), 101–105.

[Cl90] L. Clozel, The fundamental lemma for stable base change, *Duke Math. J.* **61**, no. 1, (1990), 255–302.

[Fer] A. Ferrari, Théorème de l'indice et formule des traces, *Manuscripta Math.* **124**, (2007), 363–390.

[Ga] D. Gaitsgory, Construction of central elements in the affine Hecke algebra via nearby cycles, *Invent. Math.* **144**, no. 2, (2001) 253–280.

[G] U. Görtz, Alcove walks and nearby cycles on affine flag manifolds, *J. Alg. Comb.* **26**, no. 4 (2007), 415–430.

[GH] U. Görtz, T. Haines, The Jordan-Hoelder series for nearby cycles on some Shimura varieties and affine flag varieties, *J. Reine Angew. Math.* **609** (2007), 161–213.

[GR] B. H. Gross, M. Reeder, Arithmetic invariants of discrete Langlands parameters *Duke Math. J.* **154**, no. 3, (2010), 431–508.

[H01] T. Haines, Test functions for Shimura varieties: the Drinfeld case, *Duke Math. J.* **106** no. 1, (2001), 19–40.

[H05] T. Haines, Introduction to Shimura varieties with bad reduction of parahoric type, *Clay Math. Proc.* **4**, (2005), 583–642.

[H07] T. Haines, Intertwiners for unramified groups, expository note (2007). Available at www.math.umd.edu/~tjh.

[H09] T. Haines, The base change fundamental lemma for central elements in parahoric Hecke algebras, *Duke Math. J.*, **149**, no. 3 (2009), 569–643.

[H11] T. Haines, Endoscopic transfer of the Bernstein center, IAS/Princeton Number theory seminar, April 6, 2011. Slides available at www.math.umd.edu/~tjh.

[H12] T. Haines, Base change for Bernstein centers of depth zero principal series blocks, *Ann. Scient. École Norm. Sup.* 4e t. **45**, 2012, 681–718.

[HKP] T. Haines, R. Kottwitz, A. Prasad, Iwahori-Hecke algebras, *J. Ramanujan Math. Soc.* **25**, no. 2 (2010), 113–145.

[HN02a] T. Haines, B.C. Ngô, Nearby cycles for local models of some Shimura varieties, *Compo. Math.* **133**, (2002), 117–150.

[HN02b] T. Haines, B. C. Ngô, Alcoves associated to special fibers of local models, *Amer. J. Math.* **124** (2002), 1125–1152.

[HP] T. Haines, A. Pettet, Formulae relating the Bernstein and Iwahori-Matsumoto presentations of an affine Hecke algebra, *J. Alg.* **252** (2002), 127–149.

[HRa1] T. Haines, M. Rapoport, Shimura varieties with $\Gamma_1(p)$-level via Hecke algebra isomorphisms: the Drinfeld case, *Ann. Scient. École Norm. Sup.* 4e série, t. **45**, (2012), 719–785.

[HRo] T. Haines, S. Rostami, The Satake isomorphism for special maximal parahoric Hecke algebras, *Represent. Theory* **14** (2010), 264–284.

[Hal] T. Hales, On the fundamental lemma for standard endoscopy: reduction to unit elements, *Canad. J. Math.*, **47**(5), (1995), 974–994.

[Kal] T. Kaletha, Epipelagic *L*-packets and rectifying characters. Preprint (2012). arXiv:1209.1720.

[Kaz] D. Kazhdan, Cuspidal geometry of *p*-adic groups, *J. Analyse Math.*, **47**, (1986), 1–36.

[KL] D. Kazhdan and G. Lusztig, Representations of Coxeter groups and Hecke algebras, *Invent. Math.* **53** (1979), 165–184.

[KV] D. Kazhdan, Y. Varshavsky, On endoscopic transfer of Deligne-Lusztig functions, *Duke Math. J.* **161**, no. 4, (2012), 675–732.

[Ko84a] R. Kottwitz, Stable trace formula: cuspidal tempered terms, *Duke Math. J.* **51**, no. 3, (1984), 611–650.

[Ko84b] R. Kottwitz, Shimura varieties and twisted orbital integrals, *Math. Ann.* **269** (1984), 287–300.

[Ko86] R. Kottwitz, Base change for unit elements of Hecke algebras, *Comp. Math.* **60**, (1986), 237–250.

[Ko88] R. Kottwitz, Tamagawa numbers, *Ann. of Math.* **127** (1988), 629–646.

[Ko90] R. Kottwitz, Shimura varieties and λ-adic representations, in *Automorphic forms, Shimura varieties and L-functions*. Proceedings of the Ann Arbor conference, Academic Press, 1990.

[Ko92a] R. Kottwitz, Points of some Shimura varieties over finite fields, *J. Amer. Math. Soc.* **5**, (1992), 373–444.

[Ko92b] R. Kottwitz, On the λ-adic representations associated to some simple Shimura varieties, *Invent. Math.* **108**, (1992), 653–665.

[Ko97] R. Kottwitz, Isocrystals with additional structure. II, *Compositio Math.* **109**, (1997), 255-339.

[KR] R. Kottwitz, M. Rapoport, Minuscule alcoves for GL_n and GSP_{2n}, *Manuscripta Math.* **102**, no. 4, (2000), 403–428.

[Land] E. Landvogt, *A compactification of the Bruhat-Tits building*, Lecture Notes in Mathematics 1619, Springer, 1996.

[L1] R. P. Langlands, Shimura varieties and the Selberg trace formula, *Can. J. Math.* **29**, (1977), 1292–1299.

[L2] R. P. Langlands, On the zeta-functions of some simple Shimura varieties, *Can. J. Math.* **31**, (1979), 1121–1216.

[Mil] J. S. Milne, The points on a Shimura variety modulo a prime of good reduction, In: *The zeta functions of Picard modular surfaces*, ed. R. P. Landlands and D. Ramakrishnan, CRM, 1992, 151–253.

[MV] I. Mirković, K. Vilonen, Geometric Langlands duality and representations of algebraic groups over commutative rings, *Ann. of Math.* (2) **166**, no. 1, (2007), 95–143.

[Mor] S. Morel, *On the cohomology of certain non-compact Shimura varieties*, Annals of Math. Studies 173, Princeton University Press, 2010.

[Ngo] B. C. Ngô, Le lemme fondamental pour les algèbres de Lie, *Publ. Math. IHÉS* **111**, (2010), 1–169.

[PRS] G. Pappas, M. Rapoport, B. Smithling, Local models of Shimura varieties, I. Geometry and combinatorics, to appear in the *Handbook of Moduli*, 84 pp.

[PZ] G. Pappas, X. Zhu, Local models of Shimura varieties and a conjecture of Kottwitz, *Invent. Math.* **194**, no. 1 (2013), 147–254.

[PrRa] Prasad, D., Ramakrishnan, D. Self-dual representations of division algebras and Weil groups: a contrast, *Amer. J. Math.* **134**, no. 3, (2012), 749–767.

[Ra90] M. Rapoport, On the bad reduction of Shimura varieties, in *Automorphic forms, Shimura varieties and L-functions*. Proceedings of the Ann Arbor conference, Academic Press, 1990.

[Ra05] M. Rapoport: A guide to the reduction modulo p of Shimura varieties. *Astérisque* **298**, (2005), 271–318.

[RZ]   M. Rapoport, T. Zink, *Period spaces for p-divisible groups*, Annals of Math. Studies 141, Princeton University Press, 1996.

[Ren] D. Renard, *Représentations des groupes réductifs p-adiques*. Cours Spécialisés, 17. Société Mathématique de France, 2010.

[Ri]   T. Richarz, On the Iwahori-Weyl group, available at www.math.uni-bonn.de/people/richarz/.

[Roc] A. Roche, The Bernstein decomposition and the Bernstein centre, In: *Ottawa Lectures on admissible representations of reductive p-adic groups*, Fields Inst. Monogr. 26, Amer. Math. Soc., 2009, pp. 3–52.

[Ro]   S. Rostami, The Bernstein presentation for general connected reductive groups, arXiv:1312.7374.

[Sch1] P. Scholze, The Langlands-Kottwitz approach for the modular curve, *IMRN* 2011, no. 15, 3368–3425.

[Sch2] P. Scholze, The Langlands-Kottwitz approach for some simple Shimura varieties. *Invent. Math.* **192**, no. 3, 627–661.

[Sch3] P. Scholze, The local Langlands correspondence for GL_n over p-adic fields. *Invent. Math.* **192**, no. 3, 663–715.

[Sch4] P. Scholze, The Langlands-Kottwitz method and deformation spaces of p-divisible groups, *J. Amer. Math. Soc.* **26** (2013), 227-259.

[SS]   P. Scholze, S. W. Shin, On the cohomology of compact unitary group Shimura varieties at ramified split places, *J. Amer. Math. Soc.* **26** (2013), 261–294.

[Sh]   S. W. Shin, Galois representations arising from some compact Shimura varieties, *Ann. of Math.* (2) **173**, no. 3, (2011), 1645–1741.

[Sm1] B. Smithling, Topological flatness of orthogonal local models in the split, even case, *I. Math. Ann.* **35**, no. 2, (2011), 381–416.

[Sm2] B. Smithling, Admissibility and permissibility for minuscule cocharacters in orthogonal groups, *Manuscripta Math.* **136**, no. 3–4, (2011), 295–314.

[Sm3] B. Smithling, Topological flatness of local models for ramified unitary groups. I. The odd dimensional case, *Adv. Math.* **226**, no. 4, (2011), 3160–3190.

[St]   R. Steinberg, Endomorphisms of linear algebraic groups, *Mem. Amer. Math. Soc.* **80** (1968), 1–108.

[Str] B. Stroh, Sur une conjecture de Kottwitz au bord, *Ann. Sci. Ec. Norm. Sup.* (4) **45**, no. 1, (2012), 143–165.

[Var] S. Varma, On certain elements in the Bernstein center of $\mathbb{G}L_2$, *Represent. Theory* **17**, (2013), 99–119.

[Vo]   D. Vogan, The local Langlands conjecture, In: *Representation theory of groups and algebras*, Contemp. Math. 145, 1993, pp. 305–379.

[Wal97] J.-L. Waldspurger, Le lemme fondamental implique le transfert, *Compositio Math.* **105**, (1997), 153–236.

[Wal04] J.-L. Waldspurger, Endoscopie et changement de caractéristique, *J. Inst. Math. Jussieu* **5**, no. 3, (2006), 423–525.

[Wal08] J.-L. Waldspurger, L'endoscopie tordue n'est pas si tordue, *Mem. Amer. Math. Soc.* **194**, no. 198, 2008.

[Zhu] X. Zhu, The geometric Satake correspondence for ramified groups, arXiv:1107.5762.

5

Conditional results on the birational section conjecture over small number fields

Yuichiro Hoshi

Abstract

In the present chapter, we give necessary and sufficient conditions for a birational Galois section of a projective smooth curve over either the *field of
rational numbers* or an *imaginary quadratic field* to be *geometric*. As a consequence, we prove that, over such a small number field, to prove the birational
section conjecture for projective smooth curves, it suffices to verify that,
roughly speaking, for any birational Galois section of the projective line, the
local points associated to the birational Galois section *avoid three distinct
rational points*, and, moreover, a certain Galois representation determined by
the birational Galois section is *unramified at all but finitely many primes*.

Contents

2000 *Mathematics Subject Classification. 14H30.*

Automorphic Forms and Galois Representations, ed. Fred Diamond, Payman L. Kassaei and
Minhyong Kim. Published by Cambridge University Press. © Cambridge University Press 2014.

Introduction

Let k be a field of characteristic 0, $\overline{k}$ an algebraic closure of k, and X a *projective smooth geometrically connected curve* over k. Write $G_k \overset{\text{def}}{=} \text{Gal}(\overline{k}/k)$ for the absolute Galois group of k determined by the fixed algebraic closure $\overline{k}$ of k. Now we have a natural surjection

$$\pi_1(X) \twoheadrightarrow G_k$$

from the étale fundamental group $\pi_1(X)$ of X to G_k induced by the structure morphism of X. Then Grothendieck's section conjecture may be stated as follows: If k is *finitely generated over the field of rational numbers*, and X is of *genus* ≥ 2, then any section of this surjection $\pi_1(X) \twoheadrightarrow G_k$ arises from a k-rational point of X, i.e., the image of any section of this surjection coincides with, or, equivalently, is contained in, a decomposition subgroup of $\pi_1(X)$ associated to a k-rational point of X. In the present chapter, we discuss the *birational version* of this conjecture, i.e., the *birational section conjecture*. Denote by $k(X)$ the function field of X. Fix an algebraic closure $\overline{k(X)}$ of $k(X)$ containing $\overline{k}$ and write $G_{k(X)} \overset{\text{def}}{=} \text{Gal}(\overline{k(X)}/k(X))$. Then the natural inclusions $k \hookrightarrow k(X)$, $\overline{k} \hookrightarrow \overline{k(X)}$ determine a surjection

$$G_{k(X)} \twoheadrightarrow G_k \,,$$

which factors through the above surjection $\pi_1(X) \twoheadrightarrow G_k$. We shall refer to a section of this surjection $G_{k(X)} \twoheadrightarrow G_k$ as a [*pro-$\mathfrak{P}$rimes*] *birational Galois section* of X/k [cf. Definition 5.2]. In the present chapter, we discuss the *geometricity of birational Galois sections*.

Let x be a closed point of X and $D_x \subseteq G_{k(X)}$ a decomposition subgroup of $G_{k(X)}$ associated to x. Then, as is well-known, the image of the composite $D_x \hookrightarrow G_{k(X)} \twoheadrightarrow G_k$ coincides with the open subgroup $G_{k(x)} \subseteq G_k$ of G_k corresponding to the residue field $k(x)$ of X at x, and, moreover, the resulting *surjection* $D_x \twoheadrightarrow G_{k(x)}$ admits a [not necessarily unique] *section*. In particular, if, moreover, $k(x) = x$, i.e., $x \in X(k)$, then the closed subgroup $D_x \subseteq G_{k(X)}$ of $G_{k(X)}$ contains the image of a [not necessarily unique] birational Galois section of X/k. We shall say that a birational Galois section of X/k is *geometric* if its image is contained in a decomposition subgroup of $G_{k(X)}$ associated to a [necessarily k-rational] closed point of X [cf. Definition 5.3].

The birational section conjecture over *local fields* has been solved affirmatively. In [8], Koenigsmann proved that if k is either a *p-adic local field* for some prime number p [i.e., a finite extension of the p-adic completion of the field of rational numbers] or the *field of real numbers*, then any birational Galois section of X/k is *geometric* [cf. [8] Proposition 2.4, (2)]. Moreover, in [12], Pop obtained, by a refined discussion of Koenigsmann's discussion, a result concerning birational Galois sections over p-adic local fields [cf. [12], Theorem A], which leads naturally to a proof of the *geometrically pro-p version* of Koenigsmann's result over p-adic local fields [cf. Proposition 5.10]. In [18], Wickelgren proved a strong version of the birational section conjecture over the *field of real numbers* [cf. [18], Corollary 1.2].

In the remainder of the Introduction, we discuss the geometricity of birational Galois sections over *number fields*; suppose that k is a *number field* [i.e., a finite extension of the field of rational numbers].

First, let us recall that, in [1], Esnault and Wittenberg proved that if the Shafarevich–Tate group of the Jacobian variety of X over k is *finite*, then the existence of a birational Galois section of X/k implies the existence of a *divisor of degree* 1 on X; more precisely, the existence of a section of the natural surjection $G_{k(X)}/[G_{\overline{k} \cdot k(X)}, G_{\overline{k} \cdot k(X)}] \twoheadrightarrow G_k$, where we write $G_{\overline{k} \cdot k(X)} \overset{\text{def}}{=} \mathrm{Gal}(\overline{k(X)}/\overline{k} \cdot k(X))$ and $[G_{\overline{k} \cdot k(X)}, G_{\overline{k} \cdot k(X)}]$ for the closure of the commutator subgroup of $G_{\overline{k} \cdot k(X)}$, is equivalent to the existence of a *divisor of degree* 1 on X [cf. [1], Theorem 2.1]. Next, let us recall that, in [4], Harari and Stix proved, as a consequence of results obtained by Stoll in [16], that if there exist an abelian variety A over k and a nonconstant morphism $X \to A$ over k such that both the Mordell–Weil group and the Shafarevich–Tate group of A/k are *finite*, then any birational Galois section of X/k is *geometric* [cf. [4], Theorem 17]. This result of Harari and Stix gives us some examples of X/k for which any birational Galois section is *geometric* [cf. [4], Remark 18, (1)].

To state our main results, let us discuss *local points associated to a birational Galois section*. Write $\mathfrak{P}_k^f$ for the set of nonarchimedean primes of k. For each $\mathfrak{p} \in \mathfrak{P}_k^f$, fix an algebraic closure $\overline{k}_\mathfrak{p}$ of the $\mathfrak{p}$-adic completion $k_\mathfrak{p}$ of k containing $\overline{k}$ and write $G_\mathfrak{p} \overset{\text{def}}{=} \mathrm{Gal}(\overline{k}_\mathfrak{p}/k_\mathfrak{p}) \subseteq G_k$. Finally, write $\mathbb{A}_k^f \subseteq \prod_{\mathfrak{p} \in \mathfrak{P}_k^f} k_\mathfrak{p}$ for the finite part of the adele ring of k, i.e., the subring of $\prod_{\mathfrak{p} \in \mathfrak{P}_k^f} k_\mathfrak{p}$ consisting of elements $(a_\mathfrak{p})_{\mathfrak{p} \in \mathfrak{P}_k^f} \in \prod_{\mathfrak{p} \in \mathfrak{P}_k^f} k_\mathfrak{p}$ such that $a_\mathfrak{p}$ is contained in the ring of integers of $k_\mathfrak{p}$ for all but finitely many $\mathfrak{p} \in \mathfrak{P}_k^f$. Then it follows from a result obtained in [8], as well as [12], that, for each $\mathfrak{p} \in \mathfrak{P}_k^f$, a birational Galois section s of X/k *uniquely determines* a $k_\mathfrak{p}$-valued point $x_\mathfrak{p}$ of X such that, for any open subscheme $U \subseteq X$ of X, the image of the homomorphism $G_\mathfrak{p} \to \pi_1(U \otimes_k k_\mathfrak{p})$ naturally determined by the isomorphism $\pi_1(U \otimes_k k_\mathfrak{p}) \overset{\sim}{\to} \pi_1(U) \times_{G_k} G_\mathfrak{p}$ and the composite $G_\mathfrak{p} \hookrightarrow G_k \overset{s}{\to} G_{k(X)} \twoheadrightarrow \pi_1(U)$ is contained in a decomposition subgroup of $\pi_1(U \otimes_k k_\mathfrak{p})$ associated to $x_\mathfrak{p}$ [cf. Proposition 5.24]; we shall refer to the $k_\mathfrak{p}$-valued point $x_\mathfrak{p}$ as the $k_\mathfrak{p}$-*valued point of* X *associated to* s [cf. Definition 5.20]. In particular, [since X is *projective* over k] the birational Galois section s *uniquely determines* an $\mathbb{A}_k^f$-valued point

$x_{\mathbb{A}} \overset{\text{def}}{=} (x_{\mathfrak{p}})_{\mathfrak{p} \in \mathfrak{P}_k^f} \in X(\mathbb{A}_k^f) \subseteq \prod_{\mathfrak{p} \in \mathfrak{P}_k^f} X(k_{\mathfrak{p}})$ of X; we shall refer to the $\mathbb{A}_k^f$-valued point $x_{\mathbb{A}}$ as the $\mathbb{A}_k^f$-*valued point of X associated to s* [cf. Definition 5.20]. Note that if the birational Galois section s is *geometric*, then there exists a [necessarily unique] k-rational point $x \in X(k)$ of X such that, for each $\mathfrak{p} \in \mathfrak{P}_k^f$, the $k_{\mathfrak{p}}$-valued point of X determined by x is the $k_{\mathfrak{p}}$-valued point of X associated to s [cf. Remark 5.21].

The following result is the main result, which gives necessary and sufficient conditions for a birational Galois section of a projective smooth geometrically connected curve over a *small number field*, i.e., either the *field of rational numbers* or an *imaginary quadratic field*, to be *geometric* [cf. Theorem 5.40 in the case where $\mathcal{C}$ consists of all finite groups].

Theorem A. *Let k be either the* **field of rational numbers** *or an* **imaginary quadratic field**, *X a* **projective smooth geometrically connected curve** *over k, and s a [pro-$\mathfrak{Primes}$]* **birational Galois section** *of X/k [cf. Definition 5.2]. Then the following conditions are equivalent:*

(1) *s is* **geometric** *[cf. Definition 5.3].*
(2) *The following two conditions are satisfied:*
   (2-i) *There exists a finite morphism $\phi \colon X \to \mathbb{P}_k^1$ over k such that, for each $\mathfrak{p} \in \mathfrak{P}_k^f$, the composite*

$$\operatorname{Spec} k_{\mathfrak{p}} \xrightarrow{x_{\mathfrak{p}}} X \xrightarrow{\phi} \mathbb{P}_k^1$$

   determines a $k_{\mathfrak{p}}$-valued point of $\mathbb{P}_k^1 \setminus \{0, 1, \infty\} \subseteq \mathbb{P}_k^1$.
   (2-ii) *For each open subscheme $U \subseteq X$ of X which is a* **hyperbolic curve** *over k [cf. §0], there exists a prime number l_U such that the pro-l_U Galois section of U/k [cf. Definition 5.2] naturally determined by s is either* **cuspidal** *[cf. Definition 5.37, (i)] or* **unramified almost everywhere** *[cf. Definition 5.37, (ii)].*
(3) *There exists a finite morphism $\phi \colon X \to \mathbb{P}_k^1$ over k such that the composite*

$$\operatorname{Spec} \mathbb{A}_k^f \xrightarrow{x_{\mathbb{A}}} X \xrightarrow{\phi} \mathbb{P}_k^1$$

   determines an $\mathbb{A}_k^f$-valued point of $\mathbb{P}_k^1 \setminus \{0, 1, \infty\} \subseteq \mathbb{P}_k^1$.
(4) *There exist a* **finite** *subset $T \subseteq \mathfrak{P}_k^f$ of $\mathfrak{P}_k^f$ and a closed subscheme $Z \subseteq X$ of X which is* **finite** *over k such that, for each $\mathfrak{p} \in \mathfrak{P}_k^f \setminus T$, [the image of] the $k_{\mathfrak{p}}$-valued point $x_{\mathfrak{p}}$ of X is contained in $Z \subseteq X$.*

The outline of the proof of Theorem A is as follows: The implications (1) $\Rightarrow$ (2) $\Rightarrow$ (3) follow immediately from the various definitions involved, together with some results that are proved in Appendix and derived from the discussion given in [7]. Next, to verify the implications (3) $\Rightarrow$ (4) $\Rightarrow$ (1), let us observe that since k is either the *field of rational numbers* or an *imaginary quadratic field*, the following assertion (†) holds [cf. Lemma 5.32]:

(†): for every [pro-$\mathfrak{Primes}$] birational Galois section of $\mathbb{P}^1_k/k$, if the associated $\mathbb{A}^f_k$-valued point of $\mathbb{P}^1_k$ determines an $\mathbb{A}^f_k$-valued point of $\mathbb{G}_{m,k} \overset{\text{def}}{=} \mathbb{P}^1_k \setminus \{0, \infty\}$, then the induced [pro-$\mathfrak{Primes}$] Galois section of $\mathbb{G}_{m,k}/k$ arises from a k-rational point of $\mathbb{G}_{m,k}$.

Then the implication (3) $\rightarrow$ (4) follows immediately from (†). Thus, it remains to verify the implication (4) $\Rightarrow$ (1). Since X is projective, we have a closed immersion $X \hookrightarrow \mathbb{P}^N_k$ for some positive integer N. Now by condition (4), we may assume without loss of generality that, for every nonarchimedean prime $\mathfrak{p}$ of k, the $k_\mathfrak{p}$-valued point of $\mathbb{P}^N_k$ determined by this closed immersion $X \hookrightarrow \mathbb{P}^N_k$ and the $k_\mathfrak{p}$-valued point of X associated to the given birational Galois section s lies on the open subscheme $\mathbb{G}_{m,k} \times_k \cdots \times_k \mathbb{G}_{m,k} \subseteq \mathbb{A}^N_k \subseteq \mathbb{P}^N_k$ of $\mathbb{P}^N_k$; in particular, we have a $k_\mathfrak{p}$-valued point of $\mathbb{G}_{m,k} \times_k \cdots \times_k \mathbb{G}_{m,k}$. Moreover, again by condition (4), together with the above assertion (†), one verifies easily that, for each $\mathfrak{p}$, any one of the N factors of the *coordinate* of the $k_\mathfrak{p}$-valued point of $\mathbb{G}_{m,k} \times_k \cdots \times_k \mathbb{G}_{m,k}$ is *k-rational*. In particular, it follows that X admits a k-rational point. Thus, by applying this observation to the various open subgroups of $G_{k(X)}$ that contain the image of s, we conclude that s is *geometric*. This completes the explanation of the outline of the proof of Theorem A.

Note that Theorem A is a result *without any assumption on the finiteness of a Shafarevich–Tate group*. Next, let us observe that the equivalence (1) $\Leftrightarrow$ (3) of Theorem A may be regarded as a *tripod analogue* of the result due to Harari and Stix discussed above, i.e., [4], Theorem 17. The condition that k is either the field of rational numbers or an imaginary quadratic field [i.e., the assumption that the group of units of the ring of integers of k is *finite*] in the statement of Theorem A may be regarded as an analogue of the finiteness condition on the Mordell–Weil group in the statement of [4], Theorem 17; on the other hand, since any abelian variety is *proper*, in the case of [4], Theorem 17, the condition corresponding to our condition that the birational Galois section determines [not only a $\left(\prod_{\mathfrak{p} \in \mathfrak{P}^f_k} k_\mathfrak{p} \right)$-valued point but also] an $\mathbb{A}^f_k$-valued point of the tripod $\mathbb{P}^1_k \setminus \{0, 1, \infty\}$ in Theorem A is *automatically satisfied*.

Let us also observe that our main result leads naturally to some examples – which, however, were already essentially obtained by Stoll – of projective smooth curves for which any *prosolvable* birational Galois section is *geometric* [cf. Remarks 5.35; 5.41, (iv)].

As a corollary of Theorem A, we prove the following result [cf. Corollary 5.42].

Theorem B. *Let k be either the* **field of rational numbers** *or an* **imaginary quadratic field**. *Then the following assertions are equivalent:*

(1) *Any [pro-𝔓rimes] birational Galois section [cf. Definition 5.2] of any projective smooth geometrically connected curve over k is* **geometric** *[cf. Definition 5.3].*

(2) *Any [pro-𝔓rimes] birational Galois section of* $\mathbb{P}_k^1 / k$ *is* **geometric**.

(3) *Any [pro-𝔓rimes] birational Galois section s of* $\mathbb{P}_k^1 / k$ *satisfies the following two conditions:*

  (3-i) *There exist three distinct elements a, b, c $\in$ $\mathbb{P}_k^1(k)$ of $\mathbb{P}_k^1(k)$ such that, for any nonarchimedean prime $\mathfrak{p}$ of k, the $k_{\mathfrak{p}}$-valued point of $\mathbb{P}_k^1$ associated to s is $\notin \{a, b, c\} \subseteq (\mathbb{P}_k^1(k) \subseteq) \mathbb{P}_k^1(k_{\mathfrak{p}})$.*

  (3-ii) *There exists a prime number l such that the pro-l Galois section of $\mathbb{P}_k^1 \setminus \{0, 1, \infty\}$ [cf. Definition 5.2] naturally determined by s is either* **cuspidal** *[cf. Definition 5.37, (i)] or* **unramified almost everywhere** *[cf. Definition 5.37, (ii)].*

(4) *Any [pro-𝔓rimes] birational Galois section s of* $\mathbb{P}_k^1 / k$ *satisfies the following two conditions:*

  (4-i) *There exist three distinct elements a, b, c $\in$ $\mathbb{P}_k^1(k)$ of $\mathbb{P}_k^1(k)$ such that, for any nonarchimedean prime $\mathfrak{p}$ of k, the $k_{\mathfrak{p}}$-valued point of $\mathbb{P}_k^1$ associated to s is $\notin \{a, b, c\} \subseteq (\mathbb{P}_k^1(k) \subseteq) \mathbb{P}_k^1(k_{\mathfrak{p}})$.*

  (4-ii) *Write $s^{\mathbb{P}}$ for the pro-𝔓rimes Galois section of $\mathbb{P}_k^1 \setminus \{0, 1, \infty\}$ [cf. Definition 5.2] naturally determined by s. Then it holds either that $s^{\mathbb{P}}$ is* **cuspidal** *[cf. Definition 5.37, (i)], or that there exists a prime number l such that the l-adic Galois representation*

$$G_k \xrightarrow{s^{\mathbb{P}}} \pi_1(\mathbb{P}_k^1 \setminus \{0, 1, \infty\}) \longrightarrow GL_2(\mathbb{Z}_l)$$

  – where the second arrow $\pi_1(\mathbb{P}_k^1 \setminus \{0, 1, \infty\}) \to GL_2(\mathbb{Z}_l)$ is the l-adic representation of $\pi_1(\mathbb{P}_k^1 \setminus \{0, 1, \infty\})$ determined by the **Legendre family of elliptic curves** *over $\mathbb{P}_k^1 \setminus \{0, 1, \infty\}$, i.e., the elliptic curve over $\mathbb{P}_k^1 \setminus \{0, 1, \infty\} = \mathrm{Spec}\, k[u^{\pm 1}, (1 - u)^{-1}]$ determined by the equation "$y^2 = x(x - 1)(x - u)$" – is* **unramified at all but finitely many primes** *of k.*

As a consequence [cf. the equivalences (1) $\Leftrightarrow$ (3) and (1) $\Leftrightarrow$ (4) of Theorem B], for a number field k which is either the *field of rational numbers* or an *imaginary quadratic field*, to prove the *birational section conjecture* over k [i.e., assertion (1) of Theorem B], it suffices to verify that, roughly speaking, for any birational Galois section of the projective line over k, the local points associated to the birational Galois section *avoid three distinct rational points* [cf. conditions (3-i), (4-i)], and, moreover, a certain Galois representation determined by the birational Galois section is *unramified at all but finitely many primes* [cf. conditions (3-ii), (4-ii)]. However, it is not clear to the author at the time of writing whether or not these are always satisfied.

Finally, the author should mention that the referee pointed out that, after the present paper, Stix presented, in [15], results that are similar to and stronger than some results of the present paper. In fact, for instance, a similar result to the equivalence (1) $\Leftrightarrow$ (2) of Theorem A (respectively, the equivalence (1) $\Leftrightarrow$ (4) of Theorem B; Lemma 5.32; Theorem 5.33; Corollary 5.34) of the present paper may be found as [15], Theorem A (respectively, [15], Theorem C; [15], Proposition 4; [15], Corollary 9; [15], Corollary 11).

Acknowledgements

The author would like to thank Akio Tamagawa, Takahiro Tsushima, and Seidai Yasuda for their helpful comments and, especially, discussions concerning §3. The author also would like to thank the referee for some comments and suggestions. This research was supported by Grant-in-Aid for Young Scientists (B), No. 22740012, Japan Society for the Promotion of Science.

0. Notations and conventions

Numbers: The notation $\mathfrak{Primes}$ will be used to denote the set of all prime numbers. The notation $\mathbb{Z}$ will be used to denote the ring of rational integers. If $\Sigma \subseteq \mathfrak{Primes}$, then we shall refer to a nonzero integer whose prime divisors are contained in Σ as a Σ-*integer*, and we shall write $\widehat{\mathbb{Z}}^{\Sigma}$ for the pro-Σ completion of $\mathbb{Z}$, i.e., $\widehat{\mathbb{Z}}^{\Sigma} \overset{\text{def}}{=} \varprojlim \mathbb{Z}/n\mathbb{Z}$, where the projective limit is over all positive Σ-integers n. We shall refer to a finite (respectively, finitely generated) extension of the field of rational numbers as a *number field* (respectively, *finitely generated field of characteristic* 0). If $p \in \mathfrak{Primes}$, then the notation $\mathbb{Z}_p$ will be used to denote the p-adic completion of $\mathbb{Z}$, and we shall refer to a finite extension of the p-adic completion of the field of rational numbers as a *p-adic local field*.

Profinite groups: Let G be a profinite group and $H \subseteq G$ a closed subgroup of G. Then we shall denote by $Z_G(H)$, $N_G(H)$, $Z_G^{\text{loc}}(H)$ the *centralizer, normalizer, local centralizer* of H in G, respectively, i.e.,

$$Z_G(H) \stackrel{\text{def}}{=} \{\, g \in G \mid ghg^{-1} = h \text{ for any } h \in H \,\},$$

$$N_G(H) \stackrel{\text{def}}{=} \{\, g \in G \mid g \cdot H \cdot g^{-1} = H \,\},$$

$$Z_G^{\text{loc}}(H) \stackrel{\text{def}}{=} \varinjlim_U Z_G(U)$$

– where the injective limit is over all open subgroups $U \subseteq H$ of H. We shall refer to $Z(G) \stackrel{\text{def}}{=} Z_G(G)$, $Z^{\text{loc}}(G) \stackrel{\text{def}}{=} Z_G^{\text{loc}}(G)$ as the *center*, *local center* of G, respectively. We shall say that G is *center-free*, *slim* if $Z(G) = \{1\}$, $Z^{\text{loc}}(G) = \{1\}$, respectively.

Let $\Sigma \subseteq \mathfrak{Primes}$ be a nonempty subset of $\mathfrak{Primes}$ [where we refer to the discussion entitled "Numbers" concerning the set $\mathfrak{Primes}$]. Then we shall say that a finite group G is a Σ-*group* if the cardinality of G is a Σ-integer [where we refer to the discussion entitled "Numbers" concerning the term "Σ-integer"].

Let C be a full formation [i.e., a family of finite groups that is closed under taking quotients, subgroups, and extensions]. We shall say that a finite group is a C-*group* if [a finite group which is isomorphic to] the finite group is contained in C. We shall say that a profinite group is a *pro-C group* if every finite quotient of the profinite group is a C-group. We shall write $\Sigma(C) \subseteq \mathfrak{Primes}$ for the set of prime numbers $p \in \mathfrak{Primes}$ such that $\mathbb{Z}/p\mathbb{Z}$ is a C-group. Here, we note that one verifies easily that $\Sigma(C) = \mathfrak{Primes}$ if and only if C contains all finite solvable groups. If C consists of all Σ-groups for some nonempty subset $\Sigma \subseteq \mathfrak{Primes}$, then we shall refer to a pro-C group as a *pro-Σ group*.

Let G be a profinite group. Then we shall write $\text{Aut}(G)$ for the group of [continuous] automorphisms of G, $\text{Inn}(G) \subseteq \text{Aut}(G)$ for the group of inner automorphisms of G, and

$$\text{Out}(G) \stackrel{\text{def}}{=} \text{Aut}(G)/\text{Inn}(G).$$

If, moreover, G is *topologically finitely generated*, then one verifies easily that the topology of G admits a basis of *characteristic open subgroups*, which thus induces a *profinite topology* on the group $\text{Aut}(G)$, hence also a *profinite topology* on the group $\text{Out}(G)$.

Curves: Let S be a scheme and X a scheme over S. Then we shall say that X is a *smooth curve* over S if there exist a scheme X^{cpt} which is smooth, proper, geometrically connected, and of relative dimension 1 over S and a closed subscheme $D \subseteq X^{\text{cpt}}$ of X^{cpt} which is finite and étale over S such that the complement $X^{\text{cpt}} \setminus D$ of D in X^{cpt} is isomorphic to X over S. Note that, as is well-known, if X is a smooth curve over [the spectrum of] a field k, then the pair "(X^{cpt}, D)" is *uniquely determined up to canonical isomorphism over k*; we shall refer to X^{cpt} as the *smooth compactification* of X over k and to a geometric point of X^{cpt} whose image lies on D as a *cusp* of X.

Let S be a scheme. Then we shall say that a smooth curve X over S is a *hyperbolic curve* (respectively, *tripod*) over S if there exist a pair (X^{cpt}, D) satisfying the condition in the above definition of the term "smooth curve" and a pair (g, r) of nonnegative integers such that $2g - 2 + r > 0$ (respectively, $(g, r) = (0, 3)$), any geometric fiber of $X^{\mathrm{cpt}} \to S$ is [a necessarily smooth, proper, and connected curve] of genus g, and the degree of $D \subseteq X^{\mathrm{cpt}}$ over S is r.

Let S be a scheme, $U \subseteq S$ an open subscheme of S, and X a hyperbolic curve over U. Then we shall say that X admits *good reduction* over S if there exists a hyperbolic curve X_S over S such that $X_S \times_S U$ is isomorphic to X over U.

1. Birational Galois sections and their geometricity

In the present §1, we discuss the notion of a *birational Galois section*. In the present §1, let $\mathcal{C}$ be a full formation, k a field of characteristic 0, and $\overline{k}$ an algebraic closure of k. For a finite extension k' ($\subseteq \overline{k}$) of k, write $G_{k'} \overset{\text{def}}{=} \mathrm{Gal}(\overline{k}/k')$.

Definition 5.1. Let X be a *quasi-compact* scheme which is *geometrically integral* over k.

(i) We shall write

$$k(X)$$

for the function field of X.

(ii) We shall write

$$\Delta^{\mathcal{C}}_{X/k}$$

for the *pro-$\mathcal{C}$ geometric fundamental group* of X, i.e., the maximal pro-$\mathcal{C}$ quotient of $\pi_1(X \otimes_k \overline{k})$, and

$$\Pi^{\mathcal{C}}_{X/k}$$

for the *geometrically pro-$\mathcal{C}$ fundamental group* of X, i.e., the quotient of $\pi_1(X)$ by the kernel of the natural surjection $\pi_1(X \otimes_k \overline{k}) \twoheadrightarrow \Delta^{\mathcal{C}}_{X/k}$. If X is the spectrum of a k-algebra R, then we shall write

$$\Delta_{R/k}^{\mathcal{C}} \overset{\text{def}}{=} \Delta_{X/k}^{\mathcal{C}} \; ; \quad \Pi_{R/k}^{\mathcal{C}} \overset{\text{def}}{=} \Pi_{X/k}^{\mathcal{C}} \, .$$

Thus, we have a commutative diagram of profinite groups

$$
\begin{array}{ccccccccc}
1 & \longrightarrow & \Delta_{k(X)/k}^{\mathcal{C}} & \longrightarrow & \Pi_{k(X)/k}^{\mathcal{C}} & \longrightarrow & G_k & \longrightarrow & 1 \\
 & & \downarrow & & \downarrow & & \| & & \\
1 & \longrightarrow & \Delta_{X/k}^{\mathcal{C}} & \longrightarrow & \Pi_{X/k}^{\mathcal{C}} & \longrightarrow & G_k & \longrightarrow & 1
\end{array}
$$

[cf. (i)] – where the horizontal sequences are *exact* [cf. [3], Exposé IX, Théorème 6.1].

If $\mathcal{C}$ consists of all Σ-*groups* [cf. §0] for some nonempty subset $\Sigma \subseteq \mathfrak{Primes}$ [cf. §0], then we shall write

$$\Delta_{X/k}^{\Sigma} \overset{\text{def}}{=} \Delta_{X/k}^{\mathcal{C}} \; ; \quad \Pi_{X/k}^{\Sigma} \overset{\text{def}}{=} \Pi_{X/k}^{\mathcal{C}} \, .$$

Definition 5.2. Let X be a *quasi-compact* scheme which is *geometrically integral* over k. Then we shall refer to a section of the upper (respectively, lower) exact sequence of the commutative diagram of Definition 5.1, (ii), as a *pro-$\mathcal{C}$ birational Galois section* (respectively, *pro-$\mathcal{C}$ Galois section*) of X/k. The $\Delta_{k(X)/k}^{\mathcal{C}}$-conjugacy (respectively, $\Delta_{X/k}^{\mathcal{C}}$-conjugacy) class of a pro-$\mathcal{C}$ birational Galois section (respectively, pro-$\mathcal{C}$ Galois section) of X/k as the *conjugacy class* of the pro-$\mathcal{C}$ birational Galois section (respectively, pro-$\mathcal{C}$ Galois section).

If $\mathcal{C}$ consists of all Σ-*groups* for some nonempty subset $\Sigma \subseteq \mathfrak{Primes}$, then we shall refer to a pro-$\mathcal{C}$ birational Galois section (respectively, pro-$\mathcal{C}$ Galois section) of X/k as a *pro-Σ birational Galois section* (respectively, *pro-Σ Galois section*) of X/k.

Definition 5.3. Let X be a *smooth curve* over k [cf. §0] and s a pro-$\mathcal{C}$ birational Galois section (respectively, pro-$\mathcal{C}$ Galois section) of X/k [cf. Definition 5.2]. Then we shall say that s is *geometric* if the image of s is contained in a decomposition subgroup of $\Pi_{k(X)/k}^{\mathcal{C}}$ (respectively, $\Pi_{X/k}^{\mathcal{C}}$) associated to a [necessarily k-rational] closed point of the [uniquely determined] smooth compactification of X over k.

Remark 5.4. Let X be a *smooth curve* over k. Then it follows immediately from the various definitions involved that the geometricity of a pro-$\mathcal{C}$ birational Galois section (respectively, pro-$\mathcal{C}$ Galois section) of X/k *depends only* on its conjugacy class [cf. Definition 5.2].

Remark 5.5. Let X, Y be *smooth curves* over k and $Y \rightarrow X$ a *dominant* morphism over k, which thus determines a finite extension $k(X) \hookrightarrow k(Y)$ over k. If a pro-$\mathcal{C}$ birational Galois section (respectively, pro-$\mathcal{C}$ Galois section) s

of Y/k is *geometric*, then it follows immediately from the various definitions involved that the pro-$\mathcal{C}$ birational Galois section (respectively, pro-$\mathcal{C}$ Galois section) of X/k determined by s and the morphism $Y \to X$ [i.e., the pro-$\mathcal{C}$ birational Galois section (respectively, pro-$\mathcal{C}$ Galois section) of X/k obtained as the composite of s and the natural open homomorphism $\Pi^{\mathcal{C}}_{k(Y)/k} \to \Pi^{\mathcal{C}}_{k(X)/k}$ (respectively, $\Pi^{\mathcal{C}}_{Y/k} \to \Pi^{\mathcal{C}}_{X/k}$) induced by $Y \to X$] is *geometric*.

Remark 5.6. Let X be a *projective smooth curve* over k, $U \subseteq X$ an open subscheme of X, and s a pro-$\mathcal{C}$ birational Galois section of X/k. Then it follows immediately from the various definitions involved that if s is *geometric*, then the pro-$\mathcal{C}$ Galois section of U/k naturally determined by s [i.e., the pro-$\mathcal{C}$ Galois section of U/k obtained as the composite of s and the natural surjection $\Pi^{\mathcal{C}}_{k(X)/k} \twoheadrightarrow \Pi^{\mathcal{C}}_{U/k}$] is *geometric*.

Lemma 5.7. *Let X be a **hyperbolic curve** over k [cf. §0] and x, y closed points of the [uniquely determined] smooth compactification of X. Suppose that k is **generalized sub-p-adic** [i.e., k is isomorphic to a subfield of a finitely generated extension of the p-adic completion of the maximal unramified extension of the p-adic completion of the field of rational numbers – cf. [11], Definition 4.11] for some $p \in \Sigma(\mathcal{C})$ [cf. §0]. Then the following conditions are equivalent:*

(1) *$x = y$.*
(2) *There exist respective decomposition subgroups D_x, $D_y \subseteq \Pi^{\mathcal{C}}_{X/k}$ of $\Pi^{\mathcal{C}}_{X/k}$ associated to x, y such that the image of the composite*

$$D_x \cap D_y \hookrightarrow \Pi^{\mathcal{C}}_{X/k} \twoheadrightarrow G_k$$

*is **open**.*

Proof. The implication (1) $\Rightarrow$ (2) is immediate. Next, we verify the implication (2) $\Rightarrow$ (1). Suppose that condition (2) is satisfied. Then it is immediate that, to verify the implication (2) $\Rightarrow$ (1), by replacing $\Pi^{\mathcal{C}}_{X/k}$ by an open subgroup of $\Pi^{\mathcal{C}}_{X/k}$, we may assume without loss of generality that X is of *genus* ≥ 2, and, moreover, the displayed composite of condition (2) is *surjective*, hence also that x and y are k-*rational*. Thus, to verify the implication (2) $\Rightarrow$ (1), by replacing X by its smooth compactification, we may assume without loss of generality that x, $y \in X(k)$. Then, by considering the quotient $\Pi^{\mathcal{C}}_{X/k} \twoheadrightarrow \Pi^{\{p\}}_{X/k}$ of $\Pi^{\mathcal{C}}_{X/k}$, the implication (2) $\Rightarrow$ (1) follows immediately from [11], Theorem 4.12 [cf. also [11], Remark following Theorem 4.12], together with a similar argument to the argument used in the proof of [10], Theorem C. This completes the proof of the implication (2) $\Rightarrow$ (1), hence also of Lemma 5.7. $\square$

198   Yuichiro Hoshi

Lemma 5.8. *Let X be a* **hyperbolic curve** *over k, s a pro-$\mathcal{C}$ Galois section of X/k [cf. Definition 5.2], and k' ($\subseteq \bar{k}$) a finite extension of k. Suppose that k is either*

(a) *a* **finitely generated field of characteristic 0** *[cf. §0] or*
(b) *a* **p-adic local field** *for some $p \in \Sigma(\mathcal{C})$ [cf. §0].*

Then the following conditions are equivalent:

(1) *s is* **geometric** *[cf. Definition 5.3].*
(2) *The pro-$\mathcal{C}$ Galois section $s|_{G_{k'}}$ of $X \otimes_k k'/k'$ determined by s is* **geometric.**
(3) *For any open subgroup $H \subseteq \Pi^{\mathcal{C}}_{X/k}$ of $\Pi^{\mathcal{C}}_{X/k}$ containing the image of s, the [uniquely determined] smooth compactification of the finite étale covering of X corresponding to $H \subseteq \Pi^{\mathcal{C}}_{X/k}$ admits a k'-valued point.*

Proof. The equivalence (1) $\Leftrightarrow$ (2) follows immediately from a similar argument to the argument used in the proof of [7], Lemma 54. Here, we note that if k satisfies the condition (b), then, in order to apply a similar argument to the argument used in the proof of [7], Lemma 54, we have to replace "[Moc99, Theorem C]" (respectively, the *finiteness* of the set "$(X^n)^{\mathrm{cpt}}(k)$" obtained by Mordell–Faltings's theorem) in the proof of [7], Lemma 54, by Lemma 5.7 (respectively, the *compactness* of the set "$(X^n)^{\mathrm{cpt}}(k)$" obtained by the consideration of a suitable model of "$(X^n)^{\mathrm{cpt}}$" over the ring of integers of k). The equivalence (2) $\Leftrightarrow$ (3) follows immediately from a similar argument to the argument applied in the proof of [17], Proposition 2.8, (iv). This completes the proof of Lemma 5.8. □

Lemma 5.9. *Let X be a* **smooth curve** *over k, s a pro-$\mathcal{C}$ birational Galois section of X/k [cf. Definition 5.2], and k' ($\subseteq \bar{k}$) a finite extension of k. Suppose that k is either*

(a) *a* **finitely generated field of characteristic 0** *[cf. §0] or*
(b) *a* **p-adic local field** *[cf. §0] for some $p \in \Sigma(\mathcal{C})$ [cf. §0].*

Then the following conditions are equivalent:

(1) *s is* **geometric** *[cf. Definition 5.3].*
(2) *The pro-$\mathcal{C}$ birational Galois section $s|_{G_{k'}}$ of $X \otimes_k k'/k'$ determined by s is* **geometric.**
(3) *For any open subgroup $H \subseteq \Pi^{\mathcal{C}}_{k(X)/k}$ of $\Pi^{\mathcal{C}}_{k(X)/k}$ containing the image of s, the [uniquely determined] smooth compactification of the normalization of X in the finite extension of $k(X)$ corresponding to $H \subseteq \Pi^{\mathcal{C}}_{k(X)/k}$ admits a k'-valued point.*

Proof. This follows immediately from a similar argument to the argument applied in the proof of Lemma 5.8.   □

The following result was essentially proved in [12] by a refined discussion of the discussion given in [8].

Proposition 5.10. *Let p be a prime number and X a **smooth curve** over k. Suppose that $p \in \Sigma(\mathcal{C})$ [cf. §0], and that k is a **p-adic local field**. Then any pro-$\mathcal{C}$ birational Galois section of X/k [cf. Definition 5.2] is **geometric** [cf. Definition 5.3].*

Proof. It follows from the equivalence (1) ⇔ (3) of Lemma 5.9 that, to verify Proposition 5.10, it suffices to verify that, for any open subgroup $H \subseteq \Pi^{\mathcal{C}}_{k(X)/k}$ of $\Pi^{\mathcal{C}}_{k(X)/k}$ containing the image of s, the [uniquely determined] smooth compactification of the normalization of X in the finite extension of $k(X)$ corresponding to $H \subseteq \Pi^{\mathcal{C}}_{k(X)/k}$ admits a $k(\zeta_p)$-valued point, where we use the notation $\zeta_p \in \bar{k}$ to denote a *primitive p-th root of unity*. On the other hand, this follows immediately from [12], Theorem A, (2). This completes the proof of Proposition 5.10.   □

2. Local geometricity of birational Galois sections

In the present §2, we discuss the notion of the *local geometricity* of birational Galois sections of smooth curves over number fields. In the present §2, let $\mathcal{C}$ be a full formation, k a *number field* [cf. §0], $\bar{k}$ an algebraic closure of k, and X a *smooth curve* over k [cf. §0]. Write

$$\mathfrak{o}_k \subseteq k$$

for the ring of integers of k,

$$\mathfrak{P}^f_k$$

for the set of all nonarchimedean primes of k, and

$$X^{\mathrm{cpt}}$$

for the [uniquely determined] smooth compactification of X over k. Moreover, for each $\mathfrak{p} \in \mathfrak{P}^f_k$, write

$$k_{\mathfrak{p}}$$

for the $\mathfrak{p}$-adic completion of k and

$$\mathfrak{o}_{\mathfrak{p}} \subseteq k_{\mathfrak{p}}$$

for the ring of integers of $k_{\mathfrak{p}}$. For each $\mathfrak{p} \in \mathfrak{P}_k^f$, let us fix an algebraic closure $\overline{k}_{\mathfrak{p}}$ of $k_{\mathfrak{p}}$ containing $\overline{k}$ and write

$$G_{\mathfrak{p}} \overset{\text{def}}{=} \mathrm{Gal}(\overline{k}_{\mathfrak{p}}/k_{\mathfrak{p}}) \subseteq G_k \overset{\text{def}}{=} \mathrm{Gal}(\overline{k}/k).$$

Definition 5.11. Let s be a pro-$\mathcal{C}$ Galois section of X/k [cf. Definition 5.2]. For a nonarchimedean prime $\mathfrak{p} \in \mathfrak{P}_k^f$ of k, we shall say that s is *geometric at* $\mathfrak{p}$ if the pro-$\mathcal{C}$ Galois section of $X \otimes_k k_{\mathfrak{p}}/k_{\mathfrak{p}}$ naturally determined by s [i.e., the pro-$\mathcal{C}$ Galois section of $X \otimes_k k_{\mathfrak{p}}/k_{\mathfrak{p}}$ determined by the natural isomorphism

$$\Pi^{\mathcal{C}}_{X \otimes_k k_{\mathfrak{p}}/k_{\mathfrak{p}}} \overset{\sim}{\longrightarrow} \Pi^{\mathcal{C}}_{X/k} \times_{G_k} G_{\mathfrak{p}}$$

and the composite

$$G_{\mathfrak{p}} \hookrightarrow G_k \overset{s}{\to} \Pi^{\mathcal{C}}_{X/k}]$$

is *geometric* [cf. Definition 5.3]. For a subset $S \subseteq \mathfrak{P}_k^f$ of $\mathfrak{P}_k^f$, we shall say that s is *geometric at* S if, for each $\mathfrak{p} \in S$, s is geometric at $\mathfrak{p}$. Finally, we shall say that s is *locally geometric* if s is geometric at $\mathfrak{P}_k^f$.

Remark 5.12. In the notation of Definition 5.11, it is immediate that if s is *geometric* [cf. Definition 5.3], then s is *locally geometric*.

Definition 5.13. Let $S \subseteq \mathfrak{P}_k^f$ be a subset of $\mathfrak{P}_k^f$. Then we shall write

$$\widetilde{\mathbb{A}}_k^f|_S \overset{\text{def}}{=} \prod_{\mathfrak{p} \in S} k_{\mathfrak{p}};$$

$$\mathbb{A}_k^f|_S \overset{\text{def}}{=} \left\{ (a_{\mathfrak{p}})_{\mathfrak{p} \in S} \in \widetilde{\mathbb{A}}_k^f|_S \,\middle|\, a_{\mathfrak{p}} \in \mathfrak{o}_{\mathfrak{p}} \text{ for all but finitely many } \mathfrak{p} \in S \right\};$$

$$\widetilde{\mathbb{A}}_k^f \overset{\text{def}}{=} \widetilde{\mathbb{A}}_k^f|_{\mathfrak{P}_k^f}; \quad \mathbb{A}_k^f \overset{\text{def}}{=} \mathbb{A}_k^f|_{\mathfrak{P}_k^f}.$$

Remark 5.14. Since X^{cpt} is *proper* over k, for any subset $S \subseteq \mathfrak{P}_k^f$ of $\mathfrak{P}_k^f$, the natural injection $X^{\mathrm{cpt}}(\mathbb{A}_k^f|_S) \hookrightarrow X^{\mathrm{cpt}}(\widetilde{\mathbb{A}}_k^f|_S)$ is *bijective*.

Definition 5.15. Let s be a pro-$\mathcal{C}$ Galois section of X/k [cf. Definition 5.2]. If s is geometric at a nonarchimedean prime $\mathfrak{p} \in \mathfrak{P}_k^f$ of k [cf. Definition 5.11], i.e., there exists a $k_{\mathfrak{p}}$-valued point $x_{\mathfrak{p}} \in X^{\mathrm{cpt}}(k_{\mathfrak{p}}) = (X^{\mathrm{cpt}} \otimes_k k_{\mathfrak{p}})(k_{\mathfrak{p}})$ of X^{cpt} such that the image of the pro-$\mathcal{C}$ Galois section of $X \otimes_k k_{\mathfrak{p}}/k_{\mathfrak{p}}$ naturally determined by s is contained in a decomposition subgroup of $\Pi^{\mathcal{C}}_{X \otimes_k k_{\mathfrak{p}}/k_{\mathfrak{p}}}$ associated to $x_{\mathfrak{p}}$, then we shall refer to such a $k_{\mathfrak{p}}$-valued point "$x_{\mathfrak{p}}$" of X^{cpt} as a $k_{\mathfrak{p}}$-*valued point of X^{cpt} associated to* s. If s is geometric at a subset $S \subseteq \mathfrak{P}_k^f$ of $\mathfrak{P}_k^f$, then we shall refer to an $\widetilde{\mathbb{A}}_k^f|_S$-valued point, or, equivalently [cf. Remark 5.14], an $\mathbb{A}_k^f|_S$-valued point, of X^{cpt} determined by $k_{\mathfrak{p}}$-valued points of X^{cpt} associated to s – where $\mathfrak{p}$ ranges over elements of S – as an $\widetilde{\mathbb{A}}_k^f|_S$-*valued point*, or, equivalently, an $\mathbb{A}_k^f|_S$-*valued point, of* X^{cpt} *associated to* s.

Remark 5.16. In the notation of Definition 5.15, suppose that s is *geometric* [cf. Definition 5.3], hence also *locally geometric* [cf. Definition 5.11; Remark 5.12]. Then it is immediate that there exists a k-rational point $x \in X^{\mathrm{cpt}}(k)$ of X^{cpt} such that, for each $\mathfrak{p} \in \mathfrak{P}_k^f$, the $k_{\mathfrak{p}}$-valued point of X^{cpt} determined by x is a $k_{\mathfrak{p}}$-valued point of X^{cpt} associated to s. In particular, the $\mathbb{A}_k^f$-valued point of X^{cpt} determined by x is an $\mathbb{A}_k^f$-valued point of X^{cpt} associated to s.

Note that if $\mathcal{C}$ contains all finite *solvable* groups, and X is a *hyperbolic curve* over k [cf. §0], then it follows from Theorem 5.33 below that the converse holds, i.e., if s is *locally geometric*, and there exists a k-rational point $x \in X^{\mathrm{cpt}}(k)$ of X^{cpt} such that, for each $\mathfrak{p} \in \mathfrak{P}_k^f$, the $k_{\mathfrak{p}}$-valued point of X^{cpt} determined by x is a $k_{\mathfrak{p}}$-valued point of X^{cpt} associated to s, then s is *geometric*.

Lemma 5.17. *Let s be a pro-$\mathcal{C}$ birational Galois section of X/k [cf. Definition 5.2] and $\mathfrak{p} \in \mathfrak{P}_k^f$. For an open subscheme $U \subseteq X^{\mathrm{cpt}}$ of X^{cpt}, write*

$$s[U]$$

for the pro-$\mathcal{C}$ Galois section of U/k [cf. Definition 5.2] naturally determined by s [i.e., the pro-$\mathcal{C}$ Galois section of U/k obtained as the composite of s and the natural surjection $\Pi_{k(X)/k}^{\mathcal{C}} \twoheadrightarrow \Pi_{U/k}^{\mathcal{C}}$];

$$s[U, \mathfrak{p}]$$

for the pro-$\mathcal{C}$ Galois section of $U \otimes_k k_{\mathfrak{p}}/k_{\mathfrak{p}}$ naturally determined by s [i.e., the pro-$\mathcal{C}$ Galois section of $U \otimes_k k_{\mathfrak{p}}/k_{\mathfrak{p}}$ determined by the natural isomorphism

$$\Pi_{U \otimes_k k_{\mathfrak{p}}/k_{\mathfrak{p}}}^{\mathcal{C}} \xrightarrow{\sim} \Pi_{U/k}^{\mathcal{C}} \times_{G_k} G_{\mathfrak{p}}$$

and the composite

$$G_{\mathfrak{p}} \hookrightarrow G_k \xrightarrow{s} \Pi_{k(X)/k}^{\mathcal{C}} \twoheadrightarrow \Pi_{U/k}^{\mathcal{C}}].$$

Then the following conditions are equivalent:

(1) *There exists a $k_{\mathfrak{p}}$-valued point $x_{\mathfrak{p}} \in X^{\mathrm{cpt}}(k_{\mathfrak{p}}) = (X^{\mathrm{cpt}} \otimes_k k_{\mathfrak{p}})(k_{\mathfrak{p}})$ of X^{cpt} such that, for any open subscheme $U \subseteq X^{\mathrm{cpt}}$ of X^{cpt}, the image of the pro-$\mathcal{C}$ Galois section $s[U, \mathfrak{p}]$ of $U \otimes_k k_{\mathfrak{p}}/k_{\mathfrak{p}}$ is contained in a decomposition subgroup of $\Pi_{U \otimes_k k_{\mathfrak{p}}/k_{\mathfrak{p}}}^{\mathcal{C}}$ associated to $x_{\mathfrak{p}}$.*

(2) *For any open subscheme $U \subseteq X^{\mathrm{cpt}}$ of X^{cpt}, the pro-$\mathcal{C}$ Galois section $s[U]$ of U/k is* **geometric at $\mathfrak{p}$** *[cf. Definition 5.11], i.e., the pro-$\mathcal{C}$ Galois section $s[U, \mathfrak{p}]$ of $U \otimes_k k_{\mathfrak{p}}/k_{\mathfrak{p}}$ is* **geometric** *[or, equivalently, the image of $s[U, \mathfrak{p}]$ is contained in a decomposition subgroup of $\Pi^{\mathcal{C}}_{U \otimes_k k_{\mathfrak{p}}/k_{\mathfrak{p}}}$ associated to a $k_{\mathfrak{p}}$-rational point of $X^{\mathrm{cpt}} \otimes_k k_{\mathfrak{p}}$].*

(3) *For any open subgroup $H \subseteq \Pi^{\mathcal{C}}_{k(X)/k}$ of $\Pi^{\mathcal{C}}_{k(X)/k}$ containing the image of s, the [uniquely determined] smooth compactification of the normalization of X in the finite extension of $k(X)$ corresponding to $H \subseteq \Pi^{\mathcal{C}}_{k(X)/k}$ admits a $k_{\mathfrak{p}}$-valued point.*

(4) *The image of the homomorphism $G_{\mathfrak{p}} \to \Pi^{\mathcal{C}}_{k(X)/k} \times_{G_k} G_{\mathfrak{p}}$ induced by s is contained in the image of a decomposition subgroup of $\Pi^{\mathcal{C}}_{k_{\mathfrak{p}}(X \otimes_k k_{\mathfrak{p}})/k_{\mathfrak{p}}}$ associated to a [necessarily $k_{\mathfrak{p}}$-rational] closed point of $X^{\mathrm{cpt}} \otimes_k k_{\mathfrak{p}}$ by the natural surjection $\Pi^{\mathcal{C}}_{k_{\mathfrak{p}}(X \otimes_k k_{\mathfrak{p}})/k_{\mathfrak{p}}} \twoheadrightarrow \Pi^{\mathcal{C}}_{k(X)/k} \times_{G_k} G_{\mathfrak{p}}$.*

Proof. The implications $(4) \Rightarrow (1) \Rightarrow (2) \Rightarrow (3)$ are immediate. Finally, the implication $(3) \Rightarrow (4)$ follows immediately from a similar argument to the argument applied in the proof of [17], Proposition 2.8, (iv). This completes the proof of Lemma 5.17. $\qquad\qquad\square$

Definition 5.18. Let s be a pro-$\mathcal{C}$ birational Galois section of X/k [cf. Definition 5.2]. For a nonarchimedean prime $\mathfrak{p} \in \mathfrak{P}^f_k$ of k, we shall say that s is *geometric at $\mathfrak{p}$* if the pair $(s, \mathfrak{p})$ satisfies equivalent conditions (1), (2), (3), and (4) of Lemma 5.17. For a subset $S \subseteq \mathfrak{P}^f_k$ of $\mathfrak{P}^f_k$, we shall say that s is *geometric at S* if, for each $\mathfrak{p} \in S$, s is geometric at $\mathfrak{p}$. Finally, we shall say that s is *locally geometric* if s is geometric at $\mathfrak{P}^f_k$.

Remark 5.19. In the notation of Definition 5.18, it is immediate that if s is *geometric* [cf. Definition 5.3], then s is *locally geometric*.

Definition 5.20. Let s be a pro-$\mathcal{C}$ birational Galois section of X/k [cf. Definition 5.2]. If s is geometric at a nonarchimedean prime $\mathfrak{p} \in \mathfrak{P}^f_k$ of k [cf. Definition 5.18], i.e., the pair $(s, \mathfrak{p})$ satisfies condition (1) of Lemma 5.17, then we shall refer to a $k_{\mathfrak{p}}$-valued point "$x_{\mathfrak{p}}$" of X^{cpt} appearing in condition (1) of Lemma 5.17 as a $k_{\mathfrak{p}}$-*valued point of X^{cpt} associated to s*. If s is geometric at a subset $S \subseteq \mathfrak{P}^f_k$ of $\mathfrak{P}^f_k$, then we shall refer to an $\widetilde{\mathbb{A}}^f_k|_S$-valued point, or, equivalently [cf. Remark 5.14], an $\mathbb{A}^f_k|_S$-valued point, of X^{cpt} determined by $k_{\mathfrak{p}}$-valued points of X^{cpt} associated to s – where $\mathfrak{p}$ ranges over elements of S – as an $\widetilde{\mathbb{A}}^f_k|_S$-*valued point*, or, equivalently, an $\mathbb{A}^f_k|_S$-*valued point*, of X^{cpt} *associated to s*.

Remark 5.21. In the notation of Definition 5.20, suppose that s is *geometric* [cf. Definition 5.3], hence also *locally geometric* [cf. Definition 5.18;

Remark 5.19]. Then it is immediate that there exists a k-rational point $x \in X^{\mathrm{cpt}}(k)$ of X^{cpt} such that, for each $\mathfrak{p} \in \mathfrak{P}_k^f$, the $k_{\mathfrak{p}}$-valued point of X^{cpt} determined by x is a $k_{\mathfrak{p}}$-valued point of X^{cpt} associated to s. In particular, the $\mathbb{A}_k^f$-valued point of X^{cpt} determined by x is an $\mathbb{A}_k^f$-valued point of X^{cpt} associated to s.

Note that if $\mathcal{C}$ contains all finite *solvable* groups, then it follows from Theorem 5.33 below that the converse holds, i.e., if s is *locally geometric*, and there exists a k-rational point $x \in X^{\mathrm{cpt}}(k)$ of X^{cpt} such that, for each $\mathfrak{p} \in \mathfrak{P}_k^f$, the $k_{\mathfrak{p}}$-valued point of X^{cpt} determined by x is a $k_{\mathfrak{p}}$-valued point of X^{cpt} associated to s, then s is *geometric*.

Lemma 5.22. *Let s be a pro-$\mathcal{C}$ birational Galois section (respectively, pro-$\mathcal{C}$ Galois section) of X/k [cf. Definition 5.2] and $S \subseteq \mathfrak{P}_k^f$ a subset of $\mathfrak{P}_k^f$. Suppose that s is **geometric at** S [cf. Definition 5.18 (respectively, Definition 5.11)], and that, for each $\mathfrak{p} \in S$, the residue characteristic of $\mathfrak{p}$ is $\in \Sigma(\mathcal{C})$ [cf. §0]. Suppose, moreover, that if s is a pro-$\mathcal{C}$ Galois section of X/k, then X is a **hyperbolic curve** over k [cf. §0]. Then an $\mathbb{A}_k^f|_S$-valued point of X^{cpt} associated to s [cf. Definition 5.20 (respectively, Definition 5.15)] is **uniquely determined** by s.*

Proof. Observe that, to verify Lemma 5.22, by replacing S by a subset of S of *cardinality* 1, we may assume without loss of generality that $S = \{\mathfrak{p}\}$ for some $\mathfrak{p} \in \mathfrak{P}_k^f$. Then the uniqueness in question follows immediately from Lemma 5.7. This completes the proof of Lemma 5.22. $\square$

Lemma 5.23. *Let s be a pro-$\mathcal{C}$ birational Galois section of X/k [cf. Definition 5.2], $S \subseteq \mathfrak{P}_k^f$ a subset of $\mathfrak{P}_k^f$, and $x_{\mathbb{A}} \in X^{\mathrm{cpt}}(\mathbb{A}_k^f|_S)$ an $\mathbb{A}_k^f|_S$-valued point of X^{cpt}. Suppose that s is **geometric at** S [cf. Definition 5.18]. Write $s[X]$ for the pro-$\mathcal{C}$ Galois section of X/k [cf. Definition 5.2] naturally determined by s. Then the following hold:*

(i) *$s[X]$ is **geometric at** S [cf. Definition 5.11].*

(ii) *If $x_{\mathbb{A}} \in X^{\mathrm{cpt}}(\mathbb{A}_k^f|_S)$ is an $\mathbb{A}_k^f|_S$-valued point of X^{cpt} associated to s [cf. Definition 5.20], then $x_{\mathbb{A}} \in X^{\mathrm{cpt}}(\mathbb{A}_k^f|_S)$ is an $\mathbb{A}_k^f|_S$-valued point of X^{cpt} associated to $s[X]$ [cf. Definition 5.15; assertion (i)].*

(iii) *Suppose, moreover, that, for each $\mathfrak{p} \in S$, the residue characteristic of $\mathfrak{p}$ is $\in \Sigma(\mathcal{C})$ [cf. §0], and that X is a **hyperbolic curve** over k. Then it holds that $x_{\mathbb{A}} \in X^{\mathrm{cpt}}(\mathbb{A}_k^f|_S)$ is an $\mathbb{A}_k^f|_S$-valued point of X^{cpt} associated to s if and only if $x_{\mathbb{A}} \in X^{\mathrm{cpt}}(\mathbb{A}_k^f|_S)$ is an $\mathbb{A}_k^f|_S$-valued point of X^{cpt} associated to $s[X]$ [cf. assertion (i)].*

Proof. Assertions (i), (ii) follow immediately from the various definitions involved. Assertion (iii) follows immediately from Lemma 5.22, together with assertion (ii). This completes the proof of Lemma 5.23. □

The following result was essentially proved in [12] by a refined discussion of the discussion given in [8].

Proposition 5.24. *Let s be a pro-$\mathcal{C}$ birational Galois section of X/k [cf. Definition 5.2] and $S \subseteq \mathfrak{P}_k^f$ a subset of $\mathfrak{P}_k^f$ such that, for each $\mathfrak{p} \in S$, the residue characteristic of $\mathfrak{p}$ is $\in \Sigma(\mathcal{C})$ [cf. §0]. Then s is* **geometric at S** *[cf. Definition 5.18]. In particular, s determines a* **unique** *$\mathbb{A}_k^f|_S$-valued point of X^{cpt} [cf. Definition 5.20].*

Proof. If s is *geometric at S*, then the uniqueness of an $\mathbb{A}_k^f|_S$-valued point of X^{cpt} associated to s follows from Lemma 5.22. Thus, to verify Proposition 5.24, it suffices to verify that s is *geometric at S*. Moreover, it follows immediately from the various definitions involved that, to verify that s is *geometric at S*, by replacing S by a subset of S of *cardinality* 1, we may assume without loss of generality that $S = \{\mathfrak{p}\}$ for some $\mathfrak{p} \in \mathfrak{P}_k^f$, whose residue characteristic we denote by p. Thus, it follows from Lemma 5.25 below [cf. condition (5) of Lemma 5.25 below] that, to verify Proposition 5.24, it suffices to verify that, for any open subgroup $H \subseteq \Pi_{k(X)/k}^{\mathcal{C}}$ of $\Pi_{k(X)/k}^{\mathcal{C}}$ containing the image of s, the [uniquely determined] smooth compactification Y of the normalization of X in the finite extension of $k(X)$ corresponding to $H \subseteq \Pi_{k(X)/k}^{\mathcal{C}}$ admits a $k_{\mathfrak{p}}(\zeta_p)$-valued point, where we use the notation $\zeta_p \in \bar{k}$ to denote a *primitive p-th root of unity*. On the other hand, by considering the restriction of the pro-$\mathcal{C}$ birational Galois section of Y/k naturally determined by s to the closed subgroup $\mathrm{Gal}(\bar{k}/k(\zeta_p)^{\mathrm{h}})$ of G_k, where we write $k(\zeta_p)^{\mathrm{h}} \subseteq \bar{k}$ for the algebraic closure of $k(\zeta_p)$ in $k_{\mathfrak{p}}(\zeta_p)$, we conclude from [12], Theorem B, (2), that $Y(k(\zeta_p)^{\mathrm{h}}) \neq \emptyset$, hence also that $Y(k_{\mathfrak{p}}(\zeta_p)) \neq \emptyset$. This completes the proof of Proposition 5.24. □

Lemma 5.25. *In the notation of Lemma 5.17, suppose, moreover, that the residue characteristic of $\mathfrak{p}$ is $\in \Sigma(\mathcal{C})$ [cf. §0]. Let $k_{\mathfrak{p}}'$ $(\subseteq \bar{k}_{\mathfrak{p}})$ be a finite extension of $k_{\mathfrak{p}}$. Then equivalent conditions (1), (2), (3), and (4) of Lemma 5.17 are equivalent to the following conditions:*

(5) *For any open subgroup $H \subseteq \Pi_{k(X)/k}^{\mathcal{C}}$ of $\Pi_{k(X)/k}^{\mathcal{C}}$ containing the image of s, the [uniquely determined] smooth compactification of the normalization of X in the finite extension of $k(X)$ corresponding to $H \subseteq \Pi_{k(X)/k}^{\mathcal{C}}$ admits a $k_{\mathfrak{p}}'$-valued point.*

(6) *The image of the composite of the natural inclusion* $\mathrm{Gal}(\overline{k}_{\mathfrak{p}}/k'_{\mathfrak{p}}) \hookrightarrow G_{\mathfrak{p}}$
and the homomorphism $G_{\mathfrak{p}} \to \Pi^{\mathcal{C}}_{k(X)/k} \times_{G_k} G_{\mathfrak{p}}$ *induced by s is contained*
in the image of a decomposition subgroup of $\Pi^{\mathcal{C}}_{k_{\mathfrak{p}}(X \otimes_k k_{\mathfrak{p}})/k_{\mathfrak{p}}}$ *associated to*
a closed point of $X^{\mathrm{cpt}} \otimes_k k_{\mathfrak{p}}$ *[necessarily defined over a subfield of* $k'_{\mathfrak{p}}$*] by*
the natural surjection $\Pi^{\mathcal{C}}_{k_{\mathfrak{p}}(X \otimes_k k_{\mathfrak{p}})/k_{\mathfrak{p}}} \twoheadrightarrow \Pi^{\mathcal{C}}_{k(X)/k} \times_{G_k} G_{\mathfrak{p}}$.

Proof. The implication (3) $\Rightarrow$ (5) is immediate. Moreover, by applying the
implication (3) $\Rightarrow$ (4) of Lemma 5.17 to the restriction of s to a suitable open
subgroup of G_k, we conclude that the implication (5) $\Rightarrow$ (6) holds. Finally,
the implication (6) $\Rightarrow$ (4) follows immediately from a similar argument to
the argument applied in the proof of [7], Lemma 54, by replacing "[Moc99,
Theorem C]" (respectively, the *finiteness* of the set "$(X^n)^{\mathrm{cpt}}(k)$" obtained by
Mordell–Faltings's theorem) in the proof of [7], Lemma 54, by Lemma 5.7
(respectively, the *compactness* of the set "$(X^n)^{\mathrm{cpt}}(k)$" obtained by the consideration of a suitable model of "$(X^n)^{\mathrm{cpt}}$" over the ring of integers of k). This
completes the proof of Lemma 5.25. $\square$

Proposition 5.26. *Suppose that X is a* **hyperbolic curve** *over k. Let s be a*
pro-$\mathcal{C}$ Galois section of X/k [cf. Definition 5.2] and $S \subseteq \mathfrak{P}^f_k$ a subset of
$\mathfrak{P}^f_k$ *such that, for each $\mathfrak{p} \in S$, the residue characteristic of $\mathfrak{p}$ is $\in \Sigma(\mathcal{C})$ [cf.*
§0]. Suppose that s arises from a pro-$\mathcal{C}$ birational Galois section of X/k [cf.
Definition 5.2]. Then s is **geometric at S** *[cf. Definition 5.11]. In particular, s*
determines a **unique** $\mathbb{A}^f_k|_S$*-valued point of X^{cpt} [cf. Definition 5.15].*

Proof. The fact that s is *geometric at S* follows immediately from Proposition 5.24, together with Lemma 5.23, (i). The fact that s determines a
unique $\mathbb{A}^f_k|_S$-valued point of X^{cpt} follows immediately from Lemma 5.22. This
completes the proof of Proposition 5.26. $\square$

3. Galois sections of tori that locally arise from points

In the present §3, we discuss *Galois sections of tori that locally arise from
points*. We maintain the notation of the preceding §2. Let $\Sigma \subseteq \mathfrak{P}\mathrm{rimes}$ be a
nonempty subset of $\mathfrak{P}\mathrm{rimes}$ [cf. §0]. Write

$$\mathrm{Div}(\mathfrak{o}_k) \stackrel{\mathrm{def}}{=} \bigoplus_{\mathfrak{p} \in \mathfrak{P}^f_k} \mathbb{Z} \;\;\to\;\; \mathrm{Pic}(\mathfrak{o}_k) \stackrel{\mathrm{def}}{=} \mathrm{Pic}(\mathrm{Spec}\, \mathfrak{o}_k);$$

d_k for the minimal positive integer such that $d_k \cdot \mathrm{Pic}(\mathfrak{o}_k) = \{0\}$;

$$\mathbb{G}_{m,\mathbb{Z}} \stackrel{\mathrm{def}}{=} \mathbb{P}^1_{\mathbb{Z}} \setminus \{0, \infty\} = \mathrm{Spec}\, \mathbb{Z}[u^{\pm 1}];$$

for each $\mathfrak{p} \in \mathfrak{P}_k^f$,

$$v_\mathfrak{p} : k^\times \longrightarrow \mathbb{Z}$$

for the $\mathfrak{p}$-adic valuation which induces a *surjection* $k_\mathfrak{p}^\times \twoheadrightarrow \mathbb{Z}$;

$$\begin{aligned} \mathrm{div}_k : \quad k^\times \quad &\longrightarrow \quad \mathrm{Div}(\mathfrak{o}_k) \\ a \quad &\mapsto \quad \textstyle\sum_{\mathfrak{p} \in \mathfrak{P}_k^f} v_\mathfrak{p}(a) \cdot \mathfrak{p}\,. \end{aligned}$$

[Thus, we have an *exact* sequence of modules

$$0 \longrightarrow \mathfrak{o}_k^\times \longrightarrow k^\times \xrightarrow{\mathrm{div}_k} \mathrm{Div}(\mathfrak{o}_k) \longrightarrow \mathrm{Pic}(\mathfrak{o}_k) \longrightarrow 0\,.]$$

Write, moreover, for a ring R,

$$\mathbb{G}_{m,R} \overset{\text{def}}{=} \mathbb{G}_{m,\mathbb{Z}} \otimes_\mathbb{Z} R\,.$$

Let us identify $\mathbb{G}_{m,R}(R)$ with $R^\times$ by the invertible function $u \in R[u^{\pm 1}]^\times$:

$$\mathbb{G}_{m,R}(R) \simeq R^\times\,.$$

Definition 5.27. Let M be a module. Then we shall write

$$M[\Sigma] \overset{\text{def}}{=} \varprojlim M/nM$$

– where n ranges over positive Σ-integers [cf. §0].

Lemma 5.28.

(i) *For every [not necessarily algebraic] extension k' of k, if we write*

$$\mathrm{GS}^\Sigma(\mathbb{G}_{m,k'}/k')$$

for the set of conjugacy classes of pro-Σ Galois sections of $\mathbb{G}_{m,k'}/k'$ [cf. Definition 5.2], then the map

$$\mathrm{GS}^\Sigma(\mathbb{G}_{m,k'}/k') \longrightarrow H^1(k', \Delta^\Sigma_{\mathbb{G}_{m,k}/k})$$

determined by the natural isomorphism

$$\Delta^\Sigma_{\mathbb{G}_{m,k'}/k'} \xrightarrow{\sim} \Delta^\Sigma_{\mathbb{G}_{m,k}/k}$$

induced by $k \hookrightarrow k'$ and the map

$$\mathrm{GS}^\Sigma(\mathbb{G}_{m,k'}/k') \longrightarrow H^1(k', \Delta^\Sigma_{\mathbb{G}_{m,k'}/k'})$$

*given by mapping an element $s \in \mathrm{GS}^\Sigma(\mathbb{G}_{m,k'}/k')$ to the element of $H^1(k', \Delta^\Sigma_{\mathbb{G}_{m,k'}/k'})$ obtained by considering the difference of s and the element of $\mathrm{GS}^\Sigma(\mathbb{G}_{m,k'}/k')$ arising from the k-rational point $1 \in (k')^\times \simeq \mathbb{G}_{m,k'}(k')$ is **bijective**.*

(ii) *There exists a natural isomorphism $\Delta^{\Sigma}_{\mathbb{G}_{m,k}/k} \overset{\sim}{\to} \widehat{\mathbb{Z}}^{\Sigma}(1)$ [where "(1)" denotes a Tate twist] such that, for every [not necessarily algebraic] extension k' of k, the following diagram of sets commutes:*

$$
\begin{array}{ccccc}
\mathbb{G}_{m,k}(k') & \longrightarrow & \mathrm{GS}^{\Sigma}(\mathbb{G}_{m,k'}/k') & \overset{\sim}{\longrightarrow} & H^1(k', \Delta^{\Sigma}_{\mathbb{G}_{m,k}/k}) \\
\wr\downarrow & & & & \wr\downarrow \\
(k')^{\times} & \longrightarrow & (k')^{\times}[\Sigma] & \overset{\sim}{\longrightarrow} & H^1(k', \widehat{\mathbb{Z}}^{\Sigma}(1)).
\end{array}
$$

Here, the left-hand upper horizontal arrow is the natural map given by mapping a k'-rational point of $\mathbb{G}_{m,k'}$ to the conjugacy class of a pro-Σ Galois section of $\mathbb{G}_{m,k'}/k'$ associated to the k'-rational point, the right-hand upper horizontal arrow is the bijection of (i), *the left-hand vertical arrow is the natural identification by the fixed invertible function u, the right-hand vertical arrow is the isomorphism induced by the isomorphism in question $\Delta^{\Sigma}_{\mathbb{G}_{m,k}/k} \overset{\sim}{\to} \widehat{\mathbb{Z}}^{\Sigma}(1)$, the left-hand lower horizontal arrow is the natural homomorphism [cf. Definition 5.27], and the right-hand lower horizontal arrow is the natural isomorphism given by the Kummer theory.*

(iii) *Let $S \subseteq \mathfrak{P}^f_k$ be a subset of $\mathfrak{P}^f_k$. Then there exists a natural isomorphism between the commutative diagram of sets*

$$
\begin{array}{ccccc}
\mathbb{G}_{m,k}(k) & = & \mathbb{G}_{m,k}(k) & \longrightarrow & \mathrm{GS}^{\Sigma}(\mathbb{G}_{m,k}/k) \\
\downarrow & & \downarrow & & \downarrow \\
\mathbb{G}_{m,k}(\mathbb{A}^f_k|_S) & \longrightarrow & \mathbb{G}_{m,k}(\widetilde{\mathbb{A}}^f_k|_S) & \longrightarrow & \prod_{\mathfrak{p}\in S} \mathrm{GS}^{\Sigma}(\mathbb{G}_{m,k_{\mathfrak{p}}}/k_{\mathfrak{p}})
\end{array}
$$

and the commutative diagram of modules

$$
\begin{array}{ccccc}
k^{\times} & = & k^{\times} & \longrightarrow & k^{\times}[\Sigma] \\
\downarrow & & \downarrow & & \downarrow \\
(\mathbb{A}^f_k|_S)^{\times} & \longrightarrow & (\widetilde{\mathbb{A}}^f_k|_S)^{\times} & \longrightarrow & \prod_{\mathfrak{p}\in S}(k^{\times}_{\mathfrak{p}}[\Sigma]).
\end{array}
$$

Proof. Assertion (i) follows immediately from the various definitions involved. Next, we verify assertion (ii). For a positive integer n, write $\mu_n \subseteq \overline{k}$ for the group of n-th roots of unity. Then, as is well-known, for a positive Σ-integer n, there exist natural isomorphisms

$$
H^1(\Delta^{\Sigma}_{\mathbb{G}_{m,k}/k}, \mu_n) \overset{\sim}{\longrightarrow} H^1(\mathbb{G}_{m,\overline{k}}, \mu_n) \overset{\sim}{\longleftarrow} \overline{k}[u^{\pm 1}]^{\times}/(\overline{k}[u^{\pm 1}]^{\times})^n .
$$

Thus, the invertible function $u \in \overline{k}[u^{\pm 1}]^{\times}$ determines an element of

$$
H^1(\Delta^{\Sigma}_{\mathbb{G}_{m,k}/k}, \widehat{\mathbb{Z}}^{\Sigma}(1)) \overset{\sim}{\longrightarrow} \mathrm{Hom}(\Delta^{\Sigma}_{\mathbb{G}_{m,k}/k}, \widehat{\mathbb{Z}}^{\Sigma}(1)).
$$

On the other hand, one verifies easily that the resulting homomorphism $\Delta_{\mathbb{G}_{m,k}/k}^{\Sigma} \to \widehat{\mathbb{Z}}^{\Sigma}(1)$ is an *isomorphism* and satisfies the condition in the statement of assertion (ii). This completes the proof of assertion (ii). Assertion (iii) follows immediately from assertion (ii). This completes the proof of Lemma 5.28. $\qquad\square$

Lemma 5.29. *The following hold:*

(i) *The exact sequence of modules*

$$1 \longrightarrow \mathfrak{o}_k^{\times} \longrightarrow k^{\times} \xrightarrow{\mathrm{div}_k} \mathrm{Div}(\mathfrak{o}_k)$$

determines an exact sequence of modules

$$1 \longrightarrow \mathfrak{o}_k^{\times}[\Sigma] \longrightarrow k^{\times}[\Sigma] \xrightarrow{\mathrm{div}_k[\Sigma]} \mathrm{Div}(\mathfrak{o}_k)[\Sigma].$$

(ii) *There is **no nontrivial** element of the cokernel of the natural homomorphism $k^{\times} \to k^{\times}[\Sigma]$ which is **annihilated** by a Σ-integer.*

Proof. First, we verify assertion (i). Write $M \overset{\text{def}}{=} \mathrm{Im}(\mathrm{div}_k) \subseteq \mathrm{Div}(\mathfrak{o}_k)$ for the image of div_k. Then since M is a *free* $\mathbb{Z}$-module, there exists a section of the natural surjection $k^{\times} \twoheadrightarrow M$; thus, we obtain a noncanonical isomorphism $\mathfrak{o}_k^{\times} \times M \overset{\sim}{\to} k^{\times}$. In particular, the natural homomorphism $\mathfrak{o}_k^{\times}[\Sigma] \to k^{\times}[\Sigma]$ is *injective*. Thus, to verify assertion (i), it suffices to verify that the kernel of $\mathrm{div}_k[\Sigma]$ is contained in $\mathfrak{o}_k^{\times}[\Sigma] \subseteq k^{\times}[\Sigma]$, or, equivalently [by the existence of the noncanonical isomorphism $\mathfrak{o}_k^{\times} \times M \overset{\sim}{\to} k^{\times}$], the natural homomorphism $M[\Sigma] \to \mathrm{Div}(\mathfrak{o}_k)[\Sigma]$ is *injective*.

For an element $\sum a_i \cdot \mathfrak{p}_i \in \mathrm{Div}(\mathfrak{o}_k)$, where $a_i \in \mathbb{Z}$ and $\mathfrak{p}_i \in \mathfrak{P}_k^f$, write $[\sum a_i \cdot \mathfrak{p}_i] \in \mathrm{Pic}(\mathfrak{o}_k) = \mathrm{Div}(\mathfrak{o}_k)/M$ for the element of the cokernel of div_k determined by $\sum a_i \cdot \mathfrak{p}_i \in \mathrm{Div}(\mathfrak{o}_k)$. Now, for an element $a \in \mathrm{Pic}(\mathfrak{o}_k)$, let us fix a nonarchimedean prime $\mathfrak{q}_a \in \mathfrak{P}_k^f$ of k such that $a = [1 \cdot \mathfrak{q}_a]$. [Note that it follows immediately from Chebotarev's density theorem that the subset of $\mathfrak{P}_k^f$ consisting of $\mathfrak{p} \in \mathfrak{P}_k^f$ such that $a = [1 \cdot \mathfrak{p}]$ is of *density* $1/\sharp\mathrm{Pic}(\mathfrak{o}_k)$; in particular, such a $\mathfrak{q}_a$ always exists.] Write $T \overset{\text{def}}{=} \{\mathfrak{q}_a \in \mathfrak{P}_k^f \mid a \in \mathrm{Pic}(\mathfrak{o}_k)\}$. Moreover, for each $\mathfrak{p} \in \mathfrak{P}_k^f \setminus T$, write $x_{\mathfrak{p}} \overset{\text{def}}{=} 1 \cdot \mathfrak{p} - 1 \cdot \mathfrak{q}_{[1 \cdot \mathfrak{p}]} \in \mathrm{Div}(\mathfrak{o}_k)$. Then one verifies easily that the $\mathbb{Z}$-submodule $N \subseteq \mathrm{Div}(\mathfrak{o}_k)$ generated by $\{x_{\mathfrak{p}} \mid \mathfrak{p} \in \mathfrak{P}_k^f \setminus T\}$ is contained in M and determines a section of the natural projection $\mathrm{Div}(\mathfrak{o}_k) \twoheadrightarrow \bigoplus_{\mathfrak{p} \in \mathfrak{P}_k^f \setminus T} \mathbb{Z}$. In particular, we obtain a commutative diagram of *free* $\mathbb{Z}$-modules

$$0 \longrightarrow N \longrightarrow M \longrightarrow M/N \longrightarrow 0$$

$$0 \longrightarrow N \longrightarrow \mathrm{Div}(\mathfrak{o}_k) \longrightarrow \mathrm{Div}(\mathfrak{o}_k)/N \longrightarrow 0$$

– where the horizontal sequences are *exact*, and the vertical arrows are *injective*. On the other hand, since $\mathrm{Pic}(\mathfrak{o}_k)$, hence also T, is *finite*, M/N and $\mathrm{Div}(\mathfrak{o}_k)/N$ are *finitely generated free* $\mathbb{Z}$-modules. In particular, one verifies easily that the natural homomorphism $(M/N)[\Sigma] \to (\mathrm{Div}(\mathfrak{o}_k)/N)[\Sigma]$ is *injective*. Thus, it follows immediately that the natural homomorphism in question $M[\Sigma] \to \mathrm{Div}(\mathfrak{o}_k)[\Sigma]$ is *injective*. This completes the proof of assertion (i).

Next, we verify assertion (ii). Now let us observe that one verifies easily that there is *no nontrivial* element of the cokernel of the natural homomorphism $\mathbb{Z} \to \widehat{\mathbb{Z}}^{\Sigma}$ which is *annihilated* by a Σ-integer. Thus, assertion (ii) follows immediately from the existence of the [noncanonical] isomorphism $\mathfrak{o}_k^{\times} \times M \overset{\sim}{\to} k^{\times}$ obtained in the proof of assertion (i), together with the well-known fact that $\mathfrak{o}_k^{\times}$ is *finitely generated*. This completes the proof of assertion (ii). $\square$

Remark 5.30. The observation given in the proof of Lemma 5.29 was related to the author by *A. Tamagawa* and *S. Yasuda*.

Lemma 5.31. *By applying Lemma 5.28, (iii), let us identify* $\mathbb{G}_{m,k}(k)$ *(respectively,* $\mathbb{G}_{m,k}(\mathbb{A}_k^f)$*;* $\mathrm{GS}^{\Sigma}(\mathbb{G}_{m,k}/k)$*;* $\prod_{\mathfrak{p} \in \mathfrak{P}_k^f} \mathrm{GS}^{\Sigma}(\mathbb{G}_{m,k_{\mathfrak{p}}}/k_{\mathfrak{p}}))$ *with* $k^{\times}$ *(respectively,* $(\mathbb{A}_k^f)^{\times}$*;* $k^{\times}[\Sigma]$*;* $\prod_{\mathfrak{p} \in \mathfrak{P}_k^f}(k_{\mathfrak{p}}^{\times}[\Sigma]))$*. Suppose that* k *is either the* **field of rational numbers** *or an* **imaginary quadratic field***. Let*

$$(a_{\mathfrak{p}})_{\mathfrak{p} \in \mathfrak{P}_k^f} \in (\mathbb{A}_k^f)^{\times} \simeq \mathbb{G}_{m,k}(\mathbb{A}_k^f)$$

$$a \in k^{\times}[\Sigma] \simeq \mathrm{GS}^{\Sigma}(\mathbb{G}_{m,k}/k)$$

be such that their images in $\prod_{\mathfrak{p} \in \mathfrak{P}_k^f}(k_{\mathfrak{p}}^{\times}[\Sigma]) \simeq \prod_{\mathfrak{p} \in \mathfrak{P}_k^f} \mathrm{GS}^{\Sigma}(\mathbb{G}_{m,k_{\mathfrak{p}}}/k_{\mathfrak{p}})$ *[cf. the diagrams of Lemma 5.28, (iii)] coincide. Then the following hold:*

(i) $a^{d_k} \in k^{\times}[\Sigma] \simeq \mathrm{GS}^{\Sigma}(\mathbb{G}_{m,k}/k)$ *is contained in the image of the natural homomorphism* $k^{\times} \simeq \mathbb{G}_{m,k}(k) \to k^{\times}[\Sigma] \simeq \mathrm{GS}^{\Sigma}(\mathbb{G}_{m,k}/k)$.

(ii) *If* d_k *is a* Σ**-integer***, then* $a \in k^{\times}[\Sigma] \simeq \mathrm{GS}^{\Sigma}(\mathbb{G}_{m,k}/k)$ *is contained in the image of the natural homomorphism* $k^{\times} \simeq \mathbb{G}_{m,k}(k) \to k^{\times}[\Sigma] \simeq \mathrm{GS}^{\Sigma}(\mathbb{G}_{m,k}/k)$.

(iii) *If* d_k *is a* Σ**-integer***, and we fix an element* $\tilde{a} \in k^{\times} \simeq \mathbb{G}_{m,k}(k)$ *whose image in* $k^{\times}[\Sigma] \simeq \mathrm{GS}^{\Sigma}(\mathbb{G}_{m,k}/k)$ *coincides with* a *[cf. (ii)], then, for each* $\mathfrak{p} \in \mathfrak{P}_k^f$ *whose residue characteristic is* $\in \Sigma$*, the difference* $a_{\mathfrak{p}} \cdot \tilde{a}^{-1} \in k_{\mathfrak{p}}^{\times}$ *is a* **root of unity** *whose order is a* $(\mathfrak{Primes} \setminus \Sigma)$*-integer.*

Proof. First, we verify assertion (i). Since the image of $a \in k^\times[\Sigma] \simeq \mathrm{GS}^\Sigma(\mathbb{G}_{m,k}/k)$ in $\prod_{\mathfrak{p} \in \mathfrak{P}_k^f}(k_\mathfrak{p}^\times[\Sigma]) \simeq \prod_{\mathfrak{p} \in \mathfrak{P}_k^f} \mathrm{GS}^\Sigma(\mathbb{G}_{m,k_\mathfrak{p}}/k_\mathfrak{p})$ is contained in the image of the natural homomorphism $(\mathbb{A}_k^f)^\times \simeq \mathbb{G}_{m,k}(\mathbb{A}_k^f) \to \prod_{\mathfrak{p} \in \mathfrak{P}_k^f}(k_\mathfrak{p}^\times[\Sigma]) \simeq \prod_{\mathfrak{p} \in \mathfrak{P}_k^f} \mathrm{GS}^\Sigma(\mathbb{G}_{m,k_\mathfrak{p}}/k_\mathfrak{p})$, one verifies easily that the image of $a \in k^\times[\Sigma]$ by the homomorphism $\mathrm{div}_k[\Sigma] \colon k^\times[\Sigma] \to \mathrm{Div}(\mathfrak{o}_k)[\Sigma]$ is contained in the $\mathbb{Z}$-submodule $\mathrm{Div}(\mathfrak{o}_k) \subseteq \mathrm{Div}(\mathfrak{o}_k)[\Sigma]$. Thus, it follows immediately from the definition of d_k that there exists $\widetilde{b} \in k^\times$ such that the images $\widetilde{b}$ and a^{d_k} in $\mathrm{Div}(\mathfrak{o}_k)[\Sigma]$ *coincide*. On the other hand, since k is either the *field of rational numbers* or an *imaginary quadratic field*, it holds that $\mathfrak{o}_k^\times$ is *finite*, which thus implies that $\mathfrak{o}_k^\times \to \mathfrak{o}_k^\times[\Sigma]$ is *surjective*. Thus, it follows immediately from Lemma 5.29, (i), that, by replacing $\widetilde{b}$ by a suitable element of $k^\times$, we conclude that a^{d_k} coincides with the image of $\widetilde{b} \in k^\times$ in $k^\times[\Sigma]$. This completes the proof of assertion (i).

Assertion (ii) follows immediately from Lemma 5.29, (ii), together with assertion (i); our assumption that d_k is a Σ-*integer*. Finally, we verify assertion (iii). One verifies easily that, for each $\mathfrak{p} \in \mathfrak{P}_k^f$ whose residue characteristic is $\in \Sigma$, the kernel of the natural homomorphism $k_\mathfrak{p}^\times \to k_\mathfrak{p}^\times[\Sigma]$ consists of roots of unity in $k_\mathfrak{p}$ whose orders are $(\mathfrak{Primes} \setminus \Sigma)$-integers. Thus, assertion (iii) follows immediately from assertion (ii). This completes the proof of assertion (iii). $\qquad\square$

Lemma 5.32. *In the notation of Lemma 5.31, if $\Sigma = \mathfrak{Primes}$, then the commutative diagram of sets*

$$
\begin{array}{ccc}
\mathbb{G}_{m,k}(k) & \longrightarrow & \mathrm{GS}^\Sigma(\mathbb{G}_{m,k}/k) \\
\downarrow & & \downarrow \\
\mathbb{G}_{m,k}(\mathbb{A}_k^f) & \longrightarrow & \prod_{\mathfrak{p} \in \mathfrak{P}_k^f} \mathrm{GS}^\Sigma(\mathbb{G}_{m,k_\mathfrak{p}}/k_\mathfrak{p})
\end{array}
$$

is **cartesian**.

Proof. This follows immediately from Lemma 5.31, (ii), (iii). $\qquad\square$

4. Conditional results on the birational section conjecture

In the present §4, we prove conditional results on the birational section conjecture for projective smooth curves over number fields. We maintain the notation of the preceding §3.

First, let us recall the following result that was essentially proved in [16]. It seems to the author that [at least, a similar result to] the following result is likely to be well-known to experts. Since, however, the result could not be found in the literature, the author decided to give a proof.

Theorem 5.33. *Let C be a full formation that contains all finite **solvable groups**, k a **number field** [cf. §0], X a **projective smooth curve** (respectively, **hyperbolic curve**) over k [cf. §0], and s a pro-C **birational Galois section** (respectively, **locally geometric** pro-C **Galois section**) of X/k [cf. Definition 5.2 (respectively, Definitions 5.2; 5.11)]. Write $\mathfrak{P}_k^f$ for the set of nonarchimedean primes of k and X^{cpt} for the [uniquely determined] smooth compactification of X over k. For each $\mathfrak{p} \in \mathfrak{P}_k^f$, write $k_{\mathfrak{p}}$ for the $\mathfrak{p}$-adic completion of k. Then the following conditions are equivalent:*

(1) *s is **geometric** [cf. Definition 5.3].*
(2) *There exist a subset $T \subseteq \mathfrak{P}_k^f$ of $\mathfrak{P}_k^f$ of **density 0** and a closed subscheme $Z \subseteq X^{\mathrm{cpt}}$ of X^{cpt} which is **finite** over k such that, for each $\mathfrak{p} \in \mathfrak{P}_k^f \setminus T$, the [image of the uniquely determined – cf. Lemma 5.22] $k_{\mathfrak{p}}$-valued point of X^{cpt} associated to s [cf. Definition 5.20; Proposition 5.24 (respectively, Definition 5.15)] is contained in $Z \subseteq X^{\mathrm{cpt}}$.*

Proof. First, we verify Theorem 5.33 in the case where s is a *locally geometric pro-C Galois section*. The implication (1) $\Rightarrow$ (2) is immediate [cf. also Remark 5.16]. Next, we verify the implication (2) $\Rightarrow$ (1). Now observe that it follows from the equivalence (1) $\Leftrightarrow$ (2) of Lemma 5.8 that, to verify the implication (2) $\Rightarrow$ (1), by replacing k by a suitable finite extension of k, we may assume without loss of generality that k is *totally imaginary*. Next, observe that, for each open subgroup $H \subseteq \Pi_{X/k}^C$ of $\Pi_{X/k}^C$ containing the image of s, if we write Y for the connected finite étale covering of X corresponding to $H \subseteq \Pi_{X/k}^C$ [thus, $\Pi_{Y/k}^C = H \subseteq \Pi_{X/k}^C$], then since the morphism $Y \to X$ is *finite*, one verifies easily that the pro-C Galois section of Y/k naturally determined by s [which is necessarily *locally geometric* by the various definitions involved] satisfies condition (2). Thus, to verify the implication (2) $\Rightarrow$ (1), by replacing X by such a suitable Y, we may assume without loss of generality that X is of *genus* ≥ 2; moreover, it follows from the equivalence (1) $\Leftrightarrow$ (3) of Lemma 5.8 that, to verify the implication (2) $\Rightarrow$ (1), by applying the conclusion to various open subgroups of $\Pi_{X/k}^C$ containing the image of s, it suffices to verify that $X^{\mathrm{cpt}}(k) \neq \emptyset$. In particular, since X is of *genus* ≥ 2, and [one verifies easily that] the pro-C Galois section of X^{cpt}/k naturally determined by s is *locally geometric* and satisfies condition (2), to verify that $X^{\mathrm{cpt}}(k) \neq \emptyset$, by replacing X by X^{cpt}, we may assume without loss of generality that $X^{\mathrm{cpt}} = X$.

Now since s is *locally geometric*, and k is *totally imaginary*, it follows immediately from the definition of "$X(\mathbb{A}_k)_{\bullet}^{\mathrm{f\text{-}ab}}$" [cf. [16], Definition 5.4, (3)] that the [uniquely determined] $k_{\mathfrak{p}}$-valued points of X associated to s – where $\mathfrak{p}$ ranges over nonarchimedean primes of k – form a part of an element of $X(\mathbb{A}_k)_{\bullet}^{\text{f-ab}}$. Thus, it follows immediately from [16], Theorem 8.2, together with condition (2), that $Z(k) \neq \emptyset$, hence also $X(k) \neq \emptyset$. This completes the proof of the implication (2) $\Rightarrow$ (1), hence also of Theorem 5.33 in the case where s is a *locally geometric pro-$\mathcal{C}$ Galois section*.

Next, we verify Theorem 5.33 in the case where s is a *pro-$\mathcal{C}$ birational Galois section*. The implication (1) $\Rightarrow$ (2) is immediate [cf. also Remark 5.21]. Next, we verify the implication (2) $\Rightarrow$ (1). First, observe that it follows immediately from a similar argument to the argument applied in the proof of Theorem 5.33 in the case where s is a *locally geometric pro-$\mathcal{C}$ Galois section* that, to verify Theorem 5.33 in the case where s is a *pro-$\mathcal{C}$ birational Galois section*, by replacing $\Pi_{k(X)/k}^{\mathcal{C}}$ by an open subgroup of $\Pi_{k(X)/k}^{\mathcal{C}}$ containing the image of s, we may assume without loss of generality that X is of *genus* ≥ 2; moreover, it follows from the equivalence (1) $\Leftrightarrow$ (3) of Lemma 5.9 that, to verify the implication (2) $\Rightarrow$ (1), by applying the conclusion to various open subgroups of $\Pi_{k(X)/k}^{\mathcal{C}}$ containing the image of s, it suffices to verify that $X(k) \neq \emptyset$. On the other hand, since X is of *genus* ≥ 2, in light of Proposition 5.26, by applying Theorem 5.33 in the case where s is a *locally geometric pro-$\mathcal{C}$ Galois section* to the pro-$\mathcal{C}$ Galois section of X/k naturally determined by s, we conclude that $X(k) \neq \emptyset$. This completes the proof of the implication (2) $\Rightarrow$ (1), hence also of Theorem 5.33 in the case where s is a *pro-$\mathcal{C}$ birational Galois section*. $\square$

Theorem 5.33 naturally leads to the following corollary that was essentially proved by Stoll [cf., e.g., [16], Theorem 8.6].

Corollary 5.34. *Let $\mathcal{C}$ be a full formation that contains all finite **solvable** groups, k a **number field** [cf. §0], and X a **projective smooth curve** (respectively, **hyperbolic curve**) over k [cf. §0]. Suppose that there exist an **abelian variety** A over k and a **nonconstant** morphism $X \to A$ over k such that both the Mordell–Weil group and the Shafarevich–Tate group of A/k are **finite**. Then any pro-$\mathcal{C}$ **birational Galois section** (respectively, any **locally geometric** pro-$\mathcal{C}$ **Galois section**) of X/k [cf. Definition 5.2 (respectively, Definitions 5.2; 5.11)] is **geometric** [cf. Definition 5.3].*

Proof. Write

$$\widehat{\text{Sel}}^f(A/k) \stackrel{\text{def}}{=} \varprojlim_n \text{Ker}\Big(H^1(k, A(\overline{k})[n]) \to \prod_{\mathfrak{p} \in \mathfrak{P}_k^f} H^1(k_{\mathfrak{p}}, A(\overline{k}))\Big)$$

– where the projective limit is over all positive integers n, and $A(\overline{k})[n]$ is the subgroup of $A(\overline{k})$ consisting of elements of $A(\overline{k})$ that are annihilated by n. Then the well-known natural G_k-equivariant isomorphism $A(\overline{k})[n] \xrightarrow{\sim} \Delta_{A/k}^{\mathfrak{Primes}}/n\Delta_{A/k}^{\mathfrak{Primes}}$ induces a natural injection $\widehat{\mathrm{Sel}}^f(A/k) \hookrightarrow H^1(k, \Delta_{A/k}^{\mathfrak{Primes}})$; moreover, it follows immediately from the various definitions involved that the pro-$\mathfrak{Primes}$ Kummer homomorphism $A(k) \to H^1(k, \Delta_{A/k}^{\mathfrak{Primes}})$ associated to A [cf., e.g., [6], Remark 1.1.4, (iii)] factors through $\widehat{\mathrm{Sel}}^f(A/k) \subseteq H^1(k, \Delta_{A/k}^{\mathfrak{Primes}})$, which thus implies that we have a natural *injection* $A(k) \hookrightarrow \widehat{\mathrm{Sel}}^f(A/k)$. [Here, this *injectivity* is a formal consequence of the well-known fact that there is *no nontrivial divisible* element of $A(k)$.] On the other hand, since the Shafarevich–Tate group of A/k is *finite*, in light of the fact that the absolute Galois group of the completion of k at an *archimedean prime* is either $\simeq \mathbb{Z}/2\mathbb{Z}$ or $\simeq \{1\}$, one verifies easily that, for each positive integer n, the cokernel of the natural homomorphism

$$A(k)/nA(k) \longrightarrow \mathrm{Ker}\Big(H^1(k, A(\overline{k})[n]) \to \prod_{\mathfrak{p}\in\mathfrak{P}_k^f} H^1(k_{\mathfrak{p}}, A(\overline{k}))\Big)$$

is *annihilated by a positive integer which does not depend on* n. Thus, since the Mordell–Weil group of A/k is *finite*, it follows immediately that the resulting injection $A(k) \hookrightarrow \widehat{\mathrm{Sel}}^f(A/k)$ is an *isomorphism*.

Let s be a pro-$\mathcal{C}$ birational Galois section (respectively, *locally geometric* pro-$\mathcal{C}$ Galois section) of X/k. Write s^A for the pro-$\mathfrak{Primes}$ Galois section of A/k obtained as the composite

$$G_k \xrightarrow{s} \Pi_{k(X)/k}^{\mathcal{C}} \longrightarrow \Pi_{X/k}^{\mathcal{C}} \longrightarrow \Pi_{A/k}^{\mathcal{C}} = \Pi_{A/k}^{\mathfrak{Primes}}$$

$$\text{(respectively, } G_k \xrightarrow{s} \Pi_{X/k}^{\mathcal{C}} \longrightarrow \Pi_{A/k}^{\mathcal{C}} = \Pi_{A/k}^{\mathfrak{Primes}})$$

– where the third (respectively, second) arrow is the homomorphism over G_k induced by the *nonconstant* morphism $X \to A$ over k. Then s^A naturally determines an element of $H^1(k, \Delta_{A/k}^{\mathfrak{Primes}})$ [cf., e.g., [6], Remark 1.1.4, (ii)]; moreover, it follows immediately from Proposition 5.24 (respectively, our assumption that s is *locally geometric*), together with the various definitions involved, that this element is contained in $A(k) \xrightarrow{\sim} \widehat{\mathrm{Sel}}^f(A/k) \subseteq H^1(k, \Delta_{A/k}^{\mathfrak{Primes}})$. In particular, since $X \to A$ is *nonconstant*, and the Mordell–Weil group $A(k)$ is *finite*, it follows immediately from the *injectivity* of the pro-$\mathfrak{Primes}$ Kummer homomorphism associated to $A \otimes_k k_{\mathfrak{p}}$ [that is a formal consequence of the well-known fact that there is *no nontrivial divisible* element of $A(k_{\mathfrak{p}})$], together with [6], Remark 1.1.4, (iii), that s satisfies condition (2) of Theorem 5.33. Thus, it follows from the implication

(2) $\Rightarrow$ (1) of Theorem 5.33 that s is *geometric*. This completes the proof of Corollary 5.34. □

Remark 5.35. As in the cases of [4], Theorem 17; [16], Theorem 8.6, one may apply Corollary 5.34 to obtain some examples of projective smooth curves over number fields for which any *prosolvable* birational Galois section [i.e., any pro-$\mathcal{C}$ birational Galois section in the case where $\mathcal{C}$ consists of all finite solvable groups] is *geometric* [cf., e.g., the discussions in [4], Remark 18, (1); [16], Example 8.7; [16], Corollary 8.8].

Remark 5.36. The observation given in the proof of Corollary 5.34 was related to the author by A. *Tamagawa* and S. *Yasuda*.

Definition 5.37. Suppose that X is a *hyperbolic curve* over [the number field] k. Let s be a pro-$\mathcal{C}$ Galois section of X/k [cf. Definition 5.2].

(i) We shall say that s is *cuspidal* if the image of s is contained in a decomposition subgroup of $\Pi^{\mathcal{C}}_{X/k}$ associated to a cusp of X/k.

(ii) We shall say that s is *unramified almost everywhere* if the composite

$$G_k \xrightarrow{s} \Pi^{\mathcal{C}}_{X/k} \longrightarrow \mathrm{Aut}(\Delta^{\mathcal{C}}_{X/k})$$

– where the second arrow is the action of $\Pi^{\mathcal{C}}_{X/k}$ on $\Delta^{\mathcal{C}}_{X/k}$ obtained by conjugation – is unramified for all but finitely many $\mathfrak{p} \in \mathfrak{P}^f_k$.

Remark 5.38. In the notation of Definition 5.37, it is immediate that if s is *cuspidal* [cf. Definition 5.37, (i)], then s is *geometric* [cf. Definition 5.3].

Proposition 5.39. *Suppose that Σ is **finite**. Then any **geometric** [cf. Definition 5.3] pro-Σ Galois section [cf. Definition 5.2] of a hyperbolic curve over a number field is either **cuspidal** [cf. Definition 5.37, (i)] or **unramified almost everywhere** [cf. Definition 5.37, (ii)].*

Proof. This follows immediately from Proposition 5.A.50. □

Next, we prove the main result of the chapter.

Theorem 5.40. *Let $\mathcal{C}$ be a full formation, k either the **field of rational numbers** or an **imaginary quadratic field**, X a **projective smooth curve** over k [cf. §0], and s a pro-$\mathcal{C}$ **birational Galois section** of X/k [cf. Definition 5.2]. Write $\mathfrak{o}_k$ for the ring of integers of k and $\mathfrak{P}^f_k$ for the set of nonarchimedean primes of k. For each $\mathfrak{p} \in \mathfrak{P}^f_k$, write $k_\mathfrak{p}$ for the $\mathfrak{p}$-adic completion of k and $\mathfrak{o}_\mathfrak{p}$ for the ring of integers of $k_\mathfrak{p}$. Write, moreover, $\mathbb{A}^f_k$ for the finite part of the adele ring of k, i.e.,*

$$\mathbb{A}_k^f \overset{\text{def}}{=} \left\{ (a_{\mathfrak{p}})_{\mathfrak{p} \in \mathfrak{P}_k^f} \in \prod_{\mathfrak{p} \in \mathfrak{P}_k^f} k_{\mathfrak{p}} \,\middle|\, a_{\mathfrak{p}} \in \mathfrak{o}_{\mathfrak{p}} \text{ for all but finitely many } \mathfrak{p} \right\}.$$

Suppose that the following three conditions are satisfied:

(a) *The pro-$\mathcal{C}$ birational Galois section s is* **locally geometric** *[cf. Definition 5.18].*

(b) $\Sigma(\mathcal{C})$ *[cf. §0] is* **cofinite**, *i.e.,* $\mathfrak{Primes} \setminus \Sigma(\mathcal{C})$ *[cf. §0] is* **finite**.

(c) $\mathrm{Pic}(\mathfrak{o}_k) \overset{\text{def}}{=} \mathrm{Pic}(\mathrm{Spec}\,\mathfrak{o}_k)$ *is* **annihilated** *by a $\Sigma(\mathcal{C})$-integer [cf. §0].*

[Note that it follows from Proposition 5.24 that if $\Sigma(\mathcal{C}) = \mathfrak{Primes}$, or, equivalently [cf. §0], $\mathcal{C}$ contains all finite solvable groups, then the above three conditions are satisfied.] Then the following conditions are equivalent:

(1) *The pro-$\mathcal{C}$ birational Galois section s is* **geometric** *[cf. Definition 5.3].*

(2) *The following two conditions are satisfied:*

   (2-i) *There exist a finite morphism $\phi \colon X \to \mathbb{P}_k^1$ over k and, for each $\mathfrak{p} \in \mathfrak{P}_k^f$, a $k_{\mathfrak{p}}$-valued point $x_{\mathfrak{p}}$ of X associated to s [cf. Definition 5.20; condition (a)] [note that if the residue characteristic of $\mathfrak{p}$ is $\in \Sigma(\mathcal{C})$, then the $k_{\mathfrak{p}}$-valued point $x_{\mathfrak{p}}$ of X associated to s is uniquely determined – cf. Lemma 5.22] such that the composite*

$$\mathrm{Spec}\,k_{\mathfrak{p}} \overset{x_{\mathfrak{p}}}{\longrightarrow} X \overset{\phi}{\longrightarrow} \mathbb{P}_k^1$$

   determines a $k_{\mathfrak{p}}$-valued point of $\mathbb{P}_k^1 \setminus \{0, 1, \infty\} \subseteq \mathbb{P}_k^1$.

   (2-ii) *For each open subscheme $U \subseteq X$ of X which is a* **hyperbolic curve** *over k [cf. §0], there exists a prime number $l_U \in \Sigma(\mathcal{C})$ contained in $\Sigma(\mathcal{C})$ such that the pro-l_U Galois section of U/k [cf. Definition 5.2] naturally determined by s is either* **cuspidal** *[cf. Definition 5.37, (i)] or* **unramified almost everywhere** *[cf. Definition 5.37, (ii)].*

(3) *There exist a finite morphism $\phi \colon X \to \mathbb{P}_k^1$ over k and an $\mathbb{A}_k^f$-valued point $x_{\mathbb{A}}$ of X associated to s [cf. Definition 5.20; condition (a)] [note that if $\Sigma(\mathcal{C}) = \mathfrak{Primes}$, then the $\mathbb{A}_k^f$-valued point $x_{\mathbb{A}}$ of X associated to s is uniquely determined – cf. Lemma 5.22] such that the composite*

$$\mathrm{Spec}\,\mathbb{A}_k^f \overset{x_{\mathbb{A}}}{\longrightarrow} X \overset{\phi}{\longrightarrow} \mathbb{P}_k^1$$

determines an $\mathbb{A}_k^f$-valued point of $\mathbb{P}_k^1 \setminus \{0, 1, \infty\} \subseteq \mathbb{P}_k^1$.

(4) *There exist a* **finite** *subset $T \subseteq \mathfrak{P}_k^f$ of $\mathfrak{P}_k^f$ and a closed subscheme $Z \subseteq X$ of X which is* **finite** *over k such that, for each $\mathfrak{p} \in \mathfrak{P}_k^f \setminus T$ whose residue characteristic is $\in \Sigma(\mathcal{C})$, the [image of the uniquely determined – cf. Lemma 5.22] $k_{\mathfrak{p}}$-valued point $x_{\mathfrak{p}}$ of X associated to s [cf. Definition 5.20; condition (a)] is contained in $Z \subseteq X$.*

Proof. The implication (1) ⇒ (2) follows immediately from Proposition 5.39, together with Remark 5.21. Next, we verify the implication (2) ⇒ (3). Suppose that condition (2) is satisfied. Then, by condition (2-i), for each $\mathfrak{p} \in \mathfrak{P}_k^f$, the composite

$$\operatorname{Spec} k_\mathfrak{p} \xrightarrow{x_\mathfrak{p}} X \xrightarrow{\phi} \mathbb{P}_k^1$$

determines a $k_\mathfrak{p}$-valued point of $\mathbb{P}_k^1 \setminus \{0, 1, \infty\}$. Thus, to verify the implication (2) ⇒ (3), it suffices to verify that the above $k_\mathfrak{p}$-valued point of $\mathbb{P}_k^1 \setminus \{0, 1, \infty\}$ obtained as the composite $\phi \circ x_\mathfrak{p}$ determines an $\mathfrak{o}_\mathfrak{p}$-*valued point of* $\mathbb{P}_{\mathfrak{o}_\mathfrak{p}}^1 \setminus \{0, 1, \infty\}$ *for all but finitely many* $\mathfrak{p} \in \mathfrak{P}_k^f$. Write $U \subseteq X$ for the open subscheme of X obtained as the inverse image of $\mathbb{P}_k^1 \setminus \{0, 1, \infty\} \subseteq \mathbb{P}_k^1$ by ϕ. Then, by condition (2-ii), there exists a prime number $l_U \in \Sigma(\mathcal{C})$ contained in $\Sigma(\mathcal{C})$ such that the pro-l_U Galois section s^U of U/k obtained as the composite

$$G_k \xrightarrow{s} \Pi_{k(X)/k}^{\mathcal{C}} \longrightarrow \Pi_{U/k}^{\{l_U\}}$$

is either *cuspidal* or *unramified almost everywhere*. Write $s^\mathbb{P}$ for the pro-l_U Galois section of $\mathbb{P}_k^1 \setminus \{0, 1, \infty\}$ obtained as the composite

$$G_k \xrightarrow{s} \Pi_{k(X)/k}^{\mathcal{C}} \longrightarrow \Pi_{U/k}^{\{l_U\}} \longrightarrow \Pi_{(\mathbb{P}_k^1 \setminus \{0,1,\infty\})/k}^{\{l_U\}}$$

– where the third arrow is the homomorphism over G_k induced by ϕ. Then since the morphism $U \to \mathbb{P}_k^1 \setminus \{0, 1, \infty\}$ induced by ϕ is *finite*, one verifies easily that the homomorphism $\Pi_{U/k}^{\{l_U\}} \to \Pi_{(\mathbb{P}_k^1 \setminus \{0,1,\infty\})/k}^{\{l_U\}}$ maps *injectively* any cuspidal decomposition subgroup of $\Pi_{U/k}^{\{l_U\}}$ associated to a cusp of U/k to a cuspidal decomposition subgroup of $\Pi_{(\mathbb{P}_k^1 \setminus \{0,1,\infty\})/k}^{\{l_U\}}$ associated to a cusp of $\mathbb{P}_k^1 \setminus \{0, 1, \infty\}$. Thus, it follows immediately that if s^U is *cuspidal*, then $s^\mathbb{P}$ is *cuspidal*. On the other hand, by applying Lemma 5.7 [to "$\phi \circ x_\mathfrak{p}$" for $\mathfrak{p} \in \mathfrak{P}_k^f$ whose residue characteristic is $= l_U$], it follows immediately from condition (2-i) that $s^\mathbb{P}$ is *not cuspidal*. Thus, we conclude that s^U is *not cuspidal*, hence also [by condition (2-ii)] *unramified almost everywhere*. In particular, it follows Proposition 5.A.55, (ii), that $s^\mathbb{P}$ is *unramified almost everywhere*. Therefore, it follows immediately from Proposition 5.A.50, together with condition (2-i), that the $k_\mathfrak{p}$-valued point of $\mathbb{P}_k^1$ obtained as the composite $\phi \circ x_\mathfrak{p}$ determines an $\mathfrak{o}_\mathfrak{p}$-*valued point of* $\mathbb{P}_{\mathfrak{o}_\mathfrak{p}}^1 \setminus \{0, 1, \infty\}$ *for all but finitely many* $\mathfrak{p} \in \mathfrak{P}_k^f$. This completes the proof of the implication (2) ⇒ (3).

Next, we verify the implication (3) ⇒ (4). Suppose that condition (3) is satisfied. Write $s^\mathbb{G}$ for the pro-$\Sigma(\mathcal{C})$ Galois section of $\mathbb{G}_{m,k} \overset{\text{def}}{=} \mathbb{P}_k^1 \setminus \{0, \infty\}$ over k obtained as the composite

$$G_k \xrightarrow{s} \Pi^{\mathcal{C}}_{k(X)/k} \to \Pi^{\mathcal{C}}_{k(\mathbb{P}^1_k)/k} \twoheadrightarrow \Pi^{\mathcal{C}}_{\mathbb{G}_{m,k}/k} = \Pi^{\Sigma(\mathcal{C})}_{\mathbb{G}_{m,k}/k}$$

– where the second arrow is the homomorphism over G_k induced by ϕ – and $t^{\mathbb{G}}$ for the pro-$\Sigma(\mathcal{C})$ Galois section of $\mathbb{G}_{m,k}$ over k obtained as the composite

$$G_k \xrightarrow{s} \Pi^{\mathcal{C}}_{k(X)/k} \to \Pi^{\mathcal{C}}_{k(\mathbb{P}^1_k)/k} \xrightarrow{\sim} \Pi^{\mathcal{C}}_{k(\mathbb{P}^1_k)/k} \twoheadrightarrow \Pi^{\mathcal{C}}_{\mathbb{G}_{m,k}/k} = \Pi^{\Sigma(\mathcal{C})}_{\mathbb{G}_{m,k}/k}$$

– where the second arrow is the homomorphism over G_k induced by ϕ, and the third arrow is the automorphism over G_k induced by the automorphism of $\mathbb{P}^1_k$ over k given by "$u \mapsto 1 - u$". Then it follows immediately from condition (3) that there exists an element $(a_{\mathfrak{p}})_{\mathfrak{p} \in \mathfrak{P}^f_k} \in (\widetilde{\mathbb{A}}^f_k)^{\times}$ such that $(a_{\mathfrak{p}})_{\mathfrak{p} \in \mathfrak{P}^f_k}, (1 - a_{\mathfrak{p}})_{\mathfrak{p} \in \mathfrak{P}^f_k} \in (\mathbb{A}^f_k)^{\times} \simeq \mathbb{G}_{m,k}(\mathbb{A}^f_k)$, and, moreover, the respective images of the pro-$\Sigma(\mathcal{C})$ Galois sections $s^{\mathbb{G}}, t^{\mathbb{G}} \in \mathrm{GS}^{\Sigma(\mathcal{C})}(\mathbb{G}_{m,k}/k) \simeq k^{\times}[\Sigma(\mathcal{C})]$ [cf. Lemma 5.28, (i), (ii)] in the set $\prod_{\mathfrak{p} \in \mathfrak{P}^f_k} \mathrm{GS}^{\Sigma(\mathcal{C})}(\mathbb{G}_{m,k_{\mathfrak{p}}}/k_{\mathfrak{p}}) \simeq \prod_{\mathfrak{p} \in \mathfrak{P}^f_k} (k^{\times}_{\mathfrak{p}}[\Sigma(\mathcal{C})])$ [cf. the diagrams of Lemma 5.28, (iii)] coincide with the respective images of the elements $(a_{\mathfrak{p}})_{\mathfrak{p} \in \mathfrak{P}^f_k}, (1 - a_{\mathfrak{p}})_{\mathfrak{p} \in \mathfrak{P}^f_k} \in (\mathbb{A}^f_k)^{\times} \simeq \mathbb{G}_{m,k}(\mathbb{A}^f_k)$ in the set $\prod_{\mathfrak{p} \in \mathfrak{P}^f_k} (k^{\times}_{\mathfrak{p}}[\Sigma(\mathcal{C})]) \simeq \prod_{\mathfrak{p} \in \mathfrak{P}^f_k} \mathrm{GS}^{\Sigma(\mathcal{C})}(\mathbb{G}_{m,k_{\mathfrak{p}}}/k_{\mathfrak{p}})$. Thus, it follows from Lemma 5.31, (ii), (iii); together with condition (c), that

(*): there exist $\widetilde{a}_s, \widetilde{a}_t \in k^{\times}$ such that, for $\mathfrak{p} \in \mathfrak{P}^f_k$, if we write $u_{\mathfrak{p}} \overset{\mathrm{def}}{=} a_{\mathfrak{p}} \cdot \widetilde{a}^{-1}_s$, $v_{\mathfrak{p}} \overset{\mathrm{def}}{=} (1 - a_{\mathfrak{p}}) \cdot \widetilde{a}^{-1}_t \in k^{\times}_{\mathfrak{p}}$, and the residue characteristic of $\mathfrak{p}$ is $\in \Sigma(\mathcal{C})$, then $u_{\mathfrak{p}}, v_{\mathfrak{p}}$ are *roots of unity* of $k_{\mathfrak{p}}$ whose orders are ($\mathfrak{Primes} \setminus \Sigma(\mathcal{C})$)-integers.

Now let us observe that, for $\mathfrak{p} \in \mathfrak{P}^f_k$, the pair $(u_{\mathfrak{p}}, v_{\mathfrak{p}})$ satisfies the equation

$$1 = \widetilde{a}_s \cdot u_{\mathfrak{p}} + \widetilde{a}_t \cdot v_{\mathfrak{p}}.$$

Thus, it follows immediately from [2], Theorem 1.1, together with condition (b), that the set $\{(u_{\mathfrak{p}}, v_{\mathfrak{p}})\}_{\mathfrak{p} \in \mathfrak{P}^f_k}$, hence also the set $\{u_{\mathfrak{p}}\}_{\mathfrak{p} \in \mathfrak{P}^f_k}$, is *finite*. In particular, since $a_{\mathfrak{p}} = \widetilde{a}_s \cdot u_{\mathfrak{p}}$ [cf. (*)], it follows immediately that the pro-$\mathcal{C}$ birational Galois section of $\mathbb{P}^1_k/k$ obtained as the composite

$$G_k \xrightarrow{s} \Pi^{\mathcal{C}}_{k(X)/k} \longrightarrow \Pi^{\mathcal{C}}_{k(\mathbb{P}^1_k)/k}$$

– where the second arrow is the homomorphism over G_k induced by ϕ – satisfies condition (4). Therefore, since ϕ is *finite*, one verifies easily that the pro-$\mathcal{C}$ birational Galois section s satisfies condition (4). This completes the proof of the implication (3) $\Rightarrow$ (4).

Finally, we verify the implication (4) $\Rightarrow$ (1). Suppose that condition (4) is satisfied. Let us fix an element $\mathfrak{p}_0 \in \mathfrak{P}^f_k \setminus T$ of $\mathfrak{P}^f_k \setminus T$ such that the residue characteristic of $\mathfrak{p}_0$ is $\in \Sigma(\mathcal{C})$ [note that, by condition (b), such a $\mathfrak{p}_0$ always exists] and write $r(\mathfrak{p}_0)$ for the cardinality of the set of roots of unity of $k_{\mathfrak{p}_0}$. Now observe that, for any open subgroup $H \subseteq \Pi^{\mathcal{C}}_{k(X)/k}$ of $\Pi^{\mathcal{C}}_{k(X)/k}$ containing the image of s, if we write Y for the normalization of X in the finite extension of $k(X)$ corresponding to $H \subseteq \Pi^{\mathcal{C}}_{k(X)/k}$ [thus, $\Pi^{\mathcal{C}}_{k(Y)/k} = H \subseteq \Pi^{\mathcal{C}}_{k(X)/k}$], then since the morphism $Y \to X$ is *finite*, the pro-$\mathcal{C}$ birational Galois section of Y/k determined by s satisfies condition (4) relative to the finite subset "T" $\subseteq \mathfrak{P}^f_k$ appearing in condition (4). Thus, it follows from the equivalence (1) $\Leftrightarrow$ (3) of Lemma 5.9 that, to verify condition (1), by applying the conclusion to various such H's, it suffices to verify that X admits a $k(\zeta_{r(\mathfrak{p}_0)})$-valued point – where we use the notation $\zeta_{r(\mathfrak{p}_0)} \in \overline{k}$ to denote a *primitive $r(\mathfrak{p}_0)$-th root of unity*.

For each $\mathfrak{p} \in \mathfrak{P}^f_k$, let us fix a $k_{\mathfrak{p}}$-valued point $x_{\mathfrak{p}}$ of X associated to s [cf. condition (a)]. Now since X is *projective*, there exists a closed immersion $X \hookrightarrow \mathbb{P}^N_k$ over k for some positive integer N. Then it follows immediately from condition (4) that there exists a hyperplane $H \subseteq \mathbb{P}^N_k$ defined over k such that, for any $\mathfrak{p} \in \mathfrak{P}^f_k$, [the image of] the fixed $k_{\mathfrak{p}}$-valued point $x_{\mathfrak{p}}$ of X is contained in $X \setminus (X \cap H) \subseteq \mathbb{P}^N_k \setminus H \simeq \mathbb{A}^N_k$. Moreover, again by condition (4) – by considering a suitable automorphism of $\mathbb{A}^1_k$ over k – we may assume without loss of generality that, for each $i \in \{1, \cdots, N\}$ and $\mathfrak{p} \in \mathfrak{P}^f_k$, the $k_{\mathfrak{p}}$-valued point of $\mathbb{A}^1_k$ obtained as the composite

$$\operatorname{Spec} k_{\mathfrak{p}} \xrightarrow{x_{\mathfrak{p}}} X \setminus (X \cap H) \hookrightarrow \mathbb{P}^N_k \setminus H \simeq \mathbb{A}^N_k \xrightarrow{\operatorname{pr}_i} \mathbb{A}^1_k$$

factors through $\mathbb{G}_{m,k} \overset{\text{def}}{=} \mathbb{A}^1_k \setminus \{0\} \subseteq \mathbb{A}^1_k$. Therefore, we conclude that there exist an open subscheme $U \subseteq X$ of X and a closed immersion $U \hookrightarrow \mathbb{G}_{m,k} \times_k \cdots \times_k \mathbb{G}_{m,k}$ over k such that the $\widetilde{\mathbb{A}}^f_k$-valued point $x_{\mathbb{A}} \overset{\text{def}}{=} (x_{\mathfrak{p}})_{\mathfrak{p} \in \mathfrak{P}^f_k}$ of X determined by the fixed $k_{\mathfrak{p}}$-valued points $x_{\mathfrak{p}}$ lies on U. On the other hand, again by condition (4), one verifies easily that, for each $i \in \{1, \cdots, N\}$, the $\widetilde{\mathbb{A}}^f_k$-valued point of $\mathbb{G}_{m,k}$ obtained as the composite

$$\operatorname{Spec} \widetilde{\mathbb{A}}^f_k \xrightarrow{x_{\mathbb{A}}} U \hookrightarrow \mathbb{G}_{m,k} \times_k \cdots \times_k \mathbb{G}_{m,k} \xrightarrow{\operatorname{pr}_i} \mathbb{G}_{m,k}$$

determines an $\mathbb{A}^f_k$-*valued point of* $\mathbb{G}_{m,k}$. Thus, it follows immediately from Lemma 5.31, (ii), (iii); condition (c), that, for each $i \in \{1, \cdots, N\}$, the $k_{\mathfrak{p}_0}$-valued point of $\mathbb{G}_{m,k}$ obtained as the composite

$$\operatorname{Spec} k_{\mathfrak{p}_0} \xrightarrow{x_{\mathfrak{p}_0}} U \hookrightarrow \mathbb{G}_{m,k} \times_k \cdots \times_k \mathbb{G}_{m,k} \xrightarrow{\operatorname{pr}_i} \mathbb{G}_{m,k}$$

determines a $k(\zeta_{r(\mathfrak{p}_0)})$-*valued point of* $\mathbb{G}_{m,k}$. In particular, since $U \hookrightarrow \mathbb{G}_{m,k} \times_k \cdots \times_k \mathbb{G}_{m,k}$ is a *closed immersion*, one verifies easily that the $k_{\mathfrak{p}_0}$-valued

ок8

---

point $x_{\mathfrak{p}_0}$ of U, hence also X, determines a $k(\zeta_{r(\mathfrak{p}_0)})$-valued point. This completes the proof of the implication (4) $\Rightarrow$ (1), hence also of Theorem 5.40. $\square$

Remark 5.41.

(i) Theorem 5.40 is a result *without any assumption on the finiteness of a Shafarevich–Tate group*.

(ii) The equivalence (1) $\Leftrightarrow$ (3) of Theorem 5.40 may be regarded as a *tripod analogue* of [4], Theorem 17. The condition that k is either the field of rational numbers or an imaginary quadratic field [i.e., the assumption that $\mathfrak{o}_k^\times$ is finite] in the statement of Theorem 5.40 may be regarded as an analogue of the finiteness condition on the Mordell–Weil group in the statement of [4], Theorem 17; the condition that $\mathrm{Pic}(\mathfrak{o}_k)$ is annihilated by a $\Sigma(\mathcal{C})$-integer in the statement of Theorem 5.40 may be regarded as an analogue of the finiteness condition on the Shafarevich–Tate group in the statement of [4], Theorem 17. On the other hand, since any abelian variety is *proper*, in the case of [4], Theorem 17, the condition corresponding to our condition that the birational Galois section determines [not only an $\widetilde{\mathbb{A}}_k^f$-valued point but also] an $\mathbb{A}_k^f$-valued point of the tripod $\mathbb{P}_k^1 \setminus \{0, 1, \infty\}$ in Theorem 5.40 is *automatically satisfied*.

(iii) If $\mathcal{C}$ contains all finite *solvable* groups, then Theorem 5.33 implies the equivalence (1) $\Leftrightarrow$ (4) of Theorem 5.40.

(iv) One verifies easily that the proof of the equivalence (1) $\Leftrightarrow$ (4) of Theorem 5.40 gives us an *alternative proof* of Corollary 5.34 in the case where s is a pro-$\mathcal{C}$ birational Galois section, and k is either the *field of rational numbers* or an *imaginary quadratic field*. Indeed, in the notation of Corollary 5.34, it follows from the argument given in the proof of Corollary 5.34 that every pro-$\mathcal{C}$ birational Galois section s of X/k satisfies condition (4) of Theorem 5.40. Thus, it follows from the equivalence (1) $\Leftrightarrow$ (4) of Theorem 5.40 that s is *geometric*.

Corollary 5.42. *Let k be either the* **field of rational numbers** *or an* **imaginary quadratic field** *and $\overline{k}$ an algebraic closure of k. Write $G_k \stackrel{\text{def}}{=} \mathrm{Gal}(\overline{k}/k)$ and $\mathfrak{P}_k^f$ for the set of nonarchimedean primes of k. For each $\mathfrak{p} \in \mathfrak{P}_k^f$, write $k_\mathfrak{p}$ for the $\mathfrak{p}$-adic completion of k. Then the following assertions are equivalent:*

(1) *Any pro-$\mathfrak{Primes}$ birational Galois section [cf. Definition 5.2] of any projective smooth curve over k [cf. §0] is* **geometric** *[cf. Definition 5.3].*

(2) *Any pro-$\mathfrak{Primes}$ birational Galois section of $\mathbb{P}_k^1/k$ is* **geometric**.

(3) *Any pro-$\mathfrak{Primes}$ birational Galois section s of $\mathbb{P}_k^1/k$ satisfies the following two conditions:*

(3-i) *There exist three distinct elements a, b, $c \in \mathbb{P}_k^1(k)$ of $\mathbb{P}_k^1(k)$ such that, for any nonarchimedean prime $\mathfrak{p}$ of k, the [uniquely determined – cf. Lemma 5.22] $k_{\mathfrak{p}}$-valued point of $\mathbb{P}_k^1$ associated to s [cf. Definition 5.20; Proposition 5.24] is $\notin \{a, b, c\} \subseteq (\mathbb{P}_k^1(k) \subseteq) \mathbb{P}_k^1(k_{\mathfrak{p}})$.*

(3-ii) *There exists a prime number l such that the pro-l Galois section of $\mathbb{P}_k^1 \setminus \{0, 1, \infty\}$ [cf. Definition 5.2] naturally determined by s is either* **cuspidal** *[cf. Definition 5.37, (i)] or* **unramified almost everywhere** *[cf. Definition 5.37, (ii)].*

(4) *Any pro-$\mathfrak{P}\mathrm{rimes}$ birational Galois section s of $\mathbb{P}_k^1/k$ satisfies the following two conditions:*

(4-i) *There exist three distinct elements a, b, $c \in \mathbb{P}_k^1(k)$ of $\mathbb{P}_k^1(k)$ such that, for any nonarchimedean prime $\mathfrak{p}$ of k, the [uniquely determined – cf. Lemma 5.22] $k_{\mathfrak{p}}$-valued point of $\mathbb{P}_k^1$ associated to s [cf. Definition 5.20; Proposition 5.24] is $\notin \{a, b, c\} \subseteq (\mathbb{P}_k^1(k) \subseteq) \mathbb{P}_k^1(k_{\mathfrak{p}})$.*

(4-ii) *Write $s^{\mathbb{P}}$ for the pro-$\mathfrak{P}\mathrm{rimes}$ Galois section of $\mathbb{P}_k^1 \setminus \{0, 1, \infty\}$ naturally determined by s. Then it holds either that $s^{\mathbb{P}}$ is* **cuspidal**, *or that there exists a prime number l such that the l-adic Galois representation*

$$G_k \xrightarrow{\ s^{\mathbb{P}}\ } \Pi^{\mathfrak{P}\mathrm{rimes}}_{(\mathbb{P}_k^1 \setminus \{0,1,\infty\})/k} \longrightarrow \mathrm{GL}_2(\mathbb{Z}_l)$$

– where we refer to Definition 5.1, (ii), concerning the profinite group $\Pi^{\mathfrak{P}\mathrm{rimes}}_{(\mathbb{P}_k^1 \setminus \{0,1,\infty\})/k}$; the second arrow $\Pi^{\mathfrak{P}\mathrm{rimes}}_{(\mathbb{P}_k^1 \setminus \{0,1,\infty\})/k} \to \mathrm{GL}_2(\mathbb{Z}_l)$ is the l-adic representation determined by the **Legendre family of elliptic curves** *over $\mathbb{P}_k^1 \setminus \{0, 1, \infty\}$, i.e., the elliptic curve over $\mathbb{P}_k^1 \setminus \{0, 1, \infty\} = \mathrm{Spec}\, k[u^{\pm 1}, (1 - u)^{-1}]$ determined by the equation "$y^2 = x(x - 1)(x - u)$" – is* **unramified at all but finitely many** $\mathfrak{p} \in \mathfrak{P}_k^f$.

Proof. The implications (1) $\Rightarrow$ (2) $\Rightarrow$ (4) are immediate [cf. also Remark 5.21]. On the other hand, the implication (2) $\Rightarrow$ (1) follows immediately from the fact that any projective smooth curve over k may be obtained as the normalization of $\mathbb{P}_k^1$ in the finite extension of $k(\mathbb{P}_k^1)$ corresponding to an open subgroup of $\Pi^{\mathfrak{P}\mathrm{rimes}}_{k(\mathbb{P}_k^1)/k}$. Moreover, let us observe that it follows immediately from Proposition 5.39, together with Remark 5.21, that the implications (2) $\Rightarrow$ (3) holds.

Finally, we verify the implication (3) $\Rightarrow$ (2) (respectively, (4) $\Rightarrow$ (2)). Suppose that assertion (3) (respectively, assertion (4)) holds. Let s be a pro-$\mathfrak{P}\mathrm{rimes}$ birational Galois section of $\mathbb{P}_k^1/k$. For each nonarchimedean prime $\mathfrak{p}$ of k, write $x_{\mathfrak{p}}$ for the [uniquely determined] $k_{\mathfrak{p}}$-valued point of $\mathbb{P}_k^1$ associated to s. Then it follows from condition (3-i) (respectively, condition (4-i)) that, by considering a suitable automorphism of $\mathbb{P}_k^1$ over k, we may assume without loss of generality that, for any $\mathfrak{p} \in \mathfrak{P}_k^f$, $x_\mathfrak{p} \in \mathbb{P}_k^1(k_\mathfrak{p})$ is $\notin \{0, 1, \infty\} \subseteq \mathbb{P}_k^1(k_\mathfrak{p})$. Thus, for any prime number l, by applying Lemma 5.7 [to "$x_\mathfrak{p}$" for $\mathfrak{p} \in \mathfrak{P}_k^f$ whose residue characteristic is $= l$], it follows immediately that the pro-l Galois section $s^{\mathbb{P},\{l\}}$ of $\mathbb{P}_k^1 \setminus \{0, 1, \infty\}$ obtained as the composite

$$G_k \xrightarrow{s} \Pi_{k(\mathbb{P}_k^1)/k}^{\mathfrak{P}\text{rimes}} \longrightarrow \Pi_{(\mathbb{P}_k^1\setminus\{0,1,\infty\})/k}^{\{l\}},$$

hence also the pro-$\mathfrak{P}$rimes Galois section $s^{\mathbb{P}}$ of $\mathbb{P}_k^1 \setminus \{0, 1, \infty\}$ obtained as the composite

$$G_k \xrightarrow{s} \Pi_{k(\mathbb{P}_k^1)/k}^{\mathfrak{P}\text{rimes}} \longrightarrow \Pi_{(\mathbb{P}_k^1\setminus\{0,1,\infty\})/k}^{\mathfrak{P}\text{rimes}},$$

is *not cuspidal*. Thus, by condition (3-ii) (respectively, condition (4-ii)), we conclude that there exists a prime number l_0 such that $s^{\mathbb{P},\{l_0\}}$ is *unramified almost everywhere* (respectively, the l_0-adic Galois representation obtained as the displayed composite of condition (4-ii) is *unramified at all but finitely many* $\mathfrak{p} \in \mathfrak{P}_k^f$). Thus, since [we have assumed that] for any $\mathfrak{p} \in \mathfrak{P}_k^f$, $x_\mathfrak{p} \in \mathbb{P}_k^1(k_\mathfrak{p})$ is $\notin \{0, 1, \infty\} \subseteq \mathbb{P}_k^1(k_\mathfrak{p})$, it follows immediately from Proposition 5.A.50 (respectively, [14], Theorem 1) that the birational pro-$\mathfrak{P}$rimes Galois section s of $\mathbb{P}_k^1/k$ satisfies condition (3) of Theorem 5.40, hence also [by the equivalence (1) $\Leftrightarrow$ (3) of Theorem 5.40] that s is *geometric*. This completes the proof of the implication (3) $\Rightarrow$ (2) (respectively, (4) $\Rightarrow$ (2)), hence also of Corollary 5.42. $\qquad\square$

Appendix A. Ramification of Galois sections

In the present §A, we discuss the *ramification of Galois sections* of hyperbolic curves over p-adic local fields. In the present §A, let $\Sigma \subseteq \mathfrak{P}$rimes be a nonempty subset of $\mathfrak{P}$rimes [cf. §0], k a *p-adic local field* for some prime number p [cf. §0], $\bar{k}$ an algebraic closure of k, and X a *hyperbolic curve* over k [cf. §0]. For a finite extension k' ($\subseteq \bar{k}$) of k, write

$$G_{k'} \overset{\text{def}}{=} \text{Gal}(\bar{k}/k'),$$

$$I_{k'} \subseteq G_{k'}$$

for the inertia subgroup of $G_{k'}$, and

$$\mathfrak{o}_{k'} \subseteq k'$$

for the ring of integers of k'. Write, moreover,

$$X^{\mathrm{cpt}}$$

for the [uniquely determined] smooth compactification of X over k;

$$\Delta^{\Sigma}_{X/k}$$

for the *pro-Σ geometric fundamental group* of X, i.e., the maximal pro-Σ quotient of $\pi_1(X \otimes_k \overline{k})$;

$$\Pi^{\Sigma}_{X/k}$$

for the *geometrically pro-Σ fundamental group* of X, i.e., the quotient of $\pi_1(X)$ by the kernel of the natural surjection $\pi_1(X \otimes_k \overline{k}) \twoheadrightarrow \Delta^{\Sigma}_{X/k}$. Thus, we have an exact sequence of profinite groups [cf. [3], Exposé IX, Théorème 6.1]

$$1 \longrightarrow \Delta^{\Sigma}_{X/k} \longrightarrow \Pi^{\Sigma}_{X/k} \longrightarrow G_k \longrightarrow 1 \,.$$

Let s be a pro-Σ Galois section of X/k [cf. [6], Definition 1.1, (i)], i.e., a section of the above exact sequence of profinite groups.

Definition 5.A.43. We shall say that s is *unramified* (respectively, *potentially unramified*) if the image of the composite

$$I_k \longrightarrow G_k \overset{s}{\longrightarrow} \Pi^{\Sigma}_{X/k} \longrightarrow \mathrm{Aut}(\Delta^{\Sigma}_{X/k})$$

– where the third arrow is the action of $\Pi^{\Sigma}_{X/k}$ on $\Delta^{\Sigma}_{X/k}$ obtained by conjugation – is trivial (respectively, finite).

Proposition 5.A.44. *The following hold:*

(i) *If $p \in \Sigma$, then any pro-Σ Galois section of X/k [cf. [6], Definition 1.1, (i)] is* **not potentially unramified**, *hence also* **not unramified** *[cf. Definition 5.A.43].*

(ii) *If X does* **not admit good reduction** *over $\mathfrak{o}_k$ [cf. §0], then any pro-Σ Galois section of X/k is* **not unramified**.

Proof. Let s be a pro-Σ Galois section of X/k. First, we verify assertion (i). Now one verifies easily that there exists a *characteristic* open subgroup $H \subseteq \Delta^{\Sigma}_{X/k}$ of $\Delta^{\Sigma}_{X/k}$ such that the connected finite étale covering of X corresponding to the open subgroup $H \cdot \mathrm{Im}(s)$ of $\Pi^{\Sigma}_{X/k}$ topologically generated by H and $\mathrm{Im}(s)$ is of *genus* ≥ 1. On the other hand, since $H \subseteq \Delta^{\Sigma}_{X/k}$ is *characteristic*, and [as is well-known] $\Delta^{\Sigma}_{X/k}$ is *slim* [cf. §0], it follows from [7], Lemma 5, that we have a natural *injection* $\mathrm{Aut}(\Delta^{\Sigma}_{X/k}) \hookrightarrow \mathrm{Aut}(H)$. Thus, to verify assertion (i), by replacing $\Pi^{\Sigma}_{X/k}$ by the open subgroup $H \cdot \mathrm{Im}(s)$, we may assume without loss of generality that X^{cpt} is of *genus* ≥ 1. Next, let us observe that, as is well-known, since $p \in \Sigma$, and X^{cpt} is of *genus* ≥ 1, there exist G_k-equivariant isomorphisms

$$H^2(\Delta^\Sigma_{X^{\mathrm{cpt}}/k}, \mathbb{Z}_p) \simeq H^2(X^{\mathrm{cpt}} \otimes_k \overline{k}, \mathbb{Z}_p) \simeq \mathbb{Z}_p(1)$$

– where "(1)" denotes a *Tate twist*. In particular, [the restriction to I_k of] the p-adic cyclotomic representation $\chi_p \colon I_k \to \mathrm{Aut}(\mathbb{Z}_p(1))$ factors through the displayed composite of Definition 5.A.43. On the other hand, one may verify easily that the image of χ_p is *infinite*. Thus, s is *not potentially unramified*. This completes the proof of assertion (i).

Next, we verify assertion (ii). Suppose that X does *not admit good reduction* over $\mathfrak{o}_k$. Now it follows from assertion (i) that, to verify assertion (ii), we may assume without loss of generality that $p \notin \Sigma$. Then it follows immediately from [17], Theorem 0.8, that the image of the composite

$$I_k \longrightarrow \mathrm{Aut}(\Delta^\Sigma_{X/k}) \longrightarrow \mathrm{Out}(\Delta^\Sigma_{X/k})$$

– where the first arrow is the displayed composite of Definition 5.A.43 – is *nontrivial*, hence that s is *not unramified*. This completes the proof of assertion (ii). $\square$

Definition 5.A.45. If X admits *good reduction* $\mathcal{X}$ over $\mathfrak{o}_k$ [cf. §0], then we shall write

$$(\pi_1(X) \twoheadrightarrow \Pi^\Sigma_{X/k} \twoheadrightarrow) \Pi^{\Sigma\text{-ét}}_{X/k}$$

for the quotient of $\pi_1(X)$ by the normal closed subgroup topologically normally generated by the kernels of the natural surjections $\pi_1(X) \twoheadrightarrow \Pi^\Sigma_{X/k}$, $\pi_1(X) \twoheadrightarrow \pi_1(\mathcal{X})$. Thus, the natural surjection $\Pi^\Sigma_{X/k} \twoheadrightarrow G_k$ determines a surjection $\Pi^{\Sigma\text{-ét}}_{X/k} \twoheadrightarrow G_k/I_k$. We shall write

$$\Delta^{\Sigma\text{-ét}}_{X/k}$$

for the kernel of the surjection $\Pi^{\Sigma\text{-ét}}_{X/k} \twoheadrightarrow G_k/I_k$. Thus, we have a commutative diagram of profinite groups

$$
\begin{array}{ccccccccc}
1 & \longrightarrow & \Delta^\Sigma_{X/k} & \longrightarrow & \Pi^\Sigma_{X/k} & \longrightarrow & G_k & \longrightarrow & 1 \\
 & & \downarrow & & \downarrow & & \downarrow & & \\
1 & \longrightarrow & \Delta^{\Sigma\text{-ét}}_{X/k} & \longrightarrow & \Pi^{\Sigma\text{-ét}}_{X/k} & \longrightarrow & G_k/I_k & \longrightarrow & 1
\end{array}
$$

– where the horizontal sequences are *exact*.

Remark 5.A.46. In the notation of Definition 5.A.45, as is well-known, if $p \notin \Sigma$, then the left-hand vertical arrow $\Delta^\Sigma_{X/k} \to \Delta^{\Sigma\text{-ét}}_{X/k}$ of the commutative diagram of Definition 5.A.45 is an *isomorphism*. In particular, the right-hand upper horizontal arrow $\Pi^\Sigma_{X/k} \to G_k$ induces an *isomorphism* $\mathrm{Ker}(\Pi^\Sigma_{X/k} \twoheadrightarrow \Pi^{\Sigma\text{-ét}}_{X/k}) \xrightarrow{\sim} I_k$, and the right-hand square of the commutative diagram of Definition 5.A.45 is *cartesian*.

Proposition 5.A.47. *The following conditions are equivalent:*

(1) *s is* **unramified** *[cf. Definition 5.A.43].*
(2) *$p \notin \Sigma$, X admits* **good reduction** *over $\mathfrak{o}_k$ [cf. §0], and the image of the composite*

$$I_k \hookrightarrow G_k \xrightarrow{s} \Pi^\Sigma_{X/k} \twoheadrightarrow \Pi^{\Sigma\text{-ét}}_{X/k}$$

[cf. Definition 5.A.45] is **trivial**.
(3) *$p \notin \Sigma$, X admits* **good reduction** *over $\mathfrak{o}_k$, and the composite*

$$I_k \hookrightarrow G_k \xrightarrow{s} \Pi^\Sigma_{X/k}$$

determines an **isomorphism**

$$I_k \xrightarrow{\sim} \mathrm{Ker}(\Pi^\Sigma_{X/k} \twoheadrightarrow \Pi^{\Sigma\text{-ét}}_{X/k}).$$

(4) *$p \notin \Sigma$, and, for any open subgroup $H \subseteq \Pi^\Sigma_{X/k}$ of $\Pi^\Sigma_{X/k}$ containing the image of s, the connected finite étale covering of X corresponding to $H \subseteq \Pi^\Sigma_{X/k}$ admits* **good reduction** *over $\mathfrak{o}_k$.*

Proof. First, we verify the equivalence (1) ⇔ (2). It follows immediately from Proposition 5.A.44 that both (1) and (2) imply that $p \notin \Sigma$, and that X admits *good reduction* over $\mathfrak{o}_k$. Thus, suppose that these conditions are satisfied. Write J for the image of the displayed composite of condition (2). Then it follows immediately from the existence of the commutative diagram of Definition 5.A.45 that $J \subseteq \Delta^{\Sigma\text{-ét}}_{X/k} \subseteq \Pi^{\Sigma\text{-ét}}_{X/k}$. Thus, it follows immediately from Remark 5.A.46 that the displayed composite $I_k \to \mathrm{Aut}(\Delta^\Sigma_{X/k})$ of Definition 5.A.43 factors as

$$I_k \twoheadrightarrow J \hookrightarrow \Delta^{\Sigma\text{-ét}}_{X/k} \to \mathrm{Aut}(\Delta^{\Sigma\text{-ét}}_{X/k}) \xleftarrow{\sim} \mathrm{Aut}(\Delta^\Sigma_{X/k})$$

– where the third arrow is the action of $\Delta^{\Sigma\text{-ét}}_{X/k}$ on $\Delta^{\Sigma\text{-ét}}_{X/k}$ obtained by conjugation. Now since, as is well-known, $\Delta^\Sigma_{X/k} \xrightarrow{\sim} \Delta^{\Sigma\text{-ét}}_{X/k}$ is *center-free*, the third arrow of this composite is *injective*. Therefore, it follows immediately that the condition that s is *unramified* is equivalent to the condition that $J = \{1\}$. This completes the proof of the equivalence (1) ⇔ (2).

The equivalence (2) $\Leftrightarrow$ (3) follows immediately from Remark 5.A.46. Next, we verify the implication (3) $\Rightarrow$ (4). Suppose that condition (3) is satisfied. Then it is immediate that if an open subgroup of $\Pi_{X/k}^\Sigma$ contains the image of s, then it arises from an open subgroup of $\Pi_{X/k}^{\Sigma\text{-ét}}$; thus, it follows immediately from the various definitions involved that the corresponding connected finite étale covering of X admits *good reduction* over $\mathfrak{o}_k$. This completes the proof of the implication (3) $\Rightarrow$ (4).

Finally, we verify the implication (4) $\Rightarrow$ (3). Suppose that condition (4) is satisfied. Let $H \subseteq \Pi_{X/k}^\Sigma$ be an open subgroup of $\Pi_{X/k}^\Sigma$ containing the image of s. Write $Y \to X$ for the connected finite étale covering of X corresponding to $H \subseteq \Pi_{X/k}^\Sigma$; thus, $\Pi_{Y/k}^\Sigma = H \subseteq \Pi_{X/k}^\Sigma$. Then it follows from condition (4) that Y admits *good reduction* over $\mathfrak{o}_k$. Thus, it follows from [9], Lemma 8.3, that the morphism $Y \to X$ *extends to a morphism between their [uniquely determined] smooth models*. In particular, it follows immediately from the definitions of $\Pi_{X/k}^{\Sigma\text{-ét}}$ and $\Pi_{Y/k}^{\Sigma\text{-ét}}$ that the inclusion $\Pi_{Y/k}^\Sigma \subseteq \Pi_{X/k}^\Sigma$ determines an inclusion $\mathrm{Ker}(\Pi_{Y/k}^\Sigma \twoheadrightarrow \Pi_{Y/k}^{\Sigma\text{-ét}}) \subseteq \mathrm{Ker}(\Pi_{X/k}^\Sigma \twoheadrightarrow \Pi_{X/k}^{\Sigma\text{-ét}})$. Thus, it follows immediately from Remark 5.A.46 that $\mathrm{Ker}(\Pi_{Y/k}^\Sigma \twoheadrightarrow \Pi_{Y/k}^{\Sigma\text{-ét}}) = \mathrm{Ker}(\Pi_{X/k}^\Sigma \twoheadrightarrow \Pi_{X/k}^{\Sigma\text{-ét}})$, hence that $\mathrm{Ker}(\Pi_{X/k}^\Sigma \twoheadrightarrow \Pi_{X/k}^{\Sigma\text{-ét}}) \subseteq \Pi_{Y/k}^\Sigma = H$. Therefore, by considering the intersection of such H's, we obtain that $\mathrm{Ker}(\Pi_{X/k}^\Sigma \twoheadrightarrow \Pi_{X/k}^{\Sigma\text{-ét}}) \subseteq \mathrm{Im}(s)$. Thus, again by Remark 5.A.46, we conclude that condition (3) holds. This completes the proof of the implication (4) $\Rightarrow$ (3), hence also of Proposition 5.A.47. $\qquad\square$

Lemma 5.A.48. *Suppose that* $p \notin \Sigma$, *and that X admits* **good reduction** *over $\mathfrak{o}_k$ [cf. §0]. Let* $\Pi \subseteq \Pi_{X/k}^{\Sigma\text{-ét}}$ *be an open subgroup of* $\Pi_{X/k}^{\Sigma\text{-ét}}$. *Write* $s_\mathfrak{o}$ *for the composite* $G_k \xrightarrow{s} \Pi_{X/k} \twoheadrightarrow \Pi_{X/k}^{\Sigma\text{-ét}}$, k' $(\subseteq \bar{k})$ *for the [necessarily unramified] finite extension of k corresponding to the image of the composite* $\Pi \hookrightarrow \Pi_{X/k}^{\Sigma\text{-ét}} \twoheadrightarrow G_k/I_k$, $Y \to X$ *for the connected finite étale covering of X corresponding to the open subgroup* $\Pi \subseteq \Pi_{X/k}^{\Sigma\text{-ét}}$, *and Y^{cpt} for the [uniquely determined] smooth compactification of Y over k'. [Here, it follows immediately from the various definitions involved that Y is a* **hyperbolic curve** *over k'; Y, hence also Y^{cpt}, admits* **good reduction** *over $\mathfrak{o}_{k'}$;* $\Pi_{Y/k'}^{\Sigma\text{-ét}} = \Pi \subseteq \Pi_{X/k}^{\Sigma\text{-ét}}$.] *Suppose, moreover, that Y is of* **genus** ≥ 2. *Then the image of the composite*

$$s_\mathfrak{o}(I_k) \cap \Delta_{Y/k'}^{\Sigma\text{-ét}} \hookrightarrow \Delta_{Y/k'}^{\Sigma\text{-ét}} \twoheadrightarrow \Delta_{Y^{\mathrm{cpt}}/k'}^{\Sigma\text{-ét}} \twoheadrightarrow (\Delta_{Y^{\mathrm{cpt}}/k'}^{\Sigma\text{-ét}})^{\mathrm{ab}}$$

– where the second arrow is the surjection induced by the open immersion $Y \hookrightarrow Y^{\mathrm{cpt}}$ – is **trivial**.

Proof. This follows immediately from a similar argument to the argument used in the proof of assertion (ii) in the proof of [6], Lemma 3.3. $\qquad\square$

Proposition 5.A.49. *Suppose that* $p \notin \Sigma$, *and that* X *admits* **good reduction** *over* $\mathfrak{o}_k$ [*cf.* §0]. *Then the following conditions are equivalent:*

(1) s *is* **ramified**, *i.e.*, **not unramified** [*cf. Definition 5.A.43*].

(2) *The image of the composite*

$$I_k \hookrightarrow G_k \overset{s}{\to} \Pi^{\Sigma}_{X/k} \twoheadrightarrow \Pi^{\Sigma\text{-ét}}_{X/k}$$

is a **nontrivial closed subgroup of a cuspidal inertia subgroup** *of* $\Pi^{\Sigma\text{-ét}}_{X/k}$ *associated to a cusp of* X/k.

(3) *The image of the composite*

$$I_k \hookrightarrow G_k \overset{s}{\to} \Pi^{\Sigma}_{X/k} \twoheadrightarrow \Phi^{\Sigma}_{X/k}$$

– where $\Phi^{\Sigma}_{X/k}$ *is the quotient of* $\Pi^{\Sigma}_{X/k}$ *defined in* [7], *Definition 1*, (iv), *i.e.*, *the quotient of* $\Pi^{\Sigma}_{X/k}$ *by the kernel* $Z_{\Pi^{\Sigma}_{X/k}}(\Delta^{\Sigma}_{X/k})$ *of the homomorphism* $\Pi^{\Sigma}_{X/k} \to \mathrm{Aut}(\Delta^{\Sigma}_{X/k})$ *obtained by conjugation – is a* **nontrivial closed subgroup of a cuspidal inertia subgroup** *of* $\Phi^{\Sigma}_{X/k}$ *associated to a cusp of* X/k.

(4) *There exists an element* $l \in \Sigma$ *of* Σ *such that the pro-*l *Galois section of* X/k [*cf.* [6], *Definition 1.1*, (i)] *naturally determined by* s *is* **ramified**.

Proof. First, we verify the equivalence (1) ⇔ (2). It follows from the equivalence (1) ⇔ (2) of Lemma 5.A.47 that s is *ramified* if and only if the image of the composite of condition (2) is *nontrivial*. On the other hand, it follows immediately from the existence of the commutative diagram of Definition 5.A.45, together with Remark 5.A.46, that the composite of condition (2) factors through the *maximal pro-*Σ *quotient of* I_k, which is, as is well-known, *procyclic*. Thus, it follows immediately from Lemma 5.A.48, together with [5], Lemma 1.6, that the equivalence (1) ⇔ (2) holds. This completes the proof of the equivalence (1) ⇔ (2). Next, let us observe that the implication (3) ⇒ (1) follows immediately from the various definitions involved [cf. also the definition of the quotient $\Phi^{\Sigma}_{X/k}$]. Next, we verify the implication (2) ⇒ (3). Since, as is well-known, $\Delta^{\Sigma}_{X/k}$ is *slim* [cf. §0], it follows from [7], Proposition 6, (ii), together with Remark 5.A.46, that we have a sequence of natural surjections $\Pi^{\Sigma}_{X/k} \twoheadrightarrow \Pi^{\Sigma\text{-ét}}_{X/k} \twoheadrightarrow \Phi^{\Sigma}_{X/k}$, which induces an *injection* $\Delta^{\Sigma}_{X/k} \overset{\sim}{\to} \Delta^{\Sigma\text{-ét}}_{X/k} \hookrightarrow \Phi^{\Sigma}_{X/k}$. Thus, one verifies easily that the implication (2) ⇒ (3) holds. This completes the proof of the implication (2) ⇒ (3). Finally, we verify the equivalence (1) ⇔ (4). For each nonempty subset $\Sigma' \subseteq \Sigma$ of Σ, write $J_{\Sigma'}$ for the image of the composite

$$I_k \hookrightarrow G_k \xrightarrow{s} \Pi^{\Sigma}_{X/k} \twoheadrightarrow \Pi^{\Sigma'}_{X/k} \twoheadrightarrow \Pi^{\Sigma'\text{-ét}}_{X/k}.$$

Then it follows immediately from the verified equivalence (1) ⇔ (2), together with the well-known structure of the maximal pro-Σ quotient of the fundamental group of a smooth curve over an algebraically closed field of characteristic $\notin \Sigma$, that, for each nonempty subset $\Sigma' \subseteq \Sigma$ of Σ, the image $J_{\Sigma'}$ is *procyclic*, and, moreover, $J_{\Sigma'}$ is the *maximal pro-Σ' quotient* of J_{Σ} [relative to the natural surjection $J_{\Sigma} \twoheadrightarrow J_{\Sigma'}$]. In particular, we conclude that $J_{\Sigma} = \{1\}$ if and only if $J_{\{l\}} = \{1\}$ for any $l \in \Sigma$, i.e., the equivalence (1) ⇔ (4) holds. This completes the proof of the equivalence (1) ⇔ (4), hence also of Proposition 5.A.49. □

Proposition 5.A.50. *Suppose that $p \notin \Sigma$, and that s is* **geometric** *[cf. [6], Definition 1.1, (iii)]. Let $x \in X^{\mathrm{cpt}}(k)$ be a k-rational point of X^{cpt} such that a decomposition subgroup of $\Pi^{\Sigma}_{X/k}$ associated to x contains the image of s. Consider the following three conditions:*

(1) *s is* **unramified** *[cf. Definition 5.A.43].*
(2) *$x \in X(k)$, and the hyperbolic curve $X \setminus \{x\}$, hence also the hyperbolic curve X, over k admits* **good reduction** *over $\mathfrak{o}_k$ [cf. §0].*
(3) *$x \notin X(k)$.*

Then we have implications

$$(2) \implies (1) \implies \textit{either (2) or (3)}.$$

In particular, if $x \in X(k)$, then we have an equivalence

$$(1) \iff (2).$$

Proof. To verify Proposition 5.A.50, it is immediate that it suffices to verify that if $x \in X(k)$, then condition (1) is equivalent to condition (2). Thus, suppose that $x \in X(k)$. Now let us observe that it follows immediately from Proposition 5.A.44, (ii), that both (1) and (2) imply that X admits *good reduction* over $\mathfrak{o}_k$. Thus, we may assume without loss of generality that X admits *good reduction* over $\mathfrak{o}_k$. Moreover, observe that it follows immediately from the equivalence (1) ⇔ (4) of Proposition 5.A.49 that, to verify the equivalence (1) ⇔ (2), by considering the pro-l Galois section of X/k naturally determined by s – where l ranges over elements of Σ – we may assume without loss of generality that Σ is of *cardinality* 1. On the other hand, since Σ is of *cardinality* 1, it follows immediately from [7], Proposition 19, (ii), that the kernel of the composite $G_k \xrightarrow{s} \Pi^{\Sigma}_{X/k} \to \mathrm{Aut}(\Delta^{\Sigma}_{X/k})$ coincides with the kernel of the pro-Σ outer Galois representation associated to the hyperbolic curve $X \setminus \{x\}$ over k. Thus, the equivalence (1) ⇔ (2) follows immediately from [17], Theorem 0.8. This completes the proof of Proposition 5.A.50. □

Proposition 5.A.51. *Suppose that $p \notin \Sigma$, that X admits* **good reduction** *over* $\mathfrak{o}_k$ *[cf. §0], and that X is* **proper** *over k. Then any pro-Σ Galois section of X/k [cf. [6], Definition 1.1, (i)] is* **unramified** *[cf. Definition 5.A.43].*

Proof. This follows immediately from the equivalence (1) $\Leftrightarrow$ (2) of Proposition 5.A.49. $\qquad\square$

Remark 5.A.52. Proposition 5.A.51 may be regarded as a *Galois section version* of the *valuative criterion for properness* of morphisms of schemes.

Remark 5.A.53. In [13], Saïdi proved the existence of a *nongeometric* pro-Σ Galois section in the situation of Proposition 5.A.51 [cf. [13], Proposition 4.2.1].

Proposition 5.A.54. *Suppose that $p \notin \Sigma$, and that X admits* **good reduction** *over $\mathfrak{o}_k$ [cf. §0]. Then s is* **unramified** *if and only if s is* **potentially unramified** *[cf. Definition 5.A.43].*

Proof. This follows immediately from the equivalence (1) $\Leftrightarrow$ (2) of Proposition 5.A.49, together with the well-known fact that any cuspidal inertia subgroup of $\Pi_{X/k}^{\Sigma\text{-ét}}$ associated to a cusp of X/k is isomorphic to $\widehat{\mathbb{Z}}^\Sigma$ as an abstract profinite group. $\qquad\square$

Proposition 5.A.55. *Let Y be a hyperbolic curve over k and $X \to Y$ a* **dominant** *morphism over k. Write s_Y for the pro-Σ Galois section of Y/k [cf. [6], Definition 1.1, (i)] determined by s, i.e., the composite $G_k \xrightarrow{s} \Pi_{X/k}^\Sigma \to \Pi_{Y/k}^\Sigma$. Then the following hold:*

(i) *Write $\Phi_{X/k}^\Sigma$, $\Phi_{Y/k}^\Sigma$ for the respective quotients of $\Pi_{X/k}^\Sigma$, $\Pi_{Y/k}^\Sigma$ defined in [7], Definition 1, (iv) [cf. also the statement of condition (3) of Proposition 5.A.49]. Then the natural homomorphism $\Pi_{X/k}^\Sigma \to \Pi_{Y/k}^\Sigma$ induces a homomorphism $\Phi_{X/k}^\Sigma \to \Phi_{Y/k}^\Sigma$.*

(ii) *If s is* **unramified** *(respectively,* **potentially unramified**) *[cf. Definition 5.A.43], then s_Y is* **unramified** *(respectively,* **potentially unramified**).*

(iii) *Suppose that $X \to Y$ is* **finite**, *and that X and Y admit* **good reduction** *over $\mathfrak{o}_k$ [cf. §0]. Then s is* **unramified** *if and only if s_Y is* **unramified**.

Proof. First, we verify assertion (i). Now since, as is well-known, the profinite group $\Delta_{Y/k}^\Sigma$ is *slim* [cf. §0], for any open subgroup $H \subseteq \Delta_{Y/k}^\Sigma$ of $\Delta_{Y/k}^\Sigma$, it follows immediately from [7], Lemma 5, that $N_{\Pi_{Y/k}^\Sigma}(H) \cap Z_{\Pi_{Y/k}^\Sigma}(\Delta_{Y/k}^\Sigma) = Z_{\Pi_{Y/k}^\Sigma}(H)$. Thus, it follows immediately from the fact that the natural homomorphism $\Delta_{X/k}^\Sigma \to \Delta_{Y/k}^\Sigma$ is *open* that the natural homomorphism $\Pi_{X/k}^\Sigma \to$

$\Pi_{Y/k}^{\Sigma}$ induces a homomorphism $\Phi_{X/k}^{\Sigma} \to \Phi_{Y/k}^{\Sigma}$. This completes the proof of assertion (i). Assertion (ii) follows immediately from the various definitions involved, together with assertion (i) [cf. also the definitions of the quotients $\Phi_{X/k}^{\Sigma}$, $\Phi_{Y/k}^{\Sigma}$]. Finally, we verify assertion (iii). It follows from Proposition 5.A.44, (i), that both the condition that s is *unramified* and the condition that s_Y is *unramified* imply that $p \notin \Sigma$. Thus, suppose that $p \notin \Sigma$. On the other hand, since $X \to Y$ is *finite*, one verifies easily that the restriction of the natural homomorphism $\Delta_{X/k}^{\Sigma} \to \Delta_{Y/k}^{\Sigma}$ to any cuspidal inertia subgroup of $\Pi_{X/k}^{\Sigma}$ associated to a cusp of X/k is *injective*. Thus, assertion (iii) follows immediately from assertions (i), (ii); the equivalence (1) $\leftrightarrow$ (3) of Proposition 5.A.49, together with the fact that the sequences of natural surjections $\Pi_{X/k}^{\Sigma} \twoheadrightarrow \Pi_{X/k}^{\Sigma\text{-ét}} \twoheadrightarrow \Phi_{X/k}^{\Sigma}$, $\Pi_{Y/k}^{\Sigma} \twoheadrightarrow \Pi_{Y/k}^{\Sigma\text{-ét}} \twoheadrightarrow \Phi_{Y/k}^{\Sigma}$ induce *injections* $\Delta_{X/k}^{\Sigma} \xrightarrow{\sim} \Delta_{X/k}^{\Sigma\text{-ét}} \hookrightarrow \Phi_{X/k}^{\Sigma}$, $\Delta_{Y/k}^{\Sigma} \xrightarrow{\sim} \Delta_{Y/k}^{\Sigma\text{-ét}} \hookrightarrow \Phi_{Y/k}^{\Sigma}$, respectively [cf. the proof of the implication (2) $\Rightarrow$ (3) of Proposition 5.A.49]. This completes the proof of assertion (iii). $\qquad\square$

References

[1] H. Esnault and O. Wittenberg, On abelian birational sections, *J. Amer. Math. Soc.* **23** (2010), no. 3, 713–724.

[2] J.-H. Evertse, H. P. Schlickewei, and W. M. Schmidt, Linear equations in variables which lie in a multiplicative group, *Ann. of Math.* (2) **155** (2002), no. 3, 807–836.

[3] A. Grothendieck et al., *Revêtements Étales et Groupe Fondamental* (SGA 1), Séminaire de géométrie algébrique du Bois Marie 1960–61, Documents Mathématiques, 3. Société Mathématique de France, Paris, 2003.

[4] D. Harari and J. Stix, Descent obstruction and fundamental exact sequence, *The Arithmetic of Fundamental Groups - PIA 2010*, 147–166, Contributions in Mathematical and Computational Sciences, vol. 2, Springer-Verlag, Berlin Heidelberg, 2012.

[5] Y. Hoshi and S. Mochizuki, On the combinatorial anabelian geometry of nodally nondegenerate outer representations, *Hiroshima Math. J.* **41** (2011), no. 3, 275–342.

[6] Y. Hoshi, Existence of nongeometric pro-p Galois sections of hyperbolic curves, *Publ. Res. Inst. Math. Sci.* **46** (2010), no. 4, 829–848.

[7] Y. Hoshi, On monodromically full points of configuration spaces of hyperbolic curves, *The Arithmetic of Fundamental Groups - PIA 2010*, 167–207, Contributions in Mathematical and Computational Sciences, vol. 2, Springer-Verlag, Berlin Heidelberg, 2012.

[8] J. Koenigsmann, On the 'section conjecture' in anabelian geometry, *J. Reine Angew. Math.* **588** (2005), 221–235.

[9] S. Mochizuki, The profinite Grothendieck conjecture for closed hyperbolic curves over number fields, *J. Math. Sci. Univ. Tokyo* **3** (1996), no. 3, 571–627.

[10] S. Mochizuki, The local pro-p anabelian geometry of curves, *Invent. Math.* **138** (1999), no. 2, 319–423.

[11] S. Mochizuki, Topics surrounding the anabelian geometry of hyperbolic curves, *Galois groups and fundamental groups*, 119–165, Math. Sci. Res. Inst. Publ. 41, Cambridge University Press, Cambridge, 2003.

[12] F. Pop, On the birational p-adic section conjecture, *Compos. Math.* **146** (2010), no. 3, 621–637.

[13] M. Saïdi, *Good sections of arithmetic fundamental groups*, arXiv:1010.1313, 2010.

[14] J.-P. Serre and J. Tate, Good reduction of abelian varieties, *Ann. of Math.* (2) **88** (1968), 492–517.

[15] J. Stix, *On the birational section conjecture with local conditions*, arXiv:1203.3236, 2012.

[16] M. Stoll, Finite descent obstructions and rational points on curves, *Algebra Number Theory* **1** (2007), no. 4, 349–391.

[17] A. Tamagawa, The Grothendieck conjecture for affine curves, *Compositio Math.* **109** (1997), no. 2, 135–194.

[18] K. Wickelgren, *2-nilpotent real section conjecture*, arXiv:1006.0265, 2010.

6

Blocks for mod p representations of $\mathrm{GL}_2(\mathbb{Q}_p)$

Vytautas Paškūnas

Abstract

Let π_1 and π_2 be absolutely irreducible smooth representations of $G = \mathrm{GL}_2(\mathbb{Q}_p)$ with a central character, defined over a finite extension of $\mathbb{F}_p$. We show that if there exists a non-split extension between π_1 and π_2 then they both appear as subquotients of the reduction modulo p of a unit ball in a crystalline Banach space representation of G. The results of Berger–Breuil describe such reductions and allow us to organize the irreducible representation into blocks. The result is new for $p = 2$; the proof, which works for all p, is new.

1. Introduction

Let L be a finite extension of $\mathbb{Q}_p$, with the ring of integers $\mathcal{O}$, a uniformizer ϖ, and residue field k, and let $G = \mathrm{GL}_2(\mathbb{Q}_p)$ and let B be the subgroup of upper-triangular matrices in G.

Theorem 6.1. *Let π_1, π_2 be smooth, absolutely irreducible k-representations of G with a central character. Suppose that $\mathrm{Ext}_G^1(\pi_2, \pi_1) \neq 0$ then after replacing L by a finite extension, we may find integers $(l, k) \in \mathbb{Z} \times \mathbb{N}$ and unramified characters $\chi_1, \chi_2 : \mathbb{Q}_p^\times \to L^\times$ with $\chi_2 \neq \chi_1 |\cdot|$, such that π_1 and π_2 are subquotients of $\overline{\Pi}^{ss}$, where $\overline{\Pi}^{ss}$ is the semi-simplification of the reduction modulo ϖ of an open bounded G-invariant lattice in Π, where Π is the universal unitary completion of*

$$(\mathrm{Ind}_B^G \chi_1 \otimes \chi_2 |\cdot|^{-1})_{\mathrm{sm}} \otimes \det^l \otimes \mathrm{Sym}^{k-1} L^2.$$

Automorphic Forms and Galois Representations, ed. Fred Diamond, Payman L. Kassaei and Minhyong Kim. Published by Cambridge University Press. © Cambridge University Press 2014.

The results of Berger–Breuil [3], Berger [2], Breuil–Emerton [6] and [22] describe explicitly the possibilities for $\overline{\Pi}^{ss}$, see Proposition 6.13. These results and the Theorem imply that $\text{Ext}_G^1(\pi_2, \pi_1)$ vanishes in many cases. Let us make this more precise.

Let $\text{Mod}_G^{sm}(\mathcal{O})$ be the category of smooth G-representation on $\mathcal{O}$-torsion modules. It contains $\text{Mod}_G^{sm}(k)$, the category of smooth G-representations on k-vector spaces, as a full subcategory. Every irreducible object π of $\text{Mod}_G^{sm}(\mathcal{O})$ is killed by ϖ, and hence is an object of $\text{Mod}_G^{sm}(k)$. Barthel–Livné [1] and Breuil [4] have classified the absolutely irreducible smooth representations π admitting a central character. They fall into four disjoint classes:

   (i)   characters $\delta \circ \det$;
   (ii) special series $\text{Sp} \otimes \delta \circ \det$;
   (iii) principal series $(\text{Ind}_B^G \delta_1 \otimes \delta_2)_{sm}$, $\delta_1 \neq \delta_2$;
   (iv) supersingular representations;

where Sp is the Steinberg representation, that is the locally constant functions from $\mathbb{P}^1(\mathbb{Q}_p)$ to k modulo the constant functions; $\delta, \delta_1, \delta_2 : \mathbb{Q}_p^\times \to k^\times$ are smooth characters and we consider $\delta_1 \otimes \delta_2$ as a character of B, which sends $\left(\begin{smallmatrix} a & b \\ 0 & d \end{smallmatrix}\right)$ to $\delta_1(a)\delta_2(d)$. Using their results and some easy arguments, see [25, §5.3], one may show that for an irreducible smooth representation π the following are equivalent: (1) π is admissible, which means that π^H is finite dimensional for all open subgroups H of G; (2) $\text{End}_G(\pi)$ is finite dimensional over k; (3) there exists a finite extension k' of k, such that $\pi \otimes_k k'$ is isomorphic to a finite direct sum of distinct absolutely irreducible k'-representations with a central character.

Let $\text{Mod}_G^{l.adm}(\mathcal{O})$ be the full subcategory of $\text{Mod}_G^{sm}(\mathcal{O})$, consisting of representations, which are equal to the union of their admissible subrepresentations. The categories $\text{Mod}_G^{sm}(\mathcal{O})$ and $\text{Mod}_G^{l.adm}(\mathcal{O})$ are abelian, see [15, Prop.2.2.18]. We define $\text{Mod}_G^{l.adm}(k)$ in exactly the same way with $\mathcal{O}$ replaced by k. Let Irr_G^{adm} be the set of irreducible representations in $\text{Mod}_G^{l.adm}(\mathcal{O})$, then Irr_G^{adm} is the set of irreducible representations in $\text{Mod}_G^{sm}(\mathcal{O})$ satisfying the equivalent conditions described above. We define an equivalence relation $\sim$ on Irr_G^{adm}: $\pi \sim \tau$, if there exists a sequence of irreducible admissible representations $\pi = \pi_1, \pi_2, \ldots, \pi_n = \tau$, such that for each i one of the following holds: (1) $\pi_i \cong \pi_{i+1}$; (2) $\text{Ext}_G^1(\pi_i, \pi_{i+1}) \neq 0$; (3) $\text{Ext}_G^1(\pi_{i+1}, \pi_i) \neq 0$. We note that it does not matter for the definition of $\sim$, whether we compute Ext_G^1 in $\text{Mod}_G^{sm}(\mathcal{O})$, $\text{Mod}_G^{sm}(k)$, $\text{Mod}_G^{l.adm}(\mathcal{O})$ or $\text{Mod}_G^{l.adm}(k)$, since we only care about vanishing or non-vanishing of $\text{Ext}_G^1(\pi_i, \pi_{i+1})$ for distinct irreducible representations. A block is an equivalence class of $\sim$.

Corollary 6.2. *The blocks containing an absolutely irreducible representation are given by the following:*

(i) $\mathfrak{B} = \{\pi\}$ *with* π *supersingular;*

(ii) $\mathfrak{B} = \{(\mathrm{Ind}_B^G \delta_1 \otimes \delta_2 \omega^{-1})_{\mathrm{sm}}, (\mathrm{Ind}_B^G \delta_2 \otimes \delta_1 \omega^{-1})_{\mathrm{sm}}\}$ *with* $\delta_2 \delta_1^{-1} \neq \omega^{\pm 1}, \mathbf{1}$;

(iii) $p > 2$ *and* $\mathfrak{B} = \{(\mathrm{Ind}_B^G \delta \otimes \delta\omega^{-1})_{\mathrm{sm}}\}$;

(iv) $p = 2$ *and* $\mathfrak{B} = \{\mathbf{1}, \mathrm{Sp}\} \otimes \delta \circ \det$;

(v) $p \geq 5$ *and* $\mathfrak{B} = \{\mathbf{1}, \mathrm{Sp}, (\mathrm{Ind}_B^G \omega \otimes \omega^{-1})_{\mathrm{sm}}\} \otimes \delta \circ \det$;

(vi) $p = 3$ *and* $\mathfrak{B} = \{\mathbf{1}, \mathrm{Sp}, \omega \circ \det, \mathrm{Sp} \otimes \omega \circ \det\} \otimes \delta \circ \det$;

where $\delta, \delta_1, \delta_2 : \mathbb{Q}_p^\times \to k^\times$ *are smooth characters and where* $\omega : \mathbb{Q}_p^\times \to k^\times$ *is the character* $\omega(x) = x|x| \pmod{\varpi}$.

One may view the cases (iii) to (vi) as degenerations of case (ii). A finitely generated smooth admissible representation of G is of finite length, [15, Thm.2.3.8]. This makes $\mathrm{Mod}_G^{\mathrm{l.adm}}(\mathcal{O})$ into a locally finite category. It follows from [17] that every locally finite category decomposes into blocks. In our situation we obtain:

$$\mathrm{Mod}_G^{\mathrm{l.adm}}(\mathcal{O}) \cong \prod_{\mathfrak{B} \in \mathrm{Irr}_G^{\mathrm{adm}} /\sim} \mathrm{Mod}_G^{\mathrm{l.adm}}(\mathcal{O})[\mathfrak{B}], \tag{6.1}$$

where $\mathrm{Mod}_G^{\mathrm{l.adm}}(\mathcal{O})[\mathfrak{B}]$ is the full subcategory of $\mathrm{Mod}_G^{\mathrm{l.adm}}(\mathcal{O})$ consisting of representations, with all irreducible subquotients in $\mathfrak{B}$. One can deduce a similar result for the category of admissible unitary L-Banach space representations of G, see [25, Prop.5.32].

The result has been previously known for $p > 2$. Breuil and the author [7, §8], Colmez [8, §VII], Emerton [16, §4] and the author [23] have computed $\mathrm{Ext}_G^1(\pi_2, \pi_1)$ by different characteristic p methods, which do not work in the exceptional cases, when $p = 2$. In this paper, we go via characteristic 0 and make use of a deep Theorem of Berger–Breuil. The proof is less involved, but it does not give any information about the extensions between irreducible representations lying in the same block.

The motivation for these calculations comes from the p-adic Langlands correspondence for $\mathrm{GL}_2(\mathbb{Q}_p)$. Colmez in [8] to a 2-dimensional absolutely irreducible L-representation of the absolute Galois group of $\mathbb{Q}_p$ has associated an admissible unitary absolutely irreducible non-ordinary L-Banach space representation of G. He showed that his construction induces an injection on the isomorphism classes and asked whether it is a bijection, see [8, §0.13]. This has been answered affirmatively in [25] for $p \geq 5$, where the knowledge of blocks has been used in an essential way. The results of this chapter should be useful in dealing with the remaining cases.

Let us give a rough sketch of the argument. Let $0 \to \pi_1 \to \pi \to \pi_2 \to 0$ be a non-split extension. The method of [7] allows us to embed π into Ω, such that $\Omega|_K$ is admissible and an injective object in $\mathrm{Mod}_K^{\mathrm{sm}}(k)$, where $K = \mathrm{GL}_2(\mathbb{Z}_p)$. Using the results of [24] we may lift Ω to an admissible unitary L-Banach space representation E of G, in the sense that we may find a G-invariant unit ball E^0 in E, such that $E^0/\varpi E^0 \cong \Omega$. Moreover, $E|_K$ is isomorphic to a direct summand of $\mathcal{C}(K, L)^{\oplus r}$, where $\mathcal{C}(K, L)$ is the space of continuous function with the supremum norm. This implies, using an argument of Emerton, that the K-algebraic vectors are dense in E. As a consequence we find a closed G-invariant subspace Π of E, such that the reduction of $\Pi \cap E^0$ modulo ϖ contains π as a subrepresentation, and Π contains $\oplus_{i=1}^m \frac{\text{c-Ind}_{KZ}^G \tilde{\mathbf{1}}_i}{(T-a_i)^{n_i}} \otimes \det^{l_i} \otimes \mathrm{Sym}^{k_i-1} L^2$ as a dense subrepresentation, where Z is the centre of G, $\tilde{\mathbf{1}}_i : KZ \to L^\times$ is a character, trivial on K, $a_i \in L$, and T is a certain Hecke operator in $\mathrm{End}_G(\text{c-Ind}_{KZ}^G \tilde{\mathbf{1}}_i)$, such that $\frac{\text{c-Ind}_{KZ}^G \tilde{\mathbf{1}}_i}{(T-a_i)}$ is an unramified principal series representation. Once we have this we are in a good shape to prove Theorem 6.1.

Acknowledgements. I thank the anonymous referee for the comments, which led to an improvement of the exposition, and Jochen Heinloth for a stimulating discussion.

2. Notation

Let L be a finite extension of $\mathbb{Q}_p$ with the ring of integers $\mathcal{O}$, uniformizer ϖ and residue field k. We normalize the valuation val on L so that $\mathrm{val}(p) = 1$, and the norm $|\,.\,|$, so that $|x| = p^{-\mathrm{val}(x)}$, for all $x \in L$. Let $G = \mathrm{GL}_2(\mathbb{Q}_p)$; Z the centre of G; B the subgroup of upper triangular matrices; $K = \mathrm{GL}_2(\mathbb{Z}_p)$; $I = \{g \in K : g \equiv \left(\begin{smallmatrix}* & * \\ 0 & *\end{smallmatrix}\right) \pmod{p}\}$; $I_1 = \{g \in K : g \equiv \left(\begin{smallmatrix}1 & * \\ 0 & 1\end{smallmatrix}\right) \pmod{p}\}$; let $\mathfrak{K}$ be the G-normalizer of I; let $H = \{\left(\begin{smallmatrix}[\lambda] & 0 \\ 0 & [\mu]\end{smallmatrix}\right): \lambda, \mu \in \mathbb{F}_p^\times\}$, where $[\lambda]$ is the Teichmüller lift of λ; let $\mathcal{G}$ be the subgroup of G generated by matrices $\left(\begin{smallmatrix}p & 0 \\ 0 & p\end{smallmatrix}\right)$, $\left(\begin{smallmatrix}0 & 1 \\ p & 0\end{smallmatrix}\right)$ and H. Let $G^+ = \{g \in G : \mathrm{val}(\det(g)) \equiv 0 \pmod{2}\}$. Since we are working with representations of locally pro-p groups in characteristic p, these representations will not be semi-simple in general; socle is the maximal semi-simple subobject. So for example, $\mathrm{soc}_G \tau$ means the maximal semi-simple G-subrepresentation of τ. Let $\mathrm{Ban}_G^{\mathrm{adm}}(L)$ be the category of admissible unitary L-Banach space representations of G, studied in [26]. This category is abelian. Let Π be an admissible unitary L-Banach space representation of

G, and let Θ be an open bounded G-invariant lattice in Π, then $\Theta/\varpi\Theta$ is a smooth admissible k-representation of G. If $\Theta/\varpi\Theta$ is of finite length as a G-representation, then we let $\overline{\Pi}^{ss}$ be the semi-simplification of $\Theta/\varpi\Theta$. Since any two such Θ's are commensurable, $\overline{\Pi}^{ss}$ is independent of the choice of Θ. Universal unitary completions are discussed in [11, §1].

3. Main

Let π_1, π_2 be distinct smooth absolutely irreducible k-representations of G with a central character. It follows from [1] and [4] that π_1 and π_2 are admissible. We suppose that there exists a non-split extension in $\mathrm{Mod}_G^{\mathrm{sm}}(\mathcal{O})$:

$$0 \to \pi_1 \to \pi \to \pi_2 \to 0. \tag{6.2}$$

Since π_1 and π_2 are distinct and irreducible, by examining the long exact sequence induced by multiplication with ϖ, we deduce that π is killed by ϖ. A similar argument shows that the existence of a non-split extension implies that the central character of π_1 is equal to the central character of π_2. Moreover, π also has a central character, which is then equal to the central character of π_1, see [23, Prop.8.1]. We denote this central character by $\zeta : Z \to k^\times$. After replacing L by a quadratic extension and twisting by a character we may assume that $\zeta(\left(\begin{smallmatrix} p & 0 \\ 0 & p \end{smallmatrix}\right)) = 1$.

Lemma 6.3. *If* $\pi_1^{I_1} \neq \pi^{I_1}$ *then Theorem 6.1 holds for* π_1 *and* π_2.

Proof. Since ζ is continuous, it is trivial on the pro-p group $Z \cap I_1$. We thus may extend ζ to ZI_1, by letting $\zeta(zu) = \zeta(z)$ for all $z \in Z$, $u \in I_1$. If τ is a smooth k-representation of G with a central character ζ then $\tau^{I_1} \cong \mathrm{Hom}_{I_1 Z}(\zeta, \tau) \cong \mathrm{Hom}_G(\text{c-Ind}_{KZ}^G \zeta, \tau)$. Thus τ^{I_1} is naturally an $\mathcal{H} := \mathrm{End}_G(\text{c-Ind}_{I_1 Z}^G \zeta)$ module. Taking I_1-invariants of (6.2) we get an exact sequence of $\mathcal{H}$-modules:

$$0 \to \pi_1^{I_1} \to \pi^{I_1} \to \pi_2^{I_1}. \tag{6.3}$$

Since π_2 is irreducible, $\pi_2^{I_1}$ is an irreducible $\mathcal{H}$-module by [27]. Hence, if $\pi_1^{I_1} \neq \pi^{I_1}$, then the last arrow is surjective. It is shown in [20], that if τ is a smooth k-representation of G, with a central character ζ, generated as a G-representation by its I_1-invariants, then the natural map $\tau^{I_1} \otimes_{\mathcal{H}}$ c-Ind$_{KZ}^G \zeta \to \tau$ is an isomorphism. This implies that the sequence $0 \to \pi_1^{I_1} \to \pi^{I_1} \to \pi_2^{I_1} \to 0$ is non-split, and hence defines a non-zero element of $\mathrm{Ext}^1_{\mathcal{H}}(\pi_2^{I_1}, \pi_1^{I_1})$. Since $\pi_i \cong \pi_i^{I_1} \otimes_{\mathcal{H}}$ c-Ind$_{KZ}^G \zeta$ for $i = 1, 2$, the $\mathcal{H}$-modules $\pi_1^{I_1}$ and $\pi_2^{I_1}$ are non-isomorphic. Non-vanishing of $\mathrm{Ext}^1_{\mathcal{H}}(\pi_2^{I_1}, \pi_1^{I_1})$ implies

that there exists a smooth character $\eta : G \to k^\times$ such that either ($\pi_1 \cong \eta$
and $\pi_2 \cong \mathrm{Sp} \otimes \eta$) or ($\pi_2 \cong \eta$ and $\pi_1 \cong \mathrm{Sp} \otimes \eta$), [25, Lem.5.24], where Sp is
the Steinberg representation. In both cases the universal unitary completion of
$(\mathrm{Ind}_B^G | \cdot | \otimes | \cdot |^{-1})_{\mathrm{sm}} \otimes \tilde{\eta}$, where $\tilde{\eta} : G \to \mathcal{O}^\times$ is any smooth character lifting
η, will satisfy the conditions of Theorem 6.1 by [13, 5.3.18]. $\square$

Lemma 6.3 allows to assume that $\pi^{I_1} = \pi_1^{I_1}$. We note that this implies that
$\mathrm{soc}_K \pi_1 \cong \mathrm{soc}_K \pi$, and, since I_1 is contained in G^+, the restriction of (6.2) to
G^+ is a non-split extension of G^+-representations.

Now we perform a renaming trick, the purpose of which is to get around
some technical issues, when $p = 2$. If either $p > 2$ or $p = 2$ and π_1 is
neither a special series nor a character then we let $\tau_1 = \pi_1$, $\tau = \pi$ and $\tau_2 =
\pi_2$. If $p = 2$ and π_1 is either a special series representation or a character,
then we let $0 \to \tau_1 \to \tau \to \tau_2 \to 0$ be the exact sequence obtained by
tensoring (6.2) with $\mathrm{Ind}_{G^+}^G \mathbf{1}$. In particular, $\tau \cong \pi \otimes \mathrm{Ind}_{G^+}^G \mathbf{1}$, which implies
that $\tau|_{G^+} \cong \pi|_{G^+} \oplus \pi|_{G^+}$ and $\tau_1|_{G^+} \cong \pi_1|_{G^+} \oplus \pi_1|_{G^+}$. Hence, $\tau^{I_1} = \tau_1^{I_1}$ and
$\mathrm{soc}_K \tau \cong \mathrm{soc}_K \tau_1 \cong \mathrm{soc}_K \pi_1 \oplus \mathrm{soc}_K \pi_1$. This implies that $\mathrm{soc}_G \tau \cong \mathrm{soc}_G \tau_1$.

Lemma 6.4. $\mathrm{soc}_G \tau \cong \mathrm{soc}_G \tau_1 \cong \pi_1$.

Proof. We already know that $\mathrm{soc}_G \tau \cong \mathrm{soc}_G \tau_1$ and we only need to consider
the case $p = 2$ and π_1 is either a special series or a character. The assumption
on π_1 implies that $\pi_1^{I_1}$ is one dimensional. Let $\mathfrak{K}$ be the normalizer of I_1 in G,
then $I_1 Z$ is a subgroup of $\mathfrak{K}$ of index 2. We note that $I = I_1$ as $p = 2$. Thus $\mathfrak{K}$
acts on $\pi_1^{I_1}$ by a character χ, such that the restriction of χ to $I_1 Z$ is equal to ζ.
Since $p = 2$, we have an exact non-split sequence of G-representations $0 \to
\mathbf{1} \to \mathrm{Ind}_{G^+}^G \mathbf{1} \to \mathbf{1} \to 0$. We note that G^+ and hence $Z I_1$ act trivially on all
the terms in this sequence. By tensoring with π_1 we obtain an exact sequence
$0 \to \pi_1 \to \tau_1 \to \pi_1 \to 0$ of G-representations. Taking I_1-invariants, gives us
an isomorphism of $\mathfrak{K}$-representations $\tau_1^{I_1} \cong \pi_1^{I_1} \otimes \mathrm{Ind}_{Z I_1}^{\mathfrak{K}} \mathbf{1}$. This representation
is a non-split extension of χ by itself. Thus τ_1 is a non-split extension of π_1 by
itself. Hence, $\mathrm{soc}_G \tau_1 \cong \pi_1$. $\square$

If $p = 2$ then $\tau_1^{I_1}$ is 2-dimensional and has a basis of the form $\{v, \left(\begin{smallmatrix} 0 & 1 \\ p & 0 \end{smallmatrix}\right)v\}$:
if π_1 is either a character or special series, this follows from the isomorphism
$\tau_1^{I_1} \cong \pi_1^{I_1} \otimes \mathrm{Ind}_{Z I_1}^{\mathfrak{K}} \mathbf{1}$, otherwise $\tau_1 = \pi_1$ and the assertion follows from [7, Cor.
6.4 (i)] noting that the work of Bartel–Livné [1] and Breuil [4] on classification
of irreducible representations of G implies that π^{I_1} is isomorphic as a module
of the pro-p Iwahori–Hecke algebra to $M(r, \lambda, \eta)$ defined in [7, Def.6.2]. Since
$\tau^{I_1} = \tau_1^{I_1}$, [7, Prop.9.2] implies that the inclusion $\tau^{I_1} \hookrightarrow \tau$ has a $\mathcal{G}$-equivariant
section.

Proposition 6.5. *There exists a G-equivariant injection $\tau \hookrightarrow \Omega$, where Ω is a smooth k-representation of G, such that $\Omega|_K$ is an injective envelope of $\mathrm{soc}_K \tau$ in $\mathrm{Mod}_K^{\mathrm{sm}}(k)$, $\left(\begin{smallmatrix} p & 0 \\ 0 & p \end{smallmatrix}\right)$ acts trivially on Ω and $\Omega|_{\mathfrak{K}} \cong \mathrm{Ind}_{\mathcal{G}}^{\mathfrak{K}} \Omega^{I_1}$.*

Proof. The existence of Ω satisfying the first two conditions follows from [7, Cor.9.11]. The last condition is satisfied as a byproduct of the construction of the action of $\left(\begin{smallmatrix} 0 & 1 \\ p & 0 \end{smallmatrix}\right)$ in [7, Lem.9.6]. $\qquad\square$

Corollary 6.6. *Let Ω be as above then $\mathrm{soc}_K \Omega \cong \mathrm{soc}_K \tau_1$ and $\mathrm{soc}_G \Omega \cong \pi_1$.*

Proof. Since τ is a subrepresentation of Ω, $\mathrm{soc}_K \tau$ is contained in $\mathrm{soc}_K \Omega$. Since $\Omega|_K$ is an injective envelope of $\mathrm{soc}_K \tau$, every non-zero K-invariant subspace of Ω intersects $\mathrm{soc}_K \tau$ non-trivially. This implies that $\mathrm{soc}_K \tau \cong \mathrm{soc}_K \Omega$. This implies the first assertion, as $\mathrm{soc}_K \tau \cong \mathrm{soc}_K \tau_1$. Moreover, every G-invariant non-zero subspace of Ω intersects τ non-trivially, since those are also K-invariant. This implies $\mathrm{soc}_G \Omega \cong \mathrm{soc}_G \tau \cong \pi_1$, where the last isomorphism follows from Lemma 6.4. $\qquad\square$

Lemma 6.7. *Let κ be a finite dimensional k-representation of $\mathcal{G}$ on which $\left(\begin{smallmatrix} p & 0 \\ 0 & p \end{smallmatrix}\right)$ acts trivially. There exists an admissible unitary L-Banach space representation $(E, \|\cdot\|)$ of $\mathfrak{K}$, such that $\|E\| \subset |L|$, $\left(\begin{smallmatrix} p & 0 \\ 0 & p \end{smallmatrix}\right)$ acts trivially on E, and the reduction modulo ϖ of the unit ball in E is isomorphic to $(\mathrm{Ind}_{\mathcal{G}}^{\mathfrak{K}} \kappa)_{\mathrm{sm}}$ as a $\mathfrak{K}$-representation.*

Proof. It is enough to prove the statement, when κ is indecomposable, which we now assume. Let $p^{\mathbb{Z}}$ be the subgroup of G generated by $\left(\begin{smallmatrix} p & 0 \\ 0 & p \end{smallmatrix}\right)$. Since the order of H is prime to p, and H has index 2 in $\mathcal{G}/p^{\mathbb{Z}}$, κ is either a character or an induction of a character from H to $\mathcal{G}/p^{\mathbb{Z}}$. In both cases we may lift κ to a representation $\tilde{\kappa}^0$ of $\mathcal{G}/p^{\mathbb{Z}}$ on a free $\mathcal{O}$-module of rank 1 or rank 2 respectively. Let $\tilde{\kappa} = \tilde{\kappa}^0 \otimes_{\mathcal{O}} L$ and let $\|\cdot\|$ be the gauge of $\tilde{\kappa}^0$. Then $\|\cdot\|$ is $\mathcal{G}$-invariant and $\tilde{\kappa}^0$ is the unit ball with respect to $\|\cdot\|$. Then $(\mathrm{Ind}_{\mathcal{G}/p^{\mathbb{Z}}}^{\mathfrak{K}/p^{\mathbb{Z}}} \tilde{\kappa})_{\mathrm{cont}}$ with the norm $\|f\|_1 := \sup_{g \in \mathfrak{K}/p^{\mathbb{Z}}} \|f(g)\|$ is a lift of $(\mathrm{Ind}_{\mathcal{G}/p^{\mathbb{Z}}}^{\mathfrak{K}/p^{\mathbb{Z}}} \kappa)_{\mathrm{sm}}$, where the subscript cont indicates continuous induction: the space of continuous functions with the right transformation property. $\qquad\square$

Theorem 6.8. *Let Ω be any representation given by Proposition 6.5. Then there exists an admissible unitary L-Banach space representation $(E, \|\cdot\|)$ of G, such that $\|E\| \subset |L|$, $\left(\begin{smallmatrix} p & 0 \\ 0 & p \end{smallmatrix}\right)$ acts trivially on E, and the reduction modulo ϖ of the unit ball in E is isomorphic to Ω as a G-representation.*

Proof. If $p \neq 2$ this is shown in [24, Thm.6.1]. We will observe that the renaming trick allows us to carry out essentially the same proof when $p = 2$. We make no assumption on p.

We first lift $\Omega|_K$ to characteristic 0. Let σ be the K-socle of Ω. Pontryagin duality induces an anti-equivalence of categories between $\mathrm{Mod}_K^{\mathrm{sm}}(k)$ and the category of pseudocompact $k[\![K]\!]$-modules, which we denote by $\mathrm{Mod}_K^{\mathrm{pro.aug}}(k)$. Since Ω is an injective envelope of σ in $\mathrm{Mod}_K^{\mathrm{sm}}(k)$, its Pontryagin dual $\Omega^\vee$ is a projective envelope of $\sigma^\vee$ in $\mathrm{Mod}_K^{\mathrm{pro.aug}}(k)$. Let $\widetilde{P}_{\sigma^\vee}$ be a projective envelope of $\sigma^\vee$ in the category of pseudocompact $\mathcal{O}[\![K]\!]$-modules. Then $\widetilde{P}_{\sigma^\vee}/\varpi\widetilde{P}_{\sigma^\vee}$ is a projective envelope of $\sigma^\vee$ in $\mathrm{Mod}_K^{\mathrm{pro.aug}}(k)$. Since projective envelopes are unique up to isomorphism, we obtain $\Omega^\vee \cong \widetilde{P}_{\sigma^\vee}/\varpi\widetilde{P}_{\sigma^\vee}$. Since τ_1 is admissible and $\sigma \cong \mathrm{soc}_K\, \tau_1$ by Corollary 6.6, σ is a finite dimensional k-vector space. In particular, $\sigma^\vee$ is a finitely generated $\mathcal{O}[\![K]\!]$-module, and so there exists a surjection of $\mathcal{O}[\![K]\!]$-modules $\mathcal{O}[\![K]\!]^{\oplus r} \twoheadrightarrow \sigma^\vee$. Since $\mathcal{O}[\![K]\!]^{\oplus r}$ is projective, and $\widetilde{P}_{\sigma^\vee} \twoheadrightarrow \sigma^\vee$ is essential, the surjection factors through $\mathcal{O}[\![K]\!]^{\oplus r} \twoheadrightarrow \widetilde{P}_{\sigma^\vee}$, and so $\widetilde{P}_{\sigma^\vee}$ is a finitely generated $\mathcal{O}[\![K]\!]$-module. Since $\widetilde{P}_{\sigma^\vee}$ is projective, we deduce that it is a direct summand of $\mathcal{O}[\![K]\!]^{\oplus r}$, and hence it is $\mathcal{O}$-torsion free.

Thus $\widetilde{P}_{\sigma^\vee}$ is an $\mathcal{O}$-torsion free, finitely generated $\mathcal{O}[\![K]\!]$-module, and its reduction modulo ϖ is isomorphic to $\Omega^\vee$ in $\mathrm{Mod}_K^{\mathrm{pro.aug}}(k)$. Let $E_0 = \mathrm{Hom}_{\mathcal{O}}^{cont}(\widetilde{P}_{\sigma^\vee}, L)$, and let $\|\cdot\|_0$ be the supremum norm. It follows from [26] that E_0 is an admissible unitary L-Banach space representation of K. Moreover, the unit ball E_0^0 in E_0 is $\mathrm{Hom}_{\mathcal{O}}^{cont}(\widetilde{P}_{\sigma^\vee}, \mathcal{O})$ and

$$\mathrm{Hom}_{\mathcal{O}}^{cont}(\widetilde{P}_{\sigma^\vee}, \mathcal{O}) \otimes_{\mathcal{O}} k \cong \mathrm{Hom}_{\mathcal{O}}^{cont}(\widetilde{P}_{\sigma^\vee}, k) \cong \mathrm{Hom}_k^{cont}(P_{\sigma^\vee}, k)$$
$$\cong (\Omega^\vee)^\vee \cong \Omega,$$

see [24, §5] for details. We extend the action of K on E_0 to the action of KZ by letting $\left(\begin{smallmatrix} p & 0 \\ 0 & p \end{smallmatrix}\right)$ act trivially.

Since σ is finite dimensional, it follows from [21, Lem.6.2.4] that Ω^{I_1} is a finite dimensional k-vector space. Since $\Omega|_{\mathfrak{K}} \cong (\mathrm{Ind}_{\tilde{G}}^{\mathfrak{K}}\,\Omega^{I_1})_{\mathrm{sm}}$ by Proposition 6.5, Lemma 6.7 implies that there exists a unitary L-Banach space representation $(E_1, \|\cdot\|_1)$ of $\mathfrak{K}$, such that $\|E_1\| \subseteq |L|$, $\left(\begin{smallmatrix} p & 0 \\ 0 & p \end{smallmatrix}\right)$ acts trivially on E_1 and the reduction of the unit ball E_1^0 in E_1 modulo ϖ is isomorphic to $\Omega|_{\mathfrak{K}}$. We claim that there exists an isometric, IZ-equivariant isomorphism $\varphi: E_1 \to E_0$ such that the following diagram of IZ-representations:

$$
\begin{array}{ccc}
E_1^0/\varpi E_1^0 & \xrightarrow[\mathrm{mod}\ \varpi]{\varphi} & E_0^0/\varpi E_0^0 \\
\downarrow{\scriptstyle\cong} & & \downarrow{\scriptstyle\cong} \\
\Omega & \xrightarrow{\ \mathrm{id}\ } & \Omega
\end{array}
\tag{6.4}
$$

commutes, where the left vertical arrow is the given $\mathfrak{K}$-equivariant isomorphism $E_1^0/\varpi E_1^0 \cong \Omega|_{\mathfrak{K}}$ and the right vertical arrow is the given KZ-equivariant isomorphism $E_0^0/\varpi E_0^0 \cong \Omega|_{KZ}$. Granting the claim, we may transport the action of $\mathfrak{K}$ on E_0 by using φ to obtain a unitary action of KZ and $\mathfrak{K}$ on E_0, such that the two actions agree on $KZ \cap \mathfrak{K}$, which is equal to IZ. The resulting action glues to the unitary action of G on E_0, see [21, Cor.5.5.5], which is stated for smooth representations, but the proof of which works for any representation. The commutativity of the above diagram implies that $E_0^0 \otimes_{\mathcal{O}} k \cong \Omega$ as a G-representation.

We will prove the claim now. Let $M = \mathrm{Hom}_{\mathcal{O}}^{cont}(E_1^0, \mathcal{O})$ equipped with the topology of pointwise convergence. Then M is an object of $\mathrm{Mod}_I^{\mathrm{pro.aug}}(\mathcal{O})$, and $M \otimes_{\mathcal{O}} k \cong \Omega^{\vee}$ in $\mathrm{Mod}_I^{\mathrm{pro.aug}}(k)$, see [24, Lem.5.4]. Since $\Omega|_K$ is injective in $\mathrm{Mod}_K^{\mathrm{sm}}(k)$, $\Omega|_I$ is injective in $\mathrm{Mod}_I^{\mathrm{sm}}(k)$. Since I_1 is a pro-p group, every non-zero I-invariant subspace of Ω intersects Ω^{I_1} non-trivially. Thus $\Omega|_I$ is an injective envelope of Ω^{I_1} in $\mathrm{Mod}_I^{\mathrm{sm}}(k)$. Hence, $\Omega^{\vee}$ is a projective envelope of $(\Omega^{I_1})^{\vee}$ in $\mathrm{Mod}_I^{\mathrm{pro.aug}}(k)$. Since M is $\mathcal{O}$-torsion free, and $M \otimes_{\mathcal{O}} k$ is a projective envelope of $(\Omega^{I_1})^{\vee}$ in $\mathrm{Mod}_I^{\mathrm{pro.aug}}(k)$, [24, Prop.4.6] implies that M is a projective envelope of $(\Omega^{I_1})^{\vee}$ in $\mathrm{Mod}_I^{\mathrm{pro.aug}}(\mathcal{O})$. The same holds for $\widetilde{P}_{\sigma^{\vee}}$. Since projective envelopes are unique up to isomorphism, there exists an isomorphism $\psi : \widetilde{P}_{\sigma^{\vee}} \xrightarrow{\cong} M$ in $\mathrm{Mod}_I^{\mathrm{pro.aug}}(\mathcal{O})$. It follows from [24, Cor.4.7] that the natural map $\mathrm{Aut}_{\mathcal{O}[\![I]\!]}(\widetilde{P}_{\sigma^{\vee}}) \to \mathrm{Aut}_{k[\![I]\!]}(\widetilde{P}_{\sigma^{\vee}}/\varpi \widetilde{P}_{\sigma^{\vee}})$ is surjective. Using this we may choose ψ so that the following diagram in $\mathrm{Mod}_K^{\mathrm{pro.aug}}(k)$:

$$
\begin{array}{ccc}
\widetilde{P}_{\sigma^{\vee}}/\varpi \widetilde{P}_{\sigma^{\vee}} & \xrightarrow[\mathrm{mod}\ \varpi]{\psi} & M/\varpi M \\
\Big\downarrow{\cong} & & \Big\downarrow{\cong} \\
\Omega^{\vee} & \xrightarrow{\mathrm{id}} & \Omega^{\vee}
\end{array}
\tag{6.5}
$$

commutes. Dually we obtain an isometric I-equivariant isomorphism of unitary L-Banach space representations of I, $\psi^d : \mathrm{Hom}_{\mathcal{O}}^{cont}(M, L) \to \mathrm{Hom}_{\mathcal{O}}^{cont}(\widetilde{P}_{\sigma^{\vee}}, L)$. It follows from [26, Thm.1.2] that $(E_1, \|\cdot\|_1)$ is naturally and isometrically isomorphic to $\mathrm{Hom}_{\mathcal{O}}^{cont}(M, L)$ with the supremum norm. This gives our φ. The commutativity of (6.5) implies the commutativity of (6.4). $\qquad\square$

Corollary 6.9. *The Banach space representation $(E, \|\cdot\|)$ constructed in Theorem 6.8 is isometrically, K-equivariantly isomorphic to a direct summand of $\mathcal{C}(K, L)^{\oplus r}$, where $\mathcal{C}(K, L)$ is the space of continuous functions from K to L, equipped with the supremum norm, and r is a positive integer.*

Proof. It follows from the construction of E, that $(E, \|\cdot\|)$ is isometrically, K-equivariantly isomorphic to $\mathrm{Hom}_{\mathcal{O}}^{cont}(\widetilde{P}_{\sigma^{\vee}}, L)$ with the supremum norm. Moreover, it follows from the proof of Theorem 6.8 that $\widetilde{P}_{\sigma^{\vee}}$ is a direct summand of $\mathcal{O}[\![K]\!]^{\oplus r}$. It is shown in [26, Lem.2.1, Cor.2.2] that the natural map

$K \rightarrow \mathcal{O}[\![K]\!]$, $g \mapsto g$ induces an isometrical, K-equivariant isomorphism between $\mathcal{C}(K, L)$ and $\mathrm{Hom}_{\mathcal{O}}^{cont}(\mathcal{O}[\![K]\!], L)$. □

If F is a finite extension of $\mathbb{Q}_p$ then exactly the same proof works. We note that [7, Thm.9.8] is proved for $\mathrm{GL}_2(F)$. We record this as a corollary below. Let $\mathcal{O}_F$ be the ring of integers of F, ϖ_F a uniformizer, k_F the residue field, let $\mathcal{G}_F$ be the subgroup of $\mathrm{GL}_2(F)$ generated by the matrices $\left(\begin{smallmatrix} \varpi_F & 0 \\ 0 & \varpi_F \end{smallmatrix}\right)$, $\left(\begin{smallmatrix} 0 & 1 \\ \varpi_F & 0 \end{smallmatrix}\right)$ and $\left(\begin{smallmatrix} [\lambda] & 0 \\ 0 & [\mu] \end{smallmatrix}\right)$, for $\lambda, \mu \in k_F^\times$, where $[\lambda]$ is the Teichmüller lift of λ. Let I_1 be the standard pro-p Iwahori subgroup of G.

Corollary 6.10. *Let τ be an admissible smooth k-representation of $\mathrm{GL}_2(F)$, such that $\left(\begin{smallmatrix} \varpi_F & 0 \\ 0 & \varpi_F \end{smallmatrix}\right)$ acts trivially on τ and if $p = 2$ assume that the inclusion $\tau^{I_1} \hookrightarrow \tau$ has a $\mathcal{G}_F$-equivariant section. Then there exists a $\mathrm{GL}_2(F)$-equivariant embedding $\tau \hookrightarrow \Omega$, such that $\Omega|_{\mathrm{GL}_2(\mathcal{O}_F)}$ is an injective envelope of $\mathrm{GL}_2(\mathcal{O}_F)$-socle of τ in the category of smooth k-representations of $\mathrm{GL}_2(\mathcal{O}_F)$ and $\left(\begin{smallmatrix} \varpi_F & 0 \\ 0 & \varpi_F \end{smallmatrix}\right)$ acts trivially on Ω. Moreover, we may lift Ω to an admissible unitary L-Banach space representation of $\mathrm{GL}_2(F)$.*

Remark 6.11. We also note that one could work with a fixed central character throughout.

Let $V_{l,k} = \mathrm{det}^l \otimes \mathrm{Sym}^{k-1} L^2$, for $k \in \mathbb{N}$ and $l \in \mathbb{Z}$. Rather unfortunately k also denotes the residue field of L, we hope that this will not cause any confusion.

Proposition 6.12. *Let $(E, \|\,.\,\|)$ be a unitary L-Banach space representation of K isomorphic in the category of unitary admissible L-Banach space representations of K to a direct summand of $\mathcal{C}(K, L)^{\oplus r}$. The evaluation map*

$$\bigoplus_{(l,k) \in \mathbb{Z} \times \mathbb{N}} \mathrm{Hom}_K(V_{l,k}, E) \otimes V_{l,k} \rightarrow E \qquad (6.6)$$

is injective and the image is a dense subspace of E. Moreover, the subspaces $\mathrm{Hom}_K(V_{l,k}, E)$ are finite dimensional.

Proof. The argument is the same as given in the proof of [14, Prop.5.4.1]. We have provided some details in the Appendix at the request of the referee. It is enough to prove the statement for $\mathcal{C}(K, L)$, since then it is true for $\mathcal{C}(K, L)^{\oplus r}$ and by applying the idempotent, which cuts out E, we may deduce the same statement for E. In the case $E = \mathcal{C}(K, L)$, the assertion follows from Proposition 6.A.17 applied to $G = \mathrm{GL}_2$. We note that every rational irreducible representation of GL_2/L is isomorphic to $V_{l,k}$ for a unique pair $(l, k) \in \mathbb{Z} \times \mathbb{N}$. The last assertion follows from (6.A.13) below. □

Proposition 6.13. *Let* $\rho = (\mathrm{Ind}_B^G \chi_1 \otimes \chi_2 |\cdot|^{-1})_{\mathrm{sm}}$ *be a smooth principal series representation of* G, *where* $\chi_1, \chi_2 : \mathbb{Q}_p^\times \to L^\times$ *smooth characters with* $\chi_1|\cdot| \neq \chi_2$. *Let* Π *be the universal unitary completion of* $\rho \otimes V_{l,k}$. *Then* Π *is an admissible, finite length* L-*Banach space representation of* G. *Moreover, if* Π *is non-zero and we let* $\overline{\Pi}^{ss}$ *be the semi-simplification of the reduction modulo* ϖ *of an open bounded* G-*invariant lattice in* Π, *then either* $\overline{\Pi}^{ss}$ *is irreducible supersingular, or*

$$\overline{\Pi}^{ss} \subseteq (\mathrm{Ind}_B^G \delta_1 \otimes \delta_2 \omega^{-1})_{\mathrm{sm}}^{ss} \oplus (\mathrm{Ind}_B^G \delta_2 \otimes \delta_1 \omega^{-1})_{\mathrm{sm}}^{ss}, \qquad (6.7)$$

for some smooth characters $\delta_1, \delta_2 : \mathbb{Q}_p^\times \to k^\times$, *where the superscript* ss *indicates the semi-simplification.*

Proof. If $\Pi \neq 0$ then $-(k+l) \leq \mathrm{val}(\chi_1(p)) \leq -l$, $-(k+l) \leq \mathrm{val}(\chi_2(p)) \leq -l$ and $\mathrm{val}(\chi_1(p)) + \mathrm{val}(\chi_2(p)) = -(k+2l)$, [24, Lem.7.9], [11, Lem.2.1]. If both inequalities are strict and $\chi_1 \neq \chi_2$ then it is shown in [3, §5.3] that Π is non-zero, admissible and absolutely irreducible. The assertion about $\overline{\Pi}^{ss}$ then follows from [2].

If both inequalities are strict, $\chi_1 = \chi_2$ and Π is non-zero it is shown in [22, Prop.4.2] that there exist $\mathcal{O}$-lattices M in $\rho \otimes V_{l,k}$ and M' in $\rho' \otimes V_{l,k}$, where $\rho' = (\mathrm{Ind}_B^G \chi_1' \otimes \chi_2' |\cdot|^{-1})_{\mathrm{sm}}$ for some distinct smooth characters, $\chi_1', \chi_2' : \mathbb{Q}_p^\times \to L^\times$ congruent to χ_1, χ_2 modulo $1 + (\varpi)$, such that both lattices are finitely generated $\mathcal{O}[G]$-modules and their reductions modulo ϖ are isomorphic. Since M is $\mathcal{O}$-torsion free, the completion of $\rho \otimes V_{l,k}$ with respect to the gauge of M is non-zero, and since M is a finitely generated $\mathcal{O}[G]$-module, the completion is the universal unitary completion, [11, Prop.1.17], thus is isomorphic to Π. Let Π^0 be the unit ball in Π with respect to the gauge of M. Then $\Pi^0/\varpi \Pi^0 \cong M/\varpi M \cong M'/\varpi M'$. Now by the same argument the completion of $\rho' \otimes V_{l,k}$ with respect to the gauge of M' is the universal unitary completion of of $\rho' \otimes V_{l,k}$. Since $\chi_1' \neq \chi_2'$ we may apply the results of Berger–Breuil [3] to conclude that the semi-simplification of $M'/\varpi M'$ has the desired form.

Suppose that either $\mathrm{val}(\chi_1(p)) = -l$ or $\mathrm{val}(\chi_2(p)) = -l$. If $\chi_1 = \chi_2 |\cdot|$ then this forces $k = 1$, so that $V_{l,k}$ is a character and $\rho \otimes V_{l,k} \cong (\mathrm{Ind}_B^G |\cdot| \otimes |\cdot|^{-1})_{\mathrm{sm}} \otimes \eta$, where $\eta : G \to L^\times$ is a unitary character. It follows from [13, Lem.5.3.18] that the universal unitary completion of $\rho \otimes V_{l,k}$ is admissible and of length 2. Moreover, $\overline{\Pi}^{ss} \cong \overline{\eta} \oplus \mathrm{Sp} \otimes \overline{\eta} \cong (\mathrm{Ind}_B^G \overline{\eta} \otimes \overline{\eta})_{\mathrm{sm}}^{ss}$. If $\chi_1 \neq \chi_2 |\cdot|$ then it follows from [6, Lem.2.2.1] that the universal unitary completion of $\rho \otimes V_{l,k}$ is isomorphic to a continuous induction of a unitary character. Hence $\overline{\Pi}^{ss}$ is isomorphic to the semi-simplification of a principal series representation. $\qquad\square$

Proof of Theorem 6.1. Let $(E, \| \cdot \|)$ be the unitary L-Banach space representation of G constructed in the proof of Theorem 6.8. Let E^0 be the unit ball in E, then by construction we have $E^0/\varpi E^0 \cong \Omega$, where Ω is a smooth k-representation of G, satisfying the conditions of Proposition 6.5. Let $V = \oplus \operatorname{Hom}_K(V_{l,k}, E) \otimes V_{l,k}$, where the sum is taken over all $(l, k) \in \mathbb{Z} \times \mathbb{N}$. It follows from Corollary 6.9 and Proposition 6.12 that the natural map $V \to E$ is injective and the image is dense. Let $\{V^i\}_{i \geq 0}$ be any increasing, exhaustive filtration of V by finite dimensional K-invariant subspaces. Then $V^i \cap E^0$ is a K-invariant $\mathcal{O}$-lattice in V^i, and we denote by $\overline{V}^i$ its reduction modulo ϖ. It follows from [24, Lem.5.5] that the reduction modulo ϖ induces a K-equivariant injection $\overline{V}^i \hookrightarrow \Omega$. The density of V in E implies that $\{\overline{V}^i\}_{i \geq 0}$ is an increasing, exhaustive filtration of Ω by finite dimensional, K-invariant subspaces. Recall that Ω contains τ as a subrepresentation, see Proposition 6.5. Now τ is finitely generated as a G-representation, since it is of finite length. Thus we may conclude, that there exists a finite dimensional K-invariant subspace W of V, such that τ is contained in the G-subrepresentation of Ω generated by $\overline{W}$.

Let $\varphi : V_{l,k} \to E$ be a non-zero K-equivariant, L-linear homomorphism. Let $R(\varphi)$ be the G-subrepresentation of E in the category of (abstract) G-representations on L-vector spaces, generated by the image of φ. Frobenius reciprocity gives us a surjection c-$\operatorname{Ind}_{KZ}^G \tilde{1} \otimes V_{l,k} \twoheadrightarrow R(\varphi)$, where $\tilde{1} : KZ \to L^\times$ is an unramified character, such that $\left(\begin{smallmatrix} p & 0 \\ 0 & p \end{smallmatrix}\right)$ acts trivially on $V_{l,k} \otimes \tilde{1}$. Now $\operatorname{End}_G(\text{c-}\operatorname{Ind}_{KZ}^G \tilde{1})$ is isomorphic to the ring of polynomials over L in one variable T. It follows from the proof of [24, Cor.7.4] that the surjection factors through $\frac{\text{c-}\operatorname{Ind}_{KZ}^G \tilde{1}}{P(T)} \otimes V_{l,k} \twoheadrightarrow R(\varphi)$, for some non-zero $P(T) \in L[T]$.

Let R be the (abstract) G-subrepresentation of E generated by W, and let Π be the closure of R in E. Since W is isomorphic to a finite direct sum of $V_{l,k}$'s, we deduce that if we replace L by a finite extension there exists a surjection:

$$\bigoplus_{i=1}^m \frac{\text{c-}\operatorname{Ind}_{KZ}^G \tilde{1}_i}{(T - a_i)^{n_i}} \otimes V_{l_i,k_i} \twoheadrightarrow R, \tag{6.8}$$

for some $a_i \in L$, $n_i \in \mathbb{N}$ and $(l_i, k_i) \in \mathbb{Z} \times \mathbb{N}$. Let $\rho_i = \frac{\text{c-}\operatorname{Ind}_{KZ}^G \tilde{1}_i}{T - a_i}$, then using (6.8) we may construct a finite, increasing, exhaustive filtration $\{R^j\}_{j \geq 0}$ of R by G-invariant subspaces, such that for each j there exists a surjection $\rho_i \otimes V_{l_i,k_i} \twoheadrightarrow R^j/R^{j-1}$, for some $1 \leq i \leq m$. Moreover, by choosing n_i and m in (6.8) to be minimal, we may assume that $\operatorname{Hom}_G(\rho_i \otimes V_{l_i,k_i}, R)$ is non-zero for all $1 \leq i \leq m$. Let Π^j be the closure of R^j in E. We note that since E is admissible, Π^j is an admissible unitary L-Banach space representation of

G, moreover the category $\mathrm{Ban}_G^{\mathrm{adm}}(L)$ is abelian. Since R^j is dense in Π^j, its image is dense in Π^j/Π^{j-1}. Hence, for each j there exists a G-equivariant map $\varphi_j : \rho_i \otimes V_{l_i,k_i} \to \Pi^j/\Pi^{j-1}$ with a dense image. Let Π_i be the universal unitary completion of $\rho_i \otimes V_{l_i,k_i}$. Since the target of φ_j is unitary, we can extend it to a continuous G-equivariant map $\tilde{\varphi}_j : \Pi_i \to \Pi^j/\Pi^{j-1}$. Moreover, since the target of φ_j is admissible and the image is dense, $\tilde{\varphi}_j$ is surjective.

For each closed subspace U of E, we let $\overline{U}$ be the reduction of $(U \cap E^0)$ modulo ϖ. It follows from [24, Lem.5.5] that the reduction modulo ϖ induces an injection $\overline{U} \hookrightarrow \Omega$. Since Π contains W, $\overline{\Pi}$ will contain $\overline{W}$. Since $\overline{\Pi}$ is G-invariant, it will contain τ. Now $\{\overline{\Pi}^j\}_{j\geq 0}$ defines a finite, increasing, exhaustive filtration of $\overline{\Pi}$ by G-invariant subspaces. Since π_2 is an irreducible subquotient of τ, there exists j, such that π_2 is an irreducible subquotient of $\overline{\Pi}^j/\overline{\Pi}^{j-1}$.

Each representation ρ_i is an unramified principal series representation, considered in Proposition 6.13, see [5, Prop.3.2.1]. Hence, Π_i is an admissible, finite length L-Banach space representation of G, moreover $\overline{\Pi}_i^{ss}$ is of finite length as described in Proposition 6.13. The surjection $\tilde{\varphi}_j : \Pi_i \twoheadrightarrow \Pi^j/\Pi^{j-1}$ induces a surjection $\overline{\Pi}_i^{ss} \twoheadrightarrow \overline{(\Pi^j/\Pi^{j-1})}^{ss}$. It follows from [24, Lem.5.5] that the semi-simplification of $\overline{\Pi}^j/\overline{\Pi}^{j-1}$ is isomorphic to $\overline{(\Pi^j/\Pi^{j-1})}^{ss}$. Thus π_2 is a subquotient of $\overline{\Pi}_i^{ss}$.

Since $\mathrm{Hom}_G(\rho_i \otimes V_{l_i,k_i}, \Pi)$ is non-zero, there exists a non-zero continuous G-invariant homomorphism $\varphi : \Pi_i \to \Pi$. Let Σ be the image of φ. Since Π_i and Π are admissible, we have a surjection $\Pi_i \twoheadrightarrow \Sigma$ and an injection $\Sigma \hookrightarrow \Pi$ in the abelian category $\mathrm{Ban}_G^{\mathrm{adm}}(L)$. The surjection induces a surjection $\overline{\Pi}_i^{ss} \twoheadrightarrow \overline{\Sigma}^{ss}$. The injection induces an injection $\overline{\Sigma} \hookrightarrow \overline{\Pi} \hookrightarrow \Omega$. Since $\mathrm{soc}_G \Omega \cong \pi_1$ by Corollary 6.6 and $\overline{\Sigma}$ is non-zero, we deduce that $\pi_1 \cong \mathrm{soc}_G \overline{\Sigma}$. Hence, π_1 is a subquotient of $\overline{\Pi}_i^{ss}$. $\qquad\square$

Lemma 6.14. *Let κ and λ be smooth k-representations of G and let l be a finite extension of k. Then $\mathrm{Ext}_G^i(\kappa, \lambda) \otimes_k l \cong \mathrm{Ext}_G^i(\kappa \otimes_k l, \lambda \otimes_k l)$, for all $i \geq 0$, where the Ext groups are computed in $\mathrm{Mod}_G^{\mathrm{sm}}(k)$ and $\mathrm{Mod}_G^{\mathrm{sm}}(l)$, respectively.*

Proof. The assertion for $i = 0$ follows from [25, Lem.5.1]. Hence, it is enough to find an injective resolution of λ in $\mathrm{Mod}_G^{\mathrm{sm}}(k)$, which remains injective after tensoring with l. Such resolution may be obtained by considering $(\mathrm{Ind}_{\{1\}}^G V)_{\mathrm{sm}}$, where $\{1\}$ is the trivial subgroup of G and V is a k-vector space. We note that $(\mathrm{Ind}_{\{1\}}^G V)_{\mathrm{sm}} \otimes_k l \cong (\mathrm{Ind}_{\{1\}}^G V \otimes_k l)_{\mathrm{sm}}$, since l is finite over k. $\qquad\square$

Proof of Corollary 6.2. Lemma 6.14 implies that replacing L by a finite extension does not change the blocks. It follows from Proposition 6.13 and Theorem 6.1 that an irreducible supersingular representation is in a block on its own. Let $\pi\{\delta_1, \delta_2\}$ be the semi-simple representation defined by (6.7), where $\delta_1, \delta_2 : \mathbb{Q}_p^\times \to k^\times$ are smooth characters. We have to show that all irreducible subquotients of $\pi\{\delta_1, \delta_2\}$ lie in the same block. We adopt an argument used in [8]. It follows from [5, 5.3.3.1, 5.3.3.2, 5.3.4.1] that there exists an irreducible unitary L-Banach space representation Π of G, such that $\overline{\Pi}^{ss} \cong \pi\{\delta_1, \delta_2\}$, then [8, Prop.VII.4.5(i)] asserts that we may choose an open bounded G-invariant lattice Θ in Π such that $\Theta/\varpi\Theta$ is indecomposable. It follows from (6.1) that all the irreducible subquotients of $\Theta/\varpi\Theta$ lie in the same block.

We will list explicitly the irreducible subquotients of $\pi\{\delta_1, \delta_2\}$. It is shown in [1] that if $\delta_2\delta_1^{-1} \neq \omega$ then $(\mathrm{Ind}_B^G \delta_1 \otimes \delta_2\omega^{-1})_{sm}$ is absolutely irreducible, and there exists a non-split exact sequence

$$0 \to \delta_1 \circ \det \to (\mathrm{Ind}_B^G \delta_1 \otimes \delta_2\omega^{-1})_{sm} \to \mathrm{Sp}\otimes\delta_1 \circ \det \to 0 \qquad (6.9)$$

if $\delta_2\delta_1^{-1} = \omega$. Taking this into account there are the following possibilities for decomposing $\pi\{\delta_1, \delta_2\}$ into irreducible direct summands depending on δ_1, δ_2 and p:

(i) If $\delta_2\delta_1^{-1} \neq \omega^{\pm 1}, \mathbf{1}$ then

$$\pi\{\delta_1, \delta_2\} \cong (\mathrm{Ind}_B^G \delta_1 \otimes \delta_2\omega^{-1})_{sm} \oplus (\mathrm{Ind}_B^G \delta_2 \otimes \delta_1\omega^{-1})_{sm};$$

(ii) if $\delta_2 = \delta_1 (= \delta)$ then
  (a) if $p > 2$ then $\pi\{\delta, \delta\} \cong (\mathrm{Ind}_B^G \delta \otimes \delta\omega^{-1})_{sm}^{\oplus 2}$;
  (b) if $p = 2$ then $\pi\{\delta, \delta\} \cong (\mathrm{Sp}^{\oplus 2} \oplus \mathbf{1}^{\oplus 2}) \otimes \delta \circ \det$.

(iii) if $\delta_2\delta_1^{-1} = \omega^{\pm 1}$ then
  (a) if $p \geq 5$ then $\pi\{\delta_1, \delta_2\} \cong (\mathbf{1} \oplus \mathrm{Sp} \oplus (\mathrm{Ind}_B^G \omega \otimes \omega^{-1})_{sm}) \otimes \delta \circ \det$;
  (b) if $p = 3$ then $\pi\{\delta_1, \delta_2\} \cong (\mathbf{1} \oplus \mathrm{Sp} \oplus \omega \circ \det \oplus \mathrm{Sp} \otimes \omega \circ \det) \otimes \delta \circ \det$;
  (c) if $p = 2$ then we are in the case (ii)(b),
      where δ is either δ_1 or δ_2.

Finally, we note that in the case (ii)(b) instead of using [5, 5.3.3.2], which is stated without proof, we could have observed that since (6.9) is non-split, $\mathrm{Sp}\otimes\delta_1 \circ \det$ and $\delta_1 \circ \det$ lie in the same block. $\square$

Appendix A. Density of algebraic vectors

Let X be an affine scheme of finite type over $\mathbb{Z}_p$ and let $A = \Gamma(X, \mathcal{O}_X)$. By choosing an isomorphism $A \cong \mathbb{Z}_p[x_1, \dots, x_n]/(f_1, \dots, f_m)$ we may identify the $X(\mathbb{Z}_p)$ with a closed subset of $\mathbb{Z}_p^n$. The induced topology on $X(\mathbb{Z}_p)$ is independent of a choice of the isomorphism, see [9, Prop.2.1]. Let $\mathcal{C}(X(\mathbb{Z}_p), L)$ be the space of continuous functions from $X(\mathbb{Z}_p)$ to L. Since $X(\mathbb{Z}_p)$ is compact, $\mathcal{C}(X(\mathbb{Z}_p), L)$ equipped with the supremum norm is an L-Banach space. Recall that $X(\mathbb{Z}_p) = \mathrm{Hom}_{\mathbb{Z}_p-alg}(A, \mathbb{Z}_p)$. We denote by $C^{\mathrm{alg}}(X(\mathbb{Z}_p), L)$ the functions $f : X(\mathbb{Z}_p) \to L$, which are obtained by evaluating elements of $A \otimes_{\mathbb{Z}_p} L$ at $\mathbb{Z}_p$-valued points of X.

Lemma 6.A.15. $C^{\mathrm{alg}}(X(\mathbb{Z}_p), L)$ *is a dense subspace of* $C(X(\mathbb{Z}_p), L)$.

Proof. We first look at the special case, when $X = \mathbb{A}^n$, so that $A = \mathbb{Z}_p[x_1, \ldots, x_n]$ and $X(\mathbb{Z}_p) = \mathbb{Z}_p^n$. Since addition and multiplication in $\mathbb{Z}_p$ are continuous functions, we deduce that $C^{\mathrm{alg}}(\mathbb{A}(\mathbb{Z}_p), L)$ is a subspace of $C(\mathbb{A}(\mathbb{Z}_p), L)$. The density follows from the theory of Mahler expansions, see for example [19, III.1.2.4]. In the general case, we choose an isomorphism $A \cong \mathbb{Z}_p[x_1, \ldots, x_n]/(f_1, \ldots, f_m)$ and identify $X(\mathbb{Z}_p)$ with a closed subset of $\mathbb{A}^n(\mathbb{Z}_p) = \mathbb{Z}_p^n$. The restriction of functions to $X(\mathbb{Z}_p)$ induces a surjective map $r : C(\mathbb{A}^n(\mathbb{Z}_p), L) \to C(X(\mathbb{Z}_p), L)$, see for example [10, Thm.3.1(1)]. Since $C^{\mathrm{alg}}(\mathbb{A}^n(\mathbb{Z}_p), L)$ is dense in $C(\mathbb{A}^n(\mathbb{Z}_p), L)$ and $\sup_{x \in X(\mathbb{Z}_p)} |r(f)(x)| \le \sup_{x \in \mathbb{A}^n(\mathbb{Z}_p)} |f(x)|$ for all $f \in C(\mathbb{A}^n(\mathbb{Z}_p), L)$ we deduce that $r(C^{\mathrm{alg}}(\mathbb{A}^n(\mathbb{Z}_p), L))$ is a dense subspace. Since it is equal to $C^{\mathrm{alg}}(X(\mathbb{Z}_p), L)$ we are done. $\qquad\square$

Remark 6.A.16. If X is an affine scheme of finite type over $\mathcal{O}_F$, where $\mathcal{O}_F$ is a ring of integers in a finite field extension F over $\mathbb{Q}_p$, then there are two ways to topologize $X(\mathcal{O}_F)$: as $\mathcal{O}_F$ points of X and as $\mathbb{Z}_p$-points of the Weil restriction of X to $\mathbb{Z}_p$. However, they coincide, see [9, Ex.2.4].

Proposition 6.A.17. *Let G be an affine group scheme of finite type over $\mathbb{Z}_p$ such that G_L is a split connected reductive group over L. Then the evaluation map*

$$\bigoplus_{[V]} \mathrm{Hom}_{G(\mathbb{Z}_p)}(V, C(G(\mathbb{Z}_p), L)) \otimes V \to C(G(\mathbb{Z}_p), L), \qquad (6.A.10)$$

where the sum is taken over all the isomorphism classes of irreducible rational representations of G_L, is injective and the image is equal to $C^{\mathrm{alg}}(G(\mathbb{Z}_p), L)$. In particular, the image of (6.A.10) is a dense subspace of $C(G(\mathbb{Z}_p), L)$.

Proof. The category of rational representations of G_L is semi-simple as L is of characteristic 0, see [18, II.5.6 (6)]. Hence, the regular representation $\mathcal{O}(G_L)$ decomposes into a direct sum of irreducible representations. Since we have assumed that G_L is split, every irreducible rational representation V of G_L is absolutely irreducible [18, II.2.9]. This implies that $\mathrm{End}_{G_L}(V) = L$ for every irreducible representation V. It follows from Frobenius reciprocity [18, I.3.7 (3)] and the semi-simplicity of $\mathcal{O}(G_L)$ that we have an isomorphism of G_L-representations:

$$\mathcal{O}(G_L) \cong \bigoplus_{[V]} V^* \otimes V \qquad (6.A.11)$$

where the G_L-action on V^* is trivial. The isomorphism (6.A.11) is $G(L)$-equivariant, and hence $G(\mathbb{Z}_p)$-equivariant, which gives us an isomorphism of $G(\mathbb{Z}_p)$-representations:

$$\mathcal{C}^{\mathrm{alg}}(G(\mathbb{Z}_p), L) \cong \bigoplus_{[V]} V^* \otimes V. \qquad (6.A.12)$$

The map $\varphi \mapsto [v \mapsto \varphi(v)(1)]$ induces an isomorphism

$$\mathrm{Hom}_{G(\mathbb{Z}_p)}(V, \mathcal{C}(G(\mathbb{Z}_p), L)) \cong V^*, \qquad (6.A.13)$$

with the inverse map given by $\ell \mapsto [v \mapsto [g \mapsto \ell(gv)]]$. Since every V is a finite dimensional L-vector space, we conclude from (6.A.12), (6.A.13) that the injection

$$\mathrm{Hom}_{G(\mathbb{Z}_p)}(V, \mathcal{C}^{\mathrm{alg}}(G(\mathbb{Z}_p), L)) \hookrightarrow \mathrm{Hom}_{G(\mathbb{Z}_p)}(V, \mathcal{C}(G(\mathbb{Z}_p), L)) \quad (6.A.14)$$

is an isomorphism. Moreover, as a byproduct we obtain that $\mathrm{End}_{G(\mathbb{Z}_p)}(V) = L$ and $\mathrm{Hom}_{G(\mathbb{Z}_p)}(V, W) = 0$, if V and W are non-isomorphic irreducible representations of G_L. We conclude that the evaluation map is injective, and the image is equal to $\mathcal{C}^{\mathrm{alg}}(G(\mathbb{Z}_p), L)$. Lemma 6.A.15 implies the last assertion.
$\square$

References

[1] L. BARTHEL AND R. LIVNÉ, 'Irreducible modular representations of GL$_2$ of a local field', *Duke Math. J.* 75, (1994) 261–292.
[2] L. BERGER, 'Représentations modulaires de GL$_2(\mathbb{Q}_p)$ et représentations galoisiennes de dimension 2', *Astérisque* 330 (2010), 263–279.
[3] L. BERGER AND C. BREUIL, 'Sur quelques représentations potentiellement cristallines de GL$_2(\mathbb{Q}_p)$', *Astérisque* 330 (2010) 155–211.
[4] C. BREUIL, 'Sur quelques représentations modulaires et p-adiques de GL$_2(\mathbb{Q}_p)$. I', *Compositio* 138, (2003), 165–188.
[5] C. BREUIL, 'Sur quelques représentations modulaires et p-adiques de GL$_2(\mathbb{Q}_p)$. II', *J. Inst. Math. Jussieu* 2, (2003), 1–36.
[6] C. BREUIL AND M. EMERTON, 'Représentations p-adiques ordinaires de GL$_2(\mathbb{Q}_p)$ et compatibilité local-global', *Astérisque* 331 (2010), 255–315.
[7] C. BREUIL AND V. PAŠKŪNAS, 'Towards a modulo p Langlands correspondence for GL$_2$', *Memoirs of AMS*, 216, 2012.
[8] P. COLMEZ, 'Représentations de GL$_2(\mathbb{Q}_p)$ et (φ, Γ)-modules', *Astérisque* 330 (2010) 281–509.
[9] B. CONRAD, 'Weil and Grothendieck approaches to adelic points', *Enseign. Math. (2)* 58 (2012), no. 1–2, 61–97.
[10] R.L. ELLIS, 'Extending continuous functions on zero-dimensional spaces', *Math. Ann.*, 186, (1970), 114–122.

[11] M. EMERTON, 'p-adic L-functions and unitary completions of representations of p-adic reductive groups', *Duke Math. J.* 130 (2005), no. 2, 353–392.

[12] M. EMERTON, 'Locally analytic vectors in representations of locally p-adic analytic groups', to appear in *Memoirs of the AMS*.

[13] M. EMERTON, 'Local-global compatibility conjecture in the p-adic Langlands programme for $GL_2/\mathbb{Q}$', *Pure and Applied Math. Quarterly* 2 (2006), no. 2, 279–393.

[14] M. EMERTON, 'Local-global compatibility in the p-adic Langlands programme for $GL_2/\mathbb{Q}$', Preprint 2011, available at www.math.uchicago.edu/~emerton/preprints.html.

[15] M. EMERTON, 'Ordinary parts of admissible representations of p adic reductive groups I. Definition and first properties', *Astérisque* 331 (2010), 335–381.

[16] M. EMERTON, 'Ordinary parts of admissible representations of p-adic reductive groups II. Derived functors', *Astérisque* 331 (2010), 383–438.

[17] P. GABRIEL, 'Des catégories abéliennes', *Bull. Soc. Math. France* 90 (1962) 323–448.

[18] J.C. JANTZEN, *Representations of algebraic groups*, 2nd edn, Mathematical Surveys and Momographs, Vol. 107, AMS, 2003.

[19] M. LAZARD, 'Groupes analytiques p-adiques', *Publ. Math. IHES* 26 (1965).

[20] R. OLLIVIER, 'Le foncteur des invariants sous l'action du pro-p-Iwahori de GL(2, F)', *J. für die Reine und Angewandte Mathematik* 635 (2009) 149–185.

[21] V. PAŠKŪNAS, 'Coefficient systems and supersingular representations of $GL_2(F)$', *Mémoires de la SMF*, 99, (2004).

[22] V. PAŠKŪNAS 'On some crystalline representations of $GL_2(\mathbb{Q}_p)$', *Algebra Number Theory* 3 (2009), no. 4, 411–421.

[23] V. PAŠKŪNAS, 'Extensions for supersingular representations of $GL_2(\mathbb{Q}_p)$', *Astérisque* 331 (2010) 317–353.

[24] V. PAŠKŪNAS, 'Admissible unitary completions of locally $\mathbb{Q}_p$-rational representations of $GL_2(F)$', *Represent. Theory* 14 (2010), 324–354.

[25] V. PAŠKŪNAS, 'The image of Colmez's Montreal functor', to appear in *Publ. Math. IHES*. DOI: 10.1007/s10240-013-0049-y.

[26] P. SCHNEIDER AND J. TEITELBAUM, 'Banach space representations and Iwasawa theory', *Israel J. Math.* 127, (2002) 359–380.

[27] M.-F. VIGNÉRAS, 'Representations modulo p of the p-adic group $GL(2, F)$', *Compositio Math.* 140 (2004) 333–358.

7

From étale P_+-representations to G-equivariant sheaves on G/P

Peter Schneider, Marie-France Vigneras, and Gergely Zabradi[*]

Abstract

Let $K/\mathbb{Q}_p$ be a finite extension with ring of integers o, let G be a connected reductive split $\mathbb{Q}_p$-group of Borel subgroup $P = TN$ and let α be a simple root of T in N. We associate to a finitely generated module D over the Fontaine ring over o endowed with a semilinear étale action of the monoid T_+ (acting on the Fontaine ring via α), a $G(\mathbb{Q}_p)$-equivariant sheaf of o-modules on the compact space $G(\mathbb{Q}_p)/P(\mathbb{Q}_p)$. Our construction generalizes the representation $D \boxtimes \mathbb{P}^1$ of $GL(2, \mathbb{Q}_p)$ associated by Colmez to a (φ, Γ)-module D endowed with a character of $\mathbb{Q}_p^*$.

Contents

[*] The third author was partially supported by OTKA Research grant no. K-101291.

Automorphic Forms and Galois Representations, ed. Fred Diamond, Payman L. Kassaei and Minhyong Kim. Published by Cambridge University Press. © Cambridge University Press 2014.

1. Introduction

1.1. Notation

We fix a finite extension $K/\mathbb{Q}_p$ of ring of integers o and an algebraic closure $\overline{\mathbb{Q}}_p$ of K. We denote by $\mathcal{G}_p = \mathrm{Gal}(\overline{\mathbb{Q}}_p/\mathbb{Q}_p)$ the absolute Galois group of $\mathbb{Q}_p$, by $\Lambda(\mathbb{Z}_p) = o[[\mathbb{Z}_p]]$ the Iwasawa o-algebra of maximal ideal $\mathcal{M}(\mathbb{Z}_p)$, and by $\mathcal{O}_{\mathcal{E}}$ the Fontaine ring which is the p-adic completion of the localization of $\Lambda(\mathbb{Z}_p)$ with respect to the elements not in $p\Lambda(\mathbb{Z}_p)$. We put on $\mathcal{O}_{\mathcal{E}}$ the weak topology inducing the $\mathcal{M}(\mathbb{Z}_p)$-adic topology on $\Lambda(\mathbb{Z}_p)$, a fundamental system of neighbourhoods of 0 being $(p^n\mathcal{O}_{\mathcal{E}} + \mathcal{M}(\mathbb{Z}_p)^n)_{n\in\mathbb{N}}$. The action of $\mathbb{Z}_p - \{0\}$ by multiplication on $\mathbb{Z}_p$ extends to an action on $\mathcal{O}_{\mathcal{E}}$.

We fix an arbitrary split reductive connected $\mathbb{Q}_p$-group G and a Borel $\mathbb{Q}_p$-subgroup $P = TN$ with maximal $\mathbb{Q}_p$-subtorus T and unipotent radical N. We denote by w_0 the longest element of the Weyl group of T in G, by Φ_+ the set of roots of T in N, and by $u_\alpha : \mathbb{G}_a \to N_\alpha$, for $\alpha \in \Phi_+$, a $\mathbb{Q}_p$-homomorphism onto the root subgroup N_α of N such that $tu_\alpha(x)t^{-1} = u_\alpha(\alpha(t)x)$ for $x \in \mathbb{Q}_p$ and $t \in T(\mathbb{Q}_p)$, and $N_0 = \prod_{\alpha\in\Phi_+} u_\alpha(\mathbb{Z}_p)$ is a subgroup of $N(\mathbb{Q}_p)$. We denote by T_+ the monoid of dominant elements t in $T(\mathbb{Q}_p)$ such that $\mathrm{val}_p(\alpha(t)) \geq 0$ for all $\alpha \in \Phi_+$, by $T_0 \subset T_+$ the maximal subgroup, by T_{++} the subset of strictly dominant elements, i.e. $\mathrm{val}_p(\alpha(t)) > 0$ for all $\alpha \in \Phi_+$, and we put $P_+ = N_0T_+$, $P_0 = N_0T_0$. The natural action of T_+ on N_0 extends to an action on the Iwasawa o-algebra $\Lambda(N_0) = o[[N_0]]$. The compact set $G(\mathbb{Q}_p)/P(\mathbb{Q}_p)$ contains the open dense subset $\mathcal{C} = N(\mathbb{Q}_p)w_0P(\mathbb{Q}_p)/P(\mathbb{Q}_p)$ homeomorphic to $N(\mathbb{Q}_p)$ and the compact subset $\mathcal{C}_0 = N_0w_0P(\mathbb{Q}_p)/P(\mathbb{Q}_p)$ homeomorphic to N_0. We put $\overline{P}(\mathbb{Q}_p) = w_0P(\mathbb{Q}_p)w_0^{-1}$.

Each simple root α gives a $\mathbb{Q}_p$-homomorphism $x_\alpha : N \to \mathbb{G}_a$ with section u_α. We denote by $\ell_\alpha : N_0 \to \mathbb{Z}_p$, resp. $\iota_\alpha : \mathbb{Z}_p \to N_0$, the restriction of x_α, resp. u_α, to N_0, resp. $\mathbb{Z}_p$.

For example, $G = GL(n)$, P is the subgroup of upper triangular matrices, N consists of the strictly upper triangular matrices (1 on the diagonal), T is the diagonal subgroup, $N_0 = N(\mathbb{Z}_p)$, the simple roots are $\alpha_1, \ldots, \alpha_{n-1}$ where $\alpha_i(\mathrm{diag}(t_1, \ldots, t_n)) = t_it_{i+1}^{-1}$, x_{α_i} sends a matrix to its $(i, i+1)$-coefficient, $u_{\alpha_i}(.)$ is the strictly upper triangular matrix, with $(i, i+1)$-coefficient . and 0 everywhere else.

We denote by $C^\infty(X, o)$ the o-module of locally constant functions on a locally profinite space X with values in o, and by $C_c^\infty(X, o)$ the subspace of compactly supported functions.

1.2. General overview

Colmez established a correspondence $V \mapsto \Pi(V)$ from the absolutely irreducible K-representations V of dimension 2 of the Galois group $\mathcal{G}_p$ to the unitary admissible absolutely irreducible K-representations Π of $GL(2, \mathbb{Q}_p)$ admitting a central character [6]. This correspondence relies on the construction of a representation $D(V) \boxtimes \mathbb{P}^1$ of $GL(2, \mathbb{Q}_p)$ for any representation V (not necessarily of dimension 2) of $\mathcal{G}_p$ and any unitary character $\delta : \mathbb{Q}_p^* \to o^*$. When the dimension of V is 2 and when $\delta = (x|x|)^{-1}\delta_V$, where δ_V is the character of $\mathbb{Q}_p^*$ corresponding to the representation det V by local class field theory, then $D(V) \boxtimes \mathbb{P}^1$ is an extension of $\Pi(V)$ by its dual twisted by $\delta \circ \det$. It is a general belief that the correspondence $V \to \Pi(V)$ should extend to a correspondence from representations V of dimension d to representations Π of $GL(d, \mathbb{Q}_p)$.

We generalize here Colmez's construction of the representation $D \boxtimes \mathbb{P}^1$ of $GL(2, \mathbb{Q}_p)$, replacing $GL(2)$ by the arbitrary split reductive connected $\mathbb{Q}_p$-group G. More precisely, we denote by $\mathcal{O}_{\mathcal{E},\alpha}$ the ring $\mathcal{O}_{\mathcal{E}}$ with the action of T_+ via a simple root $\alpha \in \Delta$ (if the rank of G is 1, α is unique and we omit α). For any finitely generated $\mathcal{O}_{\mathcal{E},\alpha}$-module D with an étale semilinear action of T_+, we construct a representation of $G(\mathbb{Q}_p)$. It is realized as the space of global sections of a $G(\mathbb{Q}_p)$-equivariant sheaf on the compact quotient $G(\mathbb{Q}_p)/P(\mathbb{Q}_p)$. When the rank of G is 1, the compact space $G(\mathbb{Q}_p)/P(\mathbb{Q}_p)$ is isomorphic to $\mathbb{P}^1(\mathbb{Q}_p)$ and when $G = GL(2)$ we recover Colmez's sheaf.

We review briefly the main steps of our construction.

1. We show that the category of étale T_+-modules finitely generated over $\mathcal{O}_{\mathcal{E},\alpha}$ is equivalent to the category of étale T_+-modules finitely generated over $\Lambda_{\ell_\alpha}(N_0)$, for a topological ring $\Lambda_{\ell_\alpha}(N_0)$ generalizing the Fontaine ring $\mathcal{O}_{\mathcal{E}}$, which is better adapted to the group G, and depends on the simple root α.

2. We show that the sections over $\mathcal{C}_0 \simeq N_0$ of a $P(\mathbb{Q}_p)$-equivariant sheaf $\mathcal{S}$ of o-modules over $\mathcal{C} \simeq N$ is an étale $o[P_+]$-module $\mathcal{S}(\mathcal{C}_0)$ and that the functor $\mathcal{S} \mapsto \mathcal{S}(\mathcal{C}_0)$ is an equivalence of categories.

3. When $\mathcal{S}(\mathcal{C}_0)$ is an étale T_+-module finitely generated over $\Lambda_{\ell_\alpha}(N_0)$, and the root system of G is irreducible, we show that the $P(\mathbb{Q}_p)$-equivariant sheaf $\mathcal{S}$ on $\mathcal{C}$ extends to a $G(\mathbb{Q}_p)$-equivariant sheaf over $G(\mathbb{Q}_p)/P(\mathbb{Q}_p)$ if and only if the rank of G is 1.

4. For any strictly dominant element $s \in T_{++}$, we associate functorially to an étale T_+-module M finitely generated over $\Lambda_{\ell_\alpha}(N_0)$, a $G(\mathbb{Q}_p)$-equivariant sheaf $\mathfrak{Y}_s$ of o-modules over $G(\mathbb{Q}_p)/P(\mathbb{Q}_p)$ with sections over $\mathcal{C}_0$ a dense étale $\Lambda(N_0)[T_+]$-submodule M_s^{bd} of M. When the rank of G is 1, the sheaf

$\mathfrak{Y}_s$ does not depend on the choice of $s \in T_{++}$, and $M_s^{bd} = M$; when $G = GL(2)$ we recover the construction of Colmez. For a general G, the sheaf $\mathfrak{Y}_s$ depends on the choice of $s \in T_{++}$, the system $(\mathfrak{Y}_s)_{s \in T_{++}}$ of sheaves is compatible, and we associate functorially to M the $G(\mathbb{Q}_p)$-equivariant sheaves $\mathfrak{Y}_\cup$ and $\mathfrak{Y}_\cap$ of o-modules over $G(\mathbb{Q}_p)/P(\mathbb{Q}_p)$ with sections over C_0 equal to $\cup_{s \in T_{++}} M_s^{bd}$ and $\cap_{s \in T_{++}} M_s^{bd}$, respectively.

1.3. The rings $\Lambda_{\ell_\alpha}(N_0)$ and $\mathcal{O}_{\mathcal{E},\alpha}$

Fixing a simple root $\alpha \in \Delta$, the topological local ring $\Lambda_{\ell_\alpha}(N_0)$, generalizing the Fontaine ring $\mathcal{O}_{\mathcal{E}}$, is defined as in [11] with the surjective homomorphism $\ell_\alpha : N_0 \to \mathbb{Z}_p$.

We denote by $\mathcal{M}(N_{\ell_\alpha})$ the maximal ideal of the Iwasawa o-algebra $\Lambda(N_{\ell_\alpha}) = o[[N_{\ell_\alpha}]]$ of the kernel N_{ℓ_α} of ℓ_α. The ring $\Lambda_{\ell_\alpha}(N_0)$ is the $\mathcal{M}(N_{\ell_\alpha})$-adic completion of the localization of $\Lambda(N_0)$ with respect to the Ore subset of elements which are not in $\mathcal{M}(N_{\ell_\alpha})\Lambda(N_0)$. This is a noetherian local ring with maximal ideal $\mathcal{M}_{\ell_\alpha}(N_0)$ generated by $\mathcal{M}(N_{\ell_\alpha})$. We put on $\Lambda_{\ell_\alpha}(N_0)$ the weak topology with fundamental system of neighbourhoods of 0 equal to $(\mathcal{M}_{\ell_\alpha}(N_0)^n + \mathcal{M}(N_0)^n)_{n \in \mathbb{N}}$. The action of T_+ on N_0 extends to an action on $\Lambda_{\ell_\alpha}(N_0)$. We denote by $\mathcal{O}_{\mathcal{E},\alpha}$ the ring $\mathcal{O}_{\mathcal{E}}$ with the action of T_+ induced by $(t, x) \mapsto \alpha(t)x : T_+ \times \mathbb{Z}_p \to \mathbb{Z}_p$. The homomorphism ℓ_α and its section ι_α induce T_+-equivariant ring homomorphisms

$$\ell_\alpha : \Lambda_{\ell_\alpha}(N_0) \to \mathcal{O}_{\mathcal{E},\alpha}, \quad \iota_\alpha : \mathcal{O}_{\mathcal{E},\alpha} \to \Lambda_{\ell_\alpha}(N_0), \text{ such that } \ell_\alpha \circ \iota_\alpha = \text{id}.$$

1.4. Equivalence of categories

An étale T_+-module over $\Lambda_{\ell_\alpha}(N_0)$ is a finitely generated $\Lambda_{\ell_\alpha}(N_0)$-module M with a semi-linear action $T_+ \times M \to M$ of T_+ which is étale, i.e. the action φ_t on M of each $t \in T_+$ is injective and

$$M = \oplus_{u \in J(N_0/tN_0t^{-1})} u\varphi_t(M),$$

if $J(N_0/tN_0t^{-1}) \subset N_0$ is a system of representatives of the cosets N_0/tN_0t^{-1}; in particular, the action of each element of the maximal subgroup T_0 of T_+ is invertible. We denote by ψ_t the left inverse of φ_t vanishing on $u\varphi_t(M)$ for $u \in N_0$ not in tN_0t^{-1}. These modules form an abelian category $\mathcal{M}_{\Lambda_{\ell_\alpha}(N_0)}^{et}(T_+)$.

We define analogously the abelian category $\mathcal{M}_{\mathcal{O}_{\mathcal{E},\alpha}}^{et}(T_+)$ of finitely generated $\mathcal{O}_{\mathcal{E},\alpha}$-modules with an étale semilinear action of T_+. The action φ_t of each element $t \in T_+$ such that $\alpha(t) \in \mathbb{Z}_p^*$ is invertible. We show that the action $T_+ \times D \to D$ of T_+ on $D \in \mathcal{M}_{\mathcal{O}_{\mathcal{E},\alpha}}^{et}(T_+)$ is continuous for the weak topology on D; the canonical action of the inverse T_- of T is also continuous.

Theorem 7.1. *The base change functors* $\mathcal{O}_{\mathcal{E}} \otimes_{\ell_\alpha} -$ *and* $\Lambda_{\ell_\alpha}(N_0) \otimes_{\iota_\alpha} -$ *induce quasi-inverse isomorphisms*

$$\mathbb{D} : \mathcal{M}^{et}_{\Lambda_{\ell_\alpha}(N_0)}(T_+) \rightarrow \mathcal{M}^{et}_{\mathcal{O}_{\mathcal{E},\alpha}}(T_+), \quad \mathbb{M} : \mathcal{M}^{et}_{\mathcal{O}_{\mathcal{E},\alpha}}(T_+) \rightarrow \mathcal{M}^{et}_{\Lambda_{\ell_\alpha}(N_0)}(T_+).$$

Using this theorem, we show that the action of T_+ and of the inverse monoid T_- (given by the operators ψ) on an étale T_+-module over $\Lambda_{\ell_\alpha}(N_0)$ is continuous for the weak topology.

1.5. *P*-equivariant sheaves on $\mathcal{C}$

The o-algebra $C^\infty(N_0, o)$ is naturally an étale $o[P_+]$−module, and the monoid P_+ acts on the o-algebra $\text{End}_o M$ by $(b, F) \mapsto \varphi_b \circ F \circ \psi_b$. We show that there exists a unique $o[P_+]$-linear map

$$\text{res} : C^\infty(N_0, o) \rightarrow \text{End}_o M$$

sending the characteristic function 1_{N_0} of N_0 to the identity id_M; moreover res is an algebra homomorphism which sends $1_{b.N_0}$ to $\varphi_b \circ \psi_b$ for all $b \in P_+$ acting on $x \in N_0$ by $(b, x) \mapsto b.x$.

For the sake of simplicity, we denote now by the same letter a group defined over $\mathbb{Q}_p$ and the group of its $\mathbb{Q}_p$-rational points.

Let M^P be the $o[P]$-module induced by the canonical action of the inverse monoid P_- of P_+ on M; as a representation of N, it is isomorphic to the representation induced by the action of N_0 on M. The value at 1, denoted by $\text{ev}_0 : M^P \rightarrow M$, is P_--equivariant, and admits a P_+-equivariant splitting $\sigma_0 : M \rightarrow M^P$ sending $m \in M$ to the function equal to $n \mapsto nm$ on N_0 and vanishing on $N - N_0$. The $o[P]$-submodule M_c^P of M^P generated by $\sigma_0(M)$ is naturally isomorphic to $A[P] \otimes_{A[P_+]} M$. When $M = C^\infty(N_0, o)$ then $M_c^P = C_c^\infty(N, o)$ and $M^P = C^\infty(N, o)$ with the natural $o[P]$-module structure. We have the natural o-algebra embedding

$$F \mapsto \sigma_0 \circ F \circ \text{ev}_0 : \text{End}_o M \rightarrow \text{End}_o M^P,$$

sending id_M to the idempotent $R_0 = \sigma_0 \circ \text{ev}_0$ in $\text{End}_o M^P$.

Proposition 7.2. *There exists a unique $o[P]$-linear map*

$$\text{Res} : C_c^\infty(N, o) \rightarrow \text{End}_o M^P$$

sending 1_{N_0} to R_0; moreover Res *is an algebra homomorphism.*

The topology of N is totally disconnected and by a general argument, the functor of compact global sections is an equivalence of categories from the

P-equivariant sheaves on $N \simeq C$ to the nondegenerate modules on the skew group ring

$$C_c^\infty(N, o)\#P = \oplus_{b \in P} b C_c^\infty(N, o)$$

in which the multiplication is determined by the rule $(b_1 f_1)(b_2 f_2) = b_1 b_2 f_1^{b_2} f_2$ for $b_i \in P$, $f_i \in C_c^\infty(N, o)$ and $f_1^{b_2}(.) = f_1(b_2.)$.

Theorem 7.3. *The functor of sections over $N_0 \simeq C_0$ from the P-equivariant sheaves on $N \simeq C$ to the étale $o[P_+]$-modules is an equivalence of categories.*

The space of global sections of a P-equivariant sheaf S on C is $S(C) = S(C_0)^P$.

1.6. Generalities on G-equivariant sheaves on G/P

The functor of global sections from the G-equivariant sheaves on G/P to the modules on the skew group ring $\mathcal{A}_{G/P} = C^\infty(G/P, o)\#G$ is an equivalence of categories. We have the intermediate ring $\mathcal{A}$

$$\mathcal{A}_C = C_c^\infty(C, o)\#P \subset \mathcal{A} = \oplus_{g \in G} g C_c^\infty(g^{-1}C \cap C, o) \subset \mathcal{A}_{G/P},$$

and the o-module

$$\mathcal{Z} = \oplus_{g \in G} g C_c^\infty(C, o)$$

which is a left ideal of $\mathcal{A}_{G/P}$ and a right $\mathcal{A}$-submodule.

Proposition 7.4. *The functor*

$$Z \mapsto Y(Z) = \mathcal{Z} \otimes_{\mathcal{A}} Z$$

from the nondegenerate $\mathcal{A}$-modules to the $\mathcal{A}_{G/P}$-modules is an equivalence of categories; moreover the G-sheaf on G/P corresponding to $Y(Z)$ extends the P-equivariant sheaf on C corresponding to $Z|_{\mathcal{A}_C}$.

Given an étale $o[P_+]$-module M, we consider the problem of extending to $\mathcal{A}$ the o-algebra homomorphism

$$\mathrm{Res} : \mathcal{A}_C \to \mathrm{End}_o(M_c^P), \quad \sum_{b \in P} b f_b \mapsto b \circ \mathrm{Res}(f_b).$$

We introduce the subrings

$$\mathcal{A}_0 = 1_{C_0} \mathcal{A} 1_{C_0} = \oplus_{g \in G} g C^\infty(g^{-1}C_0 \cap C_0, o) \subset \mathcal{A},$$
$$\mathcal{A}_{C0} = 1_{C_0} \mathcal{A}_C 1_{C_0} = \oplus_{b \in P} b C^\infty(b^{-1}C_0 \cap C_0, o) \subset \mathcal{A}_C.$$

The skew monoid ring $\mathcal{A}_{\mathcal{C}_0} = C^\infty(\mathcal{C}_0, o)\#P_+ = \oplus_{b\in P_+} bC^\infty(\mathcal{C}_0, o)$ is contained in $\mathcal{A}_{\mathcal{C}0}$. The intersection $g^{-1}\mathcal{C}_0 \cap \mathcal{C}_0$ is not 0 if and only if $g \in N_0\overline{P}N_0$. The subring $\mathrm{Res}(\mathcal{A}_{\mathcal{C}0})$ of $\mathrm{End}_o(M^P)$ necessarily lies in the image of $\mathrm{End}_o(M)$.

The group P acts on $\mathcal{A}$ by $(b, y) \mapsto (b1_{G/P})y(b1_{G/P})^{-1}$ for $b \in P$, and the map $b \otimes y \mapsto (b1_{G/P})y(b1_{G/P})^{-1}$ gives $o[P]$ isomorphisms

$$o[P] \otimes_{o[P_+]} \mathcal{A}_0 \to \mathcal{A} \quad \text{and} \quad o[P] \otimes_{o[P_+]} \mathcal{A}_{\mathcal{C}0} \to \mathcal{A}_\mathcal{C}.$$

Proposition 7.5. *Let M be an étale $o[P_+]$-module. We suppose given, for any $g \in N_0\overline{P}N_0$, an element $\mathcal{H}_g \in \mathrm{End}_o(M)$. The map*

$$\mathcal{R}_0 : \mathcal{A}_0 \to \mathrm{End}_o(M), \quad \sum_{g\in N_0\overline{P}N_0} gf_g \mapsto \sum_{g\in N_0\overline{P}N_0} \mathcal{H}_g \circ \mathrm{res}(f_g)$$

is a P_+-equivariant o-algebra homomorphism which extends $\mathrm{Res}\,|_{\mathcal{A}_{\mathcal{C}0}}$ if and only if, for all $g, h \in N_0\overline{P}N_0$, $b \in P \cap N_0\overline{P}N_0$, and all compact open subsets $\mathcal{V} \subset \mathcal{C}_0$, the relations

H1. $\mathrm{res}(1_\mathcal{V}) \circ \mathcal{H}_g = \mathcal{H}_g \circ \mathrm{res}(1_{g^{-1}\mathcal{V}\cap\mathcal{C}_0})$,
H2. $\mathcal{H}_g \circ \mathcal{H}_h = \mathcal{H}_{gh} \circ \mathrm{res}(1_{h^{-1}\mathcal{C}_0\cap\mathcal{C}_0})$,
H3. $\mathcal{H}_b = b \circ \mathrm{res}(1_{b^{-1}\mathcal{C}_0\cap\mathcal{C}_0})$

hold true. In this case, the unique $o[P]$-equivariant map $\mathcal{R} : \mathcal{A} \to \mathrm{End}_\mathcal{A}(M_c^P)$ extending $\mathcal{R}_0$ is multiplicative.

When these conditions are satisfied, we obtain a G-equivariant sheaf on G/P with sections on $\mathcal{C}_0$ equal to M.

1.7. $(s, \mathrm{res}, \mathfrak{C})$-integrals $\mathcal{H}_g$

Let M be an étale T_+-module M over $\Lambda_{\ell_\alpha}(N_0)$ with the weak topology. We denote by $\mathrm{End}_o^{cont}(M)$ the o-module of continuous o-linear endomorphisms of M, and for g in $N_0\overline{P}N_0$, by $U_g \subseteq N_0$ the compact open subset such that

$$U_g w_0 P/P = g^{-1}\mathcal{C}_0 \cap \mathcal{C}_0.$$

For $u \in U_g$, we have a unique element $\alpha(g, u) \in N_0 T$ such that $g u w_0 N = \alpha(g, u)u w_0 N$. We consider the map

$\alpha_{g,0} : N_0 \to \mathrm{End}_o^{cont}(M)$

$\alpha_{g,0}(u) = \mathrm{Res}(1_{\mathcal{C}_0}) \circ \alpha(g, u) \circ \mathrm{Res}(1_{\mathcal{C}_0})$ for $u \in U_g$ and $\alpha_{g,0}(u)$

$\qquad = 0$ otherwise.

The module M is Hausdorff complete but not compact; also we introduce a notion of integrability with respect to a special family $\mathfrak{C}$ of compact subsets $C \subset M$, i.e. satisfying:

$\mathfrak{C}(1)$ Any compact subset of a compact set in $\mathfrak{C}$ also lies in $\mathfrak{C}$.

$\mathfrak{C}(2)$ If $C_1, C_2, \ldots, C_n \in \mathfrak{C}$ then $\bigcup_{i=1}^{n} C_i$ is in $\mathfrak{C}$, as well.

$\mathfrak{C}(3)$ For all $C \in \mathfrak{C}$ we have $N_0 C \in \mathfrak{C}$.

$\mathfrak{C}(4)$ $M(\mathfrak{C}) := \bigcup_{C \in \mathfrak{C}} C$ is an étale $o[P_+]$-submodule of M.

A map from $M(\mathfrak{C})$ to M is called $\mathfrak{C}$-continuous if its restriction to any $C \in \mathfrak{C}$ is continuous. The o-module $\operatorname{Hom}_o^{\mathfrak{C}ont}(M(\mathfrak{C}), M)$ of $\mathfrak{C}$-continuous o-linear homomorphisms from $M(\mathfrak{C})$ to M with the $\mathfrak{C}$-open topology, is a topological complete o-module.

For $s \in T_{++}$, the open compact subgroups $N_k = s^k N_0 s^{-k} \subset N$ for $k \in \mathbb{Z}$, form a decreasing sequence of union N and intersection $\{1\}$. A map $F \colon N_0 \to \operatorname{Hom}_A^{\mathfrak{C}ont}(M(\mathfrak{C}), M)$ is called $(s, \operatorname{res}, \mathfrak{C})$-integrable if the limit

$$\int_{N_0} F \, d\operatorname{res} := \lim_{k \to \infty} \sum_{u \in J(N_0/N_k)} F(u) \circ \operatorname{res}(1_{uN_k}),$$

where $J(N_0/N_k) \subseteq N_0$, for any $k \in \mathbb{N}$, is a set of representatives for the cosets in N_0/N_k, exists in $\operatorname{Hom}_A^{\mathfrak{C}ont}(M(\mathfrak{C}), M)$ and is independent of the choice of the sets $J(N_0/N_k)$. We denote by $\mathcal{H}_{g, J(N_0/N_k)}$ the sum in the right-hand side when $F = \alpha_{g,0}(.)|_{M(\mathfrak{C})}$.

Proposition 7.6. *For all* $g \in N_0 \overline{P} N_0$, *the map* $\alpha_{g,0}(.)|_{M(\mathfrak{C})} \colon N_0 \to \operatorname{Hom}_A^{\mathfrak{C}ont}(M(\mathfrak{C}), M)$ *is* $(s, \operatorname{res}, \mathfrak{C})$-*integrable when*

$\mathfrak{C}(5)$ *For any* $C \in \mathfrak{C}$ *the compact subset* $\psi_s(C) \subseteq M$ *also lies in* $\mathfrak{C}$.

$\mathfrak{T}(1)$ *For any* $C \in \mathfrak{C}$ *such that* $C = N_0 C$, *any open* $A[N_0]$-*submodule* $\mathcal{M}$ *of* M, *and any compact subset* $C_+ \subseteq L_+$ *there exists a compact open subgroup* $P_1 = P_1(C, \mathcal{M}, C_+) \subseteq P_0$ *and an integer* $k(C, \mathcal{M}, C_+) \geq 0$ *such that*

$$s^k (1 - P_1) C_+ \psi_s^k \subseteq E(C, \mathcal{M}) \qquad \text{for any } k \geq k(C, \mathcal{M}, C_+).$$

The integrals $\mathcal{H}_g$ *of* $\alpha_{g,0}(.)|_{M(\mathfrak{C})}$ *satisfy the relations H1, H2, H3, when they belong to* $\operatorname{End}_A(M(\mathfrak{C}))$, *and when*

$\mathfrak{C}(6)$ *For any* $C \in \mathfrak{C}$ *the compact subset* $\varphi_s(C) \subseteq M$ *also lies in* $\mathfrak{C}$.

$\mathfrak{T}(2)$ *Given a set* $J(N_0/N_k) \subset N_0$ *of representatives for cosets in* N_0/N_k, *for* $k \geq 1$, *for any* $x \in M(\mathfrak{C})$ *and* $g \in N_0 \overline{P} N_0$ *there exists a compact* A-*submodule* $C_{x,g} \in \mathfrak{C}$ *and a positive integer* $k_{x,g}$ *such that* $\mathcal{H}_{g, J(N_0/N_k)}(x) \subseteq C_{x,g}$ *for any* $k \geq k_{x,g}$.

When $\mathfrak{C}$ satisfies $\mathfrak{C}(1), \ldots, \mathfrak{C}(6)$ and the technical properties $\mathfrak{T}(1), \mathfrak{T}(2)$ are true, we obtain a G-equivariant sheaf on G/P with sections on C_0 equal to $M(\mathfrak{C})$.

1.8. Main theorem

Let M be an étale T_+-module M over $\Lambda_{\ell_\alpha}(N_0)$ with the weak topology and let $s \in T_{++}$. We have the natural T_+-equivariant quotient map

$$\ell_M : M \to D = \mathcal{O}_{\mathcal{E},\alpha} \otimes_{\ell_\alpha} M \quad , \quad m \mapsto 1 \otimes m$$

from M to $D = \mathbb{D}(M) \in \mathcal{M}_{\mathcal{O}_{\mathcal{E},\alpha}}(T_+)$, of T_+-equivariant section

$$\iota_D : D \to M = \Lambda_{\ell_\alpha}(N_0) \otimes_{\iota_\alpha} D, \quad d \mapsto 1 \otimes d.$$

We note that $o[N_0]\iota_D(D)$ is dense in M. A lattice D_0 in D is a $\Lambda(\mathbb{Z}_p)$-submodule generated by a finite set of generators of D over $\mathcal{O}_{\mathcal{E}}$. When D is killed by a power of p, the o-module

$$M_s^{bd}(D_0) := \{m \in M \mid \ell_M(\psi_s^k(u^{-1}m)) \in D_0 \text{ for all } u \in N_0 \text{ and } k \in \mathbb{N}\}$$

of M is compact and is a $\Lambda(N_0)$-module. Let $\mathfrak{C}_s$ be the family of compact subsets of M contained in $M_s^{bd}(D_0)$ for some lattice D_0 of D, and let $M_s^{bd} = \cup_{D_0} M_s^{bd}(D_0)$ the union being taken over all lattices D_0 in D. In general, M is p-adically complete, $M/p^n M$ is an étale T_+-module over $\Lambda_{\ell_\alpha}(N_0)$, and $D/p^n D = \mathbb{D}(M/p^n M)$. We denote by $p_n : M \to M/p^n M$ the reduction modulo p^n, and by $\mathfrak{C}_{s,n}$ the family of compact subsets constructed above for $M/p^n M$. We define the family $\mathfrak{C}_s$ of compact subsets $C \subset M$ such that $p_n(C) \in \mathfrak{C}_{s,n}$ for all $n \geq 1$, and the o-module M_s^{bd} of $m \in M$ such that the set of $\ell_M(\psi_s^k(u^{-1}m))$ for $k \in \mathbb{N}$, $u \in N_0$ is bounded in D for the weak topology.

By reduction to the easier case where M is killed by a power of p, we show that $\mathfrak{C}_s$ satisfies $\mathfrak{C}(1), \ldots, \mathfrak{C}(6)$ and that the technical properties $\mathfrak{T}(1), \mathfrak{T}(2)$ are true.

Proposition 7.7. *Let M be an étale T_+-module M over $\Lambda_{\ell_\alpha}(N_0)$ and let $s \in T_{++}$.*

(i) *M_s^{bd} is a dense $\Lambda(N_0)[T_+]$-étale submodule of M containing $\iota_D(D)$.*

(ii) *For $g \in N_0\overline{P}N_0$, the $(s, \mathrm{res}, \mathfrak{C}_s)$-integrals $\mathcal{H}_{g,s}$ of $\alpha_{g,0}|_{M_s^{bd}}$ exist, lie in $\mathrm{End}_o(M_s^{bd})$, and satisfy the relations H1, H2, H3.*

(iii) *For $s_1, s_2 \in T_{++}$, there exists $s_3 \in T_{++}$ such that $M_{s_3}^{bd}$ contains $M_{s_1}^{bd} \cup M_{s_2}^{bd}$ and $\mathcal{H}_{g,s_1} = \mathcal{H}_{g,s_2}$ on $M_{s_1}^{bd} \cap M_{s_2}^{bd}$.*

The endomorphisms $\mathcal{H}_{g,s} \in \operatorname{End}_o(M_s^{bd})$ induce endomorphisms of $\cap_{s \in T_{++}} M_s^{bd}$ and of $\cup_{s \in T_{++}} M_s^{bd} = \sum_{s \in T_{++}} M_s^{bd}$ satisfying the relations H1, H2, H3. Moreover $\cup_{s \in T_{++}} M_s^{bd}$ and $\cap_{s \in T_{++}} M_s^{bd}$ are $\Lambda(N_0)[T_+]$-étale submodules of M containing $\iota_D(D)$. Our main theorem is the following:

Theorem 7.8. *There are faithful functors*

$$\mathbb{Y}_\cap, \ (\mathbb{Y}_s)_{s \in T_{++}}, \ \mathbb{Y}_\cup : \mathcal{M}_{\mathcal{O}_{\mathcal{E},\alpha}}^{et} (T_+) \longrightarrow \quad G\text{-equivariant sheaves on } G/P,$$

sending $D = \mathbb{D}(M)$ *to a sheaf with sections on* $\mathcal{C}_0$ *equal to the dense* $\Lambda(N_0)[T_+]$-*submodules of* M

$$\bigcap_{s \in T_{++}} M_s^{bd}, \quad (M_s^{bd})_{s \in T_{++}}, \ and \quad \bigcup_{s \in T_{++}} M_s^{bd},$$

respectively.

When $G = GL(2, \mathbb{Q}_p)$, the sheaves $\mathbb{Y}_s(D)$ are all equal to the G-equivariant sheaf on $G/P \simeq \mathbb{P}^1(\mathbb{Q}_p)$ of global sections $D \boxtimes \mathbb{P}^1$ constructed by Colmez. When the root system of G is irreducible of rank > 1, we check that $\cup_{s \in T_{++}} M_s^{bd}$ is never equal to M.

1.9. Structure of the chapter

In Section 2, we consider a general commutative (unital) ring A and A-modules M with two endomorphisms ψ, φ such that $\psi \circ \varphi = \operatorname{id}$. We show that the induction functor $\operatorname{Ind}_{\mathbb{N}, \psi}^{\mathbb{Z}} = \varprojlim_\psi$ is exact and that the module $A[\mathbb{Z}] \otimes_{\mathbb{N}, \varphi} M$ is isomorphic to the subrepresentation of $\operatorname{Ind}_{\mathbb{N}, \psi}^{\mathbb{Z}}(M) = \varprojlim_\psi M$ generated by the elements of the form $(\varphi^k(m))_{k \in \mathbb{N}}$.

In Section 3, we consider a general monoid $P_+ = N_0 \rtimes L_+$ contained in a group P with the property that N_0 is a group such that $t N_0 t^{-1} \subset N_0$ has a finite index for all $t \in L_+$ and we study the étale $A[P_+]$-modules M. We show that the inverse monoid $P_- = L_- N_0 \subset P$ acts on M, the inverse of $t \in L_+$ acting by the left inverse ψ_t of the action φ_t of t with kernel $\sum u \varphi_t(M)$ for $u \in N_0$ not in $t N_0 t^{-1}$. We add the hypothesis that L_+ contains a central element s such that the sequence $(s^k N_0 s^{-k})_{k \in \mathbb{Z}}$ is decreasing of trivial intersection, of union a group N, and that $P = N \rtimes L$ is the semi-direct product of N and of $L = \cup_{k \in \mathbb{N}} L_- s^k$. An $A[P_+]$-submodule of M is étale if and only if it is stable by ψ_s. The representation M^P of P induced by $M|_{P_-}$, restricted to N is the representation induced from $M|_{N_0}$, and restricted to $s^{\mathbb{Z}}$ is the representation $\varprojlim_{\psi_s} M$ induced from $M|_{s^{-\mathbb{N}}}$. The natural

$A[P_+]$-embedding $M \to M^P$ generates a subrepresentation M_c^P of M^P isomorphic to $A[P] \otimes_{A[P_+]} M$. When N is a locally profinite group and N_0 an open compact subgroup, we show the existence and the uniqueness of a unit-preserving $A[P_+]$-map res : $C^\infty(N_0, A) \to \text{End}_A(M)$, we extend it uniquely to an $A[P]$-map Res : $C^\infty(N, A) \to \text{End}_A(M^P)$, and we prove our first theorem: the equivalence between the P-equivariant sheaves of A-modules on N and the étale $A[P_+]$-modules on N_0.

In Section 4, we suppose that A is a linearly topological commutative ring, that P is a locally profinite group and that M is a complete linearly topological A-module with a continuous étale action of P_+ such that the action of P_- is also continuous, or equivalently ψ_s is continuous (we say that M is a topologically étale module). Then M^P is complete for the compact-open topology and Res is a measure on N with values in the algebra E^{cont} of continuous endomorphisms of M^P. We show that E^{cont} is a complete topological ring for the topology defined by the ideals $E_{\mathcal{L}}^{cont}$ of endomorphisms with image in an open A-submodule $\mathcal{L} \subset M^P$, and that any continuous map $N \to E^{cont}$ with compact support can be integrated with respect to Res.

In Section 5, we introduce a locally profinite group G containing P as a closed subgroup with compact quotient set G/P, such that the double cosets $P \backslash G/P$ admit a finite system W of representatives normalizing L, of image in $N_G(L)/L$ equal to a group, and the image $\mathcal{C} = Pw_0P/P$ in G/P of a double coset (with $w_0 \in W$) is open dense and homeomorphic to N by the map $n \mapsto nw_0P/P$. We show that any compact open subset of G/P is a finite disjoint union of $g^{-1}Uw_0P/P$ for $g \in G$ and $U \subset N$ a compact open subgroup. We prove the basic result that the G-equivariant sheaves of A-modules on G/P identify with modules over the skew group ring $C^\infty(G/P, A)\#G$, or with nondegenerate modules over a (non unital) subring $\mathcal{A}$, and that an étale $A[P_+]$-module M endowed with endomorphisms $\mathcal{H}_g \in \text{End}_A(M)$, for $g \in N_0\overline{P}N_0$, satisfying certain relations H1, H2, H3, gives rise to a nondegenerate $\mathcal{A}$-module. For $g \in G$ we denote $N_g \subset N$ such that $N_gw_0P/P = g^{-1}\mathcal{C} \cap \mathcal{C}$. We study the map α from the set of (g, u) with $g \in G$ and $u \in N_g$ to P defined by $guw_0N = \alpha(g, u)uw_0N$. In particular, we show the cocycle relation $\alpha(gh, u) = \alpha(g, h.u)\alpha(h, u)$ when each term makes sense. When M is compact, then M^P is compact and the action of P on M^P induces a continuous map $P \to E^{cont}$. We show that the A-linear map $\mathcal{A} \to E^{cont}$ given by the integrals of $\alpha(g, .)f(.)$ with respect to Res, for $f \in C_c^\infty(N_g, A)$, is multiplicative. As explained above, we obtain a G-equivariant sheaf of A-modules on G/P with sections M on $\mathcal{C}_0$.

In Section 6, we do not suppose that M is compact and we introduce the notion of $(s, \mathrm{res}, \mathfrak{C})$-integrability for a special family $\mathfrak{C}$ of compact subsets of M. We give an $(s, \mathrm{res}, \mathfrak{C})$-integrability criterion for the function $\alpha_{g,0}(u) = \mathrm{Res}(1_{N_0})\alpha(gh, u)\,\mathrm{Res}(1_{N_0})$ on the open subset $U_g \subset N_0$ such that $U_g w_0 P / P = g^{-1}\mathcal{C}_0 \cap \mathcal{C}_0$, for $g \in N_0 w_0 P w_0 N_0$, a criterion which ensures that the integrals $\mathcal{H}_g$ of $\alpha_{g,0}$ satisfy the relations H1, H2, H3, as well as a method of reduction to the case where M is killed by a power of p. When these criteria are satisfied, as explained in Section 5, one gets a G-equivariant sheaf of A-modules on G/P with sections M on $\mathcal{C}_0$.

Section 7 concerns classical (φ, Γ)-modules over $\mathcal{O}_{\mathcal{E}}$, seen as étale $o[P_+^{(2)}]$-module D, where the upper exponent indicates that $P_+^{(2)}$ is the upper triangular monoid P_+ of $GL(2, \mathbb{Q}_p)$. Using the properties of treillis we apply the method explained in Section 6 to this case and we obtain the sheaf constructed by Colmez.

In Section 8 we consider the case where N_0 is a compact p-adic Lie group endowed with a continuous non-trivial homomorphism $\ell : N_0 \to N_0^{(2)}$ with a section ι, that $L_* \subset L$ is a monoid acting by conjugation on N_0 and $\iota(N_0^{(2)})$, that ℓ extends to a continuous homomorphism $\ell : P_* = N_0 \rtimes L_* \to P_+^{(2)}$ sending L_* to $L_+^{(2)}$ and that ι is L_* equivariant. We consider the abelian categories of étale L_*-modules finitely generated over the microlocalized ring $\Lambda_\ell(N_0)$ resp. over $\mathcal{O}_{\mathcal{E}}$ (with the action of L_* induced by ℓ). Between these categories we have the base change functors given by the natural L_*-equivariant algebra homomorphisms $\ell : \Lambda_\ell(N_0) \to \mathcal{O}_{\mathcal{E}}$ and $\iota : \mathcal{O}_{\mathcal{E}} \to \Lambda_\ell(N_0)$. We show our second theorem: the base change functors are quasi-inverse equivalences of categories. When L_* contains an open topologically finitely generated pro-p-subgroup, we show that an étale L_*-module over $\mathcal{O}_{\mathcal{E}}$ is automatically topologically étale for the weak topology; the result extends to étale L_*-modules over $\Lambda_\ell(N_0)$, with the help of this last theorem.

In Section 9, we suppose that $\ell : P \to P^{(2)}(\mathbb{Q}_p)$ is a continuous homomorphism with $\ell(L) \subset L^{(2)}(\mathbb{Q}_p)$, and that $\iota : N^{(2)}(\mathbb{Q}_p) \to N$ is a L-equivariant section of $\ell|_N$ (as L acts on $N^{(2)}(\mathbb{Q}_p)$ via ℓ) sending $\ell(N_0)$ in N_0. The assumptions of Section 8 are satisfied for $L_* = L_+$. Given an étale L_+-module M over $\Lambda_\ell(N_0)$, we exhibit a special family $\mathfrak{C}_s$ of compact subsets in M which satisfies the criteria of Section 6 with $M(\mathfrak{C}_s)$ equal to a dense $\Lambda(N_0)[L_+]$-submodule $M_s^{bd} \subset M$. We obtain our third theorem: there exists a faithful functor from the étale L_+-modules over $\Lambda_\ell(N_0)$ to the G-equivariant sheaves on G/P sending M to the sheaf with sections M_s^{bd} on $\mathcal{C}_0$.

In Section 10, we check that our theory applies to the group $G(\mathbb{Q}_p)$ of rational points of a split reductive group of $\mathbb{Q}_p$, to a Borel subgroup $P(\mathbb{Q}_p)$

of maximal split torus $T(\mathbb{Q}_p) = L$ and to a natural homomorphism ℓ_α : $P(\mathbb{Q}_p) \to P^{(2)}(\mathbb{Q}_p)$ associated to a simple root α. We obtain our main theorem: there are compatible faithful functors from the étale $T(\mathbb{Q}_p)_+$-modules D over $\mathcal{O}_{\mathcal{E}}$ (where $T(\mathbb{Q}_p)_+$ acts via α) to the $G(\mathbb{Q}_p)$-equivariant sheaves on $G(\mathbb{Q}_p)/P(\mathbb{Q}_p)$ sheaves with sections $\mathbb{M}(D)_s^{bd}$ on $\mathcal{C}_0$, for all strictly dominant $s \in T(\mathbb{Q}_p)$. When the root system of G is irreducible of rank > 1, we show that $\cup_s M_s^{bd} \neq M = \mathbb{M}(D)$.

Acknowledgements: A part of the work on this article was done when the first and third authors visited the Institut Mathématique de Jussieu at the Universities of Paris 6 and Paris 7, and the second author visited the Mathematische Institut at the Universität Münster. We express our gratitude to these institutions for their hospitality. We thank heartily CIRM, IAS, the Fields Institute, as well as Durham, Cordoba and Caen Universities, for their invitations giving us the opportunity to present this work. Finally, we would like to thank the anonymous referee for a very careful reading of the manuscript and his suggestions for improving the presentation.

2. Induction Ind_H^G for monoids $H \subset G$

A monoid is supposed to have a unit.

2.1. Definition and remarks

Let A be a commutative ring, let G be a monoid and let H be a submonoid of G. We denote by $A[G]$ the monoid A-algebra of G and by $\mathfrak{M}_A(G)$ the category of left $A[G]$-modules, which has no reason to be equivalent to the category of right $A[G]$-modules. One can construct $A[G]$-modules starting from $A[H]$-modules in two natural ways, by taking the two adjoints of the restriction functor $\mathrm{Res}_H^G : \mathfrak{M}_A(G) \to \mathfrak{M}_A(H)$ from G to H. For $M \in \mathfrak{M}_A(H)$ and $V \in \mathfrak{M}_A(G)$ we have the isomorphism

$$\mathrm{Hom}_{A[G]}(A[G] \otimes_{A[H]} M, V) \xrightarrow{\simeq} \mathrm{Hom}_{A[H]}(M, V)$$

and the isomorphism

$$\mathrm{Hom}_{A[G]}(V, \mathrm{Hom}_{A[H]}(A[G], M)) \xrightarrow{\simeq} \mathrm{Hom}_{A[H]}(V, M). \quad (7.1)$$

For monoid algebras, restriction of homomorphisms induces the identification

$$\mathrm{Hom}_{A[H]}(A[G], M) = \mathrm{Ind}_H^G(M)$$

where $\mathrm{Ind}_H^G(M)$ is formed by the functions

$$f : G \to M \text{ such that } f(hg) = hf(g) \text{ for any } h \in H, g \in G;$$

the group G acts by right translations, $gf(x) = f(xg)$ for $g, x \in G$. The isomorphism (7.1) pairs ϕ of the left side and Φ of the right side satisfying ([14] I.5.7)

$$\phi(v)(g) = \Phi(gv) \quad \text{for } (v, g) \in V \times G.$$

It is well known that the left and right adjoint functors of Res_H^G are transitive (for monoids $H \subset K \subset G$), the left adjoint is right exact, the right adjoint is left exact.

We observe important differences between monoids and groups:

(a) The binary relation $g \sim g'$ if $g \in Hg'$ is not symmetric, there is no "quotient space" $H\backslash G$, no notion of function with finite support modulo H in $\mathrm{Ind}_H^G(M)$.

(b) When $hM = 0$ for some $h \in H$ such that $hG = G$, then $\mathrm{Ind}_H^G(M) = 0$. Indeed $f(hg) = hf(g)$ implies $f(hg) = 0$ for any $g \in G$.

(c) When G is a group generated, as a monoid, by H and the inverse monoid $H^{-1} := \{h \in G \mid h^{-1} \in H\}$, and when M in an $A[H]$-module such that the action of any element $h \in H$ on M is invertible, then $f(g) = gf(1)$ for all $g \in G$ and $f \in \mathrm{Ind}_H^G(M)$. This can be seen by induction on the minimal number $m \in \mathbb{N}$ such that $g = g_1 \cdots g_m$ with $g_i \in H \cup H^{-1}$. Then $g_1 \in H$ implies $f(g) = g_1 f(g_2 \cdots g_m)$, and $g_1 \in H^{-1}$ implies $f(g_2 \cdots g_m) = f(g_1^{-1} g_1 g_2 \cdots g_m) = g_1^{-1} f(g)$. The representation $\mathrm{Ind}_H^G(M)$ is isomorphic by $f \mapsto f(1)$ to the natural representation of G on M.

2.2. From $\mathbb{N}$ to $\mathbb{Z}$

An A-module with an endomorphism φ is equivalent to an $A[\mathbb{N}]$-module, φ being the action of $1 \in \mathbb{N}$, and an A-module with an automorphism φ is equivalent to an $A[\mathbb{Z}]$-module. When φ is bijective, $A[\mathbb{Z}] \otimes_{A[\mathbb{N}]} M$ and $\mathrm{Ind}_\mathbb{N}^\mathbb{Z}(M)$ are isomorphic to M.

In general, $A[\mathbb{Z}] \otimes_{A[\mathbb{N}]} M$ is the limit of an inductive system and $\mathrm{Ind}_\mathbb{N}^\mathbb{Z}(M)$ is the limit of a projective system. The first one is interesting when φ is injective, the second one when φ is surjective.

For $r \in \mathbb{N}$ let $M_r = M$. The general element of M_r is written x_r with $x \in M$. Let $\varinjlim (M, \varphi)$ be the quotient of $\sqcup_{r \in \mathbb{N}} M_r$ by the equivalence relation generated by $\varphi(x)_{r+1} \equiv x_r$, with the isomorphism induced by the maps $x_r \to \varphi(x)_r$ $M_r \to M_r$ of inverse induced by the maps $x_r \to x_{r+1}$ $M_r \to M_{r+1}$. Let $x \mapsto [x] : \mathbb{Z} \to A[\mathbb{Z}]$ be the canonical map. The maps $x_r \to [-r] \otimes x$ $M_r \to A[\mathbb{Z}] \otimes_{A[\mathbb{N}]} M$ for $r \in \mathbb{N}$ induce an isomorphism of $A[\mathbb{Z}]$-modules

$$\varinjlim M \quad \to \quad A[\mathbb{Z}] \otimes_{A[\mathbb{N}]} M.$$

Let

$$\varprojlim M := \{x = (x_m)_{m \in \mathbb{N}} \in \prod_{m \in \mathbb{N}} M : \varphi(x_{m+1}) = x_m \text{ for any } m \in \mathbb{N}\} \quad (7.2)$$

seen as an $A[\mathbb{Z}]$-module via the automorphism

$$x \mapsto (\varphi(x_0), x_0, x_1, \ldots) = (\varphi(x_0), \varphi(x_1), \varphi(x_2) \ldots)$$

of inverse $x \mapsto (x_1, x_2, \ldots)$. The map $f \mapsto (f(-m))_{m \in \mathbb{N}}$ is an isomorphism of $A[\mathbb{Z}]$-modules

$$\mathrm{Ind}_{\mathbb{N}}^{\mathbb{Z}}(M) \quad \to \quad \varprojlim M.$$

The submodules of M

$$M^{\varphi^\infty = 0} = \cup_{k \in \mathbb{N}} M^{\varphi^k = 0}, \quad \varphi^\infty(M) = \cap_{n \in \mathbb{N}} \varphi^n(M)$$

are stable by φ. The inductive limit sees only the quotient $M/M^{\varphi^\infty = 0}$ and the projective limit sees only the submodule $\varphi^\infty(M)$,

$$\varinjlim M = \varinjlim (M/M^{\varphi^\infty = 0}), \quad \varprojlim M = \varprojlim (\varphi^\infty(M)).$$

Lemma 7.9. *Let* $0 \to M_1 \to M_2 \to M_3 \to 0$ *be an exact sequence of A-modules with an endomorphism* φ.

(a) *The sequence*

$$0 \to \varinjlim M_1 \to \varinjlim M_2 \to \varinjlim M_3 \to 0$$

is exact.

(b) *When* φ *is surjective on* M_1, *the sequence*

$$0 \to \varprojlim M_1 \to \varprojlim M_2 \to \varprojlim M_3 \to 0$$

is exact.

Proof. This is a standard fact on inductive and projective limits. $\square$

2.3. (φ, ψ)-modules

Let M be an A-module with two endomorphisms ψ, φ such that $\psi \circ \varphi = 1$. Then ψ is surjective, φ is injective, the endomorphism $\varphi \circ \psi$ is a projector of M giving the direct decomposition

$$M = \varphi(M) \oplus M^{\psi=0}, \quad m = (\varphi \circ \psi)(m) + m^{\psi=0} \qquad (7.3)$$

for $m \in M$ and $m^{\psi=0} \in M^{\psi=0}$ the kernel of ψ. We consider the representation of $\mathbb{Z}$ induced by (M, ψ) as in Subsection 2.2,

$$\mathrm{Ind}_{\mathbb{N}, \psi}^{\mathbb{Z}}(M) \simeq \varprojlim_{\psi} M.$$

On the induced representation ψ is an isomorphism and we introduce $\varphi := \psi^{-1}$. As ψ is surjective on M, the map $\mathrm{ev}_0 \ \mathrm{Ind}_{\mathbb{N}, \psi}^{\mathbb{Z}}(M) \to M$, corresponding to the map

$$\varprojlim_{\psi} M \to M, \quad (x_m)_{m \in \mathbb{N}} \mapsto x_0$$

is surjective. A splitting is the map $\sigma_0 \ M \to \mathrm{Ind}_{\mathbb{N}, \psi}^{\mathbb{Z}}(M)$ corresponding to

$$M \to \varprojlim_{\psi} M, \quad x \mapsto (\varphi^m(x))_{m \in \mathbb{N}}. \qquad (7.4)$$

Obviously ev_0 is ψ-equivariant, σ_0 is φ-equivariant, $\mathrm{ev}_0 \circ \sigma_0 = \mathrm{id}_M$, and

$$R_0 \quad := \quad \sigma_0 \circ \mathrm{ev}_0 \in \mathrm{End}_A(\mathrm{Ind}_{\mathbb{N}, \psi}^{\mathbb{Z}}(M))$$

is an idempotent of image $\sigma_0(M)$.

Definition 7.10. The representation of $\mathbb{Z}$ compactly induced from (M, ψ) is the subrepresentation c-$\mathrm{Ind}_{\mathbb{N}, \psi}^{\mathbb{Z}}(M)$ of $\mathrm{Ind}_{\mathbb{N}, \psi}^{\mathbb{Z}}(M)$ generated by the image of $\sigma_0(M)$.

We note that, for any $k \geq 1$, the endomorphisms ψ^k, φ^k satisfy the same properties as ψ, φ because $\psi^k \circ \varphi^k = 1$. For any integer $k \geq 0$, the value at k is a surjective map $\mathrm{ev}_k \ \mathrm{Ind}_{\mathbb{N}, \psi}^{\mathbb{Z}}(M) \to M$, corresponding to the map

$$\varprojlim_{\psi} M \to M, \quad (x_m)_{m \in \mathbb{N}} \mapsto x_k \qquad (7.5)$$

of splitting $\sigma_k \ M \to \mathrm{Ind}_{\mathbb{N}, \psi}^{\mathbb{Z}}(M)$ corresponding to the map

$$M \to \varprojlim_{\psi} M, \quad x \mapsto (\psi^k(x), \ldots, \psi(x), x, \varphi(x), \varphi^2(x), \ldots) \ . \qquad (7.6)$$

The following relations are immediate:

$$\mathrm{ev}_k = \mathrm{ev}_0 \circ \varphi^k = \psi \circ \mathrm{ev}_{k+1} = \mathrm{ev}_{k+1} \circ \psi,$$

$$\sigma_k = \psi^k \circ \sigma_0 = \sigma_{k+1} \circ \varphi = \varphi \circ \sigma_{k+1}.$$

We deduce that $\sigma_k(M) \subset \sigma_{k+1}(M)$. Since $\sigma_k(M)$ is φ-invariant we have

$$\text{c-Ind}_{\mathbb{N},\psi}^{\mathbb{Z}}(M) = \sum_{k\in\mathbb{N}} \psi^k(\sigma_0(M)) = \sum_{k\in\mathbb{N}} \sigma_k(M) = \bigcup_{k\in\mathbb{N}} \sigma_k(M). \quad (7.7)$$

In $\varprojlim_{\psi}(M)$ the subspace of $(x_m)_{m\in\mathbb{N}}$ such that $x_{k+r} = \varphi^k(x_r)$ for all $k \in \mathbb{N}$ and for some $r \in \mathbb{N}$, is equal to $\text{c-Ind}_{\mathbb{N},\psi}^{\mathbb{Z}}(M)$. The definition of $\text{c-Ind}_{\mathbb{N},\psi}^{\mathbb{Z}}(M)$ is functorial. We get a functor $\text{c-Ind}_{\mathbb{N},\psi}^{\mathbb{Z}}$ from the category of A-modules with two endomorphisms ψ, φ such that $\psi \circ \varphi = 1$ (a morphism commutes with ψ and with φ) to the category of $A[\mathbb{Z}]$-modules.

Proposition 7.11. *The map*

$$A[\mathbb{Z}] \otimes_{A[\mathbb{N}],\varphi} M \to \text{Hom}_{A[\mathbb{N}],\psi}(A[\mathbb{Z}], M) = \text{Ind}_{\mathbb{N},\psi}^{\mathbb{Z}}(M)$$

$$[k] \otimes m \mapsto (\varphi^k \circ \sigma_0)(m)$$

induces an isomorphism from the tensor product $A[\mathbb{Z}] \otimes_{A[\mathbb{N}],\varphi} M$ to the compactly induced representation $\text{c-Ind}_{\mathbb{N},\psi}^{\mathbb{Z}}(M)$ (note that ψ and φ appear).

Proof. From (7.3) and the relations between the σ_k we have for $m \in M$, $k \in \mathbb{N}, k \geq 1$,

$$\sigma_k(m) = \sigma_{k-1}(\psi(m)) + \sigma_k(m^{\psi=0}).$$

By induction $\sum_{k\in\mathbb{N}} \sigma_k(M) = \sigma_0(M) + \sum_{k\geq1} \sigma_k(M^{\psi=0})$. Using (7.6) one checks that the sum is direct, hence by (7.7),

$$\text{c-Ind}_{\mathbb{N},\psi}^{\mathbb{Z}}(M) = \sigma_0(M) \oplus (\oplus_{k\geq1}\sigma_k(M^{\psi=0})).$$

On the other hand, one deduces from (7.3) that

$$A[\mathbb{Z}] \otimes_{A[\mathbb{N}],\varphi} M = ([0] \otimes M) \oplus (\oplus_{k\geq1}([-k] \otimes M^{\psi=0})).$$

$\square$

With Lemma 7.9 we deduce:

Corollary 7.12. *The functor $\text{c-Ind}_{\mathbb{N},\psi}^{\mathbb{Z}}$ is exact.*

We have two kinds of idempotents in $\text{End}_A(\text{Ind}_{\mathbb{N},\psi}^{\mathbb{Z}}(M))$, for $k \in \mathbb{N}$, defined by

$$R_k = \sigma_0 \circ \varphi^k \circ \psi^k \circ \text{ev}_0, \quad R_{-k} := \psi^k \circ R_0 \circ \varphi^k = \sigma_k \circ \text{ev}_k. \quad (7.8)$$

The first ones are the images of the idempotents $r_k := \varphi^k \circ \psi^k \in \text{End}_A(M)$ via the ring homomorphism

$$\text{End}_A(M) \to \text{End}_A \text{Ind}_{\mathbb{N},\psi}^{\mathbb{Z}}(M), \quad f \mapsto \sigma_0 \circ f \circ \text{ev}_0. \quad (7.9)$$

The second ones give an isomorphism from $\mathrm{Ind}_{\mathbb{N},\psi}^{\mathbb{Z}}(M)$ to the limit of the projective system $(\sigma_k(M), R_{-k} : \sigma_{k+1}(M) \to \sigma_k(M))$.

Lemma 7.13. *The map* $f \mapsto (R_{-k}(f))_{k \in \mathbb{N}}$ *is an isomorphism from* $\mathrm{Ind}_{\mathbb{N},\psi}^{\mathbb{Z}}(M)$ *to*

$$\varprojlim_{R_{-k}} (\sigma_k(M)) \quad := \quad \{(f_k)_{k \in \mathbb{N}} \mid f_k \in \sigma_k(M), \ f_k = R_{-k}(f_{k+1}) \quad \text{for } k \in \mathbb{N}\}$$

of inverse $(f_k)_{k \in \mathbb{N}} \to f$ *with* $\mathrm{ev}_k(f) = \mathrm{ev}_k(f_k)$.

Remark 7.14. As φ is injective, its restriction to $\cap_{n \in \mathbb{N}} \varphi^n(M)$ is an isomorphism and the following $A[\mathbb{Z}]$-modules are isomorphic (Section 2.2):

$$\mathrm{Ind}_{\mathbb{N},\varphi}^{\mathbb{Z}}(M) \quad \simeq \quad \varprojlim_{\varphi} M \quad \simeq \quad \cap_{n \in \mathbb{N}} \varphi^n(M).$$

As ψ is surjective, its action on the quotient $M/M^{\psi^\infty=0}$ is bijective and the following $A[\mathbb{Z}]$-modules are isomorphic (Section 2.2):

$$A[\mathbb{Z}] \otimes_{A[\mathbb{N}],\psi} M \quad \simeq \quad \varinjlim_{\psi} M \quad \simeq \quad M/M^{\psi^\infty=0}.$$

Remark 7.15. When the A-module M is noetherian, a ψ-stable submodule of M which generates M as a φ-module is equal to M.

Proof. Let N be a submodule of M. As M is noetherian there exists $k \in \mathbb{N}$ such that the φ-stable submodule of M generated by N is the submodule $N_k \subset M$ generated by $N, \varphi(N), \ldots, \varphi^k(N)$. When N is ψ-stable we have $\psi^k(N_k) = N$ and when N generates M as a φ-module we have $M = N_k$. In this case, $M = \psi^k(M) = \psi^k(N_k) = N$. $\square$

3. Etale P_+-modules

Let $P = N \rtimes L$ *be a semi-direct product of an invariant subgroup* N *and of a group* L *and let* $N_0 \subset N$ *be a subgroup of* N. *For any subgroups* $V \subset U \subset N$, *the symbol* $J(U/V) \subset U$ *denotes a set of representatives for the cosets in* U/V.

The group P acts on N by

$$(b = nt, x) \to b.x = ntxt^{-1}$$

for $n, x \in N$ and $t \in L$. The P-stabilizer $\{b \in P \mid b.N_0 \subset N_0\}$ of N_0 is a monoid

$$P_+ = N_0 L_+$$

where $L_+ \subset L$ is the L-stabilizer of N_0. Its maximal subgroup $\{b \in P \mid b.N_0 = N_0\}$ is the intersection $P_0 = N_0 \rtimes L_0$ of P_+ with the inverse monoid $P_- = L_- N_0$ where L_- is the inverse monoid of L_+ and L_0 is the maximal subgroup of L_+.

We suppose that the subgroup $t.N_0 = t N_0 t^{-1} \subset N_0$ *has a finite index, for all* $t \in L_+$. *Let A be a commutative ring and let M be an $A[P_+]$-module, equivalently an $A[N_0]$-module with a semilinear action of L_+.*

The action of $b \in P_+$ on M is denoted by φ_b. If $b \in P_0$ then φ_b is invertible and we also write $\varphi_b(m) = bm$, $\varphi_b^{-1}(m) = b^{-1}m$ for $m \in M$. The action $\varphi_t \in \mathrm{End}_A(M)$ of $t \in L_+$ is $A[N_0]$-semilinear:

$$\varphi_t(xm) = \varphi_t(x)\varphi_t(m) \quad \text{for} \quad x \in A[N_0], \ m \in M. \tag{7.10}$$

3.1. Etale module M

The group algebra $A[N_0]$ is naturally an $A[P_+]$-module. For $t \in L_+$ the map φ_t is injective of image $A[t N_0 t^{-1}]$, and

$$A[N_0] = \oplus_{u \in J(N_0/t N_0 t^{-1})} u\, A[t N_0 t^{-1}].$$

Definition 7.16. We say that M is étale if, for any $t \in L_+$, the map φ_t is injective and

$$M = \oplus_{u \in J(N_0/t N_0 t^{-1})} u\, \varphi_t(M). \tag{7.11}$$

An equivalent formulation is that, for any $t \in L_+$, the linear map

$$A[N_0] \otimes_{A[N_0],\varphi_t} M \to M, \quad x \otimes m \mapsto x\varphi_t(m)$$

is bijective. For M étale and $t \in L_+$, let $\psi_t \in \mathrm{End}_A(M)$ be the unique canonical left inverse of φ_t of kernel

$$M^{\psi_t=0} = \sum_{u \in (N_0 - t N_0 t^{-1})} u\varphi_t(M).$$

The trivial action of P_+ on M is not étale, and obviously the restriction to P_+ of a representation of P is not always étale.

Lemma 7.17. *Let M be an étale $A[P_+]$-module. For $t \in L_+$, the kernel $M^{\psi_t=0}$ is an $A[t N_0 t^{-1}]$-module, the idempotents in $\mathrm{End}_A M$*

$$(u \circ \varphi_t \circ \psi_t \circ u^{-1})_{u \in J(N_0/t N_0 t^{-1})}$$

are orthogonal of sum the identity. Any $m \in M$ can be written

$$m = \sum_{u \in J(N_0/t N_0 t^{-1})} u\varphi_t(m_{u,t}) \tag{7.12}$$

for unique elements $m_{u,t} \in M$, equal to $m_{u,t} = \psi_t(u^{-1}m)$.

Proof. The kernel $M^{\psi_t=0}$ is an $A[tN_0t^{-1}]$-module because $N_0 - tN_0t^{-1}$ is stable by left multiplication by tN_0t^{-1}. The endomorphism $\varphi_t \circ \psi_t$ is an idempotent because $\psi_t \circ \varphi_t = \mathrm{id}_M$. Then apply (7.11) and notice that $m \in M$ is equal to

$$m = \sum_{u \in J(N_0/tN_0t^{-1})} (u \circ \varphi_t \circ \psi_t \circ u^{-1})(m).$$

$\square$

Remark 7.18. (1) An $A[P_+]$-module M is étale when, for any $t \in L_+$, the action φ_t of t admits a left inverse $f_t \in \mathrm{End}_A M$ such that the idempotents $(u \circ \varphi_t \circ f_t \circ u^{-1})_{u \in J(N_0/tN_0t^{-1})}$ are orthogonal of sum the identity. The endomorphism f_t is the canonical left inverse ψ_t.

(2) The $A[P_+]$-module $A[N_0]$ is étale. As $A[N_0]$ is a left and right free $A[tN_0t^{-1}]$-module of rank $[N_0 : tN_0t^{-1}]$ we have for $x \in A[N_0]$,

$$x = \sum_{u \in J(N_0/tN_0t^{-1})} u\varphi_t(x_{u,t}) = \sum_{u \in J(N_0/tN_0t^{-1})} \varphi_t(x'_{u,t})u^{-1}$$

where $x_{u,t} = \psi_t(u^{-1}x)$, $x'_{u,t} = \psi_t(xu)$ and ψ_t is the left inverse of φ_t of kernel

$$\sum_{u \in N_0-tN_0t^{-1}} u A[tN_0t^{-1}] = \sum_{u \in N_0-tN_0t^{-1}} A[tN_0t^{-1}]u^{-1}.$$

Let M be an étale $A[P_+]$-module and $t \in L_+$. We denote $m \mapsto m^{\psi_t=0}$: $M \to M^{\psi_t=0}$ the projector $\mathrm{id}_M - \varphi_t \circ \psi_t$ along the decomposition $M = \varphi_t(M) \oplus M^{\psi_t=0}$.

Lemma 7.19. *Let $x \in A[N_0]$ and $m \in M$. We have*

$$\psi_t(\varphi_t(x)m) = x\psi_t(m), \quad \psi_t(x\varphi_t(m)) = \psi_t(x)m,$$

$$(\varphi_t(x)m)^{\psi_t=0} = \varphi_t(x)(m^{\psi_t=0}), \quad (x\varphi_t(m))^{\psi_t=0} = x^{\psi_t=0}\varphi_t(m).$$

Proof. We multiply $m = (\varphi_t \circ \psi_t)(m) + m^{\psi_t=0}$ on the left by $\varphi_t(x)$. By the $A[N_0]$-semilinearity of φ_t we have $\varphi_t(x)m = \varphi_t(x\psi_t(m)) + \varphi_t(x)(m^{\psi_t=0})$. As $M^{\psi_t=0}$ is an $A[tN_0t^{-1}]$-module, the uniqueness of the decomposition implies $\psi_t(\varphi_t(x)m) = x\psi_t(m)$ and $(\varphi_t(x)m)^{\psi_t=0} = \varphi_t(x)(m^{\psi_t=0})$.

We multiply $x = (\varphi_t \circ \psi_t)(x) + x^{\psi_t=0}$ on the right by $\varphi_t(m)$. By the semilinearity of φ_t we have $x\varphi_t(m) = \varphi_t(\psi_t(x)m) + x^{\psi_t=0}\varphi_t(m)$. As $A[N_0]^{\psi_t=0}\varphi_t(M) = M^{\psi_t=0}$ the uniqueness of the decomposition implies $\psi_t(x\varphi_t(m)) = \psi_t(x)m, (x\varphi_t(m))^{\psi_t=0} = x^{\psi_t=0}\varphi_t(m).$ $\square$

Lemma 7.20. *Let* $x \in A[N_0]$ *and* $m \in M$. *We have*

$$\psi_t(xm) = \sum_{u \in J(N_0/tN_0t^{-1})} \psi_t(xu)\psi_t(u^{-1}m).$$

Proof. Using (7.12), replace m by $\sum_{u \in J(N_0/tN_0t^{-1})} u\varphi_t(m_{u,t})$ in $\psi_t(xm)$. We get

$$\psi_t(xm) = \psi_t\left(\sum_{u \in J(N_0/tN_0t^{-1})} xu\varphi_t(m_{u,t}) \right) = \sum_{u \in J(N_0/tN_0t^{-1})} \psi_t(xu)m_{u,t}$$

$$= \sum_{u \in J(N_0/tN_0t^{-1})} \psi_t(xu)\psi_t(u^{-1}m)$$

using the first line of Lemma 7.19. $\qquad\square$

Proposition 7.21. *Let* M *be an étale* $A[P_+]$-*module. The map*

$$b^{-1} = (ut)^{-1} \mapsto \psi_b := \psi_t \circ u^{-1} \quad P_- \to \operatorname{End}_A(M) \text{ for } t \in L_+, \ u \in N_0,$$

defines a canonical action of P_- *on* M.

Proof. We check that $\psi_{b_1 b_2} = \psi_{b_2} \circ \psi_{b_1}$ for $b_1 = u_1 t_1, b_2 = u_2 t_2 \in P_+$. We have $\psi_{b_1 b_2} = \psi_{t_1 t_2} \circ (u_1 t_1 u_2 t_1^{-1})^{-1}$ and $\psi_{b_2} \circ \psi_{b_1} = \psi_{t_2} \circ u_2^{-1} \circ \psi_{t_1} \circ u_1^{-1}$. As $u_2^{-1} \circ \psi_{t_1} = \psi_{t_1} \circ t_1 u_2^{-1} t_1^{-1}$, it remains only to show $\psi_{t_2} \psi_{t_1} = \psi_{t_1 t_2}$. For the sake of simplicity, we note $\varphi_i = \varphi_{t_i}, \psi_i = \psi_{t_i}$. For $m \in M$ we have $m = \varphi_1(\varphi_2 \circ \psi_2(\psi_1(m)) + \psi_1(m)^{\psi_2=0}) + m^{\psi_1=0}$. This is also

$$m = (\varphi_{t_1 t_2} \circ \psi_2 \circ \psi_1)(m) + \varphi_1(\psi_1(m)^{\psi_2=0}) + m^{\psi_1=0}$$

because $\varphi_1 \circ \varphi_2 = \varphi_{t_1 t_2}$. By the uniqueness of the decomposition $m = (\varphi_{t_1 t_2} \circ \psi_{t_1 t_2})(m) + m^{\psi_{t_1 t_2}=0}$ we are reduced to show that

$$M^{\psi_{t_1 t_2}=0} = \varphi_1(M^{\psi_2=0}) + M^{\psi_1=0}.$$

It is enough to prove the inclusion $M^{\psi_{t_1 t_2}=0} \subset \varphi_1(M^{\psi_2=0}) + M^{\psi_1=0}$ to get the equality because $M = \varphi_{t_1 t_2}(M) \oplus V$ with V equal to any of them. Hence we want to show

$$\sum_{u \in N_0 - t_1 t_2 N_0 (t_1 t_2)^{-1}} u\varphi_{t_1 t_2}(M) \subset \varphi_1\left(\sum_{u \in N_0 - t_2 N_0 t_2^{-1}} u\varphi_2(M) \right)$$

$$+ \sum_{u \in N_0 - t_1 N_0 t_1^{-1}} u\varphi_1(M). \tag{7.13}$$

As $\varphi_1 \circ u \circ \varphi_2 = t_1 u t_1^{-1} \circ \varphi_{t_1 t_2}$ the right side of (7.13) is

$$\sum_{u \in t_1 N_0 t_1^{-1} - t_1 t_2 N_0 (t_1 t_2)^{-1}} u\varphi_{t_1 t_2}(M) + \sum_{u \in N_0 - t_1 N_0 t_1^{-1}} u\varphi_1(M).$$

As $\varphi_{t_1 t_2} = \varphi_1 \circ \varphi_2$ we have $\varphi_{t_1 t_2}(M) \subset \varphi_1(M)$. Hence (7.13) is true. $\qquad\square$

Lemma 7.22. *Let* $f : M \to M'$ *be an* A-*morphism between two étale* $A[P_+]$-*modules* M *and* M'. *Then* f *is* P_+-*equivariant if and only if* f *is* P_--*equivariant (for the canonical action of* P_-).

Proof. Let $t \in L_+$. We suppose that f is N_0-equivariant and we show that $f \circ \varphi_t = \varphi_t \circ f$ is equivalent to $f \circ \psi_t = \psi_t \circ f$. Our arguments follow the proof of ([5] Prop. II.3.4).

(a) We suppose $f \circ \varphi_t = \varphi_t \circ f$. Then $f(\varphi_t(M)) = \varphi_t(f(M))$ is contained in $\varphi_t(M')$ and $f(M^{\psi_t=0}) = \sum_{u \in N_0 - tN_0t^{-1}} u\varphi_t(f(M))$ is contained in $M'^{\psi_t=0}$. By Lemma 7.17, this implies $f \circ \varphi_t \circ \psi_t = \varphi_t \circ \psi_t \circ f$. As $f \circ \varphi_t = \varphi_t \circ f$ and φ_t is injective this is equivalent to $f \circ \psi_t = \psi_t \circ f$.

(b) We suppose $f \circ \psi_t = \psi_t \circ f$. Let $m \in M$. Then $f(\varphi_t(m))$ belongs to $\varphi_t(M)$ because $\varphi_t(M)$ is the subset of $x \in M$ such that $\psi_t(u^{-1}x) = 0$ for all $u \in N_0 - tN_0t^{-1}$ and we have

$$\psi_t(u^{-1}f(\varphi_t(m))) = f(\psi_t(u^{-1}(\varphi_t(m)))).$$

Let $x(m) \in M$ be the element such that $f(\varphi_t(m)) = \varphi_t(x(m))$. We have

$$x(m) = \psi_t\varphi_t(x(m)) = \psi_t(f(\varphi_t(m))) = f(\psi_t\varphi_t(m)) = f(m).$$

Therefore $f(\varphi_t(m)) = \varphi_t(f(m))$.

$\square$

Proposition 7.23. *The category* $\mathfrak{M}_A(P_+)^{et}$ *of étale* $A[P_+]$-*modules is abelian and has a natural fully faithful functor into the abelian category* $\mathfrak{M}_A(P_-)$ *of* $A[P_-]$-*modules.*

Proof. From Proposition 7.21 and Lemma 7.22, it suffices to show that the kernel and the image of a morphism $f : M \to M'$ between two étale modules M, M', are étale. Since the ring homomorphism φ_t is flat, for $t \in L_+$, the functor $\Phi_t := A[N_0] \otimes_{A[N_0], \varphi_t} -$ sends the exact sequence

$$(E) \qquad 0 \to \mathrm{Ker}\, f \to M \to M' \to \mathrm{Coker}\, f \to 0 \qquad (7.14)$$

to an exact sequence

$$(\Phi_t(E)) \qquad 0 \to \Phi_t(\mathrm{Ker}\, f) \to \Phi_t(M) \to \Phi_t(M') \to \Phi_t(\mathrm{Coker}\, f) \to 0, \qquad (7.15)$$

and the natural maps $j_- : \Phi_t(-) \to -$ define a map $\Phi_t(E) \to (E)$. The maps j_M and $j_{M'}$ are isomorphisms because M and M' are étale, hence the maps $j_{\mathrm{Ker}\, f}$ and $j_{\mathrm{Coker}\, f}$ are isomorphisms, i.e. $\mathrm{Ker}\, f$ and $\mathrm{Coker}\, f$ are étale. $\square$

Note that a subrepresentation of an étale representation of P_+ is not necessarily étale nor stable by P_-.

Remark 7.24. An arbitrary direct product or a projective limit of étale $A[P_+]$-modules is étale.

Proof. Since the $A[tN_0t^{-1}]$-module $A[N_0]$ is free of finite rank, for $t \in L_+$, the tensor product $A[N_0] \otimes_{A[tN_0t^{-1}]} -$ commutes with arbitrary projective limits. $\qquad\qquad\square$

3.2. Induced representation M^P

Let P be a locally profinite group, semi-direct product $P = N \rtimes L$ of closed subgroups N, L, let $N_0 \subset N$ be an open profinite subgroup, and let s be an element of the centre $Z(L)$ of L such that $L = L_-s^{\mathbb{Z}}$ (notation of Section 3) and $(N_k := s^k N_0 s^{-k})_{k \in \mathbb{Z}}$ is a decreasing sequence of union N and trivial intersection.

As the conjugation action $L \times N \to N$ of L on N is continuous and N_0 is compact open in N, the subgroups $L_0 \subset L$, $P_0 \subset P$ are open and the monoids P_+, P_- are open in P.

We have

$$P = P_-s^{\mathbb{Z}} \;=\; s^{\mathbb{Z}}P_+$$

because, for $n \in N$ and $t \in L$, there exists $k \in \mathbb{N}$ and $n_0 \in N_0$ such that $n = s^{-k}n_0s^k$ and $ts^{-k} \in L_-$. Thus $tn = ts^{-k}n_0s^k \in P_-s^k$ and $(tn)^{-1} \in s^{-k}P_+$. In particular P is generated by P_+ and by its inverse P_-.

Let M be an étale left $A[P_+]$-module. We denote by φ the action of s on M and by ψ the canonical left inverse of φ, by

$$M^P \;=\; \mathrm{Ind}_{P_-}^P(M)$$

the $A[P]$-module induced from the canonical action of P_- on M (Section 2.1).

When $f : P \to M$ is an element of M^P, the values of f on $s^{\mathbb{N}}$ determine the values of f on N and reciprocally because, for any $u \in N_0, k \in \mathbb{N}$,

$$f(s^{-k}us^k) = (\psi^k \circ u)(f(s^k)),$$

$$f(s^k) = \sum_{v \in J(N_0/N_k)} (v \circ \varphi^k)(f(s^{-k}v^{-1}s^k)). \qquad (7.16)$$

The first equality is obvious from the definition of $\mathrm{Ind}_{P_-}^P$, the second equality is obvious by the first equality as the idempotents $(v \circ \varphi^k \circ \psi^k \circ v^{-1})_{v \in J(N_0/N_k)}$ are orthogonal of sum the identity, by Lemma 7.17.

Proposition 7.25. (a) *The restriction to $s^{\mathbb{Z}}$ is an $A[s^{\mathbb{Z}}]$-equivariant isomorphism*

$$M^P \;\to\; \mathrm{Ind}_{s^{-\mathbb{N}}}^{s^{\mathbb{Z}}}(M).$$

(b) *The restriction to N is an N-equivariant bijection from M^P to $\mathrm{Ind}_{N_0}^N(M)$.*

Proof. (a) As $P = P_- s^{\mathbb{Z}}$ and $s^{-\mathbb{N}} \subset P_- \cap s^{\mathbb{Z}}$ (it is an equality if N is not trivial), the restriction to $s^{\mathbb{Z}}$ is a $s^{\mathbb{Z}}$-equivariant injective map $M^P \to \mathrm{Ind}_{s^{-\mathbb{N}}}^{s^{\mathbb{Z}}}(M)$. To show that the map is surjective, let $\phi \in \mathrm{Ind}_{s^{-\mathbb{N}}}^{s^{\mathbb{Z}}}(M)$ and $b \in P$. Then, for $b = b_- s^r$ with $b_- \in P_-, r \in \mathbb{Z}$,

$$f(b) \quad := \quad b_- \phi(s^r)$$

is well defined because the right side depends only on b, and not on the choice of (b_-, r). Indeed for two choices $b = b_- s^r = b'_- s^{r'}$ with $b_-, b'_- \in P_-, r \geq r'$ in $\mathbb{Z}$, we have

$$b_- \phi(s^r) = b'_- s^{r'-r} \phi(s^r) = b'_- \phi(s^{r'}).$$

The well defined function $b \mapsto f(b)$ on P belongs obviously to M^P and its restriction to $s^{\mathbb{Z}}$ is equal to ϕ.

(b) As $P_- \cap N = N_0$ the restriction to N is an N-equivariant map $M^P \to \mathrm{Ind}_{N_0}^N(M)$. The map is injective because the restriction to N of $f \in M^P$ determines the restriction of f to $s^{\mathbb{N}}$ by (7.16) which determines f by (a). We have the natural injective map

$$f \mapsto \phi_f \quad : \quad \mathrm{Ind}_{s^{-\mathbb{N}}}^{s^{\mathbb{Z}}}(M) \to M^P \to \mathrm{Ind}_{N_0}^N(M) \qquad (7.17)$$

$$\phi_f(s^{-k} u s^k) = (\psi^k \circ u)(f(s^k)) \quad \text{for} \quad k \in \mathbb{N}, u \in N_0,$$

and we have the map

$$\phi \mapsto f_\phi \quad : \quad \mathrm{Ind}_{N_0}^N(M) \quad \to \quad \mathrm{Ind}_{s^{-\mathbb{N}}}^{s^{\mathbb{Z}}}(M)$$

defined by

$$f_\phi(s^k) = \sum_{v \in J(N_0/N_k)} (v \circ \varphi^k)(\phi(s^{-k} v^{-1} s^k)) \quad \text{for} \quad k \in \mathbb{N}.$$

Indeed the function f_ϕ satisfies $\psi(f_\phi(s^{k+1})) = f_\phi(s^k)$: since $\psi \circ u \circ \varphi^{k+1} = s^{-1} u s \circ \varphi^k$ when $u \in N_1$ and is 0 otherwise, we have

$$\psi(f_\phi(s^{k+1})) = \psi(\sum_{v \in J(N_0/N_{k+1})} (v \circ \varphi^{k+1})(\phi(s^{-k-1} v^{-1} s^{k+1})))$$

$$= \sum_{v \in N_1 \cap J(N_0/N_{k+1})} (s^{-1} v s \circ \varphi^k)(\phi(s^{-k-1} v^{-1} s^{k+1})).$$

The last term is

$$\sum_{v \in J(N_0/N_k)} (v \circ \varphi^k)(\phi(s^{-k} v^{-1} s^k)) = f_\phi(s^k)$$

because $s^{-1}(N_1 \cap J(N_0/N_{k+1}))s$ is a system of representatives of N_0/N_k and each term of the sum does not depend on the representative. Indeed for $u \in N_0$,

$$(vs^k us^{-k} \circ \varphi^k)(\phi(s^{-k}(vs^k us^{-k})^{-1}s^k)$$
$$= (v \circ \varphi^k \circ u)(\phi(u^{-1}s^{-k}v^{-1}s^k)) = (v \circ \varphi^k)(\phi(s^{-k}v^{-1}s^k)).$$

For $u \in N_0, k \in \mathbb{N}$, we have

$$\phi_{f_\phi}(s^{-k}us^k) = (\psi^k \circ u)f_\phi(s^k)$$
$$= \sum_{v \in J(N_0/N_k)} (\psi^k \circ uv \circ \varphi^k)(\phi(s^{-k}v^{-1}s^k)) = \phi(s^{-k}us^k)$$

where the last equality comes from $\mathrm{Ker}\,\psi^k = \sum_{u \in N_0 - N_k} u\varphi^k(M)$. Moreover, we have $f_{\phi_f} = f$ as a consequence of Lemma 7.17. □

Proposition 7.26. *The induction functor*

$$\mathrm{Ind}_{P_-}^{P} \quad : \quad \mathcal{M}_A(P_+)^{et} \to \mathcal{M}_A(P_-) \to \mathcal{M}_A(P)$$

is exact.

Proof. The canonical action of any element of P_- on an étale $A[P_+]$-module is surjective. Apply Lemma 7.9. □

Proposition 7.27. *Let $f \in M^P$. Let $n, n' \in N$ and $t \in L_+$ and denote by $k(n)$ the smallest integer $k \in \mathbb{N}$ such that $n \in N_{-k}$. We have:*

$$(nf)(s^m) = (s^m ns^{-m})(f(s^m)) \quad \text{for all } m \geq k(n),$$
$$(t^{-1}f)(s^m) = \psi_t(f(s^m)) \quad \text{and} \quad (sf)(s^m) = f(s^{m+1}) \quad \text{for all } m \in \mathbb{Z},$$
$$(s^k f)(n') = \sum_{v \in J(N_0/N_k)} v\varphi^k(f(s^{-k}v^{-1}n's^k)) \quad \text{for all } k \geq 1,$$
$$(t^{-1}f)(n') = \psi_t(f(tn't^{-1})) \quad \text{and} \quad (nf)(n') = f(n'n).$$

Proof. The formulas $(sf)(s^m) = f(s^{m+1})$, $(nf)(n') = f(n'n)$ are obvious. It is clear that

$$(t^{-1}f)(s^m) = f(s^m t^{-1}) = f(t^{-1}s^m) = t^{-1}(f(s^m)) = \psi_t(f(s^m)),$$

$$(t^{-1}f)(n') = f(nt^{-1}) = f(t^{-1}tn't^{-1}) = t^{-1}(f(tn't^{-1})) = \psi_t(f(tn't^{-1})).$$

$$nf(s^m) = f(s^m n) = f(s^m ns^{-m}s^m) = (s^m ns^{-m})f(s^m).$$

Using Lemma 7.17, we write

$$(s^k f)(n') = \sum_{v \in J(N_0/N_k)} v\varphi^k(\psi^k(v^{-1}((s^k f)(n')))),$$

$$\psi^k(v^{-1}((s^k f)(n'))) = \psi^k(v^{-1}(f(n's^k))) = \psi^k(f(v^{-1}n's^k))$$
$$= f(s^{-k}v^{-1}n's^k).$$

We obtain $(s^k f)(n') = \sum_{v \in J(N_0/N_k)} v\varphi(f(s^{-k}v^{-1}n's^k))$. $\qquad\square$

Definition 7.28. The s-model and the N-model of M^P are the spaces $\operatorname{Ind}_{s-N}^{s^{\mathbb{Z}}}(M) \simeq \varprojlim_{\psi} M$ and $\operatorname{Ind}_{N_0}^{N}(M)$, respectively, with the action of P described in Proposition 7.27.

3.3. Compactly induced representation M_c^P

The map

$$\operatorname{ev}_0 \; M^P \to M, \quad f \mapsto f(1),$$

admits a splitting

$$\sigma_0 : M \to M^P.$$

For $m \in M$, $\sigma_0(m)$ vanishes on $N - N_0$ and is equal to nm on $n \in N_0$ and to $\varphi^k(m)$ on s^k for $k \in \mathbb{N}$. In particular, by Proposition 7.25.b, σ_0 is independent of the choice of s.

Lemma 7.29. *The map* ev_0 *is* P_-*-equivariant, the map* σ_0 *is* P_+*-equivariant, the* $A[P_+]$*-modules* $\sigma_0(M)$ *and* M *are isomorphic.*

Proof. It is clear on the definition of M^P that ev_0 is P_--equivariant. We show that σ_0 is L_+-equivariant using the s-model. Let $t \in L_+$. We choose $t' \in L_+, r \in \mathbb{N}$ with $t't = s^r$. Then $\varphi_{t'}\varphi_t = \varphi^r$ and $\varphi_t = \psi_{t'}\varphi^r$. We obtain for $t\sigma_0(m)(s^k) = \sigma_0(m)(s^k t)$ the following expression

$$\sigma_0(m)(t'^{-1}s^{k+r}) = \psi_{t'}(\sigma_0(m)(s^{k+r}))$$
$$= \psi_{t'}\varphi^{r+k}(m) = \varphi_t\varphi^k(m) = \varphi^k\varphi_t(m) = \sigma_0(tm)(s^k).$$

Hence $t\sigma_0(m) = \sigma_0(tm)$. We show that σ_0 is N_0-equivariant using the N-model. Let $n_0 \in N_0$ and $m \in M$. Then $n_0\sigma_0(m) = \sigma_0(n_0m)$, because for $k \in \mathbb{N}$, $u \in N_0$,

$$n_0\sigma_0(m)(s^{-k}us^k) = \sigma_0(m)(s^{-k}us^k n_0) = \sigma_0(m)(s^{-k}us^k n_0 s^{-k}s^k)$$
$$= (\psi^k \circ us^k n_0 s^{-k} \circ \varphi^k)(m) = (\psi^k \circ u \circ \varphi^k)(n_0m)$$
$$= \sigma_0(n_0m)(s^{-k}us^k).$$

$\qquad\square$

The compact induction of M from P_- to P is defined to be the $A[P]$-submodule

$$\text{c-Ind}_{P_-}^P(M) := M_c^P$$

of M^P generated by $\sigma_0(M)$. The space M_c^P is the subspace of functions $f \in M^P$ with compact restriction to N, equivalently such that $f(s^{k+r}) = \varphi^k(f(s^r))$ for all $k \in \mathbb{N}$ and some $r \in \mathbb{N}$. The restriction to $s^{\mathbb{Z}}$ is an $s^{\mathbb{Z}}$-isomorphism (Proposition 7.25)

$$M_c^P \quad \rightarrow \quad \text{c-Ind}_{s^{-\mathbb{N}},\psi}^{s^{\mathbb{Z}}}(M).$$

By Proposition 7.11, the map

$$A[P] \otimes_{A[P_+]} M \rightarrow \text{c-Ind}_{P_-}^P(M)$$

$$[s^{-k}] \otimes m \mapsto (\varphi^{-k} \circ \sigma_0)(m)$$

is an isomorphism.

Lemma 7.30. *The compact induction functor from P_- to P is isomorphic to*

$$\text{c-Ind}_{P_-}^P \simeq A[P] \otimes_{A[P_+]} \mathcal{M}_A(P_+)^{et} \rightarrow \mathcal{M}_A(P), \qquad (7.18)$$

and is exact.

Proof. For the exactness see Corollary 7.12. $\qquad\qquad\qquad\qquad\square$

3.4. P-equivariant map $\text{Res}: C_c^\infty(N, A) \rightarrow \text{End}_A(M^P)$

Let $C_c^\infty(N, A)$ be the A-module of locally constant compactly supported functions on N with values in A, with the usual product of functions and with the natural action of P,

$$P \times C_c^\infty(N, A) \rightarrow C_c^\infty(N, A), \quad (b, f) \mapsto (bf)(x) = f(b^{-1}.x).$$

For any open compact subgroup $U \subset N$, the subring $C^\infty(U, A) \subset C_c^\infty(N, A)$ of functions f supported in U, has a unit equal to the characteristic function 1_U of U, and is stable by the P-stabilizer P_U of U. We have $b1_U = 1_{b.U}$. The $A[P_U]$-module $C^\infty(U, A)$ and the $A[P]$-module $C_c^\infty(N, A)$ are cyclic generated by 1_U. The monoid $P_+ = N_0 L_+$ acts on $\text{End}_A(M)$ by

$$P_+ \times \text{End}_A(M) \rightarrow \text{End}_A(M)$$

$$(b, F) \mapsto \varphi_b \circ F \circ \psi_b.$$

Note that we have $\psi_{ut} = \psi_t \circ u^{-1}$.

Proposition 7.31. *There exists a unique P_+-equivariant A-linear map*

$$\text{res} \quad C^\infty(N_0, A) \quad \to \quad \text{End}_A(M)$$

respecting the unit. It is a homomorphism of A-algebras.

Proof. If the map res exists, it is unique because the $A[P_+]$-module $C^\infty(N_0, A)$ is generated by the unit 1_{N_0}. The existence of res is equivalent to Lemma 7.17 as we will show below. For $b \in P_+$ we have the idempotent

$$\text{res}(1_{b.N_0}) := \varphi_b \circ \psi_b \in \text{End}_A(M). \tag{7.19}$$

We claim that for any finite disjoint union $b.N_0 = \bigsqcup_{i \in I} b_i.N_0$ with $b_i \in P_+$, the idempotents $\text{res}(1_{b_i.N_0})$ are orthogonal of sum $\text{res}(1_{b.N_0})$. We may assume that $b = 1$, since the inclusion $b_i.N_0 \subset b.N_0$ yields $b^{-1}b_i \in P_+$. Write $b_i = u_i t_i$ with $u_i \in N_0$ and $t_i \in L_+$, and choose $t' \in L_+$ such that $t' \in t_i L_+$, say $t' = t_i l_i$ (with $l_i \in L_+$). Let $(n_{ij})_j$ be a system of representatives for $N_0/l_i.N_0$. Since M is étale, Lemma 7.17 shows that, for each i, the idempotents $(\varphi_{n_{ij}l_i} \circ \psi_{n_{ij}l_i})_j$ are orthogonal, with sum id_M. Note that $(v_{ij} := u_i t_i n_{ij} t_i^{-1})_{(i,j)}$ form a system of representatives for $N_0/t'.N_0$, so again by Lemma 7.17 the idempotents $(\varphi_{v_{ij}t'} \circ \psi_{v_{ij}t'})_{(i,j)}$ are orthogonal with sum id_M. The claim follows, since $v_{ij}t' = b_i(n_{ij}l_i)$, so

$$\varphi_{v_{ij}t'} \circ \psi_{v_{ij}t'} = \varphi_{b_i} \circ \varphi_{n_{ij}l_i} \circ \psi_{n_{ij}l_i} \circ \psi_{b_i}.$$

The claim being proved, we get an A-linear map res : $C^\infty(N_0, A) \to \text{End}_A(M)$ which is clearly P_+-equivariant and respects the unit. It respects the product because, for $f_1, f_2 \in C^\infty(N_0, A)$, there exists $t \in L_+$ such that f_1 and f_2 are constant on each coset $utN_0t^{-1} \subset N_0$. Hence $\text{res}(f_1 f_2) = \sum_{v \in J(N_0/tN_0t^{-1})} f_1(v) f_2(v) \text{res}(1_{vt.N_0}) = \text{res}(f_1) \circ \text{res}(f_2)$. $\square$

The group $P = NL$ acts on $\text{End}_A(M^P)$ by conjugation. We have the canonical injective algebra map

$$F \mapsto \sigma_0 \circ F \circ \text{ev}_0 \quad : \quad \text{End}_A M \to \text{End}_A(M^P). \tag{7.20}$$

It is P_+-equivariant since, by Lemma 7.29 for $b \in P_+$, we have

$$b \circ \sigma_0 \circ F \circ \text{ev}_0 \circ b^{-1} = \sigma_0 \circ \varphi_b \circ F \circ \psi_b \circ \text{ev}_0. \tag{7.21}$$

We consider the composite P_+-equivariant algebra homomorphism

$$C^\infty(N_0, A) \xrightarrow{\text{res}} \text{End}_A(M) \longrightarrow \text{End}_A(M^P),$$

sending 1_{N_0} to $R_0 := \sigma_0 \circ \text{ev}_0$ and, more generally, $1_{b.N_0}$ to $b \circ R_0 \circ b^{-1}$ for $b \in P_+$.

For $f \in M^P$, $R_0(f) \in M^P$ vanishes on $N - N_0$ and $R_0(f)(s^k) = \varphi^k(f(1))$. In the N-model, R_0 is the restriction to N_0.

We show now that the composite morphism extends to $C_c^\infty(N, A)$.

Proposition 7.32. *There exists a unique P-equivariant A-linear map*

$$\text{Res} \quad C_c^\infty(N, A) \rightarrow \text{End}_A(M^P)$$

such that $\text{Res}(1_{N_0}) = R_0$. *The map* Res *is an algebra homomorphism.*

Proof. If the map Res exists, it is unique because the $A[P]$-module $C_c^\infty(N, A)$ is generated by 1_{N_0}.

For $b \in P$ we define

$$\text{Res}(1_{b.N_0}) := b \circ R_0 \circ b^{-1}.$$

We prove that $b \circ R_0 \circ b^{-1}$ depends only on the subset $b.N_0 \subset N$, and that for any finite disjoint decomposition of $b.N_0 = \sqcup_{i \in I} b_i.N_0$ with $b_i \in P$, the idempotents $b_i \circ R_0 \circ b_i^{-1}$ are orthogonal of sum $b \circ R_0 \circ b^{-1}$.

The equivalence relation $b.N_0 = b'.N_0$ for $b, b' \in P$ is equivalent to $b' P_0 = b P_0$ because the normalizer of N_0 in P is P_0. We have $b \circ R_0 \circ b^{-1} = R_0$ when $b \in P_0$ because $\text{res}(1_{b.N_0}) = \text{res}(1_{N_0}) = \text{id}$ (Proposition 7.31). Hence $b \circ R_0 \circ b^{-1}$ depends only on $b.N_0$. By conjugation by b^{-1}, we reduce to prove that the idempotents $b_i \circ R_0 \circ b_i^{-1}$ are orthogonal of sum R_0 for any disjoint decomposition of $N_0 = \sqcup_{i \in I} b_i.N_0$ and $b_i \in P$. The b_i belong to P_+, and Proposition 7.31 implies the equality.

To prove that the A-linear map Res respects the product it suffices to check that, for any $t \in L_+, k \in \mathbb{N}$, the endomorphisms $\text{Res}(1_{vt N_0 t^{-1}}) \in \text{End}_A(M^P)$ are orthogonal idempotents, for $v \in J(N_{-k}/t N_0 t^{-1})$. We already proved this for $k = 0$ and for all $t \in L_+$, and $s^k J(N_{-k}/t N_0 t^{-1}) s^{-k} = J(N_0/s^k t N_0 t^{-1} s^{-k})$. Hence we know that

$$(s^k \circ \text{Res}(1_{vt N_0 t^{-1}}) \circ s^{-k})_{v \in J(N_{-k}/t N_0 t^{-1})}$$

are orthogonal idempotents. This implies that $(\text{Res}(1_{vt N_0 t^{-1}}))_{v \in J(N_{-k}/t N_0 t^{-1})}$ are orthogonal idempotents. □

Remark 7.33. (i) The map Res is the restriction of an algebra homomorphism

$$C^\infty(N, A) \rightarrow \text{End}_A(M^P),$$

where $C^\infty(N, A)$ is the algebra of all locally constant functions on N. For this we observe

1. The $A[P_+]$-module $C^\infty(N_0, A)$ is étale. For $t \in L_+$, the corresponding ψ_t satisfies $(\psi_t f)(x) = f(txt^{-1})$.
2. The map $(f, m) \mapsto \mathrm{res}(f)(m)$ $C^\infty(N_0, A) \times M \to M$ is ψ_t-equivariant, hence induces to a pairing $C^\infty(N_0, A)^P \times M^P \to M^P$.
3. The $A[P]$-module $C^\infty(N_0, A)^P$ is canonically isomorphic to $C^\infty(N, A)$.

(ii) The monoid $P_+ \times P_+$ acts on $\mathrm{End}_A(M)$ by $\varphi_{(b_1,b_2)} F := \varphi_{b_1} \circ F \circ \psi_{b_2}$. For this action $\mathrm{End}_A(M)$ is an étale $A[P_+ \times P_+]$-module, and we have $\psi_{(b_1,b_2)} F = \psi_{b_1} \circ F \circ \varphi_{b_2}$.

Definition 7.34. For any compact open subsets $V \subset U \subset N_0$ and $m \in M$, we denote

$$\mathrm{res}_U := \mathrm{res}(1_U), \quad M_U := \mathrm{res}_U(M), \quad m_U := \mathrm{res}_U(m),$$
$$\mathrm{res}_V^U := \mathrm{res}_V |_{M_U} : M_U \to M_V.$$

For any compact open subsets $V \subset U \subset N$ and $f \in M^P$

$$\mathrm{Res}_U := \mathrm{Res}(1_U), \quad M_U := \mathrm{Res}_U(M^P), \quad f_U := \mathrm{Res}_U(f),$$
$$\mathrm{Res}_V^U := \mathrm{Res}_V |_{M_U} : M_U \to M_V.$$

Remark 7.35. The notation is coherent for $U \subset N_0$, as follows from the following properties. For $b \in P_+$ we have

- $\mathrm{res}_{b.U} = \varphi_b \circ \mathrm{res}_U \circ \psi_b$ (Proposition 7.31);
- $b \circ \mathrm{Res}_U = \sigma_0 \circ \varphi_b \circ \mathrm{res}_U \circ \mathrm{ev}_0$ and $\mathrm{Res}_U \circ b^{-1} = \sigma_0 \circ \mathrm{res}_U \circ \psi_b \circ \mathrm{ev}_0$;
- $(\mathrm{Res}_U f)(1) = \mathrm{res}_U(f(1))$.

We note also that Proposition 7.32 implies:

Corollary 7.36. *For any compact open subset $U \subset N$ equal to a finite disjoint union $U = \sqcup_{i \in I} U_i$ of compact open subsets $U_i \subset N$, the idempotents Res_{U_i} are orthogonal of sum Res_U.*

Corollary 7.37. *For $u \in N$, the projector Res_{uN_0} is the restriction to N_0u^{-1} in the N-model.*

Proof. We have $\mathrm{Res}_{uN_0} = u \circ \mathrm{Res}_{N_0} \circ u^{-1}$ and Res_{N_0} is the restriction to N_0 in the N-model. Hence for $x \in N$, $(\mathrm{Res}_{uN_0} f)(x) = (\mathrm{Res}_{N_0} u^{-1} f)(xu)$ vanishes for $x \in N - N_0u^{-1}$ and for $v \in N_0$, $(\mathrm{Res}_{uN_0} f)(vu^{-1}) = (u^{-1}f)(v) = f(vu^{-1})$. $\square$

The constructions are functorial. A morphism $f : M \to M'$ of $A[P_+]$-modules, being also $A[P_-]$-equivariant induces a morphism $\mathrm{Ind}_{P_-}^P(f)$:

$M^P \to M'^P$ of $A[P]$-modules. On the other hand, M^P is a module over the non unital ring $C_c^\infty(N, A)$ through the map Res. The morphism $\mathrm{Ind}_{P_-}^P(f)$ is $C_c^\infty(N, A)$-equivariant. Since Res is P-equivariant, it suffices to prove that $\mathrm{Ind}_{P_-}^P(f)$ respects $R_0 = \sigma_0 \circ \mathrm{ev}_0$ which is obvious.

3.5. *P*-equivariant sheaf on *N*

We formulate now Proposition 7.32 in the language of sheaves.

Theorem 7.38. *One can associate to an étale $A[P_+]$-module M, a P-equivariant sheaf $\mathcal{S}_M$ of A-modules on the compact open subsets $U \subset N$, with*

- *sections M_U on U,*
- *restrictions Res_V^U for any open compact subset $V \subset U$,*
- *action $f \mapsto bf = \mathrm{Res}_{b.U}(bf)\ M_U \to M_{b.U}$ of $b \in P$.*

Proof. (a) Res_U^U is the identity on $M_U = \mathrm{Res}_U(M)$ because Res_U is an idempotent.

  (b) $\mathrm{Res}_W^V \circ \mathrm{Res}_V^U = \mathrm{Res}_W^U$ for compact open subsets $W \subset V \subset U \subset N$. Indeed, we have $\mathrm{Res}_W \circ \mathrm{Res}_V = \mathrm{Res}_W$ on M_U.

  (c) If U is the union of compact open subsets $U_i \subset U$ for $i \in I$, and $f_i \in M_{U_i}$ satisfying $\mathrm{Res}_{U_i \cap U_j}^{U_i}(f_i) = \mathrm{Res}_{U_i \cap U_j}^{U_j}(f_j)$ for $i, j \in I$, there exists a unique $f \in M_U$ such that $\mathrm{Res}_{U_i}^U(f) = f_i$ for all $i \in I$.

  (c1) True when $(U_i)_{i \in I}$ is a partition of U because I is finite and Res_U is the sum of the orthogonal idempotents Res_{U_i}.

  (c2) True when I is finite because the finite covering defines a finite partition of U by open compact subsets V_j for $j \in J$, such that $V_j \cap U_i$ is empty or equal to V_j for all $i \in I$, $j \in J$. By hypothesis on the f_i, if $V_j \subset U_i$, then the restriction of f_i to V_j does not depend on the choice of i, and is denoted by ϕ_j. Applying (c1), there is a unique $f \in M_U$ such that $\mathrm{Res}_{V_j}(f) = \phi_j$ for all $j \in J$. Note also that the V_j contained in U_i form a finite partition of U_i and that f_i is the unique element of M_{U_i} such that $\mathrm{Res}_{V_j}(f_i) = \phi_j$ for those j. We deduce that f is the unique element of M_U such that $\mathrm{Res}_{U_i}(f) = f_i$ for all $i \in I$.

  (c3) In general, U being compact, there exists a finite subset $I' \subset I$ such that U is covered by U_i for $i \in I'$. By (c2), there exists a unique $f_{I'} \in M_U$ such that $f_i = \mathrm{Res}_{U_i}(f_{I'})$ for all $i \in I'$. Let $i' \in I$ not belonging to I'. Then the nonempty intersections $U_{i'} \cap U_j$ for $j \in I'$ form a finite covering of $U_{i'}$ by compact open subsets. By (c2), $f_{i'}$ is the unique element of $M_{U_{i'}}$ such that $\mathrm{Res}_{U_{i'} \cap U_j}(f_j) = \mathrm{Res}_{U_{i'} \cap U_j}(f_{i'})$ for all nonempty $U_{i'} \cap U_j$. The element $\text{Res}_{U_{i'}}(f_{I'})$ has the same property, we deduce by uniqueness that $f_{i'} = \text{Res}_{U_{i'}}(f_{I'})$.

(d) Let $f \in M_U$. When $b = 1$ we have clearly $1(f) = f$. For $b, b' \in P$, we have $(bb')(f) = \text{Res}_{(bb').U}((bb')f) = \text{Res}_{b.(b'.U)}(b(b'f)) = b(b'f)$. For a compact open subset $V \subset U$, we have $b \circ \text{Res}_V \circ \text{Res}_U = \text{Res}_{bV} \circ b \circ \text{Res}_U$ in $\text{End}_A M^P$ hence $b \, \text{Res}_V^U = \text{Res}_{b.V} \, b$. $\qquad\square$

Proposition 7.39. *Let H be a topological group acting continuously on a locally compact totally disconnected space X. Any H-equivariant sheaf $\mathcal{F}$ (of A-modules) on the compact open subsets of X extends uniquely to a H-equivariant sheaf on the open subsets of X.*

Proof. This is well known. See [4] §9.2.3 Prop. 1. $\qquad\square$

Remark 7.40. The space of sections on an open subset $U \subset X$ is the projective limit of the sections $\mathcal{F}(V)$ on the compact open subsets V of U for the restriction maps $\mathcal{F}(V) \to \mathcal{F}(V')$ for $V' \subset V$.

By this general result, the P-equivariant sheaf defined by M on the compact open subsets of N (Theorem 7.38), extends uniquely to a P-equivariant sheaf $\mathcal{S}_M$ on (arbitrary open subsets of) N. We extend the definitions 7.34 to arbitrary open subsets $U \subset N$. We denote by Res_V^U the restriction maps for open subsets $V \subset U$ of N, by $\text{Res}_U = \text{Res}_U^N$ and by $M_U = \text{Res}_U(M^P)$. In this way we obtain an exact functor $M \to (M_U)_U$ from $\mathcal{M}_A(P_+)^{et}$ to the category of P-equivariant sheaves of A-modules on N. Note that for a compact open subset U even the functor $M \to M_U$ is exact.

Proposition 7.41. *The representation of P on the global sections of the sheaf $\mathcal{S}_M$ is canonically isomorphic to M^P.*

Proof. We have the obvious P-equivariant homomorphism

$$M^P \xrightarrow{\;(\text{Res}_U)_U\;} M_N = \varprojlim_U M_U.$$

The group N is the union of $s^{-k}.N_0 = s^{-k} N_0 s^k$ for $k \in \mathbb{N}$. Hence $M_N = \varprojlim_k M_{N_{-k}}$. In the s-model of M^P we have $\text{Res}_{s^{-k}.N_0} = R_{-k}$ and by the lemma 7.13 the morphism

$$f \mapsto (\text{Res}_{s^{-k}.N_0}(f))_{k \in \mathbb{N}} \quad M^P \quad \to \quad M_N$$

is bijective. $\qquad\square$

Corollary 7.42. *The restriction* $\mathrm{Res}_U^N : M_N \to M_U$ *from the global sections to the sections on an open compact subset* $U \subset N$ *is surjective with a natural splitting.*

Proof. It corresponds to an idempotent $\mathrm{Res}_U = \mathrm{Res}(1_U) \in \mathrm{End}_A(M^P)$. $\square$

3.6. Independence of N_0

Let $U \subset N$ be a compact open subgroup. For $n \in N$ and $t \in L$, the inclusion $ntUt^{-1} \subset U$ is obviously equivalent to $n \in U$ and $tUt^{-1} \subset U$. Hence the P-stabilizer $P_U := \{b \in P \mid b.U \subset U\}$ of U is the semi-direct product of U by the L-stabilizer L_U of U. As the decreasing sequence $(N_k = s^k N_0 s^{-k})_{k \in \mathbb{N}}$ forms a basis of neighbourhoods of 1 in N and $N = \cup_{r \in \mathbb{Z}} N_{-r}$, the compact open subgroup $U \subset N$ contains some N_k and is contained in some N_{-r}. This implies that the intersection $L_U \cap s^{\mathbb{N}}$ is not empty hence is equal to $s_U^{\mathbb{N}}$ where $s_U = s^{k_U}$ for some $k_U \geq 1$. The monoid $P_U = UL_U$ and the central element s_U of L satisfy the same conditions as $(P_+ = N_0 L_+, s)$, given at the beginning of Section 3.2. Our theory associates to each étale $A[P_U]$-module a P-equivariant sheaf on N.

The subspace $M_U \subset M^P$ (Definition 7.34) is stable by P_U because $b \circ \mathrm{Res}_U = \mathrm{Res}_{b.U} \circ b$ for $b \in P$ and $M_{b.U} = \mathrm{Res}_{b.U}(M) \subset \mathrm{Res}_U(M) = M_U$. As $M_U = \oplus_{u \in J(U/t.U)} u M_{t.U}$ for $t \in L_U$ the $A[P_U]$-module M_U is étale.

Proposition 7.43. *The P-equivariant sheaf S_M on N associated to the étale $A[P_+]$-module M is equal to the P-equivariant sheaf on N associated to the étale $A[P_U]$-module M_U.*

Proof. For $b \in P_U$ we denote by $\varphi_{U,b}$ the action of b on M_U and by $\psi_{U,b}$ the left inverse of $\varphi_{U,b}$ with kernel $M_{U-b.U}$. We have $M_U = M_{b.U} \oplus M_{U-b.U}$ and for $f_U \in M_U$,

$$\varphi_{U,b}(f_U) = bf_U, \quad \psi_{U,b}(f_U) = b^{-1} \mathrm{Res}_{b.U}(f_U), \quad (\varphi_{U,b} \circ \psi_{U,b})(f_U)$$
$$= \mathrm{Res}_{b.U}(f_U). \tag{7.22}$$

By the last formula and Remark 7.35, the sections on $b.U$ and the restriction maps from M_U to $M_{b.U}$ in the two sheaves are the same for any $b \in P_U$. This implies that the two sheaves are equal on (the open subsets of) U. By symmetry they are also equal on (the open subsets of) N_0. The same arguments for arbitrary compact open subgroups $U, U' \subset N$ imply that the P-equivariant sheaves on N associated to the étale $A[P_U]$-module M_U and to the étale $A[P_{U'}]$-module $M_{U'}$ are equal on (the open subsets of) U and on (the open subsets of) U'. Hence all these sheaves are equal on (the open subsets of) the compact open subsets of N and also on (the open subsets of) N.   □

3.7. Etale $A[P_+]$-module and P-equivariant sheaf on N

Proposition 7.44. *Let M be an $A[P_+]$-module such that the action φ of s on M is étale. Then M is an étale $A[P_+]$-module.*

Proof. Let $t \in L_+$. We have to show that the action φ_t of t on M is étale. As $L = L_+ s^{-\mathbb{N}}$ with s central in L, there exists $k \in \mathbb{N}$ such that $s^k t^{-1} \in L_+$. This implies $\varphi^k = \varphi_{s^k t^{-1}} \circ \varphi_t$ in $\mathrm{End}_A(M)$ and $s^k N_0 s^{-k} \subset t N_0 t^{-1}$. As φ is injective, φ_t is also injective. For any representative system $J(t N_0 t^{-1}/s^k N_0 s^{-k})$ of $t N_0 t^{-1}/s^k N_0 s^{-k}$ and any representative system $J(N_0/t N_0 t^{-1})$ of $N_0/t N_0 t^{-1}$, the set of uv for $u \in J(N_0/t N_0 t^{-1})$ and $v \in J(t N_0 t^{-1}/s^k N_0 s^{-k})$ is a representative system $J(N_0/s^k N_0 s^{-k})$ of $N_0/s^k N_0 s^{-k}$. Let ψ be the canonical left inverse of φ. We have

$$\mathrm{id} = \sum_{u \in J(N_0/t N_0 t^{-1})} u \circ \sum_{v \in J(t N_0 t^{-1}/s^k N_0 s^{-k})} v \circ \varphi^k \circ \psi^k \circ v^{-1} \circ u^{-1}$$

$$= \sum_{u \in J(N_0/t N_0 t^{-1})} u \circ \sum_{v \in J(t N_0 t^{-1}/s^k N_0 s^{-k})} v \circ \varphi_t \circ \varphi_{t^{-1} s^k} \circ \psi^k \circ v^{-1} \circ u^{-1}$$

$$= \sum_{u \in J(N_0/t N_0 t^{-1})} u \circ \varphi_t \circ \left(\sum_{v \in J(N_0/t^{-1} s^k N_0 s^{-k} t)} v \circ \varphi_{t^{-1} s^k} \circ \psi^k \circ v^{-1} \right) \circ u^{-1}.$$

We deduce that φ_t is étale of canonical left inverse ψ_t the expression between parentheses.   □

Corollary 7.45. *An $A[P_+]$-submodule $M' \subset M$ of an étale $A[P_+]$-module M is étale if and only if it is stable by the canonical inverse ψ of φ.*

Proof. If M' is ψ-stable, for $m' \in M'$ every $m'_{u,s}$ belongs to M' in (7.12). Hence the action of s on M' is étale, and M' is étale by Proposition 7.44.   □

Corollary 7.46. *The space $\mathcal{S}(N_0)$ of global sections of a P_+-equivariant sheaf $\mathcal{S}$ on N_0 is an étale representation of P_+, when the action φ of s on $\mathcal{S}(N_0)$ is injective.*

Proof. By Proposition 7.44 it suffices to show that $\mathcal{S}(N_0) = \oplus_{u \in J(N_0/s N_0 s^{-1})} us(\mathcal{S}(N_0))$. But this equality is true because N_0 is the disjoint sum of the open subsets $us N_0 s^{-1} = us.N_0$ and $\mathcal{S}(us.N_0) = us(\mathcal{S}(N_0))$.   □

The canonical left inverse ψ of the action φ of s on $\mathcal{S}(N_0)$ vanishes on $\mathcal{S}(us N_0 s^{-1})$ for $u \neq 1$ and on $\mathcal{S}(s N_0 s^{-1})$ is equal to the isomorphism $\mathcal{S}(s N_0 s^{-1}) \to \mathcal{S}(N_0)$ induced by s^{-1}.

Theorem 7.47. *The functor $M \mapsto \mathcal{S}_M$ is an equivalence of categories from the abelian category of étale $A[P_+]$-modules to the abelian category of P-equivariant sheaves of A-modules on N, of inverse the functor $\mathcal{S} \mapsto \mathcal{S}(N_0)$ of sections over N_0.*

Proof. Let $\mathcal{S}$ be a P-equivariant sheaf on N. By Corollary 7.46, the space $\mathcal{S}(N_0)$ of sections on N_0 is an étale representation of P_+ because the action φ of s on $\mathcal{S}(N_0)$ is injective.

We show now that the representation of P on the space $\mathcal{S}(N)_c$ of compact sections on N depends uniquely the representation of P_+ on $\mathcal{S}(N_0)$. The representation of N on $\mathcal{S}(N)_c$ is defined by the representation of N_0 on $\mathcal{S}(N_0)$, because $\mathcal{S}(N)_c = \oplus_{u \in J(N/N_0)} \mathcal{S}(u N_0)$ and $\mathcal{S}(u N_0) = u \mathcal{S}(N_0)$ for $u \in N$. The group P is generated by N and L_+. For $t \in L_+$, the action of t on $\mathcal{S}(N)_c$ is defined by the action of N on $\mathcal{S}(N)_c$ and by the action of t on $\mathcal{S}(N_0)$, because $t \mathcal{S}(u N_0) = t u t^{-1} t \mathcal{S}(N_0)$ with $t u t^{-1} \in N$ for $u \in N$.

We deduce that the $A[P]$-module $\mathcal{S}(N)_c$ is equal to the compact induced representation $\mathcal{S}(N_0)_c^P$, and that the sheaves $\mathcal{S}$ and $\mathcal{S}_{\mathcal{S}(N_0)}$ are equal.

Conversely, let M be an étale $A[P_+]$-module. The $A[P_+]$-module $\mathcal{S}_M(N_0)$ of sections on N_0 of the sheaf $\mathcal{S}_M$ is equal to M (Theorem 7.38). $\qquad\square$

4. Topology

4.1. Topologically étale $A[P_+]$-modules

We add to the hypothesis of Section 3.2 the following

(a) *A is a linearly topological commutative ring (the open ideals form a basis of neighbourhoods of 0).*

(b) *M is a linearly topological A-module (the open A-submodules form a basis of neighbourhoods of 0), with a continuous action of P_+*

$$P_+ \times M \to M$$

$$(b, x) \mapsto \varphi_b(x).$$

We call such an M a continuous $A[P_+]$-module. If M is also étale in the algebraic sense (Definition 7.16) and the maps ψ_t, for $t \in L_+$, are continuous we call M a topologically étale $A[P_+]$-module.

Lemma 7.48. *Let M be a continuous $A[P_+]$-module which is algebraically étale, then:*

(i) *The maps ψ_t for $t \in L_+$ are open.*

(ii) *If $\psi = \psi_s$ is continuous then M is topologically étale.*

Proof. (i) The projection of $M = M_0 \oplus M_1$ onto the algebraic direct summand M_0 (with the submodule topology) is open. Indeed let $V \subset M$ be an open subset, then $M_0 \cap (V + M_1)$ is open in M_0 and is equal to the projection of V. We apply this to $M = \varphi_t(M) \oplus \operatorname{Ker} \psi_t$ and to the projection $\varphi_t \circ \psi_t$. Then we note that $\psi_t(V) = \varphi_t^{-1}((\varphi_t \circ \psi_t)(V))$.

(ii) Given any $t \in L_+$ we find $t' \in L_+$ and $n \in \mathbb{N}$ such that $t't = s^n$. Hence $\psi_{t't} = \psi_t \circ \psi_{t'} = \psi^n$ is continuous by assumption. As $\psi_{t'}$ is surjective and open, for any open subset $V \subset M$ we have $\psi_t^{-1}(V) = \psi_{t'}((\psi_t \circ \psi_{t'})^{-1}(V))$ which is open. $\qquad\square$

Lemma 7.49. (i) *A compact algebraically étale $A[P_+]$-module is topologically étale.*

(ii) *Let M be a topologically étale $A[P_+]$-module. The P_--action $(b^{-1}, m) \mapsto \psi_b(m)$ $P_- \times M \to M$ on M is continuous.*

Proof. (i) The compactness of M implies that

$$M = \varphi_t(M) \oplus \bigoplus_{u \in (N_0 - tN_0 t^{-1})} u\varphi_t(M)$$

is a topological decomposition of M as the direct sum of finitely many closed submodules. It suffices to check that the restriction of ψ_t to each summand is continuous. On all summands except the first one ψ_t is zero. By compactness of M the map φ_t is a homeomorphism between M and the closed submodule $\varphi_t(M)$. We see that $\psi_t|\varphi_t(M)$ is the inverse of this homeomorphism and hence is continuous.

(ii) Since P_0 is open in $P_- = L_+^{-1} P_0$ we only need to show that the restriction of the P_--action to $t^{-1}P_0 \times M \to M$, for any $t \in L_+$, is continuous. We contemplate the commutative diagram

$$
\begin{array}{ccc}
t^{-1}P_0 \times M & \longrightarrow & M \\
{\scriptstyle t \cdot \times \, \mathrm{id}} \downarrow & & \uparrow {\scriptstyle \psi_t} \\
P_0 \times M & \longrightarrow & M
\end{array}
$$

where the horizontal arrows are given by the P_--action. The P_0-action on M induced by P_- coincides with the one induced by P_+-action. Therefore the bottom horizontal arrow is continuous. The left vertical arrow is trivially continuous, and ψ_t is continuous by assumption. $\qquad\square$

Lemma 7.50. *For any compact subgroup* $C \subset P_+$, *the open C-stable A-submodules of M form a basis of neighbourhoods of* 0.

Proof. We have to show that any open A-submodule $\mathcal{M}$ of M contains an open C-stable A-submodule. By continuity of the action of P_+ on M, there exists for each $c \in C$, an open A-submodule $\mathcal{M}_c$ of M and an open neighbourhood $H_c \subset P_+$ of c such that $\varphi_x(\mathcal{M}_c) \subset \mathcal{M}$ for all $x \in H_c$. By the compactness of C, there exists a finite subset $I \subset C$ such that $C = \cup_{c \in I}(H_c \cap C)$. By finiteness of I, the intersection $\mathcal{M}'' := \cap_{c \in I}\mathcal{M}_c \subset M$ is an open A-submodule such that $\mathcal{M}' := \sum_{c \in C} \varphi_c(\mathcal{M}'') \subset \mathcal{M}$. The A-submodule $\mathcal{M}'$ is C-stable and, since $\mathcal{M}'' \subset \mathcal{M}' \subset \mathcal{M}$, also open. $\qquad\square$

Let M be a topologically étale $A[P_+]$-module. Since P_0 is open in P the A-module M^P is a submodule of the A-module $C(P, M)$ of all continuous maps from P to M. We equip $C(P, M)$ with the compact-open topology which makes it a linear-topological A-module. A basis of neighbourhoods of zero is given by the submodules $\mathcal{C}(C, \mathcal{M}) := \{f \in C(P, M) \mid f(C) \subset \mathcal{M}\}$ with C and $\mathcal{M}$ running over all compact subsets in P and over all open submodules in M, respectively. With M also $C(P, M)$ is Hausdorff. It is well known that the regular action of P on $C(P, M)$ is continuous (see for instance Proposition 7.52(ii) for a proof). Therefore M^P is characterized inside $C(P, M)$ by closed conditions and hence is a closed submodule. Similarly, $\mathrm{Ind}_{s^{-\mathbb{N}}}^{s^{\mathbb{Z}}}(M)$ and $\mathrm{Ind}_{N_0}^N(M)$ are closed submodules of $C(s^{\mathbb{Z}}, M)$ and $C(N, M)$, respectively, for the compact-open topologies. Clearly the homomorphisms of restricting maps (Proposition 7.25) $M^P \to \mathrm{Ind}_{s^{-\mathbb{N}}}^{s^{\mathbb{Z}}}(M)$ and $M^P \to \mathrm{Ind}_{N_0}^N(M)$ are continuous.

Lemma 7.51. *The restriction maps* $M^P \to \mathrm{Ind}_{s^{-\mathbb{N}}}^{s^{\mathbb{Z}}}(M)$ *and* $M^P \to \mathrm{Ind}_{N_0}^N(M)$ *are topological isomorphisms.*

Proof. The topology on M^P induced by the compact-open topology on the s-model $\mathrm{Ind}_{s^{-\mathbb{N}}}^{s^{\mathbb{Z}}} M$ is the topology with basis of neighbourhoods of zero

$$B_{k,\mathcal{M}} = \{f \in M^P \mid f(s^m) \in \mathcal{M} \text{ for all } -k \leq m \leq k\},$$

for all $k \in \mathbb{N}$ and all open A-submodules $\mathcal{M}$ of M. One can replace $B_{k,\mathcal{M}}$ by

$$C_{k,\mathcal{M}} = \{f \in M^P \mid f(s^k) \in \mathcal{M}\},$$

because $B_{k,\mathcal{M}} \subset C_{k,\mathcal{M}}$ and conversely given $(k, \mathcal{M})$ there exists an open A-submodule $\mathcal{M}' \subset \mathcal{M}$ such that $\psi^m(\mathcal{M}') \subset \mathcal{M}$ for all $0 \leq m \leq 2k$ as ψ is continuous (Lemma 7.49), hence $C_{k,\mathcal{M}'} \subset B_{k,\mathcal{M}}$.

The topology on M^P induced by the compact-open topology on the N-model $\mathrm{Ind}_{N_0}^N M$ is the topology with basis of neighbourhoods of zero

$$D_{k,\mathcal{M}} = \{f \in M^P \mid f(N_{-k}) \subset \mathcal{M}\},$$

for all $(k, \mathcal{M})$ as above.

We fix an auxiliary compact open subgroup $P_0' \subset P_0$. It then suffices, by Lemma 7.50, to let $\mathcal{M}$ run, in the above families, over the open $A[P_0']$-submodules $\mathcal{M}$ of M.

Let $C \subset P$ be any compact subset and let $\mathcal{M}$ be an open $A[P_0']$-submodule of M. We choose $k \in \mathbb{N}$ large enough so that $Cs^{-k} \subset P_-$. Since Cs^{-k} is compact and P_0' is an open subgroup of P we find finitely many $b_1, \ldots, b_m \in P_+$ such that $Cs^{-k} \subset b_1^{-1} P_0' \cup \ldots \cup b_m^{-1} P_0'$. The continuity of the maps ψ_{b_i} implies the existence of an open $A[P_0']$-submodule $\mathcal{M}'$ of M such that $\psi_{b_i}(\mathcal{M}') \subset \mathcal{M}$ for any $1 \leq i \leq m$. We deduce that

$$C_{k,\mathcal{M}'} \subset C\left(\bigcup_i b_i^{-1} P_0' s^k, \mathcal{M}\right) \subset C(C, \mathcal{M}).$$

Furthermore, by the continuity of the action of P_+ on M, there exists an open submodule $\mathcal{M}''$ such that $\sum_{v \in J(N_0/N_k)} v\varphi^k(\mathcal{M}'') \subset \mathcal{M}'$. The second part of the formula (7.16) then implies that

$$D_{k,\mathcal{M}''} \subset C_{k,\mathcal{M}}.$$

$\square$

The maps $\mathrm{ev}_0 : M^P \to M$ and $\sigma_0 : M \to M^P$ are continuous (Section 3.3). We denote by $\mathrm{End}_A^{cont}(M) \subset \mathrm{End}_A(M)$ and $E^{cont} \subset E := \mathrm{End}_A(M^P)$ the subalgebra of continuous endomorphisms. We have the canonical injective algebra map (7.20)

$$f \mapsto \sigma_0 \circ f \circ \mathrm{ev}_0 \quad : \quad \mathrm{End}_A^{cont}(M) \to E^{cont}.$$

Proposition 7.52. *Let M be a topologically étale $A[P_+]$-module.*

(i) *If M is complete, resp. compact, the A-module M^P is complete, resp. compact.*

(ii) *The natural map $P \times M^P \to M^P$ is continuous.*

(iii) *$\mathrm{Res}(f) \in E^{cont}$ for each $f \in C_c^\infty(N, A)$ (Proposition 7.32).*

Proof. (i) If M is complete, by [3] TG X.9 Cor. 3 and TG X.25 Th. 2, the compact-open topology on $C(P, M)$ is complete because P is locally compact. Hence, M^P as a closed submodule is complete as well.

　　If M is compact, the s-model of M^P is compact as a closed subset of the compact space $M^{\mathbb{N}}$. Hence by Lemma 7.51, M^P is compact.

(ii) It suffices to show that the right translation action of P on $C(P, M)$ is continuous. This is well known: the map in question is the composite of the following three continuous maps

$$P \times C(P, M) \longrightarrow P \times C(P \times P, M)$$
$$(b, f) \longmapsto (b, (x, y) \mapsto f(yx)),$$

$$P \times C(P \times P, M) \longrightarrow P \times C(P, C(P, M))$$
$$(b, F) \longmapsto (b, x \mapsto [y \mapsto F(x, y)]),$$

and

$$P \times C(P, C(P, M)) \longrightarrow C(P, M)$$
$$(b, \Phi) \longmapsto \Phi(b),$$

where the continuity of the latter relies on the fact that P is locally compact.

(iii) It suffices to consider functions of the form $f = 1_{b.N_0}$ for some $b \in P$. But then $\mathrm{Res}(f) = b \circ \sigma_0 \circ \mathrm{ev}_0 \circ b^{-1}$ is the composite of continuous endomorphisms.

$\square$

4.2. Integration on N with value in $\mathrm{End}_A^{cont}(M^P)$

We suppose that M is a complete topologically étale $A[P_+]$-module.

　　We denote by E^{cont} the ring of continuous A-endomorphisms of the complete A-module M^P with the topology defined by the right ideals

$$E_{\mathcal{L}}^{cont} = \mathrm{Hom}_A^{cont}(M^P, \mathcal{L})$$

for all open A-submodules $\mathcal{L} \subset M^P$.

Lemma 7.53. *E^{cont} is a complete topological ring.*

Proof. It is clear that the maps $(x, y) \mapsto x - y$ and $(x, y) \mapsto x \circ y$ from $E^{cont} \times E^{cont}$ to E^{cont} are continuous, i.e. that E^{cont} is a topological ring. The composite of the natural morphisms

$$E^{cont} \rightarrow \varprojlim_{\mathcal{L}} E^{cont}/E_{\mathcal{L}}^{cont} \rightarrow \varprojlim_{\mathcal{L}} \mathrm{Hom}_A^{cont}(M^P, M^P/\mathcal{L})$$

is an isomorphism (the natural map $M^P \rightarrow \varprojlim_{\mathcal{L}} M^P/\mathcal{L}$ is an isomorphism), hence the two morphisms are isomorphisms since the kernel of the map $E^{cont} \rightarrow \operatorname{Hom}_A^{cont}(M^P, M^P/\mathcal{L})$ is $E_{\mathcal{L}}^{cont}$. We deduce that E^{cont} is complete. $\qquad\qquad\qquad\qquad\qquad\qquad\qquad\qquad\qquad\qquad\qquad\qquad\qquad\square$

Definition 7.54. An A-linear map $C_c^{\infty}(N, A) \rightarrow E^{cont}$ is called a measure on N with values in E^{cont}.

The map Res is a measure on N with values in E^{cont} (Proposition 7.52).

Let $C_c(N, E^{cont})$ be the space of compactly supported **continuous** maps from N to E^{cont}. We will prove that one can "integrate" a function in $C_c(N, E^{cont})$ with respect to a measure on N with values in E^{cont}.

Proposition 7.55. *There is a natural bilinear map*

$$C_c(N, E^{cont}) \times \operatorname{Hom}_A(C_c^{\infty}(N, A), E^{cont}) \rightarrow E^{cont}$$

$$(f, \lambda) \mapsto \int_N f \, d\lambda.$$

Proof. (a) Every compact subset of N is contained in a compact open subset. It follows that $C_c(N, E^{cont})$ is the union of its subspaces $C(U, E^{cont})$ of functions with support contained in U, for all compact open subsets $U \subset N$.

(b) For any open A-submodule $\mathcal{L}$ of M^P, a function in $C(U, E^{cont}/E_{\mathcal{L}}^{cont})$ is locally constant because $E^{cont}/E_{\mathcal{L}}^{cont}$ is discrete. An upper index ∞ means that we consider locally constant functions hence

$$C(U, E^{cont}/E_{\mathcal{L}}^{cont}) = C^{\infty}(U, E^{cont}/E_{\mathcal{L}}^{cont}) = C^{\infty}(U, A) \otimes_A E^{cont}/E_{\mathcal{L}}^{cont}.$$

There is a natural linear pairing

$$(C^{\infty}(U, A) \otimes_A E^{cont}/E_{\mathcal{L}}^{cont}) \times \operatorname{Hom}_A(C^{\infty}(U, A), E^{cont}) \rightarrow E^{cont}/E_{\mathcal{L}}^{cont}$$

$$(f \otimes x, \lambda) \mapsto x\lambda(f).$$

Note that $E^{cont}/E_{\mathcal{L}}^{cont}$ is a right E^{cont}-module.

(c) Let $f \in C_c(N, E^{cont})$ and let $\lambda \in \operatorname{Hom}_A(C_c^{\infty}(N, A), E^{cont})$. Let $U \subset N$ be an open compact subset containing the support of f. For any open A-submodule L of M^P let $f_{\mathcal{L}} \in C_c^{\infty}(U, E^{cont}/E_{\mathcal{L}}^{cont})$ be the map induced by f. Let

$$\int_U f_{\mathcal{L}} \, d\lambda \quad \in E^{cont}/E_{\mathcal{L}}^{cont}$$

be the image of $(f_{\mathcal{L}}, \lambda)$ by the natural pairing of (b). The elements $\int_U f_{\mathcal{L}} \, d\lambda$ combine in the projective limit $E^{cont} = \varprojlim_{\mathcal{L}} E^{cont}/E_{\mathcal{L}}^{cont}$ to give an element $\int_U f \, d\lambda \in E^{cont}$. One checks easily that $\int_U f \, d\lambda$ does not depend on the choice of U. We define

$$\int_N f \, d\lambda \quad := \quad \int_U f \, d\lambda.$$

□

We recall that $J(N/V)$ is a system of representatives of N/V when $V \subset N$ is a compact open subgroup.

Corollary 7.56. *Let* $f \in C_c(N, E^{cont})$ *and let* λ *be a measure on* N *with values in* E^{cont}. *Then*

$$\lim_{V \to \{1\}} \sum_{v \in J(N/V)} f(v) \, \lambda(1_{vV}) \quad = \quad \int_N f \, d\lambda$$

with the limit over compact open subgroups $V \subset N$ *shrinking to* $\{1\}$.

Proof. We choose an open compact subset $U \subset N$ containing the support of f. Let L be an open o-submodule of M^P and a compact open subgroup $V \subset N$ such that $uV \subset U$ and $f_{\mathcal{L}}$ (proof of Proposition 7.55) is constant on uV for all $u \in U$. Then $\int_U f_{\mathcal{L}} \, d\lambda$ is the image of

$$\sum_{v \in J(N/V)} f(v) \, \lambda(1_{vV})$$

by the quotient map $E^{cont} \to E^{cont}/E_{\mathcal{L}}^{cont}$. □

Lemma 7.57. *Let* $f \in C_c(N, E^{cont})$ *be a continuous map with support in the compact open subset* $U \subset N$, *let* λ *be a measure on* N *with values in* E^{cont}, *and let* $\mathcal{L} \subset M^P$ *be any open A-submodule. There is a compact open subgroup* $V_{\mathcal{L}} \subset N$ *such that* $UV_{\mathcal{L}} = U$ *and*

$$\int_N f 1_{uV} d\lambda - f(u)\lambda(1_{uV}) \in E_{\mathcal{L}}^{cont}$$

for any open subgroup $V \subset V_{\mathcal{L}}$ *and any* $u \in U$.

Proof. The integral in question is the limit (with respect to open subgroups $V' \subset V$) of the net

$$\sum_{v \in J(V/V')} (f(uv) - f(u))\lambda(1_{uvV'}).$$

Since $E_{\mathcal{L}}^{cont}$ is a right ideal it therefore suffices to find a compact open subgroup $V_{\mathcal{L}} \subset N$ such that $UV_{\mathcal{L}} = U$ and

$$f(uv) - f(u) \in E_{\mathcal{L}}^{cont} \qquad \text{for any } u \in U \text{ and } v \in V_{\mathcal{L}}.$$

We certainly find a compact open subgroup $\tilde{V} \subset N$ such that $U\tilde{V} = U$. The map

$$U \times \tilde{V} \to E^{cont}$$
$$(u, v) \mapsto f(uv) - f(u)$$

is continuous and maps any $(u, 1)$ to zero. Hence, for any $u \in U$, there is an open neighbourhood $U_u \subset U$ of u and a compact open subgroup $V_u \subset \tilde{V}$ such that $U_u \times V_u$ is mapped to $E_{\mathcal{L}}^{cont}$. Since U is compact we have $U = U_{u_1} \cup \ldots \cup U_{u_s}$ for finitely many appropriate $u_i \in U$. The compact open subgroup $V_{\mathcal{L}} := V_{u_1} \cap \ldots \cap V_{u_s}$ then is such that $U \times V_{\mathcal{L}}$ is mapped to $E_{\mathcal{L}}^{cont}$. $\qquad\square$

Let $C(N, E^{cont})$ be the space of continuous functions from N to E^{cont}. For any continuous function $f \in C(N, E^{cont})$, for any compact open subset $U \subset N$ and for any measure λ on N with values in E^{cont} we denote

$$\int_U f \, d\lambda = \int_N f \, 1_U \, d\lambda$$

where $1_U \in C^\infty(U, A)$ is the characteristic function of U hence $f1_U \in C_c(N, E^{cont})$ is the restriction of f to U. The "integral of f on U" (with respect to the measure λ) is equal to the "integral of the restriction of f to U".

Remark 7.58. For $f \in C_c(N, E^{cont})$ and $\phi \in C_c^\infty(N, A)$ we have

$$\int_N f\phi d\,\mathrm{Res} = \int_N \phi f d\,\mathrm{Res} = \int_N f d\,\mathrm{Res} \circ \mathrm{Res}(\phi).$$

Proof. This is immediate from the construction of the integral and the multiplicativity of Res. $\qquad\square$

5. G-equivariant sheaf on G/P

Let G be a locally profinite group containing $P = N \rtimes L$ as a closed subgroup satisfying the assumptions of Section 3.2 such that

(a) G/P is compact.
(b) There is a subset W in the G-normalizer $N_G(L)$ of L such that
 – the image of W in $N_G(L)/L$ is a subgroup,
 – G is the disjoint union of PwP for $w \in W$.
 We note that $PwP = NwP = PwN$.

(c) *There exists $w_0 \in W$ such that $N w_0 P$ is an open dense subset of G. We call*

$$\mathcal{C} := N w_0 P / P$$

the open cell of G/P.

(d) *The map $(n, b) \mapsto n w_0 b$ from $N \times P$ onto $N w_0 P$ is a homeomorphism.*

Remark 7.59. These conditions imply that

$$G = P \overline{P} P = C(w_0) C(w_0^{-1})$$

where $\overline{P} := w_0 P w_0^{-1}$ and $C(g) = PgP$ for $g \in G$.

Proof. The intersection of the two dense open subsets $g\mathcal{C}$ and $\mathcal{C}$ in G/P is open and not empty, for any $g \in G$. $\qquad\square$

The group G acts continuously on the topological space G/P,

$$G \times G/P \to G/P$$
$$(g, xP) \mapsto gxP.$$

For $n, x \in N$ and $t \in L$ we have $n t x w_0 P = n t x t^{-1} w_0 P = (n t.x) w_0 P$ hence the action of P on the open cell corresponds to the action of P on N introduced before Proposition 7.32, i.e. the homeomorphism

$$N \to \mathcal{C}, \quad u \mapsto x_u := u w_0 P$$

is P-equivariant.

When M is an étale $A[P_+]$-module, this allows us to systematically view the map Res in the following as a P-equivariant homomorphism of A-algebras

$$\mathrm{Res} : C_c^\infty(\mathcal{C}, A) \to \mathrm{End}_A(M^P)$$

and the corresponding sheaf (Theorem 7.38) as a sheaf on $\mathcal{C}$. Our purpose is to show that this sheaf extends naturally to a G-equivariant sheaf on G/P for certain étale $A[P_+]$-modules. When M is a complete topologically étale $A[P_+]$-module we note that also integration with respect to the measure Res (Proposition 7.55) will be viewed in the following as a map

$$C_c(\mathcal{C}, E^{cont}) \to E^{cont}$$
$$f \mapsto \int_{\mathcal{C}} f \, d\,\mathrm{Res}$$

on the space $C_c(\mathcal{C}, E^{cont})$ of compactly supported continuous maps from $\mathcal{C}$ to E^{cont}.

5.1. Topological G-space G/P and the map α

Definition 7.60. An open subset $\mathcal{U}$ of G/P is called standard if there is a $g \in G$ such that $g\mathcal{U}$ is contained in the open cell $\mathcal{C}$.

The inclusion $g\mathcal{U} \subset Nw_0P/P$ is equivalent to $\mathcal{U} = g^{-1}Uw_0P/P$ for a unique open subset $U \subset N$. An open subset of a standard open subset is standard. The translates by G of N_0w_0P/P form a basis of the topology of G/P.

Proposition 7.61. *A compact open subset $\mathcal{U} \subset G/P$ is a disjoint union*

$$\mathcal{U} = \bigsqcup_{g \in I} g^{-1}Vw_0P/P$$

where $V \subset N$ is a compact open subgroup and $I \subset G$ a finite subset.

Proof. We first observe that any open covering of $\mathcal{U}$ can be refined into a disjoint open covering. In our case, this implies that $\mathcal{U}$ has a finite disjoint covering by standard compact open subsets. Let $g^{-1}Uw_0P/P \subset G/P$ be a standard compact open subset. Then $U = \bigsqcup_{u \in J} uV$ (disjoint union) with a finite set $J \subset U$ and $V \subset N$ is a compact open subgroup. Then $g^{-1}Uw_0P/P = \bigsqcup_{h \in I} h^{-1}Vw_0P/P$ (disjoint union) where $I = \{u^{-1}g \mid u \in J\}$. $\qquad\square$

For $g \in G$ and x in the nonempty open subset $g^{-1}\mathcal{C} \cap \mathcal{C}$ of G/P (Remark 7.59), there is a unique element $\alpha(g,x) \in P$ such that, if $x = uw_0P/P$ with $u \in N$, then

$$guw_0N = \alpha(g,x)uw_0N.$$

We give some properties of the map α.

Lemma 7.62. *Let $g \in G$. Then*

(i) $g^{-1}\mathcal{C} \cap \mathcal{C} = \mathcal{C}$ *if and only if $g \in P$.*

(ii) *The map $\alpha(g,.) : g^{-1}\mathcal{C} \cap \mathcal{C} \to P$ is continuous.*

(iii) *We have $gx = \alpha(g,x)x$ for $x \in g^{-1}\mathcal{C} \cap \mathcal{C}$ and we have $\alpha(b,x) = b$ for $b \in P$ and $x \in \mathcal{C}$.*

Proof. (i) We have $g^{-1}\mathcal{C} \cap \mathcal{C} = \mathcal{C}$ if and only if $gNw_0P \subset Nw_0P$ if and only if $g \in P$. Indeed, the condition $hPw_0P \subset Pw_0P$ on $h \in G$ depends only on PhP and for $w \in W$, the condition $wPw_0P \subset Pw_0P$ implies $ww_0 \in Pw_0P$ hence $ww_0 \in w_0L$ by the hypothesis (b) hence $w \in L$.

(ii) Let $N_g \subset N$ be such that $N_g w_0 P/P = g^{-1}\mathcal{C} \cap \mathcal{C}$. It suffices to show that the map $u \to \alpha(g, uw_0 P)u : N_g \to P$ is continuous. This follows from the continuity of the maps $u \mapsto guw_0 N : N_g \to Pw_0 P/N = Pw_0 N/N$ and $bw_0 N \mapsto b : Pw_0 N/N \to P$.

(iii) Obvious.

$\square$

Lemma 7.63. *Let $g, h \in G$ and $x \in (gh)^{-1}\mathcal{C} \cap h^{-1}\mathcal{C} \cap \mathcal{C}$. Then $hx \in g^{-1}\mathcal{C} \cap \mathcal{C}$ and we have*

$$\alpha(gh, x) = \alpha(g, hx)\alpha(h, x).$$

Proof. The first part of the assertion is obvious. Let $x = uw_0 P$ and $hx = vw_0 P$ with $u, v \in N$. We have

$$huw_0 N = \alpha(h, x)uw_0 N, \quad gvw_0 N = \alpha(g, hx)vw_0 N,$$

$$\text{and } \alpha(gh, x)uw_0 N = ghuw_0 N.$$

The first identity implies $\alpha(h, x)uw_0 P = vw_0 P$, hence $v^{-1}\alpha(h, x)u \in P \cap w_0 Pw_0^{-1}$. Hypothesis (d) easily yields $P \cap w_0 Pw_0^{-1} = L$, hence $\alpha(h, x)u = vt$ for some $t \in L$. Multiplying the second identity on the right by $w_0^{-1}tw_0$ we obtain $gvtw_0 N = \alpha(g, hx)vtw_0 N = \alpha(g, hx)\alpha(h, x)uw_0 N$. Finally, by inserting the first identity into the right-hand side of the third identity we get

$$\alpha(gh, x)uw_0 N = g\alpha(h, x)uw_0 N = gvtw_0 N = \alpha(g, hx)\alpha(h, x)uw_0 N$$

which is the assertion.

$\square$

It will be technically convenient later to work on N instead of $\mathcal{C}$. For $g \in G$ let therefore N_g be the open subset of N such that $\mathcal{C} \cap g^{-1}\mathcal{C} = N_g w_0 P/P$. We have $N_g = N$ if and only if $g \in P$ (Lemma 7.62 (i)). We have the homeomorphism $u \mapsto x_u := uw_0 P/P : N \xrightarrow{\sim} \mathcal{C}$ and the continuous map (Lemma 7.62 (ii))

$$N_g \longrightarrow P$$
$$u \longmapsto \alpha(g, x_u)$$

such that

$$gu = \alpha(g, x_u)u\bar{n}(g, u) \quad \text{for some } \bar{n}(g, u) \in \overline{N} := w_0 N w_0^{-1},$$
$$\alpha(g, x_u)u = n(g, u)t(g, u) \quad \text{for some } n(g, u) \in N, t(g, u) \in L. \tag{7.23}$$

Lemma 7.64. *Fix $g \in G$ and let $V \subset g^{-1}\mathcal{C} \cap \mathcal{C}$ be any compact open subset. There exists a disjoint covering $V = V_1 \dot{\cup} \ldots \dot{\cup} V_m$ by compact open subsets V_i and points $x_i \in V_i$ such that*

$$\alpha(g, x_i)V_i \subset gV \qquad \text{for any } 1 \leq i \leq m.$$

Proof. We denote the inverse of the homeomorphism $u \mapsto x_u : N \xrightarrow{\sim} C$ by $x \mapsto u_x$. The image $C \subset P$ of V under the continuous map $x \mapsto \alpha(g, x)u_x$ $V \to P$ is compact. As (Lemma 7.62 (iii)) $\alpha(g, x)x = gx \in gV$ for any $x \in V$, under the continuous action of P on C, every element in the compact set C maps the point $w_0 P$ into gV. It follows that there is an open neighbourhood $V_0 \subset C$ of $w_0 P$ such that $CV_0 \subset gV$. This means that

$$\alpha(g, x)u_x V_0 \subset gV \qquad \text{for any } x \in V.$$

Using Proposition 7.61 we find, by appropriately shrinking V_0, a disjoint covering of V of the form $V = u_1 V_0 \dot{\cup} \ldots \dot{\cup} u_m V_0$ with $u_i \in N$. We put $x_i := u_i w_0 P$. $\qquad \square$

We denote by $G_X := \{x \in G \mid xX \subset X\}$ the G-stabilizer of a subset $X \subset G/P$ and by

$$G_X^\dagger := \{g \in G \mid g \in G_X, \ g^{-1} \in G_X\} = \{x \in G \mid xX = X\}$$

the subgroup of invertible elements of G_X. If G_X is open then its inverse monoid is open hence $G_X^\dagger$ is open (and conversely).

Lemma 7.65. *The G-stabilizers $G_\mathcal{U}$ and $G_\mathcal{U}^\dagger$ are open in G, for any compact open subset $\mathcal{U} \subset G/P$.*

Proof. By Proposition 7.61 it suffices to consider the case where $\mathcal{U} = U w_0 P/P$ for some compact open subgroup $U \subset N$. As $U w_0 P \subset G$ is an open subset containing w_0 there exists an open subgroup $K \subset G$ such that $Kw_0 \subset Uw_0P$. The set $U/(K \cap U)$ is finite because U is compact and $(K \cap U) \subset U$ is an open subgroup. The finite intersection $K' := \bigcap_{u \in U/(U \cap K)} uKu^{-1} = \bigcap_{u \in U} uKu^{-1}$ is an open subgroup of K which is normalized by U. But $K'U = UK'$ implies that $K'w_0P = UK'w_0P \subset U(Uw_0P)P = Uw_0P$, and hence that $K' \subset G_\mathcal{U}$. We deduce that $G_\mathcal{U}$ is open. Hence $G_\mathcal{U}^\dagger$ is open. $\qquad \square$

Remark 7.66. The G-stabilizer of the open cell C is the group P.

Proof. Lemma 7.62 (i). $\qquad \square$

For $\mathcal{U} \subset C$ the map

$$G_\mathcal{U} \times \mathcal{U} \to P \ , \quad (g, x) \mapsto \alpha(g, x) \tag{7.24}$$

is continuous because, if $\mathcal{U} = U w_0 P/P$ with U open in N, then the map $(g, u) \mapsto gu w_0 N : G_\mathcal{U} \times U \to Pw_0P/N = Pw_0N/N$ is continuous (cf. the proof of Lemma 7.62 (ii)).

5.2. Equivariant sheaves and modules over skew group rings

Our construction of the sheaf on G/P will proceed through a module theoretic interpretation of equivariant sheaves. The ring $C_c^\infty(\mathcal{C}, A)$ has no unit element. But it has sufficiently many idempotents (the characteristic functions 1_V of the compact open subsets $V \subset \mathcal{C}$). A (left) module Z over $C_c^\infty(\mathcal{C}, A)$ is called nondegenerate if for any $z \in Z$ there is an idempotent $e \in C_c^\infty(\mathcal{C}, A)$ such that $ez = z$.

It is well known that the functor

$$\text{sheaves of } A\text{-modules on } \mathcal{C} \to \text{nondegenerate } C_c^\infty(\mathcal{C}, A)\text{-modules}$$

which sends a sheaf $\mathcal{S}$ to the A-module of global sections with compact support $\mathcal{S}_c(\mathcal{C}) := \bigcup_V \mathcal{S}(V)$, with V running over all compact open subsets in $\mathcal{C}$, is an equivalence of categories. In fact, as we have discussed in the proof of Theorem 7.38 a quasi-inverse functor is given by sending the module Z to the sheaf whose sections on the compact open subset $V \subset \mathcal{C}$ are equal to $1_V Z$.

In order to extend this equivalence to equivariant sheaves we note that the group P acts, by left translations, from the right on $C_c^\infty(\mathcal{C}, A)$ which we write as $(f, b) \mapsto f^b(.) := f(b.)$. This allows to introduce the skew group ring

$$\mathcal{A}_\mathcal{C} := C_c^\infty(\mathcal{C}, A)\#P = \oplus_{b \in P} bC_c^\infty(\mathcal{C}, A)$$

in which the multiplication is determined by the rule

$$(b_1 f_1)(b_2 f_2) = b_1 b_2 f_1^{b_2} f_2 \qquad \text{for } b_i \in P \text{ and } f_i \in C_c^\infty(\mathcal{C}, A).$$

It is easy to see that the above functor extends to an equivalence of categories

$$P\text{-equivariant sheaves of } A\text{-modules on } \mathcal{C} \xrightarrow{\sim} \text{nondegenerate } \mathcal{A}_\mathcal{C}\text{-modules}.$$

We have the completely analogous formalism for the G-space G/P. The only small difference is that, since G/P is assumed to be compact, the ring $C^\infty(G/P, A)$ of locally constant A-valued functions on G/P is unital. The skew group ring

$$\mathcal{A}_{G/P} := C^\infty(G/P, A)\#G = \oplus_{g \in G} gC^\infty(G/P, A)$$

therefore is unital as well, and the equivalence of categories reads

$$G\text{-equivariant sheaves of } A\text{-modules on } G/P \xrightarrow{\sim} \text{unital } \mathcal{A}_{G/P}\text{-modules}.$$

For any open subset $\mathcal{U} \subset G/P$ the A-algebra $C_c^\infty(\mathcal{U}, A)$ of A-valued locally constant and compactly supported functions on $\mathcal{U}$ is, by extending functions by zero, a subalgebra of $C^\infty(G/P, A)$. It follows in particular that $\mathcal{A}_\mathcal{C}$ is a subring of $\mathcal{A}_{G/P}$. There is for our purposes a very important ring in between these two rings which is defined to be

$$\mathcal{A} := \mathcal{A}_{\mathcal{C}\subseteq G/P} := \oplus_{g\in G}\, gC_c^\infty(g^{-1}\mathcal{C}\cap\mathcal{C}, A).$$

That $\mathcal{A}$ indeed is multiplicatively closed is immediate from the following observation. If supp(f) denotes the support of a function $f \in C^\infty(G/P, A)$ then we have the formula

$$\mathrm{supp}(f_1^g f_2) = g^{-1}\,\mathrm{supp}(f_1)\cap\mathrm{supp}(f_2)$$
$$\text{for } g \in G \text{ and } f_1, f_2 \in C^\infty(G/P, A). \tag{7.25}$$

In particular, if $f_i \in C_c^\infty(g_i^{-1}\mathcal{C}\cap\mathcal{C}, A)$ then

$$\mathrm{supp}(f_1^{g_2} f_2) \subset g_2^{-1}(g_1^{-1}\mathcal{C}\cap\mathcal{C})\cap(g_2^{-1}\mathcal{C}\cap\mathcal{C}) \subset (g_1g_2)^{-1}\mathcal{C}\cap\mathcal{C}.$$

We also have the A-submodule

$$\mathcal{Z} := \oplus_{g\in G}\, gC_c^\infty(\mathcal{C}, A)$$

of $\mathcal{A}_{G/P}$. Using (7.25) again one sees that $\mathcal{Z}$ actually is a left ideal in $\mathcal{A}_{G/P}$ which at the same time is a right $\mathcal{A}$-submodule. This means that we have the well defined functor

$$\text{nondegenerate } \mathcal{A}\text{-modules} \ \to \ \text{unital } \mathcal{A}_{G/P}\text{-modules}$$
$$Z \mapsto \mathcal{Z}\otimes_{\mathcal{A}} Z.$$

Remark 7.67. The functor of restricting G-equivariant sheaves on G/P to the open cell $\mathcal{C}$ is faithful and detects isomorphisms.

Proof. Any sheaf homomorphism which is the zero map, resp. an isomorphism, on sections on any compact open subset of $\mathcal{C}$ has, by G-equivariance, the same property on any standard compact open subset and hence, by Proposition 7.61, on any compact open subset of G/P. □

Proposition 7.68. *The above functor $Z \mapsto \mathcal{Z}\otimes_{\mathcal{A}} Z$ is an equivalence of categories; a quasi-inverse functor is given by sending the $\mathcal{A}_{G/P}$-module Y to $\bigcup_{V\subseteq\mathcal{C}} 1_V Y$ where V runs over all compact open subsets in $\mathcal{C}$.*

Proof. We abbreviate the asserted candidate for the quasi-inverse functor by $R(Y) := \bigcup_{V\subseteq\mathcal{C}} 1_V Y$. It immediately follows from Remark 7.67 that the functor R, which in terms of sheaves is the functor of restriction, is faithful and detects isomorphisms.

By a slight abuse of notation we identify in the following a function $f \in C^\infty(G/P, A)$ with the element $1f \in \mathcal{A}_{G/P}$, where $1 \in G$ denotes the unit element. Let $V \subset \mathcal{C}$ be a compact open subset. Then $1_V \mathcal{A}_{G/P} 1_V$ is a subring of $\mathcal{A}_{G/P}$ (with the unit element 1_V), which we compute as follows:

$$
\begin{aligned}
1_V \mathcal{A}_{G/P} 1_V &= \sum_{g \in G} 1_V g C^\infty(V, A) = \sum_{g \in G} g 1_{g^{-1}V} C^\infty(V, A) \\
&= \sum_{g \in G} g C^\infty(g^{-1}V \cap V, A).
\end{aligned}
$$

We note:

- If $U \subset V \subset \mathcal{C}$ are two compact open subsets then $1_V \mathcal{A}_{G/P} 1_V \supset 1_U \mathcal{A}_{G/P} 1_U$.
- Let $f \in C_c^\infty(g^{-1}\mathcal{C} \cap \mathcal{C}, A)$ be supported on the compact open subset $U \subset g^{-1}\mathcal{C} \cap \mathcal{C}$. Then $V := U \cup gU$ is compact open in $\mathcal{C}$ as well, and $U \subset g^{-1}V \cap V$. This shows that $C_c^\infty(g^{-1}\mathcal{C} \cap \mathcal{C}, A) = \bigcup_{V \subset \mathcal{C}} C^\infty(g^{-1}V \cap V, A)$.

We deduce that

$$
\bigcup_{V \subset \mathcal{C}} 1_V \mathcal{A}_{G/P} 1_V = \mathcal{A}_{\mathcal{C} \subset G/P} = \mathcal{A}.
$$

A completely analogous computation shows that

$$
1_V \mathcal{Z} = 1_V \mathcal{A}.
$$

Given a nondegenerate $\mathcal{A}$-module Z the map

$$
1_V(\mathcal{Z} \otimes_{\mathcal{A}} Z) = (1_V \mathcal{Z}) \otimes_{\mathcal{A}} Z = (1_V \mathcal{A}) \otimes_{\mathcal{A}} Z \;\to\; 1_V Z
$$

$$
1_V a \otimes z = 1_V \otimes 1_V a z \mapsto 1_V a z
$$

therefore is visibly an isomorphism of $1_V \mathcal{A}_{G/P} 1_V$-modules. In the limit with respect to V we obtain a natural isomorphism of $\mathcal{A}$-modules

$$
R(\mathcal{Z} \otimes_{\mathcal{A}} Z) \xrightarrow{\cong} Z.
$$

On the other hand, for any unital $\mathcal{A}_{G/P}$-module Y there is the obvious natural homomorphism of $\mathcal{A}_{G/P}$-modules

$$
\mathcal{Z} \otimes_{\mathcal{A}} R(Y) \;\to\; Y
$$

$$
a \otimes z \mapsto a z.
$$

It is an isomorphism because applying the functor R, which detects isomorphisms, to it gives the identity map. □

Remark 7.69. Let Z be a nondegenerate $\mathcal{A}$-module. Viewed as an $\mathcal{A}_{\mathcal{C}}$-module it corresponds to a P-equivariant sheaf $\widetilde{Z}$ on $\mathcal{C}$. On the other hand, the $\mathcal{A}_{G/P}$-module $Y := \mathcal{Z} \otimes_{\mathcal{A}} Z$ corresponds to a G-equivariant sheaf $\widetilde{Y}$ on G/P. We have $\widetilde{Y}|\mathcal{C} = \widetilde{Z}$, i.e., the sheaf $\widetilde{Y}$ extends the sheaf $\widetilde{Z}$.

We have now seen that the step of going from $\mathcal{A}$ to $\mathcal{A}_{G/P}$ is completely formal. On the other hand, for any topologically étale $A[P_+]$-module M, the P-equivariance of Res together with Proposition 7.52 imply that Res extends to the A-algebra homomorphism

$$\text{Res} : \quad \mathcal{A}_C \rightarrow \text{End}_A^{cont}(M^P)$$
$$\sum_{b\in P} bf_b \mapsto \sum_{b\in P} b \circ \text{Res}(f_b).$$

When M is compact it is relatively easy, as we will show in the next section, to further extend this map from $\mathcal{A}_C$ to $\mathcal{A}$. This makes crucial use of the full topological module M^P and not only its submodule M_c^P of sections with compact support. When M is not compact this extension problem is much more subtle and requires more facts about the ring $\mathcal{A}$.

We introduce the compact open subset $\mathcal{C}_0 := N_0 w_0 P / P$ of $\mathcal{C}$, and we consider the unital subrings

$$\mathcal{A}_0 := 1_{\mathcal{C}_0} \mathcal{A}_{G/P} 1_{\mathcal{C}_0} = \sum_{g\in G} g C^\infty(g^{-1}\mathcal{C}_0 \cap \mathcal{C}_0, A)$$

and

$$\mathcal{A}_{C0} := 1_{\mathcal{C}_0} \mathcal{A}_C 1_{\mathcal{C}_0} = \sum_{b\in P} b C^\infty(b^{-1}\mathcal{C}_0 \cap \mathcal{C}_0, A)$$

of $\mathcal{A}$ and $\mathcal{A}_C$, respectively. Obviously $\mathcal{A}_{C0} \subseteq \mathcal{A}_0$ with the same unit element $1_{\mathcal{C}_0}$. Since $g^{-1}\mathcal{C}_0 \cap \mathcal{C}_0$ is nonempty if and only if $g \in N_0\overline{P}N_0$ we in fact have

$$\mathcal{A}_0 = \sum_{g\in N_0\overline{P}N_0} g C^\infty(g^{-1}\mathcal{C}_0 \cap \mathcal{C}_0, A).$$

The map $A[G] \longrightarrow \mathcal{A}_{G/P}$ sending g to $g1_{G/P}$ is a unital ring homomorphism. Hence we may view $\mathcal{A}_{G/P}$ as an $A[G]$-module for the adjoint action

$$G \times \mathcal{A}_{G/P} \longrightarrow \mathcal{A}_{G/P}$$
$$(g, y) \longmapsto (g1_{G/P})y(g1_{G/P})^{-1}.$$

One checks that $\mathcal{A}_C \subseteq \mathcal{A}$ are $A[P]$-submodules, that $\mathcal{A}_{C0} \subseteq \mathcal{A}_0$ are $A[P_+]$-submodules, and that the map Res : $\mathcal{A}_C \longrightarrow E^{cont}$ is a homomorphism of $A[P]$-modules.

Proposition 7.70. *The homomorphism of $A[P]$-modules*

$$A[P] \otimes_{A[P_+]} \mathcal{A}_0 \xrightarrow{\cong} \mathcal{A}$$
$$b \otimes y \longmapsto (b1_{G/P})y(b1_{G/P})^{-1}$$

is bijective; it restricts to an isomorphism $A[P] \otimes_{A[P_+]} \mathcal{A}_{C0} \xrightarrow{\cong} \mathcal{A}_C$.

Proof. Since $P = s^{-N} P_+$ the assertion amounts to the claim that

$$\mathcal{A} = \bigcup_{n \geq 0} (s^{-n} 1_{G/P}) \mathcal{A}_0 (s^n 1_{G/P})$$

and correspondingly for $\mathcal{A}_C$. But we have

$$(s^{-n} 1_{G/P}) \big(g C^\infty (g^{-1} C_0 \cap C_0, A) \big) (s^n 1_{G/P})$$
$$= s^{-n} g s^n C^\infty ((s^{-n} g^{-1} s^n) s^{-n} C_0 \cap s^{-n} C_0, A)$$

for any $n \geq 0$ and any $g \in G$. $\qquad\qquad\square$

Suppose that we may extend the map $\mathrm{Res} : \mathcal{A}_{C0} \longrightarrow \mathrm{End}_A^{cont}(M^P)$ to an $A[P_+]$-equivariant (unital) A-algebra homomorphism

$$\mathcal{R}_0 : \mathcal{A}_0 \longrightarrow \mathrm{End}_A(M^P).$$

By the above Proposition 7.70 it further extends uniquely to an $A[P]$-equivariant map $\mathcal{R} : \mathcal{A} \longrightarrow \mathrm{End}_A(M^P)$.

Lemma 7.71. *The map $\mathcal{R}$ is multiplicative.*

Proof. Using Proposition 7.70 we have that two arbitrary elements $y, z \in \mathcal{A}$ are of the form $y = (s^{-m} 1_{G/P}) y_0 (s^m 1_{G/P})$, $z = (s^{-n} 1_{G/P}) z_0 (s^n 1_{G/P})$ with $m, n \in \mathbb{N}$ and $y_0, z_0 \in \mathcal{A}_0$. We can choose $m = n$. It follows that

$$yz = (s^{-m} 1_{G/P}) y_0 z_0 (s^m 1_{G/P}) = (s^{-m} 1_{G/P}) x_0 (s^m 1_{G/P})$$

with $x_0 := y_0 z_0 \in \mathcal{A}_0$, and that

$$\begin{aligned} \mathcal{R}(yz) &= \mathcal{R}((s^{-m} 1_{G/P}) x_0 (s^m 1_{G/P})) = s^{-m} \circ \mathcal{R}_0(x_0) \circ s^m \\ &= s^{-m} \circ \mathcal{R}_0(y_0) \circ \mathcal{R}_0(z_0) \circ s^m \\ &= (s^{-m} \circ \mathcal{R}_0(y_0) \circ s^m) \circ (s^{-m} \circ \mathcal{R}_0(z_0) \circ s^m) \\ &= \mathcal{R}(y) \circ \mathcal{R}(z). \end{aligned}$$

$\qquad\qquad\square$

Note that the images $\mathrm{Res}(\mathcal{A}_{C0})$ and $\mathcal{R}_0(\mathcal{A}_0)$ necessarily lie in the image of $\mathrm{End}_A(M) = \mathrm{End}_A(\mathrm{Res}(1_{C_0})(M^P))$ by the natural embedding into $\mathrm{End}_A(M^P)$. This reduces us to search for an $A[P_+]$-equivariant (unital) A-algebra homomorphism

$$\mathcal{R}_0 : \mathcal{A}_0 \longrightarrow \mathrm{End}_A(M)$$

which extends $\mathrm{Res}\,|\mathcal{A}_{C0}$. In fact, since for $g \in N_0\overline{P}N_0$ and $f \in C^\infty(g^{-1}C_0 \cap C_0, A)$ we have $gf = (g1_{g^{-1}C_0 \cap C_0})(1f)$ with $1f \in \mathcal{A}_{C0}$ it suffices to find the elements

$$\mathcal{H}_g = \mathcal{R}_0(g1_{g^{-1}C_0 \cap C_0}) \in \mathrm{End}_A(M) \qquad \text{for } g \in N_0\overline{P}N_0.$$

Note that $P_+ = N_0L_+$ is contained in $N_0\overline{P}N_0 = N_0L\overline{N}N_0$.

Proposition 7.72. *We suppose given, for any $g \in N_0\overline{P}N_0$, an element $\mathcal{H}_g \in \mathrm{End}_A(M)$. Then the map*

$$\mathcal{R}_0: \qquad \mathcal{A}_0 \longrightarrow \mathrm{End}_A(M)$$
$$\sum_{g \in N_0\overline{P}N_0} gf_g \longmapsto \sum_{g \in N_0\overline{P}N_0} \mathcal{H}_g \circ \mathrm{res}(f_g)$$

is an $A[P_+]$-equivariant (unital) A-algebra homomorphism which extends $\mathrm{Res}\,|\mathcal{A}_{C0}$ if and only if, for all $g, h \in N_0\overline{P}N_0$, $b \in P \cap N_0\overline{P}N_0$, and all compact open subsets $V \subset C_0$, the relations

H1. $\mathrm{res}(1_V) \circ \mathcal{H}_g = \mathcal{H}_g \circ \mathrm{res}(1_{g^{-1}V \cap C_0})$,
H2. $\mathcal{H}_g \circ \mathcal{H}_h = \mathcal{H}_{gh} \circ \mathrm{res}(1_{(gh)^{-1}C_0 \cap h^{-1}C_0 \cap C_0})$,
H3. $\mathcal{H}_b = b \circ \mathrm{res}(1_{b^{-1}C_0 \cap C_0})$.

hold true. When H1 is true, H2 can equivalently be replaced by

$$\mathcal{H}_g \circ \mathcal{H}_h = \mathcal{H}_{gh} \circ \mathrm{res}(1_{h^{-1}C_0 \cap C_0}).$$

Proof. Necessity of the relations is easily checked. Vice versa, the first two relations imply that $\mathcal{R}_0$ is multiplicative. The third relation says that $\mathcal{R}_0$ extends $\mathrm{Res}\,|\mathcal{A}_{C0}$.

The last sentence of the assertion derives from the fact that we have

$$\mathcal{H}_{gh} \circ \mathrm{res}(1_{(gh)^{-1}C_0 \cap h^{-1}C_0 \cap C_0}) = \mathcal{H}_{gh} \circ \mathrm{res}(1_{(gh)^{-1}C_0 \cap C_0}) \circ \mathrm{res}(1_{h^{-1}C_0 \cap C_0})$$
$$= \mathcal{H}_{gh} \circ \mathrm{res}(1_{h^{-1}C_0 \cap C_0})$$

since $\mathcal{H}_{gh} \circ \mathrm{res}(1_{(gh)^{-1}C_0 \cap C_0}) = \mathcal{H}_{gh}$ by the first relation.

The P_+-equivariance is equivalent to the identity

$$\mathcal{R}_0((c1_{G/P})gf_g(c1_{G/P})^{-1}) = \varphi_c \circ \mathcal{R}_0(gf_g) \circ \psi_c$$

where $c \in P_+$ and f_g is any function in $C^\infty(g^{-1}C_0 \cap C_0)$. By the definition of $\mathcal{R}_0$ and the P_+-equivariance of res the left-hand side is equal to

$$\mathcal{H}_{cgc^{-1}} \circ \varphi_c \circ \mathrm{res}(f_g) \circ \psi_c$$

whereas the right-hand side is

$$\varphi_c \circ \mathcal{H}_g \circ \mathrm{res}(f_g) \circ \psi_c.$$

Since ψ_c is surjective and $\mathrm{res}(f_g) = \mathrm{res}(1_{g^{-1}C_0 \cap C_0}) \circ \mathrm{res}(f_g)$ we see that the P_+-equivariance of $\mathcal{R}_0$ is equivalent to the identity

$$\mathcal{H}_{cgc^{-1}} \circ \varphi_c \circ \mathrm{res}(1_{g^{-1}C_0 \cap C_0}) = \varphi_c \circ \mathcal{H}_g \circ \mathrm{res}(1_{g^{-1}C_0 \cap C_0}).$$

But as a special case of the first relation we have $\mathcal{H}_g \circ \mathrm{res}(1_{g^{-1}C_0 \cap C_0}) = \mathcal{H}_g$. Hence the latter identity coincides with the relation

$$\mathcal{H}_{cgc^{-1}} \circ \varphi_c \circ \mathrm{res}(1_{g^{-1}C_0 \cap C_0}) = \varphi_c \circ \mathcal{H}_g.$$

This relation holds true because $\varphi_c = \mathcal{H}_c$ and by the second relation $\mathcal{H}_{cgc^{-1}} \circ \mathcal{H}_c = \mathcal{H}_{cg}$ and $\mathcal{H}_c \circ \mathcal{H}_g = \mathcal{H}_{cg} \circ \mathrm{res}(1_{g^{-1}C_0 \cap C_0})$. $\qquad\square$

5.3. Integrating α when M is compact

Let M be a compact topologically étale $A[P_+]$-module. Then M^P is compact, hence the continuous action of P on M^P (Proposition 7.52) induces a continuous map $P \to E^{cont}$.

We will construct an extension $\widetilde{\mathrm{Res}}$ of Res to $\mathcal{A}_{C \subset G/P}$ by integration. For any $g \in G$, we consider the continuous map

$$\alpha_g \ g^{-1}C \cap C \xrightarrow{\alpha(g,.)} P \to E^{cont}.$$

We introduce the A-linear maps

$$\rho \ \mathcal{A} = \mathcal{A}_{C \subset G/P} \to C_c(C, E^{cont})$$
$$\sum_{g \in G} g f_g \mapsto \sum_{g \in G} \alpha_g f_g$$

and

$$\widetilde{\mathrm{Res}} \ \mathcal{A} = \mathcal{A}_{C \subset G/P} \to E^{cont}$$
$$a \mapsto \int_C \rho(a) d\,\mathrm{Res}.$$

For $b \in P$ the map α_b is the constant map on C with value b (Lemma 5.3 iii). It follows that

$$\widetilde{\mathrm{Res}} \,|\, \mathcal{A}_C = \mathrm{Res}$$

is an extension as we want it.

Theorem 7.73. $\widetilde{\mathrm{Res}}$ *is a homomorphism of A-algebras.*

Proof. Let $g, h \in G$ and let V_g and V_h be compact open subsets of $g^{-1}\mathcal{C} \cap \mathcal{C}$ and $h^{-1}\mathcal{C} \cap \mathcal{C}$, respectively. We have to show that

$$\widetilde{\mathrm{Res}}((g1_{V_g})(h1_{V_h})) = \widetilde{\mathrm{Res}}(g1_{V_g}) \circ \widetilde{\mathrm{Res}}(h1_{V_h})$$

holds true. This is, by definition of $\widetilde{\mathrm{Res}}$, equivalent to the identity

$$\int_{\mathcal{C}} \alpha_{gh} 1_{h^{-1}V_g \cap V_h} d\mathrm{Res} = \int_{\mathcal{C}} \alpha_g 1_{V_g} d\mathrm{Res} \circ \int_{\mathcal{C}} \alpha_h 1_{V_h} d\mathrm{Res}.$$

Let U_g, U_h be the open compact subsets of N corresponding to V_g, V_h and let f be the map $\alpha_g 1_{V_g}$, seen as a map on N with support on U_g. Let $\mathcal{L} \subset M^P$ be an open A-submodule and let $V_{\mathcal{L}}$ be chosen as in Lemma 7.57, with $\lambda = \mathrm{Res}$. If we let $N_{k,v} = \alpha(h, x_v).(vN_k)$, then the P-equivariance of Res combined with Remark 7.58 yield, for $v \in U_h$ and $k \geq 1$

$$\int_{\mathcal{C}} \alpha_g 1_{V_g} d\mathrm{Res} \circ \alpha(h, x_v) \circ \mathrm{Res}(1_{vN_k}) = \int_N f d\mathrm{Res} \circ \mathrm{Res}(1_{N_{k,v}}) \circ \alpha(h, x_v)$$

$$= \int_N (f 1_{N_{k,v}}) d\mathrm{Res} \circ \alpha(h, x_v).$$

Writing $\alpha(h, x_v) = n_v t_v$ with $n_v \in N$ and $t_v \in L$, for k large enough we have $U_g t_v N_k t_v^{-1} \subset U_g$ and $t_v N_k t_v^{-1} \subset V_{\mathcal{L}}$ for all $v \in U_h$ (by compactness of $(t_v)_{v \in U_h}$). Since $N_{k,v} = (\alpha(h, x_v).v) t_v N_k t_v^{-1}$, we deduce that $N_{k,v} \cap U_g \neq \emptyset \Leftrightarrow \alpha(h, x_v).v \in U_g \Leftrightarrow hx_v \in V_g$ and hence, by Lemma 7.57 for all sufficiently large k we have, uniformly in $v \in U_h$,

$$\int_N f 1_{N_{k,v}} d\mathrm{Res} \equiv 1_{x_v \in h^{-1}V_g \cap V_h} f(\alpha(h, x_v).v) \circ \mathrm{Res}(1_{N_{k,v}}) =$$

$$= 1_{x_v \in h^{-1}V_g \cap V_h} \alpha(g, hx_v) \circ \mathrm{Res}(1_{N_{k,v}}) \quad (\mathrm{mod}\ E_{\mathcal{L}}^{\mathrm{cont}}).$$

Combining the last two relations with Lemma 7.63, and using again the P-equivariance of Res, we obtain for k large enough and for all $v \in U_h$

$$\int_{\mathcal{C}} \alpha_g 1_{V_g} d\mathrm{Res} \circ \alpha(h, x_v) \circ \mathrm{Res}(1_{vN_k})$$

$$\equiv 1_{x_v \in h^{-1}V_g \cap V_h} \alpha(gh, x_v) \circ \mathrm{Res}(1_{vN_k}) \quad (\mathrm{mod}\ E_{\mathcal{L}}^{\mathrm{cont}}).$$

The result follows by summing over v and letting $k \to \infty$ (Corollary 7.56). □

5.4. G-equivariant sheaf on G/P

Let M be a compact topologically étale $A[P_+]$-module. We briefly survey our construction of a G-equivariant sheaf on G/P functorially associated with M.

From Proposition 7.32 we obtain an A-algebra homomorphism

$$\text{Res } C_c^\infty(\mathcal{C}, A)\#P \to E^{cont}$$

which gives rise to a P-equivariant sheaf on $\mathcal{C}$ as described in detail in Theorem 7.38. By Theorem 7.73, it extends to an A-algebra homomorphism

$$\widetilde{\text{Res}} \; \mathcal{A}_{\mathcal{C}\subset G/P} \to E^{cont}.$$

This homomorphism defines on the global sections with compact support M_c^P of the sheaf on $\mathcal{C}$ the structure of a nondegenerate $\mathcal{A}_{\mathcal{C}\subset G/P}$-module. The latter leads, by Proposition 7.68, to the unital $C_c^\infty(G/P, A)\#G$-module $\mathcal{Z} \otimes_A M_c^P$ which corresponds to a G-equivariant sheaf on G/P extending the earlier sheaf on $\mathcal{C}$ (Remark 7.39). We will denote the sections of this latter sheaf on an open subset $\mathcal{U} \subset G/P$ by $M \boxtimes \mathcal{U}$. The restriction maps in this sheaf, for open subsets $\mathcal{V} \subset \mathcal{U} \subset G/P$, will simply be written as $\text{Res}_{\mathcal{V}}^{\mathcal{U}} \; M \boxtimes \mathcal{U} \to M \boxtimes \mathcal{V}$.

We observe that for a standard compact open subset $\mathcal{U} \subset G/P$ with $g \in G$ such that $g\mathcal{U} \subset \mathcal{C}$ the action of the element g on the sheaf induces an isomorphism of A-modules $M \boxtimes \mathcal{U} \xrightarrow{\cong} M \boxtimes g\mathcal{U} = M_{g\mathcal{U}}$. Being the image of a continuous projector on M^P (Proposition 7.52), $M_{g\mathcal{U}}$ is naturally a compact topological A-module. We use the above isomorphism to transport this topology to $M \boxtimes \mathcal{U}$. The result is independent of the choice of g since, if $g\mathcal{U} = h\mathcal{U}$ for some other $h \in G$, then $h\mathcal{U} \subset (gh^{-1})^{-1}\mathcal{C} \cap \mathcal{C}$ and, by construction, the endomorphism gh^{-1} of $M \boxtimes h\mathcal{U}$ is given by the continuous map $\widetilde{\text{Res}}(gh^{-1}1_{h\mathcal{U}})$.

A general compact open subset $\mathcal{U} \subset G/P$ is the disjoint union $\mathcal{U} = \mathcal{U}_1 \dot{\cup} \ldots \dot{\cup} \mathcal{U}_m$ of standard compact open subsets $\mathcal{U}_i$ (Proposition 7.61). We equip $M \boxtimes \mathcal{U} = M \boxtimes \mathcal{U}_1 \oplus \ldots \oplus M \boxtimes \mathcal{U}_m$ with the direct product topology. One easily verifies that this is independent of the choice of the covering.

Finally, for an arbitrary open subset $\mathcal{U} \subset G/P$ we have $M \boxtimes \mathcal{U} = \varprojlim M \boxtimes \mathcal{V}$, where $\mathcal{V}$ runs over all compact open subsets $\mathcal{V} \subset \mathcal{U}$, and we equip $M \boxtimes \mathcal{U}$ with the corresponding projective limit topology.

It is straightforward to check that all restriction maps are continuous and that any $g \in G$ acts by continuous homomorphisms. We see that $(M \boxtimes \mathcal{U})_\mathcal{U}$ is a G-equivariant sheaf of compact topological A-modules.

Lemma 7.74. *For any compact open subset $\mathcal{U} \subset G/P$ the action $G_\mathcal{U}^\dagger \times (M \boxtimes \mathcal{U}) \to M \boxtimes \mathcal{U}$ of the open subgroup $G_\mathcal{U}^\dagger$ (Lemma 7.65) on the sections on $\mathcal{U}$ is continuous.*

Proof. Using Proposition 7.61, it suffices to consider the case that $\mathcal{U} \subset \mathcal{C}$. Note that $G_\mathcal{U}^\dagger$ acts by continuous automorphisms on $M \boxtimes \mathcal{U} = M_\mathcal{U}$. By (7.24) the map

$$G_{\mathcal{U}}^{\dagger} \times \mathcal{U} \;\to\; E^{cont}$$
$$(g, x) \;\mapsto\; \alpha_g(x)$$

is continuous. Hence ([3] TG X.28 Th. 3) the corresponding map

$$G_{\mathcal{U}}^{\dagger} \;\to\; C(\mathcal{U}, E^{cont})$$

is continuous, where we always equip the module $C(\mathcal{U}, E^{cont})$ of E^{cont}-valued continuous maps on $\mathcal{U}$ with the compact-open topology. On the other hand it is easy to see that, for any measure λ on C with values in E^{cont}, the map

$$\int_{\mathcal{U}} \cdot d\lambda \; C(\mathcal{U}, E^{cont}) \;\to\; E^{cont}$$

is continuous. It follows that the map

$$G_{\mathcal{U}}^{\dagger} \;\to\; E^{cont}$$
$$g \;\mapsto\; \widetilde{\mathrm{Res}}(g 1_{\mathcal{U}})$$

is continuous. The direct decomposition $M^P = M_{\mathcal{U}} \oplus M_{C-\mathcal{U}}$ gives a natural inclusion map $\mathrm{End}_A^{cont}(M_{\mathcal{U}}) \to E^{cont}$ through which the above map factorizes. The resulting map

$$G_{\mathcal{U}}^{\dagger} \;\to\; \mathrm{End}_A^{cont}(M_{\mathcal{U}})$$

is continuous and coincides with the $G_{\mathcal{U}}^{\dagger}$-action on $M_{\mathcal{U}}$. As $M_{\mathcal{U}}$ is compact this continuity implies the continuity of the action $G_{\mathcal{U}}^{\dagger} \times M_{\mathcal{U}} \to M_{\mathcal{U}}$. $\square$

The same construction can be done, starting from the compact topologically étale $A[P_U]$-module M_U, for any compact open subgroup $U \subset N$.

Proposition 7.75. *Let $U \subset N$ be a compact open subgroup. The G-equivariant sheaves on G/P associated to (N_0, M) and to (U, M_U) are equal.*

Proof. As the P-equivariant sheaves on the open cell associated to (N_0, M) and to (U, M_U) are equal by Proposition 7.43, and as the function α_g depends only on the open cell, our formal construction gives the same G-equivariant sheaf. $\square$

6. Integrating α when M is non compact

Recall that we have chosen a certain element $s \in Z(L)$ such that $L = L_- s^{\mathbb{Z}}$ and $(N_k = s^k N_0 s^{-k})_{k \in \mathbb{Z}}$ is a decreasing sequence with union N and trivial intersection. *We now suppose in addition that* $(\overline{N}_k := s^{-k} w_0 N_0 w_0^{-1} s^k)_{k \in \mathbb{Z}}$ *is a decreasing sequence with union* $\overline{N} = w_0 N w_0^{-1}$ *and trivial intersection.*

We have chosen *A* and *M* in Section 4.1. *We suppose now in addition that M is a topologically étale* $A[P_+]$-*module which is Hausdorff and complete.*

Definition 7.76. A special family of compact sets in *M* is a family $\mathfrak{C}$ of compact subsets of *M* satisfying:

$\mathfrak{C}(1)$ Any compact subset of a compact set in $\mathfrak{C}$ also lies in $\mathfrak{C}$.

$\mathfrak{C}(2)$ If $C_1, C_2, \ldots, C_n \in \mathfrak{C}$ then $\bigcup_{i=1}^{n} C_i$ is in $\mathfrak{C}$, as well.

$\mathfrak{C}(3)$ For all $C \in \mathfrak{C}$ we have $N_0 C \in \mathfrak{C}$.

$\mathfrak{C}(4)$ $M(\mathfrak{C}) := \bigcup_{C \in \mathfrak{C}} C$ is an étale $A[P_+]$-submodule of *M*.

Note that *M* is the union of its compact subsets, and that the family of all compact subsets of *M* satisfies these four properties.

Let $\mathfrak{C}$ be a special family of compact sets in *M*. A map from $M(\mathfrak{C})$ to *M* is called $\mathfrak{C}$-continuous if its restriction to any $C \in \mathfrak{C}$ is continuous. We equip the *A*-module $\mathrm{Hom}_A^{\mathfrak{C}ont}(M(\mathfrak{C}), M)$ of $\mathfrak{C}$-continuous *A*-linear homomorphisms from $M(\mathfrak{C})$ to *M* with the $\mathfrak{C}$-open topology. The *A*-submodules

$$E(C, \mathcal{M}) := \{ f \in \mathrm{Hom}_A^{\mathfrak{C}ont}(M(\mathfrak{C}), M) \colon f(C) \subseteq \mathcal{M} \},$$

for any $C \in \mathfrak{C}$ and any open *A*-submodule $\mathcal{M} \subseteq M$, form a fundamental system of open neighbourhoods of zero in $\mathrm{Hom}_A^{\mathfrak{C}ont}(M(\mathfrak{C}), M)$. Indeed, this system is closed for finite intersection by $\mathfrak{C}(2)$. Since N_0 is compact the $E(C, \mathcal{M})$ for *C* such that $N_0 C \subseteq C$ and $\mathcal{M}$ an $A[N_0]$-submodule still form a fundamental system of open neighbourhoods of zero (Lemma 7.50 and $\mathfrak{C}(3)$). We have:

- $\mathrm{Hom}_A^{\mathfrak{C}ont}(M(\mathfrak{C}), M)$ is a topological *A*-module.
- $\mathrm{Hom}_A^{\mathfrak{C}ont}(M(\mathfrak{C}), M)$ is Hausdorff, since $\mathfrak{C}$ covers $M(\mathfrak{C})$ by $\mathfrak{C}(4)$ and *M* is Hausdorff.
- $\mathrm{Hom}_A^{\mathfrak{C}ont}(M(\mathfrak{C}), M)$ is complete ([3] TG X.9 Cor.2).

6.1. (*s*, res, $\mathfrak{C}$)-integrals

We have the P_+-equivariant measure res $: C^\infty(N_0, A) \longrightarrow \mathrm{End}_A^{cont}(M)$ on N_0. If *M* is not compact then it is no longer possible to integrate any map in the *A*-module $C(N_0, \mathrm{End}_A^{cont}(M))$ of all continuous maps on N_0 with values in $\mathrm{End}_A^{cont}(M)$ against this measure. This forces us to introduce a notion of integrability with respect to a special family of compact sets in *M*.

Definition 7.77. A map $F: N_0 \to \mathrm{Hom}_A^{\mathfrak{C}ont}(M(\mathfrak{C}), M)$ is called integrable with respect to $(s, \mathrm{res}, \mathfrak{C})$ if the limit

$$\int_{N_0} F\, d\,\mathrm{res} := \lim_{k \to \infty} \sum_{u \in J(N_0/N_k)} F(u) \circ \mathrm{res}(1_{uN_k}),$$

where $J(N_0/N_k) \subseteq N_0$, for any $k \in \mathbb{N}$, is a set of representatives for the cosets in N_0/N_k, exists in $\mathrm{Hom}_A^{\mathfrak{C}ont}(M(\mathfrak{C}), M)$ and does not depend on the choice of the sets $J(N_0/N_k)$.

We suppress $\mathfrak{C}$ from the notation when $\mathfrak{C}$ is the family of all compact subsets of M.

Note that we regard $\mathrm{res}(1_{uN_k})$ as an element of $\mathrm{End}_A^{cont}(M(\mathfrak{C}))$. This makes sense as the algebraically étale submodule $M(\mathfrak{C})$ of the topologically étale module M is topologically étale.

One easily sees that the set $C^{int}(N_0, \mathrm{Hom}_A^{\mathfrak{C}ont}(M(\mathfrak{C}), M))$ of integrable maps is an A-module. The A-linear map

$$\int_{N_0} .d\,\mathrm{res} : C^{int}(N_0, \mathrm{Hom}_A^{\mathfrak{C}ont}(M(\mathfrak{C}), M)) \longrightarrow \mathrm{Hom}_A^{\mathfrak{C}ont}(M(\mathfrak{C}), M)$$

will be called the $(s, \mathrm{res}, \mathfrak{C})$-integral.

We give now a general integrability criterion.

Proposition 7.78. A map $F : N_0 \longrightarrow \mathrm{Hom}_A^{\mathfrak{C}ont}(M(\mathfrak{C}), M)$ is $(s, \mathrm{res}, \mathfrak{C})$-integrable if, for any $C \in \mathfrak{C}$ and any open A-submodule $\mathcal{M} \subseteq M$, there exists an integer $k_{C,\mathcal{M}} \geq 0$ such that

$$(F(u) - F(uv)) \circ \mathrm{res}(1_{uN_{k+1}}) \in E(C, \mathcal{M})$$

$$\text{for any } k \geq k_{C,\mathcal{M}}, \, u \in N_0, \text{ and } v \in N_k.$$

Proof. Let $J(N_0/N_k)$ and $J'(N_0/N_k)$, for $k \geq 0$, be two choices of sets of representatives. We put

$$s_k(F) := \sum_{u \in J(N_0/N_k)} F(u) \circ \mathrm{res}(1_{uN_k}) \quad \text{and} \quad s_k'(F)$$

$$:= \sum_{u' \in J'(N_0/N_k)} F(u') \circ \mathrm{res}(1_{u'N_k}).$$

Since $\mathrm{Hom}_A^{\mathfrak{C}ont}(M(\mathfrak{C}), M)$ is Hausdorff and complete it suffices to show that, given any neighbourhood of zero $E(C, \mathcal{M})$, there exists an integer $k_0 \geq 0$ such that

$$s_k(F) - s_{k+1}(F), \, s_k(F) - s_k'(F) \in E(C, \mathcal{M}) \qquad \text{for any } k \geq k_0.$$

For $u \in J(N_0/N_{k+1})$ let $\bar{u} \in J(N_0/N_k)$ and $u' \in J'(N_0/N_{k+1})$ be the unique elements such that $uN_k = \bar{u}N_k$ and $uN_{k+1} = u'N_{k+1}$, respectively. Then

$$s_k(F) = \sum_{u \in J(N_0/N_{k+1})} F(\bar{u}) \circ \mathrm{res}(1_{uN_{k+1}})$$

and hence

$$s_k(F) - s_{k+1}(F) = \sum_{u \in J(N_0/N_{k+1})} (F(u(u^{-1}\bar{u})) - F(u)) \circ \mathrm{res}(1_{uN_{k+1}}). \quad (7.26)$$

Since $u^{-1}\bar{u} \in N_k$ it follows from our assumption that the right-hand side lies in $E(C, \mathcal{M})$ for $k \geq k_{C,\mathcal{M}}$. Similarly

$$s_{k+1}(F) - s'_{k+1}(F) = \sum_{u \in J(N_0/N_{k+1})} (F(u) - F(u(u^{-1}u'))) \circ \mathrm{res}(1_{uN_{k+1}});$$

again, as $u^{-1}u' \in N_{k+1} \subseteq N_k$, the right-hand sum is contained in $E(C, \mathcal{M})$ for $k \geq k_{C,\mathcal{M}}$. $\qquad\square$

6.2. Integrability criterion for α

Let $U_g \subseteq N_0$ be the compact open subset such that $U_g w_0 P/P - g^{-1}C_0 \cap C_0$. This intersection is nonempty if and only if $g \in N_0 \bar{P} N_0$, which we therefore assume in the following. We consider the map

$$\alpha_{g,0} : N_0 \longrightarrow \mathrm{End}_A^{cont}(M)$$

$$u \longmapsto \begin{cases} \mathrm{Res}(1_{N_0}) \circ \alpha_g(x_u) \circ \mathrm{Res}(1_{N_0}) & \text{if } u \in U_g, \\ 0 & \text{otherwise} \end{cases}$$

(where we identify $\mathrm{End}_A^{cont}(M)$ with its image in E^{cont} under the natural embedding (7.20) using that $\mathrm{Res}(1_{N_0}) = \sigma_0 \circ \mathrm{ev}_0$). Restricting $\alpha_{g,0}(u) \in \mathrm{End}_A^{cont}(M)$ to $M(\mathcal{C})$ for any $u \in N_0$ we may view $\alpha_{g,0}$ as a map from N_0 to $\mathrm{End}_A^{cont}(M(\mathcal{C}))$ since $M(\mathcal{C})$ is an étale $A[P_+]$-submodule of M. However, as we do not assume $M(\mathcal{C})$ to be complete, it will be more convenient for the purpose of integration to regard $\alpha_{g,0}$ as a map into $\mathrm{Hom}_A^{cont}(M(\mathcal{C}), M)$. We want to establish a criterion for the $(s, \mathrm{res}, \mathcal{C})$-integrability of the map $\alpha_{g,0}$.

By the argument in the proof of Lemma 7.64 (applied to $V = g^{-1}C_0 \cap C_0$) we may choose an integer $k_g^{(0)} \geq 0$ such that, for any $k \geq k_g^{(0)}$, we have $U_g N_k \subseteq U_g$ and

$$\alpha(g, x_u).uN_k \subseteq gU_g \qquad \text{for any } u \in U_g. \quad (7.27)$$

Lemma 7.79. *For $u \in U_g$ and $k \geq k_g^{(0)}$ we have*

$$\alpha_{g,0}(u) \circ \mathrm{res}(1_{uN_k}) = \alpha(g, x_u) \circ \mathrm{Res}(1_{uN_k}).$$

Proof. Using the P-equivariance of Res we have

$$\alpha(g, x_u) \circ \mathrm{Res}(1_{uN_k}) = \mathrm{Res}(1_{\alpha(g,x_u).uN_k}) \circ \alpha(g, x_u) \circ \mathrm{Res}(1_{uN_k})$$
$$= \mathrm{Res}(1_{N_0}) \circ \mathrm{Res}(1_{\alpha(g,x_u).uN_k}) \circ \alpha(g, x_u) \circ \mathrm{Res}(1_{uN_k})$$
$$= \mathrm{Res}(1_{N_0}) \circ \alpha(g, x_u) \circ \mathrm{Res}(1_{N_0}) \circ \mathrm{Res}(1_{uN_k})$$
$$= \alpha_{g,0}(u) \circ \mathrm{res}(1_{uN_k})$$

where the second identity follows from (7.27).   □

For $u \in U_g$ and $k \geq k_g^{(0)}$ we put

$$\mathcal{H}_{g, J(N_0/N_k)} := \sum_{u \in U_g \cap J(N_0/N_k)} \alpha(g, x_u) \circ \mathrm{Res}(1_{uN_k}). \qquad (7.28)$$

By Lemma 7.79, each summand on the right-hand side belongs to $\mathrm{End}_A(M(\mathfrak{C}))$. If $\alpha_{g,0}$ is $(s, \mathrm{res}, \mathfrak{C})$-integrable, the limit

$$\mathcal{H}_g := \lim_{k \geq k_g^{(0)}, k \to \infty} \mathcal{H}_{g, J(N_0/N_k)} \qquad (7.29)$$

exists in $\mathrm{Hom}_A^{\mathfrak{C}ont}(M(\mathfrak{C}), M)$ and is equal to the $(s, \mathrm{res}, \mathfrak{C})$-integral of $\alpha_{g,0}$

$$\int_{N_0} \alpha_{g,0} d\, \mathrm{res} = \mathcal{H}_g. \qquad (7.30)$$

We investigate the integrability criterion of Proposition 7.78 for the function $\alpha_{g,0}$. We have to consider the elements

$$\Delta_g(u, k, v) := (\alpha_{g,0}(u) - \alpha_{g,0}(uv)) \circ \mathrm{res}(1_{uN_{k+1}}), \qquad (7.31)$$

for $u \in U_g$, $k \geq k_g^{(0)}$, and $v \in N_k$. By Lemma 7.79, we have

$$\Delta_g(u, k, v) = (\alpha_{g,0}(u) \circ \mathrm{res}(1_{uN_k}) - \alpha_{g,0}(uv) \circ \mathrm{res}(1_{uvN_k})) \circ \mathrm{res}(1_{uN_{k+1}})$$
$$= (\alpha(g, x_u) \circ \mathrm{Res}(1_{uN_k}) - \alpha(g, x_{uv}) \circ \mathrm{Res}(1_{uvN_k})) \circ \mathrm{Res}(1_{uN_{k+1}})$$
$$= (\alpha(g, x_u) - \alpha(g, x_{uv})) \circ \mathrm{Res}(1_{uN_{k+1}})$$
$$= (\alpha(g, x_u) - \alpha(g, x_{uv})) \circ u \circ \mathrm{Res}(1_{N_{k+1}}) \circ u^{-1}.$$

Recall that $N_g \subset N$ is the subset such that $N_g w_0 P / P = g^{-1}\mathcal{C} \cap \mathcal{C}$.

Lemma 7.80. *For $u \in N_g$ and $v \in N$ such that $uv \in N_g$ we have:*

i. $v \in N_{\bar{n}(g,u)}$;
ii. $\alpha(g, x_{uv}) = \alpha(g, x_u)u\alpha(\bar{n}(g, u), x_v)u^{-1}.$

Proof. i. Because of $gu = \alpha(g, x_u)u\bar{n}(g, u)$ we have

$$\alpha(g, x_u)u\bar{n}(g, u)v = guv \in \alpha(g, x_{uv})uv\overline{N}$$

and hence

$$\bar{n}(g, u)vw_0 P = u^{-1}\alpha(g, x_u)^{-1}\alpha(g, x_{uv})uvw_0 P \in Pw_0 P.$$

ii. By i. the equation $\bar{n}(g, u)vw_0 N = \alpha(\bar{n}(g, u), x_v)vw_0 N$ holds. Hence

$$guvw_0 N = \alpha(g, x_u)u\bar{n}(g, u)vw_0 N = \alpha(g, x_u)u\alpha(\bar{n}(g, u), x_v)vw_0 N$$

and therefore $\alpha(g, x_{uv})uv = \alpha(g, x_u)u\alpha(\bar{n}(g, u), x_v)v$. $\qquad\square$

Let $f: U_g \to P$ be the map $u \mapsto \alpha(g, x_u)u$. The previous computation
shows that for all $u \in U_g$ and $v \in N_k$ we have

$$\Delta_g(u, k, v) = (f(u) - f(uv)v^{-1}) \circ \mathrm{Res}(1_{N_{k+1}}) \circ u^{-1}. \tag{7.32}$$

Let $f(u) = n(g, u)t(g, u)$, with $n(g, u) \in N_0$ and $t(g, u) \in L$. Also, write
$gu = f(u)\bar{n}(g, u)$ with $\bar{n}(g, u) \in \overline{N}$. Since $t(g, U_g) \subset L$ and $\bar{n}(g, U_g) \subset \overline{N}$
are compact subsets, there is $k_g^{(1)} \geq k_g^{(0)}$ such that

$$\Lambda_g := t(g, U_g)s^{k_g^{(1)}} \subset L_+, \quad \bar{n}(g, U_g) \subset \overline{N}_{-k_g^{(1)}}. \tag{7.33}$$

Proposition 7.81. *For any compact open subgroup P_1 of P_0 there is*
$k_g^{(2)}(P_1) \geq k_g^{(1)}$ *such that for all $k \geq k_g^{(2)}(P_1)$, $u \in U_g$ and $v \in N_k$*

$$f(u) - f(uv)v^{-1} \in N_0 s^{k-k_g^{(1)}}(1 - P_1)\Lambda_g s^{-k}.$$

Proof. We abbreviate $n(u) = n(g, u)$ and similarly for $t(u)$ and $\bar{n}(u)$. Since
$f(u)\bar{n}(u)v = guv = f(uv)\bar{n}(uv)$, we have

$$f(u) - f(uv)v^{-1} = f(u)(1 - \bar{n}(u)v\bar{n}(uv)^{-1}v^{-1})$$

$$= n(u)(1 - t(u)\bar{n}(u)v\bar{n}(uv)^{-1}v^{-1}t(u)^{-1})t(u).$$

Since $n(u) \in N_0$, $t(u) \in s^{-k_g^{(1)}}\Lambda_g$ and $(t(u))_{u \in U_g}$ is compact, it suffices to prove that for any compact open subgroup P_2 of P_0 we have
$\bar{n}(u)v\bar{n}(uv)^{-1}v^{-1} \in s^k P_2 s^{-k}$ for sufficiently large k. But if $v = s^k n_0 s^{-k}$,
we can write

$$\bar{n}(u)v\bar{n}(uv)^{-1}v^{-1} = s^k(s^{-k}\bar{n}(u)s^k)n_0(s^{-k}\bar{n}(uv)^{-1}s^k)n_0^{-1}s^{-k}$$

$$\in s^k \overline{N}_{k-k_g^{(1)}}\left(\bigcup_{n_0 \in N_0} n_0\overline{N}_{k-k_g^{(1)}}n_0^{-1}\right)s^{-k}.$$

The result follows from the compactness of N_0 and the fact that the $\overline{N}_k$'s shrink to $\{1\}$ as $k \to \infty$.   □

Corollary 7.82. *For any compact open subgroup P_1 of P_0 and $k \geq k_g^{(2)}(P_1)$*

$$\Delta_g(U_g, k, N_k) \subset N_0 s^{k - k_g^{(1)}} (1 - P_1) \Lambda_g s \psi^{k+1} N_0.$$

Proof. Proposition 7.81 and relation (7.32) show that

$$\Delta_g(U_g, k, N_k) \subset N_0 s^{k - k_g^{(1)}} (1 - P_1) \Lambda_g s \circ s^{-(k+1)} \circ \mathrm{Res}(1_{N_{k+1}}) \circ N_0.$$

The P-equivariance of Res yields $s^{-(k+1)} \circ \mathrm{Res}(1_{N_{k+1}}) = \mathrm{Res}(1_{N_0}) \circ s^{-k-1}$, and this is the image of $\psi^{k+1} \in \mathrm{End}_A^{\mathrm{cont}}(M)$ in $\mathrm{End}_A^{\mathrm{cont}}(M^P)$. The result follows.   □

This leads to an integrability criterion for $\alpha_{g,0}$, which depends only on $(s, M, \mathfrak{C})$.

Proposition 7.83. *We suppose that $(s, M, \mathfrak{C})$ satisfies:*

$\mathfrak{C}(5)$ *For any $C \in \mathfrak{C}$ the compact subset $\psi(C) \subseteq M$ also lies in $\mathfrak{C}$.*

$\mathfrak{T}(1)$ *For any $C \in \mathfrak{C}$ such that $C = N_0 C$, any open $A[N_0]$-submodule $\mathcal{M}$ of M, and any compact subset $C_+ \subseteq L_+$ there exists a compact open subgroup $P_1 = P_1(C, \mathcal{M}, C_+) \subseteq P_0$ and an integer $k(C, \mathcal{M}, C_+) \geq 0$ such that*

$$s^k (1 - P_1) C_+ \psi^k(C) \subseteq \mathcal{M} \qquad \textit{for any } k \geq k(C, \mathcal{M}, C_+). \tag{7.34}$$

Then the map $\alpha_{g,0} \colon N_0 \to \mathrm{Hom}_A^{\mathfrak{C}\mathrm{ont}}(M(\mathfrak{C}), M)$ is $(s, \mathrm{res}, \mathfrak{C})$-integrable for all $g \in N_0 \overline{P} N_0$.

Proof. By the general integrability criterion of Proposition 7.78, the map $\alpha_{g,0}$ is integrable if for any $(C, \mathcal{M})$ as above, there exists $k_{C, \mathcal{M}, g} \geq 0$ such that

$$\Delta_g(u, k, v) \in E(C, \mathcal{M}) \qquad \text{for any } k \geq k_{C, \mathcal{M}, g}, u \in U_g, \text{ and } v \in N_k. \tag{7.35}$$

By Corollary 7.82, this is true if $k_{C, \mathcal{M}, g} \geq k_g^{(2)}(P_1)$ and

$$s^{k - k_g^{(1)}} (1 - P_1) \Lambda_g s \psi^{k+1}(C) \subset \mathcal{M}, \tag{7.36}$$

because $N_0 \mathcal{M} = \mathcal{M}$ and $N_0 C = C$.

We note that the set $C_+ = \Lambda_g s$ is contained in L_+ by (7.33) and is compact, that the set $C' = \psi^{k_g^{(1)}+1}(C) \subset M$ is compact and $N_0 C' = C'$ because the map ψ is continuous and $N_0 \psi(C) = \psi(s N_0 s^{-1} C) = \psi(C)$, and that (7.36) is equivalent to

$$s^{k - k_g^{(1)}} (1 - P_1) C_+ \psi^{k - k_g^{(1)}}(C) \subset E(C', \mathcal{M}).$$

By our hypothesis, there exists an open subgroup $P_1 \subset P_0$ such that this inclusion is satisfied when $k \geq k_g^{(1)} + k(C', \mathcal{M}, C_+)$. For

$$k_{C, \mathcal{M}, g} := \max(k_g^{(1)} + k(C', \mathcal{M}, C_+), k_g^{(2)}(P_1)) \qquad (7.37)$$

(7.35) is satisfied. By construction, P_1 depends on $\psi^{k_g^{(1)}+1}(C), \mathcal{M}, \Lambda_g s$, hence only on $C, \mathcal{M}, g$. □

Later, under the assumptions of Proposition 7.83, we will use the argument in the previous proof in the following slightly more general form: for $C, \mathcal{M}, C_+$ as in the proposition and an integer $k' \geq 0$ we have

$$s^{k-k'}(1 - P_1(\psi^{k'}(C), \mathcal{M}, C_+))C_+\psi^k \subseteq E(C, \mathcal{M}) \qquad (7.38)$$

for any $k \geq k' + k(\psi^{k'}(C), \mathcal{M}, C_+)$.

6.3. Extension of Res

Proposition 7.84. *Suppose that* $(s, M, \mathfrak{C})$ *satisfies the assumptions of Proposition 7.83 and that the* $(s, \mathrm{res}, \mathfrak{C})$-*integral* $\mathcal{H}_g$ *of* $\alpha_{g,0}$ *is contained in* $\mathrm{End}_A(M(\mathfrak{C}))$ *for all* $g \in N_0\overline{P}N_0$. *In addition we assume that:*

$\mathfrak{C}(6)$ *For any* $C \in \mathfrak{C}$ *the compact subset* $\varphi(C) \subseteq M$ *also lies in* $\mathfrak{C}$.

$\mathfrak{T}(2)$ *Given a set* $J(N_0/N_k) \subset N_0$ *of representatives for cosets in* N_0/N_k, *for* $k \geq 1$, *for any* $x \in M(\mathfrak{C})$ *and* $g \in N_0\overline{P}N_0$ *there exists a compact A-submodule* $C_{x,g} \in \mathfrak{C}$ *and a positive integer* $k_{x,g}$ *such that* $\mathcal{H}_{g, J(N_0/N_k)}(x) \subseteq C_{x,g}$ *for any* $k \geq k_{x,g}$.

Then the $\mathcal{H}_g$ *satisfy the relations H1, H2, H3 of Proposition 7.72.*

Remark 7.85. The properties $\mathfrak{C}(3), \mathfrak{C}(5), \mathfrak{C}(6)$ imply that for any $u \in N_0$, $k \geq 1$, and $C \in \mathfrak{C}$ also $\mathrm{res}(1_{uN_k})(C)$ lies in $\mathfrak{C}$. Indeed, $\mathrm{res}(1_{uN_k}) = u \circ \varphi^k \circ \psi^k \circ u^{-1}$.

We prove now H1 and H3, which do not use the last assumption. The proof of H2 is postponed to the next subsection.

Proof. For the proof of H1 let $V \subset C_0$ be a compact open subset and let U_1, U_2 be the compact open subsets of N_0 corresponding to V and $g^{-1}V \cap C_0$. To prove that $\mathrm{res}(1_V) \circ \mathcal{H}_g = \mathcal{H}_g \circ \mathrm{res}(1_{g^{-1}V \cap C_0})$, it suffices to verify that if k is large enough, then for all $u \in U_g$ we have

$$\mathrm{Res}(1_{U_1}) \circ \alpha(g, x_u) \circ \mathrm{Res}(1_{uN_k}) = \alpha(g, x_u) \circ \mathrm{Res}(1_{uN_k}) \circ \mathrm{Res}(1_{U_2}). \quad (7.39)$$

If $N_{k,u} = \alpha(g, x_u).(uN_k)$, then by *P*-equivariance of Res, (7.39) is equivalent to

$$\mathrm{Res}(1_{U_1 \cap N_{k,u}}) \circ \alpha(g, x_u) = \alpha(g, x_u) \circ \mathrm{Res}(1_{u N_k \cap U_2}). \qquad (7.40)$$

Write $\alpha(g, x_u) = n_u t_u$ with $n_u \in N$ and $t_u \in L$. If k is large enough, then for all $u \in U_g$ we have $U_2 N_k \subset U_2$ and $U_1 t_u N_k t_u^{-1} \subset U_1$. Since $N_{k,u} = (\alpha(g, x_u).u) t_u N_k t_u^{-1}$, we obtain

$$U_1 \cap N_{k,u} \neq \emptyset \Leftrightarrow \alpha(g, x_u).u \in U_1 \Leftrightarrow g x_u \in \mathcal{V}$$

$$\Leftrightarrow x_u \in g^{-1}\mathcal{V} \cap \mathcal{C}_0 \Leftrightarrow u \in U_2 \Leftrightarrow u N_k \subset U_2.$$

Hence (7.40) is equivalent to $0 = 0$ or to $\mathrm{Res}(1_{N_{k,u}}) \circ \alpha(g, x_u) = \alpha(g, x_u) \circ \mathrm{Res}(1_{u N_k})$, which is true as Res is P-equivariant.

H3. For $b \in P \cap N_0 \overline{P} N_0$ we have

$$\alpha_{b,0} = \text{constant map on } N_0 \text{ with value } \mathrm{res}(1_{\mathcal{C}_0}) \circ b \circ \mathrm{res}(1_{\mathcal{C}_0})$$

and hence

$$\mathcal{H}_b = \mathrm{res}(1_{\mathcal{C}_0}) \circ b \circ \mathrm{res}(1_{\mathcal{C}_0}) = b \circ \mathrm{res}(1_{b^{-1}\mathcal{C}_0 \cap \mathcal{C}_0}).$$

$\square$

6.4. Proof of the product formula

We invoke now the full set of assumptions of Proposition 7.84 and we prove the product formula

$$\mathcal{H}_g \circ \mathcal{H}_h = \mathcal{H}_{gh} \circ \mathrm{res}(1_{h^{-1}\mathcal{C}_0 \cap \mathcal{C}_0})$$

for $g, h \in N_0 \overline{P} N_0$. This suffices by Proposition 7.72.

Let $k_0 := \max(k_g^{(0)}, k_h^{(1)}, k_{gh}^{(0)}) + 1$ and let $k \geq k_0$.

As $k \geq k_h^{(0)}$ (because $k_h^{(1)} \geq k_h^{(0)}$ (7.33)), the set U_h is a disjoint union of cosets $u N_k$. We choose a set $J(N_0/N_k) \subset N_0$ of representatives of the cosets in N_0/N_k and for each $u \in J(N_0/N_k) \cap U_h$ a set $J_u(N_0/N_{k-k_0}) \subset N_0$ of representatives of the cosets in N_0/N_{k-k_0} with $n(g, u) \in J_u(N_0/N_{k-k_0})$ (see (7.23)).

We write $\mathcal{H}_g \circ \mathcal{H}_h - \mathcal{H}_{gh} \circ \mathrm{res}(1_{h^{-1}\mathcal{C}_0 \cap \mathcal{C}_0})$ as the sum over $u \in J(N_0/N_k) \cap U_h$ of

$$(\mathcal{H}_g \circ \mathcal{H}_h - \mathcal{H}_{gh} \circ \mathrm{Res}(1_{U_h})) \circ \mathrm{Res}(1_{u N_k}) = a_{k,u} + b_{k,u} + c_{k,u}, \qquad (7.41)$$

where

$$a_{k,u} := (\mathcal{H}_g \circ \mathcal{H}_h - \mathcal{H}_{g, J_u(N_0/N_{k-k_0})} \circ \mathcal{H}_{h, J(N_0/N_k)}) \circ \mathrm{Res}(1_{u N_k})$$

$$b_{k,u} := (\mathcal{H}_{g, J_u(N_0/N_{k-k_0})} \circ \mathcal{H}_{h, J(N_0/N_k)} - \mathcal{H}_{gh, J(N_0/N_k)}) \circ \mathrm{Res}(1_{U_h}) \circ \mathrm{Res}(1_{u N_k})$$

$$c_{k,u} := (\mathcal{H}_{gh, J(N_0/N_k)} - \mathcal{H}_{gh}) \circ \mathrm{Res}(1_{U_h}) \circ \mathrm{Res}(1_{u N_k}).$$

The product formula follows from the claim that $b_{k,u} = 0$ and that for an arbitrary compact subset $C \in \mathfrak{C}$ such that $N_0 C = C$, and an arbitrary open $A[N_0]$-module $\mathcal{M} \subset M$, $a_{k,u}$ and $c_{k,u}$ lie in $E(C, \mathcal{M})$ when k is very large, independently of u.

The claim results from the following three propositions.

Because $(s, M, \mathfrak{C})$ satisfies Proposition 7.83, we associate to $(C, \mathcal{M}, g)$ the integer $k_{C,\mathcal{M},g}$ defined in (7.37) which is independent of the choice of the $J(N_0/N_k)$. For the sake of simplicity, we write

$$\mathcal{H}_g^{(k)} := \mathcal{H}_{g,J(N_0/N_k)}, \quad s_g^{(k)} := \mathcal{H}_g^{(k+1)} - \mathcal{H}_g^{(k)}. \qquad (7.42)$$

By (7.26), we have, for $k \geq k_g^{(0)}$,

$$s_g^{(k)} = \sum_{u \in U_g \cap J(N_0/N_{k+1})} \Delta_g(u, k, v_u)$$

for some $v_u \in N_k$. It follows from Corollary 7.82 that, for any given compact open subgroup $P_1 \subset P_0$, we have

$$s_g^{(k)} \in \; < N_0 s^{k-k_g^{(1)}} (1 - P_1) \Lambda_g s \psi^{k+1} N_0 >_A \qquad \text{for } k \geq k_g^{(2)}(P_1), \qquad (7.43)$$

where we use the notation $< X >_A$ for the A-submodule in $\mathrm{End}_A(M)$ generated by X. We deduce from the proof of Proposition 7.83, that $s_g^{(k)} \in E(C, \mathcal{M})$ for any $k \geq k_{C,\mathcal{M},g}$.

Proposition 7.86. $(\mathcal{H}_g - \mathcal{H}_{g,J(N_0/N_k)}) \circ \mathrm{Res}(1_{uN_k}) \in E(C, \mathcal{M})$ *for any* $k \geq k_{C,\mathcal{M},g}$.

Proof. When $k \geq 0$, $k_2 \geq \max(k - 1, k_g^{(0)})$, $u' \in U_g$, $v \in N_k$ we have that $\Delta_g(u', k_2, v) \circ \mathrm{Res}(1_{uN_k})$ is equal either to $\Delta_g(u', k_2, v)$ or to 0. If follows that

$$s_g^{(k_2)} \circ \mathrm{Res}(1_{uN_k}) \subseteq E(C, \mathcal{M}) \qquad \text{for any } k_2 \geq \max(k - 1, k_{C,\mathcal{M},g}) \text{ and } k \geq 0.$$

Now we fix $k \geq k_{C,\mathcal{M},g}$. Note that $\mathrm{Res}(1_{uN_k})(C)$ is contained in $\mathfrak{C}$ by the stability of $\mathfrak{C}$ by ψ, φ, and $u^{\pm 1}$. Therefore the sequence $(\mathcal{H}_g^{(k_2)} \circ \mathrm{Res}(1_{uN_k}))_{k_2}$ converges to $\mathcal{H}_g \circ \mathrm{Res}(1_{uN_k})$ in $\mathrm{Hom}_A^{\mathfrak{C}ont}(M(\mathfrak{C}), M)$. In particular, we have

$$(\mathcal{H}_g - \mathcal{H}_g^{(k_2)}) \circ \mathrm{Res}(1_{uN_k}) \subseteq E(C, \mathcal{M})$$

$$\text{for any } k_2 \geq \max(k - 1, k_{C,\mathcal{M},g}) \text{ and } k \geq 0.$$

The statement follows by taking $k_2 = k$. $\qquad \square$

This establishes that $c_{k,u}$ lies in $E(C, \mathcal{M})$ when $k \geq k_{C,\mathcal{M},gh}$.

Note that the proposition is true also for any other system $J'(N_0/N_k) \subset N_0$ of representatives for the cosets in N_0/N_k for the same integer $k_{C,\mathcal{M},g}$. We write $\mathcal{H}_g'^{(k)}$ and $s_g'^{(k)}$ for the elements defined in (7.42) for $J'(N_0/N_k)$.

Proposition 7.87. *There exists an integer $k_{C,\mathcal{M},g,h,k_0} \in \mathbb{N}$, independent of the choices of $J(N_0/N_k)$ and $J'(N_0/N_k)$, such that:*

i. $\mathcal{H}_g'^{(k+1-k_0)} \circ \mathcal{H}_h^{(k+1)} - \mathcal{H}_g'^{(k-k_0)} \circ \mathcal{H}_h^{(k)} \in E(C,\mathcal{M})$, *for all* $k \geq k_{C,\mathcal{M},g,h,k_0}$, *and the sequence* $(\mathcal{H}_g'^{(k-k_0)} \circ \mathcal{H}_h^{(k)})$ *converges to* $\mathcal{H}_g \circ \mathcal{H}_h$ *in* $\mathrm{Hom}_A^{\mathfrak{C}ont}(M(\mathfrak{C}),M)$.

ii. $(\mathcal{H}_g \circ \mathcal{H}_h - \mathcal{H}_g'^{(k-k_0)} \circ \mathcal{H}_h^{(k)}) \circ \mathrm{Res}(1_{uN_k}) \in E(C,\mathcal{M})$, *for all* $k \geq k_{C,\mathcal{M},g,h,k_0}$.

Proof. i. To prove the first assertion, we write

$$\mathcal{H}_g'^{(k+1-k_0)} \circ \mathcal{H}_h^{(k+1)} - \mathcal{H}_g'^{(k-k_0)} \circ \mathcal{H}_h^{(k)}$$
$$= \mathcal{H}_g'^{(k+1-k_0)} \circ s_h^{(k)} + s_g'^{(k-k_0)} \circ \mathcal{H}_h^{(k)}. \tag{7.44}$$

Note that, when $k \geq k_g^{(1)}$, the endomorphisms $\mathcal{H}_g'^{(k)}$ and $\mathcal{H}_g^{(k)}$ are contained in the A-module $< N_0 s^{k-k_g^{(1)}} \Lambda_g \psi^k N_0 >_A$, because

$$\alpha(g, x_u) \circ \mathrm{Res}(1_{uN_k})$$
$$= n(g,u)t(g,u)u^{-1} u s^k \psi^k u^{-1} \subset N_0 s^{k-k_g^{(1)}} \Lambda_g \psi^k N_0 \qquad \text{for } u \in U_g.$$

We consider any compact open subgroup $P_1 \subset P_0$ and we assume $k \geq \max(k_g^{(2)}(P_1) + k_0, k_h^{(2)}(P_1))$. With (7.43) we obtain that (7.44) is contained in

$$< N_0 s^{k+1-k_0-k_g^{(1)}} \Lambda_g \psi^{k+1-k_0} N_0 s^{k-k_h^{(1)}} (1-P_1)\Lambda_h s \psi^{k+1} N_0 >_A$$
$$+ < N_0 s^{k-k_0-k_g^{(1)}} (1-P_1)\Lambda_g s \psi^{k-k_0+1} N_0 s^{k-k_h^{(1)}} \Lambda_h \psi^k N_0 >_A.$$

Recalling that $\psi^a(N_0 \varphi^{a+b}(m)) = \psi^a(N_0)\varphi^b(m) = N_0 \varphi^b(m)$ for $a,b \in \mathbb{N}$ and $m \in M$, we see that this is contained in

$$< N_0 s^{k+1-k_0-k_g^{(1)}} \Lambda_g N_0 s^{k_0-k_h^{(1)}-1} (1-P_1)\Lambda_h s \psi^{k+1} N_0 >_A$$
$$+ < N_0 s^{k-k_0-k_g^{(1)}} (1-P_1)\Lambda_g N_0 s^{k_0-k_h^{(1)}} \Lambda_h \psi^k N_0 >_A.$$

As $k + 1 - k_0 - k_g^{(1)} \geq k_g^{(2)}(P_1) + 1 - k_g^{(1)} \geq 1$ and as $\Lambda_g \subset L_+$, we have

$$N_0 s^{k+1-k_0-k_g^{(1)}} \Lambda_g N_0 \subset N_0 s^{k+1-k_0-k_g^{(1)}} \Lambda_g,$$

and this is contained in

$$< N_0 s^{k+1-k_0-k_g^{(1)}} \Lambda_g s^{k_0-k_h^{(1)}-1} (1-P_1)\Lambda_h s \psi^{k+1} N_0 >_A$$
$$+ < N_0 s^{k-k_0-k_g^{(1)}} (1-P_1)\Lambda_g s^{k_0-k_h^{(1)}} \Lambda_h \psi^k N_0 >_A.$$

We assume, as we may, that the compact open subgroup P_1 of P_0 satisfies $tP_1t^{-1} \subseteq P_1$ for all t in the compact set $\Lambda_g s^{k_0-k_h^{(1)}-1}$ of L_+. Then we finally obtain that (7.44) is contained in

$$< N_0 s^{k+1-k_0-k_g^{(1)}}(1 - P_1)\Lambda_g s^{k_0-k_h^{(1)}} \Lambda_h \psi^{k+1} N_0 >_A$$
$$+ < N_0 s^{k-k_0-k_g^{(1)}}(1 - P_1)\Lambda_g s^{k_0-k_h^{(1)}} \Lambda_h \psi^{k} N_0 >_A.$$

This subset of $\mathrm{End}_A(M)$ is contained in $E(C, \mathcal{M})$ when

$$s^{k+1-k_0-k_g^{(1)}}(1 - P_1)\Lambda_g s^{k_0-k_h^{(1)}} \Lambda_h \psi^{k+1}(C)$$

$$\text{and} \quad s^{k-k_0-k_g^{(1)}}(1 - P_1)\Lambda_g s^{k_0-k_h^{(1)}} \Lambda_h \psi^{k}(C)$$

are contained in $E(C, \mathcal{M})$ because $N_0 C = C$ and $\mathcal{M}$ is an $A[N_0]$-module. By (7.38), this is true when P_1 is contained in $P_1(\psi^{k_0+k_g^{(1)}}(C), \mathcal{M}, \Lambda_g s^{k_0-k_h^{(1)}} \Lambda_h)$ and $k \geq k_{C,\mathcal{M},g,h,k_0}$ where

$$k_{C,\mathcal{M},g,h,k_0} := \max(k_g^{(2)}(P_1) + k_0, k_h^{(2)}(P_1), k(\psi^{k_0+k_g^{(1)}}(C),$$
$$\mathcal{M}, \Lambda_g s^{k_0-k_h^{(1)}} \Lambda_h)). \tag{7.45}$$

The first assertion of i. is proved. We deduce the second assertion from the following claim and the last assumption of Proposition 7.84:

Let $(A_n)_{n \in \mathbb{N}}$ and $(B_n)_{n \in \mathbb{N}}$ be two convergent sequences in $\mathrm{Hom}_A^{\mathfrak{C}ont}$ $(M(\mathfrak{C}), M)$ with limits A and B, respectively; assume that $(B_n)_{n \in \mathbb{N}}$ and B are in $\mathrm{End}_A(M(\mathfrak{C}))$ and that, for any $x \in \mathfrak{C}$ there exists an A-submodule $C \in \mathfrak{C}$ such that $B_n(x) \in C$ for any large n. Then, if the sequence $(A_n \circ B_n)_{n \in \mathbb{N}}$ is convergent, its limit is $A \circ B$.

Let D be the limit of the sequence $(A_n \circ B_n)_n$. It suffices to show that, for any open A-submodule $\mathcal{M} \subseteq M$ and any element $x \in M(\mathfrak{C})$ we have $(D - A \circ B)(x) \in \mathcal{M}$. We write

$$D - A \circ B = (D - A_n \circ B_n) - (A - A_n) \circ B_n - A \circ (B - B_n).$$

Obviously $(D - A_n \circ B_n)(x) \in \mathcal{M}$ for large n. Secondly, the elements $B_n(x)$ for any large n are contained in some compact A-submodule $C \in \mathfrak{C}$, hence also $(B - B_n)(x)$. Moreover $A - A_n \in E(C, \mathcal{M})$ for large n. Hence $(A - A_n) \circ B_n(x) \in \mathcal{M}$ for large n. Finally, A being $\mathfrak{C}$-continuous there is an open A-submodule $\mathcal{M}' \subseteq M$ such that $A(\mathcal{M}' \cap C) \subseteq \mathcal{M}$. Furthermore $(B - B_n)(x) \in \mathcal{M}' \cap C$ for large n. Hence $A \circ (B - B_n)(x) \in \mathcal{M}$ for large n.

ii. This follows from the second assertion in i. together with Remark 7.85. $\qquad \square$

We have now proved that $a_{k,u} \in E(C, \mathcal{M})$ when $k \geq k_{C,\mathcal{M},g,h,k_0}$.

Proposition 7.88. *For $u \in J(N_0/N_k) \cap U_h$, we have*

$$\mathcal{H}_{g, J_u(N_0/N_{k-k_0})} \circ \mathcal{H}_{h, J(N_0/N_k)} \circ \mathrm{Res}(1_{uN_k}) = \mathcal{H}_{gh, J(N_0/N_k)} \circ \mathrm{Res}(1_{uN_k}). \quad (7.46)$$

Proof. The left side of (7.46) is

$$\sum_{v \in U_g \cap J_u(N_0/N_{k-k_0})} \alpha(g, x_v) \circ \mathrm{Res}(1_{vN_{k-k_0}}) \circ \alpha(h, x_u) \circ \mathrm{Res}(1_{uN_k}).$$

The right side of (7.46) is $\alpha(gh, x_u) \circ \mathrm{Res}(1_{uN_k})$ if $u \in J(N_0/N_k) \cap U_h \cap U_{gh}$ and is 0 if u does not belong to U_{gh}. We recall that

$$\alpha(h, x_u)u = n(h, u)t(h, u) \qquad \text{with } n(h, u) \in N_0 \text{ and } t(h, u) \in L_+ s^{-k_h^{(1)}}.$$

It follows that

$$\alpha(h, x_u)u N_k w_0 P \subseteq n(h, u) N_{k - k_h^{(1)}} w_0 P \subset n(h, u) N_{k - k_0} w_0 P.$$

We obtain

$$\mathrm{Res}(1_{vN_{k-k_0}}) \circ \alpha(h, x_u) \circ \mathrm{Res}(1_{uN_k})$$
$$= \begin{cases} \alpha(h, x_u) \circ \mathrm{Res}(1_{uN_k}) & \text{if } vN_{k-k_0} = n(h, u)N_{k-k_0}, \\ 0 & \text{otherwise.} \end{cases}$$

We check now that $u \in U_{gh} \cap U_h$ if and only if $n(h, u) \in U_g$. Indeed $x_u = u w_0 P/P$ belongs to $h^{-1}C_0 \cap C_0 = U_h w_0 P/P$,

$$x_u \in (gh)^{-1}C_0 \cap h^{-1}C_0 \cap C_0 \quad \text{if and only if} \quad hx_u \in g^{-1}C_0 \cap C_0$$

and $hx_u = \alpha(h, x_u)x_u = n(h, u)w_0 P/P$. It follows that $u \in U_{gh} \cap U_h$ if and only if $n(h, u) \in U_g$. As $J_u(N_0/N_{k-k_0})$ contains $n(h, u)$, we have $v = n(h, u)$ when $vN_{k-k_0} = n(h, u)N_{k-k_0}$. We deduce that the left side of (7.46) is 0 when u does not belong to U_{gh} and otherwise is equal to

$$\alpha(g, hx_u) \circ \alpha(h, x_u) \circ \mathrm{Res}(1_{uN_k}) = \alpha(gh, x_u) \circ \mathrm{Res}(1_{uN_k}),$$

where the last equality follows from the product formula for α (Lemma 7.63). $\qquad \square$

We have proved that $b_{k,u} = 0$, therefore ending the proof of the product formula.

6.5. Reduction modulo p^n

We investigate now the situation that will appear for generalized (φ, Γ)-modules M, where the reduction modulo a power of p allows us to reduce to the simpler case where M is killed by a power of p. We will use later this section to get a special family $\mathfrak{C}_s$ in M such that the $(s, \text{res}, \mathfrak{C}_s)$-integrals $\mathcal{H}_g$ exist for all $g \in N_0 \overline{P} N_0$ and satisfy the relations H1, H2, H3 of Proposition 7.72.

We assume now that (A, M) satisfies:

a. *A is a commutative ring with the p-adic topology (the ideals $p^n A$ for $n \geq 1$ form a basis of neighbourhoods of 0) and is Hausdorff.*
b. *M is a linearly topological A-module with a topology weaker than the p-adic topology (a neighbourhood of 0 contains some $p^n M$) and M is a Hausdorff and topological $A[P_+]$-module as in Section 6 (we do not suppose that M is complete).*
c. *The submodules $p^n M$, for $n \geq 1$, are closed in M.*
d. *M is p-adically complete: the linear map $M \rightarrow \varprojlim_{n \geq 1}(M/p^n M)$ is bijective.*

For all $n \geq 1$, we equip $M/p^n M$ with the quotient topology so that the quotient map $p_n : M \rightarrow M/p^n M$ is continuous. The natural homomorphism

$$M \xrightarrow{\cong} \varprojlim_{n \geq 1}(M/p^n M)$$

is a homeomorphism, and the natural homomorphism

$$\text{End}_A^{cont}(M) \xrightarrow{\cong} \varprojlim_{n \geq 1} \text{End}_A^{cont}(M/p^n M)$$

is bijective. We have:

- For a subset C of M, let $\overline{C}$ be the closure of C. Then $\overline{C} = \varprojlim_{n \geq 1} \overline{p_n(C)}$ and if C is closed, $C = \varprojlim_{n \geq 1} p_n(C)$. If C is p-compact (i.e. $p_n(C)$ are compact for all $n \geq 1$), then C is compact, and conversely ([2] I.29 Cor. and I.64 Prop.8).
- An endomorphism f of M which is p-continuous (i.e. the endomorphism f_n induced by f on $M/p^n M$ is continuous for all $n \geq 1$) is continuous, and conversely.
- An action of a topological group H on M which is p-continuous (i.e. the induced action of H on $M/p^n M$ is continuous for all $n \geq 1$) is continuous, and conversely.
- If the $M/p^n M$ are complete for all $n \geq 1$, then M is complete.
- The image $\mathfrak{C}_n$ in $M/p^n M$, for all $n \geq 1$, of a special family $\mathfrak{C}$ of compact subsets in M such that, for all positive integers n,

$$p^n M \cap M(\mathfrak{C}) = p^n M(\mathfrak{C})$$

is a special family. In this case, one has $M(\mathfrak{C}_n) = M(\mathfrak{C})/p^n M(\mathfrak{C})$.

- M is a topologically étale $A[P_+]$-module if and only if $M/p^n M$ is a topologically étale $A[P_+]$-module, for all $n \geq 1$. If we replace "topologically"
  by "algebraically", this is the same proof as for classical (φ, Γ)-modules
  (see Subsection 7.3). The canonical inverse ψ_s of the action φ_s of s is
  continuous if and only if it is p-continuous.

We introduce now our setting which will be discussed in this section.
  We suppose that:

- *M is a topologically étale $A[P_+]$-module, and $M/p^n M$ is complete for
  all $n \geq 1$.*
- *We are given, for $n \geq 1$, a special family $\mathfrak{C}_n$ of compact subsets in $M_n = M/p^n M$ such that $\mathfrak{C}_n$ contains the image of $\mathfrak{C}_{n+1}$ in M_n for all $n \geq 1$.*

Let $\mathfrak{C}$ be the set of compact subsets $C \subset M$ such that $p_n(C) \in \mathfrak{C}_n$ for all
$n \geq 1$.

Lemma 7.89. *$\mathfrak{C}$ is a special family in M and $M(\mathfrak{C}) = \varprojlim_{n \geq 1} M(\mathfrak{C}_n)$.*

Proof. $\mathfrak{C}(1)$ It is obvious that a compact subset C' of $C \in \mathfrak{C}$ is in $\mathfrak{C}$ because p_n
is continuous and $p_n(C')$ is compact.
  $\mathfrak{C}(2)$ p_n commutes with finite union hence $\mathfrak{C}$ is stable by finite union.
  $\mathfrak{C}(3)$ p_n commutes with the action of N_0 hence $C \in \mathfrak{C}$ implies $N_0 C \in \mathfrak{C}$.
  $\mathfrak{C}(4)$ By definition $x \in M(\mathfrak{C})$ if and only if $p_n(x) \in M(\mathfrak{C}_n)$ for all
$n > 1$. The compatibility of the $\mathfrak{C}_n$ implies that the $M(\mathfrak{C}_n)$ form a projective
system. We deduce $M(\mathfrak{C}) = \varprojlim_{n \geq 1} M(\mathfrak{C}_n)$. As the latter ones are topologically étale, the topological $A[P_+]$-module $M(\mathfrak{C})$ is topologically étale by
Remark 7.24.   □

We have the natural map

$$\varprojlim_n \operatorname{Hom}_A(M(\mathfrak{C}_n), M/p^n M) \to \operatorname{Hom}_A(\varprojlim_n M(\mathfrak{C}_n), \varprojlim_n M/p^n M)$$

$$= \operatorname{Hom}_A(M(\mathfrak{C}), M).$$

Lemma 7.90. *The above map induces a continuous map*

$$\varprojlim_n \operatorname{Hom}_A^{\mathfrak{C}_n cont}(M(\mathfrak{C}_n), M/p^n M) \to \operatorname{Hom}_A^{\mathfrak{C} cont}(M(\mathfrak{C}), M), \qquad (7.47)$$

for the projective limit of the $\mathfrak{C}_n$-open topologies on the left-hand side.

Proof. Let $f = \varprojlim f_n$ be a map in the image, and let $C \in \mathfrak{C}$. Then $f|_C$ is the
projective limit of the $f_n|_{p_n(C)}$ hence is continuous. This means that the map

in the assertion is well defined. For the continuity, let $C \in \mathfrak{C}$ and $\mathcal{M} \subset M$ be an open A-submodule. The preimage of $E(C, \mathcal{M})$ is equal to

$$\left(\varprojlim_n \operatorname{Hom}_A^{\mathfrak{C}_n cont}(M(\mathfrak{C}_n), M/p^n M)\right) \cap \left(\prod_n E(p_n(C), \mathcal{M} + p^n M/p^n M)\right).$$

Since $\mathcal{M}$ contains some $p^{n_0} M$, this intersection is equal to the open submodule

$$\{(f_n) \in \varprojlim_n \operatorname{Hom}_A^{\mathfrak{C}_n cont}(M(\mathfrak{C}_n), M/p^n M) : f_n \in E(p_n(C), \mathcal{M} + p^n M/p^n M)$$

for $n \leq n_0\}$.

$\square$

Proposition 7.91. *In the above setting assume that all the assumptions of Proposition 7.84 are satisfied for $(s, M/p^n M, \mathfrak{C}_n)$ and for all $n \geq 1$. Then, for all $g \in N_0 \overline{P} N_0$, the functions*

$$\alpha_{g,0} : N_0 \to \operatorname{Hom}_A^{\mathfrak{C}ont}(M(\mathfrak{C}), M)$$

are $(s, res, \mathfrak{C})$-integrable, their $(s, res, \mathfrak{C})$-integrals $\mathcal{H}_g$ belong to $\operatorname{End}_A(M(\mathfrak{C}))$ and satisfy the relations H1, H2, H3 of Proposition 7.72.

Proof. In the following we indicate with an extra index n that the corresponding notation is meant for the module $M/p^n M$ with the special family $\mathfrak{C}_n$. Then $\alpha_{g,0}(u)$ is the image of $(\alpha_{g,0,n}(u))_n$ by the map (7.47), for $u \in N_0$. It follows that $\mathcal{H}_{g,J(N_0/N_k)}$ is the image of $(\mathcal{H}_{g,J(N_0/N_k),n})_n$ for $g \in N_0 \overline{P} N_0$. By assumption the integral $\mathcal{H}_{g,n} = \lim_{k\to\infty} \mathcal{H}_{g,J(N_0/N_k),n}$ exists, lies in $\operatorname{Hom}_A^{\mathfrak{C}_n ont}(M(\mathfrak{C}_n), M/p^n M)$, and satisfies the relations H1, H2, H3 of Proposition 7.72.

The continuity of the map (7.47) implies that the image of $(\mathcal{H}_{g,n})_n$ is equal to the limit $\lim_{k\to\infty} \mathcal{H}_{g,J(N_0/N_k)}$, therefore is the integral $\mathcal{H}_g$ of $\alpha_{g,0}$. The additional properties for $\mathcal{H}_g$ are inherited from the corresponding properties of the $\mathcal{H}_{g,n}$. $\square$

Under the assumptions of Proposition 7.91, we associate to $(s, M, \mathfrak{C})$, an A-algebra homomorphism

$$\widetilde{\operatorname{Res}} \; \mathcal{A}_{C \subset G/P} \to \operatorname{End}_A(M(\mathfrak{C})^P)$$

via Propositions 7.72, 7.70, which extends the A-algebra homomorphism

$$\operatorname{Res} \; C_c^\infty(\mathcal{C}, A) \# P \to \operatorname{End}_A(M(\mathfrak{C})^P)$$

constructed in Proposition 7.32. The homomorphism Res gives rise to a P-equivariant sheaf on $\mathcal{C}$ as described in detail in Theorem 7.38. The homomorphism $\widetilde{\operatorname{Res}}$ defines on the global sections with compact support $M(\mathfrak{C})_c^P$

of the sheaf on $\mathcal{C}$ the structure of a nondegenerate $\mathcal{A}_{C \subset G/P}$-module. The latter leads, by Proposition 7.68, to the unital $C_c^\infty(G/P, A)\#G$-module $\mathcal{Z} \otimes_A M(\mathfrak{C})_c^P$ which corresponds to a G-equivariant sheaf on G/P extending the earlier sheaf on $\mathcal{C}$ (Remark 7.69).

7. Classical (φ, Γ)-modules on $\mathcal{O}_\mathcal{E}$

7.1. The Fontaine ring $\mathcal{O}_\mathcal{E}$

Let $K/\mathbb{Q}_p$ be a finite extension of ring of integers o, of uniformizer p_K and residue field k. By definition the Fontaine ring $\mathcal{O}_\mathcal{E}$ over o is the p-adic completion of the localization of the Iwasawa o-algebra $\Lambda(\mathbb{Z}_p) := o[[\mathbb{Z}_p]]$ with respect to the multiplicative set of elements which are not divisible by p. We choose a generator γ of $\mathbb{Z}_p$ of image $[\gamma]$ in $\mathcal{O}_\mathcal{E}$ and we denote $X = [\gamma] - 1 \in \mathcal{O}_\mathcal{E}$. The Iwasawa o-algebra $\Lambda(\mathbb{Z}_p)$ is a local noetherian ring of maximal ideal $\mathcal{M}(\mathbb{Z}_p)$ generated by p_K, X. It is a compact ring for the $\mathcal{M}(\mathbb{Z}_p)$-adic topology. The ring $\mathcal{O}_\mathcal{E}$ can be viewed as the ring of infinite Laurent series $\sum_{n \in \mathbb{Z}} a_n X^n$ over o in the variable X with $\lim_{n \to -\infty} a_n = 0$, and $\Lambda(\mathbb{Z}_p)$ as the subring $o[[X]]$ of Taylor series. The Fontaine ring $\mathcal{O}_\mathcal{E}$ is a local noetherian ring of maximal ideal $p_K \mathcal{O}_\mathcal{E}$ and residue field isomorphic to $k((X))$; it is a pseudocompact ring for the p-adic (= strong) topology and a complete ring (with continuous multiplication) for the weak topology. A fundamental system of open neighbourhoods of 0 for the weak topology of $\mathcal{O}_\mathcal{E}$ is given by

$$(O_{n,k} = p^n \mathcal{O}_\mathcal{E} + \mathcal{M}(\mathbb{Z}_p)^k)_{n,k \in \mathbb{N}}$$

or by

$$(B_{n,k} = p^n \mathcal{O}_\mathcal{E} + X^k \Lambda(\mathbb{Z}_p)_{n,k \in \mathbb{N}}.$$

Other fundamental systems of neighbourhoods of 0 for the weak topology are

$$(O_n := O_{n,n})_{n \geq 1} \quad \text{or} \quad (B_n := B_{n,n})_{n \geq 1}.$$

7.2. The group $GL(2, \mathbb{Q}_p)$

We consider the group $G = GL(2, \mathbb{Q}_p)$ and

$$N_0 := \begin{pmatrix} 1 & \mathbb{Z}_p \\ 0 & 1 \end{pmatrix}, \ \Gamma := \begin{pmatrix} \mathbb{Z}_p^* & 0 \\ 0 & 1 \end{pmatrix}, \ L_0 := \begin{pmatrix} \mathbb{Z}_p^* & 0 \\ 0 & \mathbb{Z}_p^* \end{pmatrix},$$

$$L_* := \begin{pmatrix} \mathbb{Z}_p - \{0\} & 0 \\ 0 & 1 \end{pmatrix},$$

$$N_k := \begin{pmatrix} 1 & p^k\mathbb{Z}_p \\ 0 & 1 \end{pmatrix}, \; L_k := \begin{pmatrix} 1 + p^k\mathbb{Z}_p & 0 \\ 0 & 1 + p^k\mathbb{Z}_p \end{pmatrix} \text{ for } k \geq 1,$$

$P_k = L_k N_k$ for $k \in \mathbb{N}$, the upper triangular subgroup P, the diagonal subgroup L, the upper unipotent subgroup N, the center Z, the mirabolic monoid $P_* = N_0 L_*$, and the monoids $L_+ = L_* Z$, $P_+ = N_0 L_+$. The subset of non invertible elements in the monoid L_* is

$$\Gamma s_P^{\mathbb{N}-\{0\}} = \{s_a := \begin{pmatrix} a & 0 \\ 0 & 1 \end{pmatrix} \text{ for } a \in p\mathbb{Z}_p - \{0\}\}.$$

An element $s \in \Gamma s_P^{\mathbb{N}-\{0\}} Z$ is called strictly dominant. In the following we identify the group $\mathbb{Z}_p$ with N_0. The action of P_+ on N_0 induces an étale ring action of P_+ (trivial on Z) on $\Lambda(N_0)$ which respects the ideal generated by p. This action extends first to the localization and then to the completion to give an étale ring action of P_+ on $\mathcal{O}_\mathcal{E}$ determined by its restriction to P_*. For the weak topology (and not for the p-adic topology), the action $P_+ \times \mathcal{O}_\mathcal{E} \to \mathcal{O}_\mathcal{E}$ of the monoid P_+ on $\mathcal{O}_\mathcal{E}$ is continuous (see Lemma 8.24.i in [12]). For $t \in L_+$ the canonical left inverse ψ_t of the action φ_t of t is continuous (this is proved in a more general setting later in Proposition 7.119).

7.3. Classical étale (φ, Γ)-modules

Let $s \in \Gamma s_P^{\mathbb{N}-\{0\}} Z$. A finitely generated étale φ_s-module D over $\mathcal{O}_\mathcal{E}$ is a finitely generated $\mathcal{O}_\mathcal{E}$-module with an étale semilinear endomorphism φ_s. These modules form an abelian category $\mathfrak{M}_{\mathcal{O}_\mathcal{E}}^{et}(\varphi_s)$. We fix such a module D.

In the following, the topology of D is its weak topology. For any surjective $\mathcal{O}_\mathcal{E}$-linear map $f : \oplus^d \mathcal{O}_\mathcal{E} \to D$, the image in D of a fundamental system of neighbourhoods of 0 in $\oplus^d \mathcal{O}_\mathcal{E}$ for the weak topology is a fundamental system of neighbourhoods of 0 in D. Finitely generated $\Lambda(N_0)$-submodules of D generating the $\mathcal{O}_\mathcal{E}$-module D will be called lattices. The map f sends $\oplus^d \Lambda(\mathbb{Z}_p)$ onto a lattice D^0 of D. We note $\mathcal{O}_{n,k} := p^n D + \mathcal{M}(\mathbb{Z}_p)^k D^0$ and $\mathcal{B}_{n,k} := p^n D + X^k D^0$. Writing $\mathcal{O}_n := \mathcal{O}_{n,n}$ and $\mathcal{B}_n := \mathcal{B}_{n,n}$, $(\mathcal{O}_n)_n$ and $(\mathcal{B}_n)_n$ are two fundamental systems of neighbourhoods of 0 in D. The topological $\mathcal{O}_\mathcal{E}$-module D is Hausdorff and complete.

A treillis D_0 in D is a compact $\Lambda(N_0)$-submodule D_0 such that the image of D_0 in the finite dimensional $k((X))$-vector space $D/p_K D$ is a $k[[X]]$-lattice ([5] Déf. I.1.1). A lattice is a treillis and a treillis contains a lattice.

For $n \geq 1$, the reduction modulo p^n of D is the finitely generated $\mathcal{O}_\mathcal{E}$-module $D/p^n D$ with the induced action of φ_s. The action remains étale, because the multiplication by p^n being a morphism in $\mathfrak{M}_{\mathcal{O}_\mathcal{E}}^{et}(\varphi_s)$ its cokernel belongs to the category. The reduction modulo p^n of ψ_s is the canonical left inverse of the reduction modulo p^n of φ_s. The reduction modulo p^n of a treillis of D is a treillis of $D/p^n D$.

Conversely, if the reduction modulo p^n of a finitely generated φ_s-module D over $\mathcal{O}_{\mathcal{E}}$ is étale for all $n \geq 1$, then D is an étale φ_s-module over $\mathcal{O}_{\mathcal{E}}$ because $D = \varprojlim_n D/p^n D$.

The weak topology of D is the projective limit of the weak topologies of $D/p^n D$.

When D is killed by a power of p and D_0 is a treillis of D, we have:

1. D_0 is open and closed in D.
2. $(\mathcal{M}(\mathbb{Z}_p)^n D_0)_{n\in\mathbb{N}}$ and $(X^n D_0)_{n\in\mathbb{N}}$ form two fundamental systems of open neighbourhoods of zero in D.
3. Any treillis of D is contained in $X^{-n} D_0$ for some $n \in \mathbb{N}$.
4. $D = \bigcup_{k\in\mathbb{N}} X^{-k} D_0$.
5. D_0 is a lattice.

The first four properties are easy; a reference is [5] Prop. I.1.2. To show that D_0 is a lattice, we pick some lattice D^0 then D_0 is contained in the lattice $X^{-n} D^0$ for some $n \in \mathbb{N}$ by property 3. Since the ring $\Lambda(N_0)$ is noetherian the assertion follows.

When D is killed by a power of p, the weak topology of D is locally compact (by properties 2 and 5).

Proposition 7.92. *Let D be a finitely generated étale φ_s-module over $\mathcal{O}_{\mathcal{E}}$. Then φ_s and its canonical inverse ψ_s are continuous.*

Proof. (a) The above $\mathcal{O}_{\mathcal{E}}$-linear surjective map $f : \oplus^d \mathcal{O}_{\mathcal{E}} \to D$ sends $(a_i)_i$ to $\sum_i a_i d_i$ for some elements $d_i \in D$. As φ_s is étale, the map $(a_i)_i \mapsto \sum_i a_i \varphi_s(d_i)$ also gives an $\mathcal{O}_{\mathcal{E}}$-linear surjective map $\oplus^d \mathcal{O}_{\mathcal{E}} \to D$. Both surjections are topological quotient maps by the definition of the topology on D, and the morphism φ_s of $\mathcal{O}_{\mathcal{E}}$ is continuous. We deduce that the morphism φ_s of D is continuous.

(b) The image of $\oplus^d \Lambda(N_0)$ by f is a lattice D_0 of D. For any $k \in \mathbb{N}$ the $\Lambda(N_0)$-submodule $D_{0,k}$ of D generated by $(\varphi_s(X^k e_i))_{1\leq i\leq d}$ also is a treillis of D because φ_s is étale. Here $\{e_i \mid 1 \leq i \leq d\}$ is a generating family of D_0. We have $\psi_s(D_{0,k}) = X^k D_0$ (cf. Lemma 7.19).

(c) When D is killed by a power of p, we deduce that ψ_s is continuous by the properties 1 and 2 of the treillis. When D is not killed by a power of p, we deduce that the reduction modulo p^n of ψ_s is continuous for all n; this implies that ψ_s is continuous because $(A = o, D)$ satisfy the properties a, b, c, d of Section 6.5, and $D/p^n D$ is a (finitely generated) étale φ_s-module over $\mathcal{O}_{\mathcal{E}}$.

$\square$

We put

$$D^+ := \{x \in D : \text{ the sequence } (\varphi_s^k(x))_{k \in \mathbb{N}} \text{ is bounded in } D\}$$

and

$$D^{++} := \{x \in D \mid \lim_{k \to \infty} \varphi_s^k(x) = 0\}. \tag{7.48}$$

Proposition 7.93. (i) *If D is killed by a power of p, then D^+ and D^{++} are lattices in D.*

(ii) *There exists a unique maximal treillis $D^\sharp$ such that $\psi_s(D^\sharp) = D^\sharp$.*

(iii) *The set of ψ_s-stable treillis in D has a unique minimal element $D^\natural$; it satisfies $\psi_s(D^\natural) = D^\natural$.*

(iv) *$X^{-k} D^\sharp$ is a treillis stable by ψ_s for all $k \in \mathbb{N}$.*

Proof. The references given in the following are stated for étale (φ_{s_p}, Γ)-modules but the proofs never use that there exists an action of Γ and they are valid for étale φ_{s_p}-modules.

(i) For $s = s_p$ this is [5] Prop. II.2.2(iii) and Lemma II.2.3. The properties of s_p which are needed for the argument are still satisfied for general s in the following form:

- $\varphi_s(X) \in \varphi_{s_p}^m(X) \Lambda(\mathbb{Z}_p)^\times$ where $s = s_0 s_p^m z$ with $s_0 \in \Gamma$, $m \geq 1$, and $z \in Z$.
- $(\varphi_s(X) X^{-1})^{p^k} \in p^{k+1} \Lambda(\mathbb{Z}_p) + X^{(p-1)p^k} \Lambda(\mathbb{Z}_p)$ for any $k \in \mathbb{N}$.

(ii) and (iii) For any finitely generated $\mathcal{O}_{\mathcal{E}}$-torsion module M we denote its Pontrjagin dual of continuous o-linear maps from M to K/o by $M^\vee := \mathrm{Hom}_o^{cont}(M, K/o)$. Obviously, $M^\vee$ again is an $\mathcal{O}_{\mathcal{E}}$-module by $(\lambda f)(x) := f(\lambda x)$ for $\lambda \in \mathcal{O}_{\mathcal{E}}$, $f \in M^\vee$, and $x \in M$. It is shown in [5] Lemma I.2.4 that:

- $M^\vee$ is a finitely generated $\mathcal{O}_{\mathcal{E}}$-torsion module,
- the topology of pointwise convergence on $M^\vee$ coincides with its weak topology as an $\mathcal{O}_{\mathcal{E}}$-module, and
- $M^{\vee\vee} = M$.

Now let D be as in the assertion but killed by a power of p. One checks that $D^\vee$ also belongs to $\mathfrak{M}_{\mathcal{O}_{\mathcal{E}}}^{et}(\varphi_s)$ with respect to the semilinear map $\varphi_s(f) := f \circ \psi_s$ for $f \in D^\vee$; moreover, the canonical left inverse is $\psi_s(f) = f \circ \varphi_s$. Next, [5] Lemma I.2.8 shows that:

– If $D_0 \subset D$ is a lattice then $D_0^{\perp} := \{d \in D^{\vee} : f(D_0) = 0\}$ is a lattice in $D^{\vee}$, and $D_0^{\vee\vee} = D_0$.

We now define $D^{\natural} := (D^{\vee})^+$ and $D^{\sharp} := (D^{\vee})^{++}$. The purely formal arguments in the proofs of [5] Prop. II.6.1, Lemma II.6.2, and Prop. II.6.3 show that $D^{\natural}$ and $D^{\sharp}$ have the asserted properties.

For a general D in $\mathfrak{M}_{\mathcal{O}_{\mathcal{E}}}^{et}(\varphi_s)$ the (formal) arguments in the proof of [5] Prop. II.6.5 show that $((D/p^n D)^{\natural})_{n\in\mathbb{N}}$ and $((D/p^n D)^{\sharp})_{n\in\mathbb{N}}$ are well defined projective systems of compact $\Lambda(\mathbb{Z}_p)$-modules (with surjective transition maps). Hence

$$D^{\natural} := \varprojlim (D/p^n D)^{\natural} \quad \text{and} \quad D^{\sharp} := \varprojlim (D/p^n D)^{\sharp}$$

have the asserted properties.

(iv) $X^{-k} D^{\sharp}$ is clearly a treillis. As X divides $\varphi_s(X) = (1 + X)^a - 1$ in $\Lambda(\mathbb{Z}_p) = o[[X]]$ (when $s \in s_a \mathbb{Z}$ for some $a \in p\mathbb{Z}_p \setminus \{0\}$), there exists $f(X) \in o[[X]]$ such that $\varphi_s(X^k) = X^k f(X)^k$. So we have

$$\psi_s(X^{-k} D^{\sharp}) = \psi_s(\varphi_s(X^{-k}) f(X)^k D^{\sharp})$$
$$= X^{-k} \psi_s(f(X)^k D^{\sharp}) \subset X^{-k} \psi_s(D^{\sharp}) \subset X^{-k} D^{\sharp}$$

since $D^{\sharp}$ is ψ_s-stable by definition. $\qquad\square$

Proposition 7.94. *Let D be a finitely generated étale φ_s-module over $\mathcal{O}_{\mathcal{E}}$ killed by a power of p. For any compact subset $C \subseteq D$, there exists an $r \in \mathbb{N}$ such that*

$$\bigcup_{k\geq 0} \psi_s^k(N_0 C) \subseteq X^{-r} D^{++}.$$

Proof. Since $N_0 C$ is compact and D^{++} and $D^{\sharp}$ are treillis, there exists $l \in \mathbb{N}$ such that $N_0 C \subset X^{-l} D^{\sharp} \subset X^{-2l} D^{++}$. By (iv) of Proposition 7.93 we obtain for all $k \geq 0$

$$\psi_s^k(N_0 C) \subset \psi_s^k(X^{-l} D^{\sharp}) \subset X^{-l} D^{\sharp} \subset X^{-2l} D^{++}$$

and we can take $r = 2l$. $\qquad\square$

Corollary 7.95. *Let D be a finitely generated étale φ_s-module over $\mathcal{O}_{\mathcal{E}}$. For any compact subset $C \subseteq D$ and any $n \in \mathbb{N}$, there exists $k_0 \in \mathbb{N}$ such that*

$$\bigcup_{k\geq k_0} \psi_s^k(N_0 C) \subseteq D^{\sharp} + p^n D.$$

Proof. We may assume that D is killed by a power of p. In view of (the proof of) Proposition 7.94 it suffices to show that for all $l > 0$ there exists a k_0 such that $\psi_s^{k_0}(X^{-l}D^\sharp) = D^\sharp$. By Proposition 7.93(ii) and (iv) we have

$$D^\sharp = \psi_s^{k+1}(D^\sharp) \subseteq \psi_s^{k+1}(X^{-l}D^\sharp) \subseteq \psi_s^k(X^{-l}D^\sharp) \subseteq X^{-l}D^\sharp$$

for any $k \geq 0$. Hence $\bigcap_k \psi_s^k(X^{-l}D^\sharp)$ is a treillis in D on which ψ_s is surjective. Therefore it coincides with $D^\sharp$ by the maximality of $D^\sharp$ (Proposition 7.93(ii)). On the other hand, the $\mathbb{Z}_p[[X]]$-module $(X^{-l}D^\sharp)/D^\sharp$ is killed by both X^l and p^n and hence is finite. So there exists a k_0 such that $\psi_s^k(X^{-l}D^\sharp) = D^\sharp$ for all $k \geq k_0$. $\qquad\square$

For any submonoid $L' \subset L_+$ containing a strictly dominant element, an étale L'-module over $\mathcal{O}_\mathcal{E}$ is a finitely generated $\mathcal{O}_\mathcal{E}$-module with an étale semilinear action of L'.

A topologically étale L'-module over $\mathcal{O}_\mathcal{E}$ will be an étale L'-module D over $\mathcal{O}_\mathcal{E}$ such that the action $L' \times D \to D$ of L' on D is continuous. This terminology is provisional since we will show later on (Corollary 7.125) in a more general context that any étale L'-module over $\mathcal{O}_\mathcal{E}$ in fact is topologically étale and, in particular, is a complete topologically étale $o[N_0 L']$-module in our previous sense.

Let D be a topologically étale L_+-module over $\mathcal{O}_\mathcal{E}$ and let $g = \begin{pmatrix} a & b \\ c & d \end{pmatrix} \in$ GL$_2(\mathbb{Q}_p)$. Denote

$$w_0 = \begin{pmatrix} 0 & 1 \\ 1 & 0 \end{pmatrix}, \quad u(b) = \begin{pmatrix} 1 & b \\ 0 & 1 \end{pmatrix}.$$

Using the formula

$$gu(r)w_0 = \begin{pmatrix} ar+b & a \\ cr+d & c \end{pmatrix},$$

one checks that the set U_g defined by $g^{-1}C_0 \cap C_0 = U_g w_0 P/P$ is

$$U_g = u(X_g) \quad \text{where} \quad X_g = \{r \in \mathbb{Z}_p | cr+d \neq 0, \ \frac{ar+b}{cr+d} \in \mathbb{Z}_p\}.$$

For each $r \in X_g$ we can write

$$gu(r)w_0 = u(g[r])t(g,r)w_0 u\left(\frac{c}{cr+d}\right),$$

where

$$g[r] = \frac{ar+b}{cr+d} \in \mathbb{Z}_p, \quad t(g,r) = \begin{pmatrix} \frac{\det g}{cr+d} & 0 \\ 0 & cr+d \end{pmatrix}.$$

We deduce that for $u(r) \in U_g$ we have

$$\alpha(g, x_{u(r)}) = u(g[r])t(g, r).$$

Let now $s = s_a z \in L_+$ be strictly dominant, with $z \in Z$ and $a \in p\mathbb{Z}_p - \{0\}$. There exists a positive integer $k_{g,s}$ such that for all $k \geq k_{g,s}$ we have $t(g, r)s^k \in L_+$. Note that $N_k = s^k N_0 s^{-k} = u(a^k \mathbb{Z}_p)$. We deduce that for $k \geq k_{g,s}$ the operators $\mathcal{H}_{g, J(N_0/N_k)}$ introduced in (7.28) are equal to the operators

$$\mathcal{H}_{g,s,J(\mathbb{Z}_p/a^k\mathbb{Z}_p)} = \sum_{r \in X_g \cap J(\mathbb{Z}_p/a^k\mathbb{Z}_p)} (1+X)^{g[r]} \varphi_{t(g,r)s^k} \psi_s^k ((1+X)^{-r}).$$

Proposition 7.96. *Let D be a topologically étale L_+-module over $\mathcal{O}_{\mathcal{E}}$. For the compact open topology in $\mathrm{End}_o^{cont}(D)$, the maps $\alpha_{g,0} : N_0 \longrightarrow \mathrm{End}_o^{cont}(D)$, for $g \in N_0 \overline{P} N_0$, are integrable with respect to s and res, for all $s \in L_+$ strictly dominant, i.e. $s = s_a z$ with $a \in p\mathbb{Z}_p - \{0\}$ and $z \in Z$, their integrals*

$$\mathcal{H}_g = \int_{N_0} \alpha_{g,0} d\,\mathrm{res} = \lim_{k \to \infty} \mathcal{H}_{g,s,J(\mathbb{Z}_p/a^k\mathbb{Z}_p)}$$

for any choices of $J(\mathbb{Z}_p/a^k\mathbb{Z}_p) \subset \mathbb{Z}_p$, do not depend on the choice of s and satisfy the relations H1, H2, H3 of Proposition 7.72.

Proof. By Proposition 7.91, we reduce to the case that D is killed by a power of p and to showing the assumptions of Proposition 7.84 for the family of all compact subsets of D. The axioms $\mathfrak{C}_i$, for $1 \leq i \leq 6$, are obviously satisfied by continuity of φ_s, ψ_s, and of the action of $n \in N_0$ on D.

i. We check first the convergence criterion of Proposition 7.83, using the theory of treillis, i.e. of lattices, in D.

Given a lattice $\mathcal{M} \subseteq D$, a compact subset $C \subseteq D$ such that $N_0 C \subseteq C$, and a compact subset $C_+ \subseteq L_+$, we want to find a compact open subgroup $P_1 \subset P_0$ and an integer $k_0 \in \mathbb{N}$ such that

$$s^k(1 - P_1)C_+ \psi_s^k(C) \subseteq \mathcal{M} \tag{7.49}$$

for all $k \geq k_0$.

We choose $r_0 \in \mathbb{N}$ with $\varphi_s^k(D^{++}) \subset \mathcal{M}$ for all $k \geq r_0$, as we may by the definition of D^{++}. We choose $r, k_0 \in \mathbb{N}$ such that $k_0 \geq r_0$ and

$$\bigcup_{k \geq k_0} \psi_s^k(C) \subseteq X^{-r} D^{++},$$

as we may by Proposition 7.94. Applying C_+ we obtain

$$\bigcup_{k \geq k_0} C_+ \psi_s^k(C) \subseteq C_+(X^{-r}D^{++}).$$

The continuity of the action of P_+ on D implies that $C_+(X^{-r}D^{++})$ is compact. Hence we can choose $r' \in \mathbb{N}$ such that $C_+(X^{-r}D^{++}) \subseteq X^{-r'}D^{++}$ and we obtain

$$\bigcup_{k \geq k_0} C_+ \psi_s^k(C) \subseteq X^{-r'}D^{++}.$$

As $X^{-r'}D^{++}$ is compact and D^{++} an open neighbourhood of 0, the continuity of the action of P_+ on D there exists a compact open subgroup $P_1 \subseteq P_+$ such that

$$(1 - P_1)X^{-r'}D^{++} \subseteq D^{++}.$$

Hence we have $s^k(1 - P_1)C_+\psi_s^k(C) \subset \varphi_s^k(D^{++}) \subset \mathcal{M}$ for all $k \geq k_0$.

ii. To obtain all the assumptions of Proposition 7.84 for the family of all compact subsets of D, it remains to prove that, given $x \in D$ and $g \in N_0 \overline{P} N_0$, $s = s_a z$ with $a \in p\mathbb{Z}_p - \{0\}$ and $z \in Z$, and $(J(\mathbb{Z}_p/a^k\mathbb{Z}_p))_k$, there exists a compact $C_{x,g,s} \subset D$ and a positive integer $k_{x,g,s}$ such that $\mathcal{H}_{g,s,J(\mathbb{Z}_p/a^k\mathbb{Z}_p)}(x) \in C_{x,g,s}$ for any $k \geq k_{x,g,s}$. This is clear because D is locally compact (by hypothesis D is killed by a power of p) and the sequence $(\mathcal{H}_{g,s,J(\mathbb{Z}_p/a^k\mathbb{Z}_p)}(x))_k$ converges.

iii. The independence of the choice of $s \in L_+$ strictly dominant results from the fact that, for $z \in Z$, $e \subset \mathbb{Z}_p^*$, and a positive integer r, we have $(zs_{p^r e})^k N_0(zs_{p^r e})^{-k} = s_p^{kr} N_0 s_p^{kr}$ and $\varphi_{zs_{p^r e}}^k \psi_{zs_{p^r e}}^k = \varphi_{s_p}^{rk} \psi_{s_p}^{kr}$ as ψ_{zs_e} is the right and left inverse of φ_{zs_e}.

$\square$

Remark 7.97. Let D be a topologically étale L_+-module over $\mathcal{O}_\mathcal{E}$, on which Z acts through a character ω. The pointwise convergence of the integrals $\int_{N_0} \alpha_{g,0} d\mathrm{res}$ is a basic theorem of Colmez, allowing him the construction of the representation of $\mathrm{GL}_2(\mathbb{Q}_p)$ that he denotes $D \boxtimes_\omega \mathbb{P}^1$. Our construction coincides with Colmez's construction because our $\mathcal{H}_g \in \mathrm{End}_o^{\mathrm{cont}}(D)$ are the same as the H_g of Colmez given in [6] lemma II.1.2 (ii). Indeed,

$$\alpha(g, x_{u(r)}) = u(g[r])t(g, r) =$$

$$\omega(cr + d)u(g[r])\begin{pmatrix} \frac{\det g}{(cr+d)^2} & 0 \\ 0 & 1 \end{pmatrix} = \omega(cr + d)\begin{pmatrix} g'[r] & g[r] \\ 0 & 1 \end{pmatrix},$$

where $g'[r] = \frac{\det g}{(cr+d)^2}$. This coincides with Colmez's formula.

The major goal of the paper is to generalize Proposition 7.96. See Proposition 7.141.

8. A generalization of (φ, Γ)-modules

We return to a general group G. We denote $G^{(2)} := GL(2, \mathbb{Q}_p)$ and the objects relative to $G^{(2)}$ will be affected with an upper index $^{(2)}$.

(a) We suppose that N_0 has the structure of a p-adic Lie group and that we have a continuous surjective homomorphism

$$\ell : N_0 \to N_0^{(2)}.$$

We choose a continuous homomorphism $\iota : N_0^{(2)} \to N_0$ which is a section of ℓ (which is possible because $N_0^{(2)} \simeq \mathbb{Z}_p$).

We have $N_0 = N_\ell \, \iota(N_0^{(2)})$ where N_ℓ is the kernel of ℓ.

We denote by $L_{\ell,+} := \{t \in L \mid tN_\ell t^{-1} \subset N_\ell, \ tN_0 t^{-1} \subset N_0\}$ the stabilizer of N_ℓ in the L-stabilizer of N_0, and by $L_{\ell,\iota} := \{t \in L \mid tN_\ell t^{-1} \subset N_\ell, \ t\iota(N_0^{(2)})t^{-1} \subset \iota(N_0^{(2)})\}$ the stabilizer of N_ℓ in the L-stabilizer of $\iota(N_0^{(2)})$. We have $L_{\ell,\iota} \subset L_{\ell,+}$.

(b) We suppose given a submonoid L_* of $L_{\ell,\iota}$ containing s and a continuous homomorphism $\ell : L_* \to L_+^{(2)}$ such that (ℓ, ι) satisfies

$$\ell(tut^{-1}) = \ell(t)\ell(u)\ell(t)^{-1}, \quad t\iota(y)t^{-1} = \iota(\ell(t)y\ell(t)^{-1}),$$

$$\text{for } u \in N_0, y \in N_0^{(2)}, t \in L_*.$$

The sequence $\ell(s^n N_0 s^{-n}) = \ell(s)^n N_0^{(2)} \ell(s)^{-n}$ in $N^{(2)}$ is decreasing with trivial intersection. The maps ℓ in (a) and (b) combine to a unique continuous homomorphism

$$\ell \ P_* := N_0 \rtimes L_* \to P_+^{(2)}.$$

8.1. The microlocalized ring $\Lambda_\ell(N_0)$

The ring $\Lambda_\ell(N_0)$, denoted by $\Lambda_{N_\ell}(N_0)$ in [12], is a generalization of the ring $\mathcal{O}_\mathcal{E}$, which corresponds to $\Lambda_{\mathrm{id}}(N_0^{(2)})$. We refer the reader to [12] for the proofs of some claims in this section.

The maximal ideal $\mathcal{M}(N_\ell)$ of the completed group o-algebra $\Lambda(N_\ell) = o[[N_\ell]]$ is generated by p_K and by the kernel of the augmentation map $o[N_\ell] \to o$.

The ring $\Lambda_\ell(N_0)$ is the $\mathcal{M}(N_\ell)$-adic completion of the localization of $\Lambda(N_0)$ with respect to the Ore subset $S_\ell(N_0)$ of elements which are not in $\mathcal{M}(N_\ell)\Lambda(N_0)$. The ring $\Lambda(N_0)$ can be viewed as the ring $\Lambda(N_\ell)[[X]]$ of skew Taylor series over $\Lambda(N_\ell)$ in the variable $X = [\gamma] - 1$ where $\gamma \in N_0$ and

$\ell(\gamma)$ is a topological generator of $\ell(N_0)$. Then $\Lambda_\ell(N_0)$ is viewed as the ring of infinite skew Laurent series $\sum_{n\in\mathbb{Z}} a_n X^n$ over $\Lambda(N_\ell)$ in the variable X with $\lim_{n\to-\infty} a_n = 0$ for the pseudocompact topology of $\Lambda(N_\ell)$.

The ring $\Lambda_\ell(N_0)$ is strict-local noetherian of maximal ideal $\mathcal{M}_\ell(N_0)$ generated by $\mathcal{M}(N_\ell)$. It is a pseudocompact ring for the $\mathcal{M}(N_\ell)$-adic topology (called the strong topology). It is a complete Hausdorff ring for the weak topology ([12] Lemma 8.2) with fundamental system of open neighbourhoods of 0 given by

$$O_{n,k} := \mathcal{M}_\ell(N_0)^n + \mathcal{M}(N_0)^k \text{ for } n \in \mathbb{N}, k \in \mathbb{N}.$$

In the computations it is sometimes better to use the fundamental systems of open neighbourhoods of 0 defined by

$$B_{n,k} := \mathcal{M}_\ell(N_0)^n + X^k \Lambda(N_0) \text{ for } n \in \mathbb{N}, k \in \mathbb{N},$$

and

$$C_{n,k} := \mathcal{M}_\ell(N_0)^n + \Lambda(N_0) X^k \text{ for } n \in \mathbb{N}, k \in \mathbb{N},$$

which are equivalent due to the two formulas

$$X^k \Lambda(N_0) \subseteq \Lambda(N_0) X^k + \mathcal{M}(N_0)^k \quad \text{and} \quad \Lambda(N_0) X^k \subseteq X^k \Lambda(N_0) + \mathcal{M}(N_0)^k,$$

We write $O_n := O_{n,n}$, $B_n := B_{n,n}$, and $C_n = C_{n,n}$. Then $(O_n)_n$, $(B_n)_n$, and $(C_n)_n$ are also fundamental system of open neighbourhoods of 0 in $\Lambda_\ell(N_0)$.

The action $(b = ut, n_0) \mapsto b.n_0 = utn_0t^{-1}$ of the monoid $P_{\ell,+} = N_0 \rtimes L_{\ell,+}$ on N_0 induces a ring action $(t, x) \mapsto \varphi_t(x)$ of $L_{\ell,+}$ on the o-algebra $\Lambda(N_0)$ respecting the ideal $\Lambda(N_0)\mathcal{M}(N_\ell)$, and the Ore set $S_\ell(N_0)$ hence defines a ring action of $L_{\ell,+}$ on the o-algebra $\Lambda_\ell(N_0)$. This action respects the maximal ideals $\mathcal{M}(N_0)$ and $\mathcal{M}_\ell(N_0)$ of the rings $\Lambda(N_0)$ and $\Lambda_\ell(N_0)$ and hence the open neighbourhoods of zero $O_{n,k}$.

Lemma 7.98. *For $t \in L_{\ell,+}$, a fundamental system of open neighbourhoods of 0 in $\Lambda_\ell(N_0)$ is given by*

$$(\varphi_t(O_{n,k})\Lambda(N_0))_{n,k\in\mathbb{N}}.$$

Proof. We have just seen $\varphi_t(O_{n,k})\Lambda(N_0) \subset O_{n,k}$. Conversely, given $n, k \in \mathbb{N}$, we have to find $n', k' \in \mathbb{N}$ such that $O_{n',k'} \subset \varphi_t(O_{n,k})\Lambda(N_0)$. This can be deduced from the following fact. Let $H' \subset H$ be an open subgroup. Then given $k' \in \mathbb{N}$, there is $k \in \mathbb{N}$ such that

$$\mathcal{M}(H')^{k'}\Lambda(H) \supset \mathcal{M}(H)^k.$$

Indeed by taking a smaller H' we can suppose that $H' \subset H$ is open normal. Then $\mathcal{M}(H')^k \Lambda(H)$ is a two-sided ideal in $\Lambda(H)$ and the factor ring $\Lambda(H)/\mathcal{M}(H')\Lambda(H)$ is an artinian local ring with maximal ideal $\mathcal{M}(H)/\mathcal{M}(H')\Lambda(H)$. It remains to observe that in any artinian local ring the maximal ideal is nilpotent.   □

Proposition 7.99. *The action of $L_{\ell,+}$ on $\Lambda_\ell(N_0)$ is étale: for any $t \in L_{\ell,+}$, the map*

$$(\lambda, x) \mapsto \lambda \varphi_t(x) : \Lambda(N_0) \otimes_{\Lambda(N_0),\varphi_t} \Lambda_\ell(N_0) \to \Lambda_\ell(N_0)$$

is bijective.

Proof. We follow ([12] Prop. 9.6, Proof, Step 1).

(1) The conjugation by t gives a natural isomorphism

$$\Lambda_\ell(N_0) \to \Lambda_{tN_\ell t^{-1}}(tN_0 t^{-1}).$$

(2) Obviously $\Lambda_{tN_\ell t^{-1}}(tN_0 t^{-1}) = \Lambda(tN_0 t^{-1}) \otimes_{\Lambda(tN_0 t^{-1})} \Lambda_{tN_\ell t^{-1}}(tN_0 t^{-1})$, and the map

$$\Lambda(tN_0 t^{-1}) \otimes_{\Lambda(tN_0 t^{-1})} \Lambda_{tN_\ell t^{-1}}(tN_0 t^{-1}) \to \Lambda(N_0) \otimes_{\Lambda(tN_0 t^{-1})}$$
$$\Lambda_{tN_\ell t^{-1}}(tN_0 t^{-1})$$

is injective because $\Lambda_{tN_\ell t^{-1}}(tN_0 t^{-1})$ is flat on $\Lambda(tN_0 t^{-1})$.

(3) The natural map

$$\Lambda(N_0) \otimes_{\Lambda(tN_0 t^{-1})} \Lambda_{tN_\ell t^{-1}}(tN_0 t^{-1}) \to \Lambda_\ell(N_0)$$

is bijective.

(4) The ring action $\varphi_t : \Lambda_\ell(N_0) \to \Lambda_\ell(N_0)$ of t on $\Lambda_\ell(N_0)$ is the composite of the maps of (1), (2), (3), hence is injective.

(5) The proposition is equivalent to (3) and φ_t injective.

     □

Remark 7.100. The proposition is equivalent to: for any $t \in L_{\ell,+}$, the map

$$(u, x) \mapsto u\varphi_t(x) : o[N_0] \otimes_{o[N_0],\varphi_t} \Lambda_\ell(N_0) \to \Lambda_\ell(N_0)$$

is bijective.

8.2. The categories $\mathfrak{M}^{et}_{\Lambda_\ell(N_0)}(L_*)$ and $\mathfrak{M}^{et}_{\mathcal{O}_{\mathcal{E},\ell}}(L_*)$

By the universal properties of localization and adic completion the continuous homomorphisms ℓ and ι between N_0 and $N_0^{(2)}$ extend to continuous o-linear homomorphisms of pseudocompact rings,

$$\ell : \Lambda_\ell(N_0) \to \mathcal{O}_\mathcal{E}, \quad \iota : \mathcal{O}_\mathcal{E} \to \Lambda_\ell(N_0), \quad \ell \circ \iota = \mathrm{id}. \tag{7.50}$$

If we view the rings as rings of Laurent series, $\ell(X) = X^{(2)}$, $\iota(X^{(2)}) = X$, and ℓ is the augmentation map $\Lambda(N_\ell) \to o$ and ι is the natural injection $o \to \Lambda(N_\ell)$, on the coefficients. We have for $n, k \in \mathbb{N}$,

$$\ell(\mathcal{M}_\ell(N_0)) = p_K \mathcal{O}_\mathcal{E}, \quad \ell(B_{n,k}) = B_{n,k}^{(2)},$$
$$\iota(p_K \mathcal{O}_\mathcal{E}) = \mathcal{M}_\ell(N_0) \cap \iota(\mathcal{O}_\mathcal{E}), \quad \iota(B_{n,k}^{(2)}) = B_{n,k} \cap \iota(\mathcal{O}_\mathcal{E}). \tag{7.51}$$

We denote by $J(N_0)$ the kernel of $\ell : \Lambda(N_0) \to \Lambda(N_0^{(2)})$ and by $J_\ell(N_0)$ the kernel of $\ell : \Lambda_\ell(N_0) \to \mathcal{O}_\mathcal{E}$. They are the closed two-sided ideals generated (as left or right ideals) by the kernel of the augmentation map $o[N_\ell] \to o$. We have

$$\Lambda_\ell(N_0) = \iota(\mathcal{O}_\mathcal{E}) \oplus J_\ell(N_0), \quad \mathcal{M}_\ell(N_0) = p_K \iota(\mathcal{O}_\mathcal{E}) \oplus J_\ell(N_0),$$
$$X^k \Lambda(N_0) = \iota((X^{(2)})^k \Lambda(N_0^{(2)})) \oplus X^k J(N_0), \tag{7.52}$$
$$B_{n,k} = \iota(B_{n,k}^{(2)}) \oplus (J_\ell(N_0) \cap B_{n,k}).$$

The maps ℓ and ι are L_*-equivariant: for $t \in L_*$,

$$\ell \circ \varphi_t = \varphi_{\ell(t)} \circ \ell, \quad \iota \circ \varphi_{\ell(t)} = \varphi_t \circ \iota, \tag{7.53}$$

thanks to the hypothesis (b) made at the beginning of this chapter. The map ι is equivariant for the canonical action of the inverse monoid L_*^{-1}, but not the map ℓ as the following lemma shows.

Lemma 7.101. *For $t \in L_*$, we have $\iota \circ \psi_{\ell(t)} = \psi_t \circ \iota$. We have $\ell \circ \psi_t = \psi_{\ell(t)} \circ \ell$ if and only if $N_\ell = t N_\ell t^{-1}$.*

Proof. Clearly $N_0 = N_\ell \rtimes \iota(N_0^{(2)})$ and $t N_0 t^{-1} = t N_\ell t^{-1} \rtimes t \iota(N_0^{(2)}) t^{-1}$ for $t \in L$. We choose, as we may, for $t \in L_{\ell,\iota}$, a system $J(N_0 / t N_0 t^{-1})$ of representatives of $N_0 / t N_0 t^{-1}$ containing 1 such that

$$J(N_0 / t N_0 t^{-1}) = \{u \iota(v) \mid u \in J(N_\ell / t N_\ell t^{-1}), \; v \in J(N_0^{(2)} / \ell(t) N_0^{(2)} \ell(t^{-1}))\}. \tag{7.54}$$

We have $\iota \circ \psi_{\ell(t)} = \psi_t \circ \iota$ because, for $\lambda \in \mathcal{O}_\mathcal{E}$, we have on one hand (7.12)

$$\lambda = \sum_{v \in J(N_0^{(2)} / \ell(t) N_0^{(2)} \ell(t)^{-1})} v \varphi_{\ell(t)}(\lambda_{v,\ell(t)}), \quad \lambda_{v,\ell(t)} = \psi_{\ell(t)}(v^{-1} \lambda),$$

$$\iota(\lambda) = \sum_{v \in J(N_0^{(2)} / \ell(t) N_0^{(2)} \ell(t)^{-1})} \iota(v) \varphi_t(\iota(\lambda_{v,\ell(t)})),$$

and on the other hand (7.12)

$$\iota(\lambda) = \sum_{u \in J(N_\ell/tN_\ell t^{-1}), v \in J(N_0^{(2)}/\ell(t)N_0^{(2)})\ell(t)^{-1})} u\iota(v)\varphi_t(\iota(\lambda)_{u\iota(v),t}),$$

where $\iota(\lambda)_{u\iota(v),t} = \psi_t(\iota(v)^{-1}u^{-1}\iota(\lambda))$. By the uniqueness of the decomposition,

$$\iota(\lambda)_{\iota(v),t} = \iota(\lambda_{v,\ell(t)}), \quad \iota(\lambda)_{u\iota(v),t} = 0 \text{ if } u \neq 1.$$

Taking $u = 1$, $v = 1$, we get $\psi_t(\iota(\lambda)) = \iota(\psi_{\ell(t)}(\lambda))$.

A similar argument shows that $\ell \circ \psi_t = \psi_{\ell(t)} \circ \ell$ if and only if $N_\ell = tN_\ell t^{-1}$. For $\lambda \in \Lambda_\ell(N_0)$,

$$\lambda = \sum_{u \in J(N_0/tN_0t^{-1})} u\varphi_t(\lambda_{u,t}), \quad \lambda_{u,t} = \psi_t(u^{-1}\lambda),$$

$$\ell(\lambda) = \sum_{u \in J(N_0/tN_0t^{-1})} \ell(u)\varphi_{\ell(t)}(\ell(\lambda_{u,t})) = \sum_{v \in J(N_0^{(2)}/\ell(t)N_0^{(2)})} v\varphi_{\ell(t)}(\ell(\lambda)_{v,\ell(t)}).$$

By the uniqueness of the decomposition,

$$\ell(\lambda)_{v,\ell(t)} = \sum_{u \in J(N_\ell/tN_\ell t^{-1})} \ell(\lambda_{u\iota(v),t}).$$

We deduce that $\ell \circ \psi_t = \psi_{\ell(t)} \circ \ell$ if and only if $N_\ell = tN_\ell t^{-1}$. $\quad\square$

Remark 7.102. $\ell \circ \psi_s \neq \psi_{\ell(s)} \circ \ell$, except in the trivial case where $\ell : N_0 \to N_0^{(2)}$ is an isomorphism, because $sN_\ell s^{-1} \neq N_\ell$ as the intersection of the decreasing sequence $s^k N_\ell s^{-k}$ for $k \in \mathbb{N}$ is trivial.

For future use, we note:

Lemma 7.103. *The left or right $o[N_0]$-submodule generated by $\iota(\mathcal{O}_\mathcal{E})$ in $\Lambda_\ell(N_0)$ is dense.*

Proof. As $o[N_0]$ is dense in $\Lambda(N_0)$ it suffices to show that the left or right $\Lambda(N_0)$-submodule generated by $\iota(\mathcal{O}_\mathcal{E})$ in $\Lambda_\ell(N_0)$ is dense. This will be shown even with respect to the $\mathcal{M}_\ell(N_0)$-adic topology.

View $\lambda \in \Lambda_\ell(N_0)$ as an infinite Laurent series $\lambda = \sum_{n \in \mathbb{Z}} \lambda_n X^n$ with $\lambda_n \in \Lambda(N_\ell)$ and $\lim_{n \to -\infty} \lambda_n = 0$ in the $\mathcal{M}(N_\ell)$-adic topology of $\Lambda(N_\ell)$. Further, note that the left, resp. right, $\Lambda(N_0)$-submodule of $\Lambda_\ell(N_0)$ generated by $\iota(\mathcal{O}_\mathcal{E})$ contains $\Lambda(N_0)X^{-m}$, resp. $X^{-m}\Lambda(N_0)$, for any positive integer m. Finally, for each $n \in \mathbb{N}$ there exists μ_n in $\Lambda(N_0)X^{-m}$, resp. $X^{-m}\Lambda(N_0)$, for some large m, such that $\lambda - \mu_n \in \mathcal{M}_\ell(N_0)^n$. $\quad\square$

Let M be a finitely generated $\Lambda_\ell(N_0)$-module and let $f : \oplus_{i=1}^n \Lambda_\ell(N_0) \to$ M be a $\Lambda_\ell(N_0)$-linear surjective map. We put on M the quotient topology of the weak topology on $\oplus_{i=1}^n \Lambda_\ell(N_0)$; this is independent of the choice of f. Then M is a Hausdorff and complete topological $\Lambda_\ell(N_0)$-module and every submodule is closed ([12] Lemma 8.22). In the same way we can equip M with the pseudocompact topology. Again M is Hausdorff and complete and every submodule is closed in the pseudocompact topology, because $\Lambda_\ell(N_0)$ is noetherian. The weak topology on M is weaker than the pseudocompact topology which is weaker than the p-adic topology. In particular the intersection of the submodules $p^n M$ for $n \in \mathbb{N}$ is 0. By [9] IV.3.Prop. 10, M is p-adically complete, i.e., the natural map $M \to \varprojlim_n M/p^n M$ is bijective.

Unless otherwise indicated, M is always understood to carry the weak topology.

Lemma 7.104. *The properties a,b,c,d of Section 6.5 are satisfied by* (o, M) *and M is complete.*

Definition 7.105. A finitely generated module M over $\Lambda_\ell(N_0)$ with an étale semilinear action of a submonoid L' of $L_{\ell,+}$ is called an étale L'-module over $\Lambda_\ell(N_0)$.

We denote by $\mathfrak{M}^{et}_{\Lambda_\ell(N_0)}(L')$ the category of étale L'-modules on $\Lambda_\ell(N_0)$.

Lemma 7.106. *The category* $\mathfrak{M}^{et}_{\Lambda_\ell(N_0)}(L')$ *is abelian.*

Proof. As in the proof of Proposition 7.23 and using that the ring $\Lambda_\ell(N_0)$ is noetherian. $\qquad\square$

The continuous homomorphism $\ell : L_* \to L_+^{(2)}$ defines an étale semilinear action of L_* on the ring $\Lambda_{id}(N_0^{(2)})$ isomorphic to $\mathcal{O}_\mathcal{E}$.

Definition 7.107. A finitely generated module D over $\mathcal{O}_\mathcal{E}$ with an étale semilinear action of L_* is called an étale L_*-module over $\mathcal{O}_\mathcal{E}$.

An element $t \in L_*$ in the kernel $L_*^{\ell=1}$ of ℓ acts trivially on $\mathcal{O}_\mathcal{E}$ hence bijectively on an étale L_*-module over $\mathcal{O}_\mathcal{E}$.

Remark 7.108. The action of $L_*^{\ell=1}$ on D extends to an action of the subgroup of L generated by $L_*^{\ell=1}$ if $L_*^{\ell=1}$ is commutative or if we assume that for each $t \in L_*^{\ell=1}$ there exists an integer $k > 0$ such that $s^k t^{-1} \in L_*$. The assumption is trivially satisfied whenever $L_* = H \cap L_+$ for some subgroup $H \subset L$.

Indeed, the subgroup generated by $L_*^{\ell=1}$ is the set of words of the form $x_1^{\pm 1} \ldots x_n^{\pm 1}$ with $x_i \in L_*^{\ell=1}$ for $i = 1, \ldots, n$. So if we have an action of all the elements and all the inverses, then we can take the products of these, as well. We need to show that this action is well defined, i.e., whenever we have a relation

$$x_1^{\pm 1} \ldots x_n^{\pm 1} = y_1^{\pm 1} \ldots y_r^{\pm 1} \qquad (7.55)$$

in the group then the action we just defined is the same using the x's or the y's. If $L_*^{\ell = 1}$ is commutative, this is easily checked. In the second case, we can choose a big enough $k = \sum_{i=1}^n k_i + \sum_{j=1}^r k_j$ such that $s^{k_i} x_i^{-1} \in L_*$ and $s^{k_j} y_j^{-1} \in L_*$. Then multiplying the relation (7.55) by s^k we obtain a relation in L_* so the two sides will define the same action on D. This shows that the actions defined using the two sides of (7.55) are equal on $\varphi_s^k(D) \subset D$. However, they are also equal on group elements $u \in N_0^{(2)}$ hence on the whole $D = \bigoplus_{u \in J(N_0^{(2)}/\varphi_s^k(N_0^{(2)}))} u\varphi_s^k(D)$.

We denote by $\mathfrak{M}^{et}_{\mathcal{O}_{\mathcal{E},\ell}}(L_*)$ the category of étale L_*-modules on $\mathcal{O}_{\mathcal{E}}$.

Lemma 7.109. *The category $\mathfrak{M}^{et}_{\mathcal{O}_{\mathcal{E},\ell}}(L_*)$ is abelian.*

Proof. As in the proof of Proposition 7.23 and using that the ring $\mathcal{O}_{\mathcal{E}}$ is noetherian. $\qquad \square$

We will prove later that the categories $\mathfrak{M}^{et}_{\mathcal{O}_{\mathcal{E},\ell}}(L_*)$ and $\mathfrak{M}^{et}_{\Lambda_\ell(N_0)}(L_*)$ are equivalent.

8.3. Base change functors

We recall a general argument of semilinear algebra (see [12]). Let A be a ring with a ring endomorphism φ_A, let B be another ring with a ring endomorphism φ_B, and let $f : A \to B$ be a ring homomorphism such that $f \circ \varphi_A = \varphi_B \circ f$. When M is an A-module with a semilinear endomorphism φ_M, its image by base change is the B-module $B \otimes_{A,f} M$ with the semilinear endomorphism $\varphi_B \otimes \varphi_M$. The endomorphism φ_M of M is called étale if the natural map

$$a \otimes m \mapsto a\varphi_M(m) : A \otimes_{A,\varphi_A} M \to M$$

is bijective.

Lemma 7.110. *When φ_M is étale, then $\varphi_B \otimes \varphi_M$ is étale.*

Proof. We have

$$B \otimes_{B,\varphi_B} (B \otimes_{A,f} M) = B \otimes_{A,\varphi_B \circ f} M = B \otimes_{f \circ \varphi_A} M$$
$$= B \otimes_{A,f} (A \otimes_{A,\varphi_A} M) \cong B \otimes_{A,f} M.$$

$\qquad \square$

Applying these general considerations to the L_*-equivariant maps ℓ : $\Lambda_\ell(N_0) \to \mathcal{O}_\mathcal{E}$ and $\iota : \mathcal{O}_\mathcal{E} \to \Lambda_\ell(N_0)$ satisfying $\ell \circ \iota = \mathrm{id}$ (see (7.50), (7.53)), we have the base change functors

$$M \mapsto \mathbb{D}(M) := \mathcal{O}_\mathcal{E} \otimes_{\Lambda_\ell(N_0),\ell} M$$

from the category of $\Lambda_\ell(N_0)$-modules to the category of $\mathcal{O}_\mathcal{E}$-modules, and

$$D \mapsto \mathbb{M}(D) := \Lambda_\ell(N_0) \otimes_{\mathcal{O}_\mathcal{E},\iota} D$$

in the opposite direction. Obviously these base change functors respect the property of being finitely generated. By the general lemma we obtain:

Proposition 7.111. *The above functors restrict to functors*

$$\mathbb{D} : \mathfrak{M}^{et}_{\Lambda_\ell(N_0)}(L_*) \to \mathfrak{M}^{et}_{\mathcal{O}_\mathcal{E},\ell}(L_*) \quad and \quad \mathbb{M} : \mathfrak{M}^{et}_{\mathcal{O}_\mathcal{E},\ell}(L_*) \to \mathfrak{M}^{et}_{\Lambda_\ell(N_0)}(L_*).$$

When $M \in \mathfrak{M}^{et}_{\Lambda_\ell(N_0)}(L_*)$, the diagonal action of L_* on $\mathbb{D}(M)$ is:

$$\varphi_t(\mu \otimes m) = \varphi_{\ell(t)}(\mu) \otimes \varphi_t(m) \quad \text{for } t \in L_*, \mu \in \mathcal{O}_\mathcal{E}, m \in M. \tag{7.56}$$

When $D \in \mathfrak{M}^{et}_{\mathcal{O}_\mathcal{E},\ell}(L_*)$, the diagonal action of L_* on $\mathbb{M}(D)$ is:

$$\varphi_t(\lambda \otimes d) = \varphi_t(\lambda) \otimes \varphi_t(d) \quad \text{for } t \in L_*, \lambda \in \Lambda_\ell(N_0), d \in D. \tag{7.57}$$

The natural map

$$\ell_M : M \to \mathbb{D}(M), \quad \ell_M(m) = 1 \otimes m$$

is surjective, L_*-equivariant, with a P_*-stable kernel $M_\ell := J_\ell(N_0)M$. The injective L_*-equivariant map

$$\iota_D : D \to \mathbb{M}(D), \quad \iota_D(d) = 1 \otimes d$$

is ψ_t-equivariant for $t \in L_*$ (same proof as Lemma 7.101).

For future use we note the following property.

Lemma 7.112. *Let $d \in D$ and $t \in L_*$. We have*

$$\psi_t(u^{-1}\iota_D(d)) = \begin{cases} \iota_D(\psi_t(v^{-1}d)) & \text{if } u = \iota(v) \text{ with } v \in N_0^{(2)}, \\ 0 & \text{if } u \in N_0 \setminus \iota(N_0^{(2)})tN_0t^{-1}. \end{cases}$$

Proof. We choose a set $J \subset N_0^{(2)}$ of representatives for the cosets in $N_0^{(2)}/\ell(t)N_0^{(2)}\ell(t)^{-1}$. The semilinear endomorphism φ_t of D is étale hence

$$d = \sum_{v \in J} v\varphi_t(d_{v,t}) \quad \text{where } d_{v,t} = \psi_t(v^{-1}d).$$

Applying ι_D we obtain

$$\iota_D(d) = \sum_v \iota(v)\iota_D(\varphi_t(d_{v,t})) = \sum_v \iota(v)\varphi_t(\iota_D(d_{v,t}))$$

$$= \sum_v \iota(v)\varphi_t(\psi_t(\iota_D(v^{-1}d))).$$

The map ι induces an injective map from J into N_0/tN_0t^{-1} with image included in a set $J(N_0/tN_0t^{-1}) \subset N_0$ of representatives for the cosets in N_0/tN_0t^{-1}. As the action φ_t of t in $\mathbb{M}(D)$ is étale, we have (7.12)

$$m = \sum_{u\in J(N_0/tN_0t^{-1})} u\varphi_t(m_{u,t})) \quad \text{where } m_{u,t} = \psi_t(u^{-1}m)$$

for any $m \in \mathbb{M}(D)$. We deduce that $\psi_t(\iota(v^{-1})\iota_D(d)) = \iota_D(d_{v,t})$ when $v \in J$ and $\psi_t(u^{-1}\iota_D(d)) = 0$ when $u \in J(N_0/tN_0t^{-1}) \setminus \iota(J)$. As any element of $N_0^{(2)}$ can belong to a set of representatives of $N_0^{(2)}/\ell(t)N_0^{(2)}\ell(t)^{-1}$, we deduce that $\psi_t(\iota(v^{-1})\iota_D(d)) = \iota_D(d_{v,t})$ for any $v \in N_0^{(2)}$. For the same reason $\psi_t(\iota(u^{-1})\iota_D(d)) = 0$ for any $u \in N_0$ which does not belong to $\iota(N_0^{(2)})tN_0t^{-1}$. $\qquad\square$

8.4. Equivalence of categories

Let $D \in \mathfrak{M}_{\mathcal{O}_{\mathcal{E}},\ell}^{et}(L_*)$. By definition $\mathbb{D}(\mathbb{M}(D)) = \mathcal{O}_{\mathcal{E}} \otimes_{\Lambda_\ell(N_0),\ell} (\Lambda_\ell(N_0) \otimes_{\mathcal{O}_{\mathcal{E}},\iota} D)$, and we have a natural map

$$\mu \otimes (\lambda \otimes d) \mapsto \mu\ell(\lambda)d : \mathcal{O}_{\mathcal{E}} \otimes_{\Lambda_\ell(N_0),\ell} (\Lambda_\ell(N_0) \otimes_{\mathcal{O}_{\mathcal{E}},\iota} D) \to D.$$

Proposition 7.113. *The natural map* $\mathbb{D}(\mathbb{M}(D)) \to D$ *is an isomorphism in* $\mathfrak{M}_{\mathcal{O}_{\mathcal{E}},\ell}^{et}(L_*)$.

Proof. The natural map is bijective because $\ell \circ \iota = \mathrm{id} : \mathcal{O}_{\mathcal{E}} \to \Lambda_\ell(N_0) \to \mathcal{O}_{\mathcal{E}}$, and L_*-equivariant because the action of $t \in L_*$ satisfies

$$\varphi_t(\mu \otimes (\lambda \otimes d)) = \varphi_{\ell(t)}(\mu) \otimes \varphi_t(\lambda \otimes d) = \varphi_{\ell(t)}(\mu) \otimes (\varphi_t(\lambda) \otimes \varphi_t(d)),$$

$$\varphi_t(\mu\ell(\lambda)d) = \varphi_{\ell(t)}(\mu(\ell(\lambda))\varphi_t(d) = \varphi_{\ell(t)}(\mu)\ell(\varphi_t(\lambda))\varphi_t(d),$$

by (7.56), (7.57). $\qquad\square$

The kernel N_ℓ of $\ell : N_0 \to \mathbb{Z}_p$ being a closed subgroup of N_0 is also a p-adic Lie group, hence contains an open pro-p-subgroup H with the following property ([11] Remark 26.9 and Thm. 27.1):

For any integer $n \geq 1$, the map $h \mapsto h^{p^n}$ is an homeomorphism of H onto an open subgroup $H_n \subseteq H$, and $(H_n)_{n\geq 1}$ is a fundamental system of open neighbourhoods of 1 in H.

The groups $s^k N_\ell s^{-k}$ for $k \geq 1$ are open and form a fundamental system of neighbourhoods of 1 in N_ℓ. For any integer $n \geq 1$ there exists a positive integer k such that any element in $s^k N_\ell s^{-k}$ is contained in H_n, hence is a p^n-th power of some element in N_ℓ. We denote by k_n the smallest positive integer such that any element in $s^{k_n} N_\ell s^{-k_n}$ is a p^n-th power of some element in N_ℓ.

Lemma 7.114. *For any positive integers n and $k \geq k_n$, we have*

$$\varphi^k(J_\ell(N_0)) \subset \mathcal{M}_\ell(N_0)^{n+1}.$$

Proof. For $u \in N_\ell$, and $j \in \mathbb{N}$, the value at u of the p^j-th cyclotomic polynomial $\Phi_{p^j}(u)$ lies in $\mathcal{M}_\ell(N_0)$ and

$$u^{p^n} - 1 = \prod_{j=0}^{n} \Phi_{p^j}(u)$$

lies in $\mathcal{M}_\ell(N_0)^{n+1}$. An element $v \in s^k N_\ell s^{-k}$ is a p^n-th power of some element in N_ℓ hence $v-1$ lies in $\mathcal{M}_\ell(N_0)^{n+1}$. The ideal $J_\ell(N_0)$ of $\Lambda_\ell(N_0)$ is generated by $u - 1$ for $u \in N_\ell$ and $\varphi^k(J_\ell(N_0))$ is contained in the ideal generated by $v - 1$ for $v \in s^k N_\ell s^{-k}$. As $\mathcal{M}_\ell(N_0)$ is an ideal of $\Lambda_\ell(N_0)$ we deduce that $\varphi^k(J_\ell(N_0)) \subset \mathcal{M}_\ell(N_0)^{n+1}$.   □

Lemma 7.115. i. *The functor $\mathbb{D}$ is faithful.*
ii. *The functor $\mathbb{M}$ is fully faithful.*

Proof. Obviously ii. follows from i. by Proposition 7.113. To prove i. let $f : M_1 \to M_2$ be a morphism in $\mathfrak{M}^{et}_{\Lambda_\ell(N_0)}(L_*)$ such that $\mathbb{D}(f) = 0$, i.e., such that $f(M_1) \subseteq J_\ell(N_0)M_2$. Since M_1 is étale we deduce that $f(M_1) \subseteq \bigcap_k \varphi^k(J_\ell(N_0))M_2$ and hence, by Lemma 7.114, in $\bigcap_n \mathcal{M}_\ell(N_0)^n M_2$. Since the pseudocompact topology on M_2 is Hausdorff we have $\bigcap_n \mathcal{M}_\ell(N_0)^n M_2 = 0$. It follows that $f = 0$.   □

Let $M \in \mathfrak{M}^{et}_{\Lambda_\ell(N_0)}(L_*)$. By definition,

$$\mathbb{MD}(M) = \Lambda_\ell(N_0) \otimes_{\mathcal{O}_{\mathcal{E},\iota}} (\mathcal{O}_\mathcal{E} \otimes_{\Lambda_\ell(N_0),\ell} M) = \Lambda_\ell(N_0) \otimes_{\Lambda_\ell(N_0),\iota\circ\ell} M.$$

In the particular case where $L_* = s^\mathbb{N}$ is the monoid generated by s, we denote the category $\mathfrak{M}^{et}_{\Lambda_\ell(N_0)}(L_*)$ (resp. $\mathfrak{M}^{et}_{\mathcal{O}_{\mathcal{E},\ell}}(L_*)$), by $\mathfrak{M}^{et}_{\Lambda_\ell(N_0)}(\varphi)$ (resp. $\mathfrak{M}^{et}_{\mathcal{O}_{\mathcal{E},\ell}}(\varphi)$). The category $\mathfrak{M}^{et}_{\Lambda_\ell(N_0)}(L_*)$ (resp. $\mathfrak{M}^{et}_{\mathcal{O}_{\mathcal{E},\ell}}(L_*)$) is a subcategory of $\mathfrak{M}^{et}_{\Lambda_\ell(N_0)}(\varphi)$ (resp. $\mathfrak{M}^{et}_{\mathcal{O}_{\mathcal{E},\ell}}(\varphi)$).

Proposition 7.116. *For any $M \in \mathfrak{M}^{et}_{\Lambda_\ell(N_0)}(\varphi)$ there is a unique morphism*

$$\Theta_M : M \to \mathbb{MD}(M) \qquad in \ \mathfrak{M}^{et}_{\Lambda_\ell(N_0)}(\varphi)$$

such that the composed map $\mathbb{D}'(\Theta_M) : \mathbb{D}(M) \xrightarrow{\mathrm{D}(\Theta_M)} \mathbb{DMD}(M) \cong \mathbb{D}(M)$ *is the identity. The morphism* Θ_M, *in fact, is an isomorphism.*

Proof. The uniqueness follows immediately from Lemma 7.115.i. The construction of such an isomorphism Θ_M will be done in three steps.

Step 1: We assume that M is free over $\Lambda_\ell(N_0)$, and we start with an arbitrary finite $\Lambda_\ell(N_0)$-basis $(\epsilon_i)_{i \in I}$ of M. By (8.2), we have

$$M = (\oplus_{i \in I} \iota(\mathcal{O}_{\mathcal{E}})\epsilon_i) \oplus (\oplus_{i \in I} J_\ell(N_0)\epsilon_i).$$

The $\Lambda_\ell(N_0)$-linear map from M to $\mathbb{MD}(M)$ sending ϵ_i to $1 \otimes (1 \otimes \epsilon_i)$ is bijective. If $\oplus_{i \in I} \iota(\mathcal{O}_{\mathcal{E}})\epsilon_i$ is φ-stable, the map is also φ-equivariant and is an isomorphism in the category $\mathfrak{M}^{et}_{\Lambda_\ell(N_0)}(\varphi)$. We will construct a $\Lambda_\ell(N_0)$-basis $(\eta_i)_{i \in I}$ of M such that $\oplus_{i \in I} \iota(\mathcal{O}_{\mathcal{E}})\eta_i$ is φ-stable.

We have

$$\varphi(\epsilon_i) = \sum_{j \in I}(a_{i,j} + b_{i,j})\epsilon_j \text{ where } a_{i,j} \in \iota(\mathcal{O}_{\mathcal{E}}), \ b_{i,j} \in J_\ell(N_0).$$

If the $b_{i,j}$ are not all 0, we will show that there exist elements $x_{i,j} \in J_\ell(N_0)$ such that $(\eta_i)_{i \in I}$ defined by

$$\eta_i := \epsilon_i + \sum_{j \in I} x_{i,j}\epsilon_j,$$

satisfies $\varphi(\eta_i) = \sum_{j \in I} a_{i,j}\eta_j$ for $i \in I$. By the Nakayama lemma ([1] II §3.2 Prop. 5), the set $(\eta_i)_{i \in I}$ is a $\Lambda_\ell(N_0)$-basis of M, and we obtain an isomorphism in $\mathfrak{M}^{et}_{\Lambda_\ell(N_0)}(\varphi)$,

$$\Theta_M \ M \to \mathbb{MD}(M), \quad \Theta(\eta_i) = 1 \otimes (1 \otimes \eta_i) \text{ for } i \in I,$$

such that $\mathbb{D}'(\Theta_M)$ is the identity morphism of $\mathbb{D}(M)$.

The conditions on the matrix $X := (x_{i,j})_{i,j \in I}$ are:

$$\varphi(\mathrm{Id} + X)(A + B) = A(\mathrm{Id} + X)$$

for the matrices $A := (a_{i,j})_{i,j \in I}$, $B := (b_{i,j})_{i,j \in I}$. The coefficients of A belong to the commutative ring $\iota(\mathcal{O}_{\mathcal{E}})$. The matrix A is invertible because the $\Lambda_\ell(N_0)$-endomorphism f of M defined by

$$f(\epsilon_i) = \varphi(\epsilon_i) \text{ for } i \in I$$

is an automorphism of M as φ is étale. We have to solve the equation

$$A^{-1}B + A^{-1}\varphi(X)(A + B) = X.$$

For any $k \geq 0$ define

$$U_k = A^{-1}\varphi(A^{-1})\ldots\varphi^{k-1}(A^{-1})\,\varphi^k(A^{-1}B)\,\varphi^{k-1}(A+B)\ldots\varphi(A+B)(A+B).$$

We have

$$A^{-1}\varphi(U_k)(A + B) = U_{k+1}.$$

Hence $X := \sum_{k \geq 0} U_k$ is a solution of our equation provided this series converges with respect to the pseudocompact topology of $\Lambda_\ell(N_0)$. The coefficients of $A^{-1}B$ belong to the two-sided ideal $J_\ell(N_0)$ of $\Lambda_\ell(N_0)$. Therefore the coefficients of U_k belong to the two-sided ideal generated by $\varphi^k(J_\ell(N_0))$. Hence the series converges (Lemma 7.114). The coefficients of every term in the series belong to $J_\ell(N_0)$ and $J_\ell(N_0)$ is closed in $\Lambda_\ell(N_0)$, hence $x_{i,j} \in J_\ell(N_0)$ for $i, j \in I$.

Step 2: We show that any module M in $\mathfrak{M}^{et}_{\Lambda_\ell(N_0)}(\varphi)$ is the quotient of another module M_1 in $\mathfrak{M}^{et}_{\Lambda_\ell(N_0)}(\varphi)$ which is free over $\Lambda_\ell(N_0)$.

Let $(m_i)_{i \in I}$ be a minimal finite system of generators of the $\Lambda_\ell(N_0)$-module M. As φ is étale, $(\varphi(m_i))_{i \in I}$ is also a minimal system of generators. We denote by $(e_i)_{i \in I}$ the canonical $\Lambda_\ell(N_0)$-basis of $\oplus_{i \in I} \Lambda_\ell(N_0)$, and we consider the two surjective $\Lambda_\ell(N_0)$-linear maps

$$f, g : \oplus_{i \in I} \Lambda_\ell(N_0) \to M, \quad f(e_i) = m_i, \ g(e_i) = \varphi(m_i).$$

In particular, we find elements $m_i' \in M$, for $i \in I$, such that $g(m_i') = \varphi(m_i)$. By the Nakayama lemma ([1] II §3.2 Prop. 5) the $(m_i')_{i \in I}$ form another $\Lambda_\ell(N_0)$-basis of $\oplus_{i \in I} \Lambda_\ell(N_0)$. The φ-linear map

$$\oplus_{i \in I} \Lambda_\ell(N_0) \to \oplus_{i \in I} \Lambda_\ell(N_0), \quad \varphi(\sum_{i \in I} \lambda_i e_i) := \sum_{i \in I} \varphi(\lambda_i) m_i'$$

therefore is étale. With this map, $M_1 := \oplus_{i \in I} \Lambda_\ell(N_0)$ is a module in $\mathfrak{M}^{et}_{\Lambda_\ell(N_0)}(\varphi)$ which is free over $\Lambda_\ell(N_0)$, and the surjective map f is a morphism in $\mathfrak{M}^{et}_{\Lambda_\ell(N_0)}(\varphi)$.

Step 3: As $\Lambda_\ell(N_0)$ is noetherian, we deduce from Step 2 that for any module M in $\mathfrak{M}^{et}_{\Lambda_\ell(N_0)}(\varphi)$ we have an exact sequence

$$M_2 \xrightarrow{f} M_1 \xrightarrow{f'} M \to 0$$

in $\mathfrak{M}^{et}_{\Lambda_\ell(N_0)}(\varphi)$ such that M_1 and M_2 are free over $\Lambda_\ell(N_0)$. We now consider the diagram

$$
\begin{array}{ccccccc}
\mathbb{MD}(M_2) & \xrightarrow{\mathbb{MD}(f)} & \mathbb{MD}(M_1) & \xrightarrow{\mathbb{MD}(f')} & \mathbb{MD}(M) & \longrightarrow & 0 \\
\Theta_{M_2} \big\uparrow \cong & & \Theta_{M_1} \big\uparrow \cong & & \Theta_M \big\uparrow & & \\
M_2 & \xrightarrow{\ \ f\ \ } & M_1 & \xrightarrow{\ \ f'\ \ } & M & \longrightarrow & 0
\end{array}
$$

Since the functors $\mathbb{M}$ and $\mathbb{D}$ are right exact both rows of the diagram are exact. By Step 1 the left two vertical maps exist and are isomorphisms. Since

$$\mathbb{D}(\mathbb{M}\mathbb{D}(f) \circ \Theta_{M_2} - \Theta_{M_1} \circ f) = \mathbb{D}(f) \circ \mathbb{D}'(\Theta_{M_2}) - \mathbb{D}'(\Theta_{M_1}) \circ \mathbb{D}(f) = 0$$

it follows from Lemma 7.115.i that the left square of the diagram commutes. Hence we obtain an induced isomorphism Θ_M as indicated, which moreover by construction satisfies $\mathbb{D}'(\Theta_M) = \mathrm{id}_{\mathbb{D}(M)}$. $\square$

Theorem 7.117. *The functors*

$$\mathbb{M} : \mathfrak{M}^{et}_{\mathcal{O}_{\mathcal{E}},\ell}(L_*) \to \mathfrak{M}^{et}_{\Lambda_\ell(N_0)}(L_*), \quad \mathbb{D} : \mathfrak{M}^{et}_{\Lambda_\ell(N_0)}(L_*) \to \mathfrak{M}^{et}_{\mathcal{O}_{\mathcal{E}},\ell}(L_*),$$

are quasi-inverse equivalences of categories.

Proof. By Proposition 7.113 and Lemma 7.115.ii it remains to show that the functor $\mathbb{M}$ is essentially surjective. Let $M \in \mathfrak{M}^{et}_{\Lambda_\ell(N_0)}(L_*)$. We have to find a $D \in \mathfrak{M}^{et}_{\mathcal{O}_{\mathcal{E}},\ell}(L_*)$ together with an isomorphism $M \cong \mathbb{M}(D)$ in $\mathfrak{M}^{et}_{\Lambda_\ell(N_0)}(L_*)$. It suffices to show that the morphism Θ_M in Proposition 7.116 is L_*-equivariant.

We want to prove that $(\Theta_M \circ \varphi_t - \varphi_t \circ \Theta_M)(m) = 0$ for any $m \in M$ and $t \in L_*$. Since $\mathbb{D}'(\Theta_M) = \mathrm{id}_{\mathbb{D}(M)}$ we certainly have $(\Theta \circ \varphi_t - \varphi_t \circ \Theta)(m) \in J_\ell(N_0)\mathbb{M}\mathbb{D}(M)$ for any $m \in M$ and $t \in L_*$. We choose for any positive integer r a set $J(N_0/N_r) \subseteq N_0$ of representatives for the cosets in N_0/N_r. Writing
(7.12)
$$m = \sum_{u \in J(N_0/N_r)} u\varphi^r(m_{u,s^r}), \quad m_{u,s^r} = \psi^r(u^{-1}m)$$

and using that $st = ts$ we see that

$$(\Theta_M \circ \varphi_t - \varphi_t \circ \Theta_M)(m) = \sum_{u \in J(N_0/N_r)} \varphi_t(u)\varphi^r((\Theta_M \circ \varphi_t - \varphi_t \circ \Theta_M)(m_{u,s^r}))$$

lies, for any r, in the $\Lambda_\ell(N_0)$-submodule of $\mathbb{M}\mathbb{D}(M)$ generated by $\varphi^r(J_\ell(N_0))\mathbb{M}\mathbb{D}(M)$. As in the proof of Lemma 7.115.ii we obtain $\bigcap_{r>0} \varphi^r(J_\ell(N_0))\mathbb{M}\mathbb{D}(M) = 0$. $\square$

Since the functors $\mathbb{M}$ and $\mathbb{D}$ are right exact they commute with the reduction modulo p^n, for any integer $n \geq 1$.

8.5. Continuity

In this section we assume that L_* contains a subgroup L_1 which is open in L_* and is a topologically finitely generated pro-p-group.

We will show that the L_*-action on any étale L_*-module over $\Lambda_\ell(N_0)$ is automatically continuous. Our proof is highly indirect so that we temporarily will have to make some definitions. But first a few partial results can be established directly.

Let M be a finitely generated $\Lambda_\ell(N_0)$-module.

Definition 7.118. A lattice in M is a $\Lambda(N_0)$-submodule of M generated by a finite system of generators of the $\Lambda_\ell(N_0)$-module M.

The lattices of M are of the form $M^0 = \sum_{i=1}^{r} \Lambda(N_0)m_i$ for a set $(m_i)_{1 \leq i \leq r}$ of generators of the $\Lambda_\ell(N_0)$-module M.

We have the three fundamental systems of neighbourhoods of 0 in M:

$$(\sum_{i=1}^{r} O_{n,k}m_i = \mathcal{M}_\ell(N_0)^n M + \mathcal{M}(N_0)^k M^0)_{n,k \in \mathbb{N}}, \tag{7.58}$$

$$(\sum_{i=1}^{r} B_{n,k}m_i = \mathcal{M}_\ell(N_0)^n M + X^k M^0)_{n,k \in \mathbb{N}}, \tag{7.59}$$

$$(\sum_{i=1}^{r} C_{n,k}m_i = \mathcal{M}_\ell(N_0)^n M + M_k^0)_{n,k \in \mathbb{N}}, \tag{7.60}$$

where M_k^0 is the lattice $\sum_{i=1}^{r} \Lambda(N_0)X^k m_i$, and is different from the set $X^k M_0$ when N_0 is not commutative.

If M is an étale L_*-module over $\Lambda_\ell(N_0)$, for any fixed $t \in L_{\ell,+}$ we have a fourth fundamental system of neighbourhoods of 0 in M:

$$(\sum_{i=1}^{r} \varphi_t(O_{n,k})\Lambda(N_0)\varphi_t(m_i))_{n,k \in \mathbb{N}},$$

given by Lemma 7.98, because $(\varphi_t(m_i))_{1 \leq i \leq r}$ is also a system of generators of the $\Lambda_\ell(N_0)$-module M.

Proposition 7.119. *Let L' be a submonoid of $L_{\ell,+}$. Let M be an étale L'-module over $\Lambda_\ell(N_0)$. Then the maps φ_t and ψ_t, for any $t \in L'$, are continuous on M.*

Proof. The ring endomorphisms φ_t of $\Lambda_\ell(N_0)$ are continuous since they preserve $\mathcal{M}(N_0)$ and $\mathcal{M}(N_\ell)$. The continuity of the φ_t on M follows as in part (a) of the proof of Proposition 7.92. The continuity of the ψ_t follows from

$$\psi_t(\sum_{i=1}^{r} \varphi_t(O_{n,k})\Lambda(N_0)\varphi_t(m_i)) = \sum_{i=1}^{r} O_{n,k}\psi_t(\Lambda(N_0))m_i = \sum_{i=1}^{r} O_{n,k}m_i.$$

$\square$

The same proof shows that, for any $D \in \mathfrak{M}^{et}_{\mathcal{O}_{\mathcal{E}},\ell}(L_*)$, the maps φ_t and ψ_t, for any $t \in L_*$, are continuous on D.

Proposition 7.120. *The L_*-action $L_* \times D \to D$ on an étale L_*-module D over $\mathcal{O}_{\mathcal{E}}$ is continuous.*

Proof. Let D be in $\mathfrak{M}^{et}_{\mathcal{O}_{\mathcal{E}},\ell}(L_*)$. Since we already know from Proposition 7.119 that each individual φ_t, for $t \in L_*$, is a continuous map on D and since L_1 is open in L_* it suffices to show that the action $L_1 \times D \to D$ of L_1 on D is continuous. As D is p-adically complete with its weak topology being the projective limit of the weak topologies on the $D/p^n D$ we may further assume that D is killed by a power of p. In this situation the weak topology on D is locally compact. By Ellis' theorem ([8] Thm. 1) we therefore are reduced to showing that the map $L_1 \times D \to D$ is separately continuous. Because of Proposition 7.119 it, in fact, remains to prove that, for any $d \in D$, the map

$$L_1 \longrightarrow D, \; g \longmapsto gd$$

is continuous at $1 \in L_1$. This amounts to finding, for any $d \in D$ and any lattice $D_0 \subset D$, an open subgroup $H \subset L_1$ such that $(H - 1)d \subset D_0$. We observe that $(X^m D_{++})_{m \in \mathbb{Z}}$ is a fundamental system of L_1-stable open neighbourhoods of zero in D such that $\bigcup_m X^m D_{++} = D$. We now choose an $m \geq 0$ large enough such that $d \in X^{-m} D_{++}$ and $X^m D_{++} \subset D_0$. The L_1 action on D induces an L_1-action on $X^{-m} D_{++}/X^m D_{++}$ which is o-linear hence given by a group homomorphism $L_1 \to \mathrm{Aut}_o(X^{-m} D_{++}/X^m D_{++})$. Since D_{++} is a finitely generated $o[[X]]$-module which is killed by a power of p we see that $X^{-m} D_{++}/X^m D_{++}$ is finite. It follows that the kernel H of the above homomorphism is of finite index in L_1. Our assumption that L_1 is a topologically finitely generated pro-p-group finally implies, by a theorem of Serre ([7] Thm. 1.17), that H is open in L_1. We obtain

$$(H - 1)d \subset (H - 1)X^{-m} D_{++} \subset X^m D_{++} \subset D_0.$$

$\square$

In the special case of classical (φ, Γ)-modules on $\mathcal{O}_{\mathcal{E}}$ the proposition is stated as Exercise 2.4.6 in [10] (with the indication of a totally different proof).

Proposition 7.121. *Let L' be a submonoid of $L_{\ell,+}$ containing an open subgroup L_2 which is a topologically finitely generated pro-p-group. Then the L'-action $L' \times \Lambda_\ell(N_0) \to \Lambda_\ell(N_0)$ on $\Lambda_\ell(N_0)$ is continuous.*

Proof. Since we know already from Propositions 7.99 and 7.119 that each individual φ_t, for $t \in L'$, is a continuous map on $\Lambda_\ell(N_0)$ and since L_2 is open in L' it suffices to show that the action $L_2 \times \Lambda_\ell(N_0) \rightarrow \Lambda_\ell(N_0)$ of L_2 on $\Lambda_\ell(N_0)$ is continuous. The ring $\Lambda_\ell(N_0)$ is $\mathcal{M}_\ell(N_0)$-adically complete with its weak topology being the projective limit of the weak topologies on the $\Lambda_\ell(N_0)/\mathcal{M}_\ell(N_0)^n \Lambda_\ell(N_0)$. It suffices to prove that the induced action of L_2 on $\Lambda' = \Lambda_\ell(N_0)/\mathcal{M}_\ell(N_0)^n$ is continuous. The weak topology on Λ' is locally compact since $(B'_k = (X^k \Lambda(N_0) + \mathcal{M}_\ell(N_0)^n)/\mathcal{M}_\ell(N_0)^n)_{k \in \mathbb{Z}}$ forms a fundamental system of compact neighbourhoods of 0. By Ellis' theorem ([8] Thm. 1) we therefore are reduced to showing that the map $L_2 \times \Lambda' \rightarrow \Lambda'$ is separately continuous. Because of Proposition 7.119 it, in fact, remains to prove that, for any $x \in \Lambda'$, the map

$$L_2 \longrightarrow \Lambda', \; g \longmapsto gx$$

is continuous at $1 \in L_2$. This amounts to finding, for any $x \in \Lambda'$ and any large $k \geq 1$, an open subgroup $H \subset L_2$ such that $(H - 1)x \subset B'_k$. We observe that the B'_k, for $k \in \mathbb{Z}$, are L_2-stable of union Λ'. We now choose an $m \geq k$ large enough such that $x \in B'_{-m}$. The L_2-action on Λ' induces an L_2-action on B'_{-m}/B'_m which is o-linear hence given by a group homomorphism $L_2 \rightarrow \text{Aut}_o(B'_{-m}/B'_m)$. Since B'_0 is isomorphic to $o[[X]] \otimes_o \Lambda(N_\ell)/\mathcal{M}(N_\ell)^n$ as an $o[[X]]$-module, and $\Lambda(N_\ell)/\mathcal{M}(N_\ell)^n$ is finite, we see that B'_{-m}/B'_m is finite. It follows that the kernel H of the above homomorphism is of finite index in L_2. Our assumption that L_2 is a topologically finitely generated pro-p-group finally implies, by a theorem of Serre ([7] Thm. 1.17), that H is open in L_2. We obtain

$$(H - 1)x \subset (H - 1)B'_{-m} \subset B'_m \subset B'_k.$$

$\square$

Lemma 7.122. i. *For any $M \in \mathfrak{M}^{et}_{\Lambda_\ell(N_0)}(L_*)$ the weak topology on $\mathbb{D}(M)$ is the quotient topology, via the surjection $\ell_M : M \twoheadrightarrow \mathbb{D}(M)$, of the weak topology on M.*

ii. *For any $D \in \mathfrak{M}^{et}_{\mathcal{O}_\mathcal{E},\ell}(L_*)$ the weak topology on $\mathbb{M}(D)$ induces, via the injection $\iota_D : D \rightarrow \mathbb{M}(D)$, the weak topology on D.*

Proof. i. If we write M as a quotient of a finitely generated free $\Lambda_\ell(N_0)$-module then we obtain an exact commutative diagram of surjective maps of the form

The horizontal maps are continuous and open by the definition of the weak topology. The left vertical map is continuous and open by direct inspection of the open zero neighbourhoods $B_{n,k}$ (see (7.51)). Hence the right vertical map ℓ_M is continuous and open.

ii. An analogous argument as for i. shows that ι_D is continuous. Moreover ι_D has the continuous left inverse $\ell_{\mathbb{M}(D)}$. Any continuous map with a continuous left inverse is a topological inclusion. $\square$

An étale L_*-module M over $\Lambda_\ell(N_0)$, resp. over $\mathcal{O}_{\mathcal{E}}$, will be called topologically étale if the L_*-action $L_* \times M \to M$ is continuous. Let $\mathfrak{M}^{et,c}_{\Lambda_\ell(N_0)}(L_*)$ and $\mathfrak{M}^{et,c}_{\mathcal{O}_{\mathcal{E}},\ell}(L_*)$ denote the corresponding full subcategories of $\mathfrak{M}^{et}_{\Lambda_\ell(N_0)}(L_*)$ and $\mathfrak{M}^{et}_{\mathcal{O}_{\mathcal{E}},\ell}(L_*)$, respectively. Note that, by construction, all morphisms in $\mathfrak{M}^{et}_{\Lambda_\ell(N_0)}(L_*)$ and in $\mathfrak{M}^{et}_{\mathcal{O}_{\mathcal{E}},\ell}(L_*)$ are automatically continuous. Also note that by Proposition 7.119 any object in these categories is a complete topologically étale $o[N_0 L_*]$-module in our earlier sense.

Proposition 7.123. *The functors $\mathbb{M}$ and $\mathbb{D}$ restrict to quasi-inverse equivalences of categories*

$$\mathbb{M} : \mathfrak{M}^{et,c}_{\mathcal{O}_{\mathcal{E}},\ell}(L_*) \to \mathfrak{M}^{et,c}_{\Lambda_\ell(N_0)}(L_*), \quad \mathbb{D} : \mathfrak{M}^{et,c}_{\Lambda_\ell(N_0)}(L_*) \to \mathfrak{M}^{et,c}_{\mathcal{O}_{\mathcal{E}},\ell}(L_*).$$

Proof. It is immediate from Lemma 7.122.i that if L_* acts continuously on $M \in \mathfrak{M}^{et}_{\Lambda_\ell(N_0)}(L_*)$ then it also acts continuously on $\mathbb{D}(M)$.

On the other hand, let $D \in \mathfrak{M}^{et}_{\mathcal{O}_{\mathcal{E}},\ell}(L_*)$ such that the action of L_* on D is continuous. We choose a lattice D_0 in D with a finite system (d_i) of generators. Given $t \in L_*$ we introduce $D_t := \sum_i \Lambda(N_0^{(2)}) t.d_i$ which is a lattice in D since the action of t on D is étale. Also $D_0 + D_t$ is a lattice in D. The $\Lambda_\ell(N_0)$-module $\mathbb{M}(D)$ is generated by $\iota_D(D_0)$ as well as by $\iota_D(D_0 + D_t)$ and both

$$(C_n \iota_D(D_0))_{n \in \mathbb{N}} \quad \text{and} \quad (C_n \iota_D(D_0 + D_t))_{n \in \mathbb{N}}$$

are fundamental systems of neighbourhoods of 0 in $\mathbb{M}(D)$ for the weak topology. To show that the action of L_* on $\mathbb{M}(D)$ is continuous, it suffices to find for any $t \in L_*, \lambda_0 \in \Lambda_\ell(N_0), d_0 \in D_0, n \in \mathbb{N}$ a neighbourhood $L_t \subset L_*$ of t and $n' \in \mathbb{N}$ such that

$$L_t.(\lambda_0 \iota_D(d_0) + C_{n'} \iota_D(D_0)) \subset t.\lambda_0 \iota_D(d_0) + C_n \iota_D(D_0 + D_t). \quad (7.61)$$

The three maps

$$\lambda \mapsto \lambda \iota_D(d_0) : \Lambda_\ell(N_0) \to \mathbb{M}(D)$$
$$d \mapsto \lambda_0 \iota_D(d) : D \to \mathbb{M}(D)$$
$$(\lambda, d) \mapsto \lambda \iota_D(d) : \Lambda_\ell(N_0 \times D) \to \mathbb{M}(D)$$

are continuous because ι_D is continuous. The action of L_* on D and on $\Lambda_\ell(N_0)$ is continuous (Proposition 7.121). Altogether this implies that we can find a small L_t such that

$$L_t.\lambda_0 \iota_D(d_0) \subset t.\lambda_0 \iota_D(d_0) + C_n \iota_D(D_0 + D_t).$$

Since ι_D is L_*-equivariant we have, for any $n' \in \mathbb{N}$,

$$L_t.C_{n'} \iota_D(D_0) = (L_t.C_{n'}) \iota_D(L_t.D_0).$$

The continuity of the action of L_* on $\Lambda_\ell(N_0)$ shows that $L_t.C_{n'} \subset C_n$ when L_t is small enough and n' is large enough.

For $d \in D_0$ we have $L_t.\Lambda(N_0^{(2)})d \subset \Lambda(N_0^{(2)})(L_t.d)$. The action of L_* on D is continuous hence, for any n', we can choose a small L_t such that $L_t.d \subset t.d + C_{n'}^{(2)} D_0$. We can choose the same L_t for each d_i and we obtain

$$L_t.D_0 \subset \sum_i \Lambda(N_0^{(2)})t.d_i + C_{n'}^{(2)} D_0.$$

Applying ι_D, we obtain

$$\iota_D(L_t.D_0) \subset \iota_D(D_t) + C_{n'} \iota_D(D_0)$$

and then

$$(L_t.C_{n'}) \iota_D(L_t.D_0) \subset C_n \iota_D(D_t) + C_n C_{n'} \iota_D(D_0).$$

We check that $C_n C_{n'} \subset C_{n,n+n'} \subset C_n$ when $n' \geq n$. Hence when n' is large enough,

$$L_t.(C_{n'} \iota_D(D_0)) \subset C_n \iota_D(D_t + D_0).$$

This ends the proof of (7.61).   □

Proposition 7.124. *We have* $\mathfrak{M}_{\mathcal{O}_\mathcal{E},\ell}^{et,c}(L_*) = \mathfrak{M}_{\mathcal{O}_\mathcal{E},\ell}^{et}(L_*)$ *and* $\mathfrak{M}_{\Lambda_\ell(N_0)}^{et,c}(L_*) = \mathfrak{M}_{\Lambda_\ell(N_0)}^{et}(L_*)$.

Proof. The first identity was shown in Proposition 7.120 and is equivalent to the second identity by Theorem 7.117 and Proposition 7.123.   □

Corollary 7.125. *Any étale L_*-module over $\Lambda_\ell(N_0)$, resp. over $\mathcal{O}_\mathcal{E}$, is a complete topologically étale $o[N_0 L_*]$-module in our sense.*

Proof. Use Propositions 7.119 and 7.124.   □

9. Convergence in L_+-modules on $\Lambda_\ell(N_0)$

In this section, we use the notation of Section 8 where we assume that N is a p-adic Lie group. We assume that ℓ and ι are continuous group homomorphisms

$$\ell : P \to P^{(2)}, \quad \iota : N^{(2)} \to N, \quad \ell \circ \iota = \mathrm{id},$$

such that $\ell(L_+) \subset L_+^{(2)}$, $\ell(N) = N^{(2)}$, $(\iota \circ \ell)(N_0) \subset N_0$, and

$$t\iota(y)t^{-1} = \iota(\ell(t)y\ell(t)^{-1}) \quad \text{for } y \in N^{(2)}, t \in L. \tag{7.62}$$

The assumptions of Section 8 are naturally satisfied with $L_* = L_+$. Indeed, the compact open subgroup N_0 of N is a compact p-adic Lie group, the group $\ell(N_0)$ is a compact non-trivial subgroup $N_0^{(2)}$ of $N^{(2)} \simeq \mathbb{Q}_p$ hence $N_0^{(2)}$ is isomorphic to $\mathbb{Z}_p$ and is open in $N^{(2)}$, the kernel of $\ell|_{N_0}$ is normalized by $L_{\ell,+}$. Note that L_+ normalizes $\iota(N_0^{(2)})$ since $\ell(L_+)$ normalises $N_0^{(2)}$ and (7.62).

Let $M \in \mathfrak{M}^{et}_{\Lambda_\ell(N_0)}(L_+)$ and $D \in \mathfrak{M}^{et}_{\mathcal{O}_\mathcal{E},\ell}(L_+)$ be related by the equivalence of categories (Theorem 7.117),

$$M = \Lambda_\ell(N_0) \otimes_{\mathcal{O}_\mathcal{E},\iota} D = \Lambda_\ell(N_0)\iota_D(D).$$

We will exhibit in this section a special family $\mathfrak{C}_s$ of compact subsets in M such that $M(\mathfrak{C}_s)$ is a dense o-submodule of M, and such that the P-equivariant sheaf on $\mathcal{C}$ associated to the étale $o[P_+]$-module $M(\mathfrak{C}_s)$ by Theorem 7.38 extends to a G-equivariant sheaf on G/P. We will follow the method explained in Subsection 6.5 which reduces the most technical part to the easier case where M is killed by a power of p.

9.1. Bounded sets

Definition 7.126. A subset A of M is called bounded if for any open neighbourhood $\mathcal{B}$ of 0 in M there exists an open neighbourhood B of 0 in $\Lambda_\ell(N_0)$ such that

$$BA \subset \mathcal{B}.$$

Compare with [12] Def. 8.5. The properties satisfied by bounded subsets of M can be proved directly or deduced from the properties of bounded subsets of $\Lambda_\ell(N_0)$ ([15] §12). Using the fundamental system (7.59) of neighbourhoods of 0, the set A is bounded if and only if for any large n there exists $n' > n$ such that

$$(\mathcal{M}_\ell(N_0)^{n'} + X^{n'}\Lambda(N_0))A \subset \mathcal{M}_\ell(N_0)^n M + X^n M^0,$$

equivalently $X^{n'-n}A \subset \mathcal{M}_\ell(N_0)^n M + M^0$. We obtain (compare with [12] Lemma 8.8):

Lemma 7.127. *A subset A of M is bounded if and only if for any large positive n there exists a positive integer n' such that*

$$A \subset \mathcal{M}_\ell(N_0)^n M + X^{-n'} M^0.$$

The following properties of bounded subsets will be used in the construction of a special family $\mathfrak{C}_s$ in the next subsection.

- Let $f : \oplus_{i=1}^r \Lambda_\ell(N_0) \to M$ be a surjective homomorphism of $\Lambda_\ell(N_0)$-modules. The image by f of a bounded subset of $\oplus_{i=1}^r \Lambda_\ell(N_0)$ is a bounded subset of M. For $1 \le i \le r$, the i-th projections $A_i \subset \Lambda_\ell(N_0)$ of a subset A of $\oplus_{i=1}^r \Lambda_\ell(N_0)$ are all bounded if and only if A is bounded.
- A compact subset is bounded.
- The $\Lambda(N_0)$-module generated by a bounded subset is bounded.
- The closure of a bounded subset is bounded.
- Given a compact subset C in $\Lambda_\ell(N_0)$ and a bounded subset A of M, the subset CA of M is bounded.
- The image of a bounded subset by $f \in \mathrm{End}_o^{cont}(M)$ is bounded. The image by ℓ_M of a bounded subset in M is bounded in D.
- A subset A of D is bounded if and only if the image A_n of A in $D/p^n D$ is bounded for all large n.
- When D is killed by a power of p, a subset A of D is bounded if and only if A is contained in a lattice, i.e. if A is contained in a compact subset (by the properties of lattices given in Section 7.3).

Lemma 7.128. *The image by ι_D of a bounded subset in D is bounded in M.*

Proof. Let $A \subset D$ be a bounded subset and let D^0 be a fixed lattice in D. For all $n \in \mathbb{N}$ there exists $n' \in \mathbb{N}$ such that $A \subset p^n D + (X^{(2)})^{-n'} D^0$ by Lemma 7.127. Applying ι_D we obtain

$$\iota_D(A) \subset p^n \iota_D(D) + X^{-n'} \iota_D(D^0) \subset \mathcal{M}_\ell(N_0)^n M + X^{-n'} M^0$$

where $M^0 = \Lambda(N_0)\iota_D(D^0)$ is a lattice in M. By the same lemma, this means that $\iota_D(A)$ is bounded in M. $\qquad\square$

9.2. The module M_s^{bd}

Definition 7.129. M_s^{bd} is the set of $m \in M$ such that the set of $\ell_M(\psi^k(u^{-1}m))$ for $k \in \mathbb{N}$, $u \in N_0$ is bounded in D.

The definition of M_s^{bd} depends on s because ψ is the canonical left inverse of the action φ of s on M. We recognize $m_{u,s^k} = \psi^k(u^{-1}m)$ appearing in the expansion (7.12).

Proposition 7.130. M_s^{bd} *is an étale $o[P_+]$-submodule of M.*

Proof. (a) We check first that M_s^{bd} is P_+-stable. As M_s^{bd} is N_0-stable and $P_+ = N_0 L_+$, it suffices to show that $tm = \varphi_t(m) \in M_s^{bd}$ when $t \in L_+$ and $m \in M_s^{bd}$. Using the expansion (7.12) of m and $st = ts$, for $k \in \mathbb{N}$ and $n_0 \in N_0$, we write $\psi^k(n_0^{-1}tm)$ as the sum over $u \in J(N_0/N_k)$ of

$$\psi^k(n_0^{-1}tu\varphi^k(m_{u,s^k})) = \psi^k(n_0^{-1}tut^{-1}\varphi^k(\varphi_t(m_{u,s^k})))$$
$$= \psi^k(n_0^{-1}tut^{-1})\varphi_t(m_{u,s^k}),$$

and $\ell_M(\psi^k(n_0^{-1}\varphi_t(m)))$ as the sum over $u \in J(N_0/N_k)$ of

$$\ell_M(\psi^k(n_0^{-1}tut^{-1})\varphi_t(m_{u,s^k})) = v_{k,n_0}\ell_M(\varphi_t(m_{u,s^k})) = v_{k,n_0}\varphi_t(\ell_M(m_{u,s^k})),$$

where $v_{k,n_0} := \ell(\psi^k(n_0^{-1}tut^{-1}))$ belongs to $N_0^{(2)}$ or is 0. As $m \in M_s^{bd}$, the set of $\ell_M(m_{u,s^k})$ for $k \in \mathbb{N}$ and $u \in N_0$ is bounded in D. Its image by the continuous map φ_t is bounded and generates a bounded $o[N_0^{(2)}]$-submodule of D. Hence $\varphi_t(m) \in M_s^{bd}$.

(b) The $o[P_+]$-module M_s^{bd} is ψ-stable (hence M_s^{bd} is étale by Corollary 7.45) because we have, for $m \in M_s^{bd}$, $u \in N_0$, $k \in \mathbb{N}$,

$$\psi^k(u^{-1}\psi(m)) = \psi^{k+1}(\varphi(u^{-1})m). \tag{7.63}$$

$\square$

The goal of this section is to show that the P-equivariant sheaf on $\mathcal{C}$ associated to the étale $o[P_+]$-module M_s^{bd} extends to a G-equivariant sheaf on G/P. We will follow the method explained in Subsection 6.5.

Let $p_n : M \to M/p^n M$ be the reduction modulo p^n for a positive integer n. Recall that M is p-adically complete.

Lemma 7.131. *The o-submodule $M_s^{bd} \subset M$ is closed for the p-adic topology, in particular*

$$M_s^{bd} = \varprojlim_n (M_s^{bd}/p^n M_s^{bd}).$$

Moreover M_s^{bd} is the set of $m \in M$ such that $p_n(m)$ belongs to $(M/p^n M)_s^{bd}$ for all $n \in \mathbb{N}$, and we have

$$M_s^{bd} = \varprojlim_n (M/p^n M)_s^{bd}.$$

Proof. (a) Let m be an element in the closure of M_s^{bd} in M for the p-adic
topology. For any $r \in \mathbb{N}$, we choose $m'_r \in M_s^{bd}$ with $m - m'_r \in p^r M$. For
each r, we choose $r' \geq 1$ such that $\ell_M(\psi^k(u^{-1} m'_r)) \in p^r D + X^{-r'} D^0$ for
all $k \in \mathbb{N}$, $u \in N_0$, applying Lemma 7.127. We have

$$\ell_M(\psi^k(u^{-1} m)) \in \ell_M(\psi^k(u^{-1} m'_r) + p^r M)$$
$$= \ell_M(\psi^k(u^{-1} m'_r)) + p^r D \subset p^r D + X^{-r'} D^0.$$

By the same lemma, $m \in M_s^{bd}$. This proves that M_s^{bd} is closed in M hence
p-adically complete.

(b) The reduction modulo p^n commutes with ℓ_M, ψ, and the action of N_0.
The following properties are equivalent:

- $m \in M_s^{bd}$,
- $\{\ell_M(\psi^k(u^{-1} m))$ for $k \in \mathbb{N}$, $u \in N_0\} \subset D$ is bounded,
- $\{\ell_{M/p^n M}(\psi^k(u^{-1} p_n(m)))$ for $k \in \mathbb{N}, u \in N_0\} \subset D/p^n D$ is bounded
  for all positive integers n,

a. $p_n(m) \in (M/p^n M)_s^{bd}$ for all positive integers n.
We deduce that $m \mapsto (p_n(m))_n : M_s^{bd} \rightarrow \varprojlim_n (M/p^n M)_s^{bd}$ is an
isomorphism.

$\square$

Proposition 7.132. $D = D_s^{bd}$ and M_s^{bd} contains $\iota_D(D)$.

Proof. (i) We show that $D = D_s^{bd}$. By Lemma 7.131, we can suppose that D
is killed by a power of p. Let $d \in D$. By Corollary 7.95, for $n \in \mathbb{N}$, there
exists $k_0 \in \mathbb{N}$ such that $\psi^k(v^{-1} d) \in D^\sharp$ for $k \geq k_0$, $v \in N_0^{(2)}$. As $D^\sharp \subset D$
is bounded, and as the set of $\psi^k(v^{-1} d)$ for all $0 \leq k < k_0$, $v \in N_0^{(2)}$, is
also bounded because the set of $v^{-1} d$ for $v \in N_0^{(2)}$ is bounded and ψ^k is
continuous, we deduce that $d \in D_s^{bd}$.

(ii) We show that M_s^{bd} contains $\iota_D(D)$ by showing

$$\{\ell_M(\psi^k(u^{-1} \iota_D(d)))$ for $k \in \mathbb{N}, u \in N_0\}$$
$$= \{\psi^k(v^{-1} d)$ for $k \in \mathbb{N}, v \in N_0^{(2)}\}$$

when $d \in D$ (the right-hand side is bounded in D by (i)). We write an
element of N_0 as $\iota(v)u$ for u in N_ℓ and $v \in N_0^{(2)}$. By Lemma 7.112,

$$\psi^k(u^{-1} \iota(v)^{-1} \iota_D(d)) = \psi^k(u^{-1} \iota_D(v^{-1} d)) = s^{-k} u^{-1} s^k \psi^k(\iota_D(v^{-1} d))$$

when $u \in s^k N_\ell s^{-k}$ and is 0 when u is not in $s^k N_\ell s^{-k}$. When $u \in$
$s^k N_\ell s^{-k}$ we have $\ell_M(s^{-k} u^{-1} s^k \psi^k(\iota_D(v^{-1} d))) = \psi^k(v^{-1} d)$ as ι_D is
ψ-equivariant.

$\square$

Proposition 7.133. M_s^{bd} *is dense in* M.

Proof. $M_s^{bd} \subset M$ is an $o[N_0]$-submodule, which by Proposition 7.132 contains $\iota_D(D)$. The $o[N_0]$-submodule of M generated by $\iota_D(D)$ is dense by Lemma 7.103.   □

We summarize: we proved that $M_s^{bd} \subset M$ is a dense $o[N_0]$-submodule, stable by L_+, and the action of L_+ on M_s^{bd} is étale.

Remark 7.134. It follows from Lemma 7.131 and the subsequent Proposition 7.135 that M_s^{bd} is a $\Lambda(N_0)$-submodule of M.

9.3. The special family $\mathfrak{C}_s$ when M is killed by a power of p

We suppose that M is killed by a power of p.

Proposition 7.135. 1. *For any lattice D_0 in D, the o-submodule*

$$M_s^{bd}(D_0) := \{m \in M \mid \ell_M(\psi^k(u^{-1}m)) \in D_0 \text{ for all } u \in N_0 \text{ and } k \in \mathbb{N}\}$$

of M is compact, and is a ψ-stable $\Lambda(N_0)$-submodule.

2. *The family $\mathfrak{C}_s$ of compact subsets of M contained in $M_s^{bd}(D_0)$ for some lattice D_0 of D, is special (Definition 7.76), satisfies $\mathfrak{C}(5)$ (Proposition 7.83) and $\mathfrak{C}(6)$ (Proposition 7.84), and $M(\mathfrak{C}_s) = M_s^{bd}$ is a $\Lambda(N_0)$-submodule of M.*

Proof. 1. (a) As ℓ and ψ are continuous (Proposition 7.119) and $D_0 \subset D$ is closed, it follows that $M_s^{bd}(D_0)$ is an intersection of closed subsets in M, hence $M_s^{bd}(D_0)$ is closed in M. As $M_s^{bd}(D_0)$ is an $o[N_0]$-submodule of M and $o[N_0]$ is dense in $\Lambda(N_0)$ we deduce that $M_s^{bd}(D_0)$ is a $\Lambda(N_0)$-submodule. It is ψ-stable by (7.63). The weak topology on M is the projective limit of the weak topologies on $M/\mathcal{M}_\ell(N_0)^n M$, and we have ([2] I.29 Corollary)

$$M_s^{bd}(D_0) = \varprojlim_{n \geq 1}(M_s^{bd}(D_0) + \mathcal{M}_\ell(N_0)^n M)/\mathcal{M}_\ell(N_0)^n M.$$

Therefore it suffices to show that

$$(M_s^{bd}(D_0) + \mathcal{M}_\ell(N_0)^n M)/\mathcal{M}_\ell(N_0)^n M$$

is compact for each large n. We will show the stronger property that it is a finitely generated $\Lambda(N_0)$-module.

(b) We prove first that $M_s^{bd}(D_0)$ is the intersection of the $\Lambda(N_0)$-modules generated by the image by φ^k of the inverse image $\ell_M^{-1}(D_0)$ of D_0 in M, for $k \in \mathbb{N}$,

$$M_s^{bd}(D_0) = \bigcap_{k \in \mathbb{N}} \Lambda(N_0)\varphi^k(\ell_M^{-1}(D_0)). \tag{7.64}$$

The inclusion from left to right follows from the expansion (7.12), as $m \in M_s^{bd}(D_0)$ is equivalent to $m_{u,s^k} = \psi^k(u^{-1}m) \in \ell_M^{-1}(D_0)$ for all $u \in N_0$ and $k \in \mathbb{N}$. The inclusion from right to left follows from

$$\ell_M \psi^k u^{-1} (\Lambda(N_0)\varphi^k(\ell_M^{-1}(D_0))) = D_0.$$

(c) We pick a lattice M_0 of M such that $\ell_M^{-1}(D_0) = M_0 + J_\ell(N_0)M$, as $J_\ell(N_0)M$ is the kernel of ℓ_M. By Lemma 7.114 we can choose for each $n \in \mathbb{N}$ a large integer r such that $\varphi^r(J_\ell(N_0)M) \subseteq \mathcal{M}_\ell(N_0)^n M$. Therefore we have

$$M_s^{bd}(D_0) \subseteq \Lambda(N_0)\varphi^r(M_0 + J_\ell(N_0)M)$$
$$\subseteq \Lambda(N_0)\varphi^r(M_0) + \mathcal{M}_\ell(N_0)^n M.$$

We deduce

$$(M_s^{bd}(D_0) + \mathcal{M}_\ell(N_0)^n M)/\mathcal{M}_\ell(N_0)^n M$$
$$\subseteq (\Lambda(N_0)\varphi^r(M_0) + \mathcal{M}_\ell(N_0)^n M)/\mathcal{M}_\ell(N_0)^n M.$$

The right term is a finitely generated $\Lambda(N_0)$-module hence the left term is finitely generated as a $\Lambda(N_0)$-module since $\Lambda(N_0)$ is noetherian.

2. The family is stable by finite union because a finite sum of lattices is a lattice. If $C \in \mathfrak{C}_s$ then $N_0 C \in \mathfrak{C}_s$ because $M_s^{bd}(D_0)$ is a $\Lambda(N_0)$-module. We have

$$M(\mathfrak{C}_s) = \cup_{D_0} M_s^{bd}(D_0) = M_s^{bd},$$

when D_0 runs over the lattices of D, the last follows from the fact that a bounded subset of D is contained in a lattice (this is the only part in the proof where the assumption that M is killed by a power of p is used). Apply Proposition 7.130.

Property $\mathfrak{C}(5)$ is immediate because $M_s^{bd}(D_0)$ is ψ-stable. Property $\mathfrak{C}(6)$ follows from $\varphi(M_s^{bd}(D_0)) \subset M_s^{bd}(D_s)$ where D_s is the lattice of D generated by $\varphi(D_0)$ (this uses part (a) of the proof of Proposition 7.130).

$\square$

Consider the lattice $M^{++} = \Lambda(N_0)i_D(D^{++})$ of M. Since $\Lambda(N_0)$ and D^{++} are φ-stable and since φ and ι_D commute, M^{++} is φ-stable and $\ell_M(M^{++}) = D^{++}$. Hence for a subset $S \subset M$ we have

$$S \subset X^r M^{++} + J_\ell(N_0)M \Leftrightarrow \ell_M(S) \subset (X^{(2)})^r D^{++}. \tag{7.65}$$

Proposition 7.136. *Let $r \in \mathbf{N}$, $C_+ \subset L_+$ a compact subset and let D_0 be a lattice in D. There is a compact open subgroup $P_1 \subset P_0$ and $k_0 \geq 0$ such that for all $k \geq k_0$*

$$s^k(1 - P_1)C_+ M_s^{bd}(D_0) \subset X^r M^{++} + \mathcal{M}_\ell(N_0)^r M.$$

Proof. Denote for simplicity $S = M_s^{bd}(D_0)$. By definition $\ell_M(S) \subset D_0$. Since $P_+^{(2)}$ acts continuously on D, $\ell(C_+)D_0$ is compact and $(X^{(2)})^r D^{++}$ is open, there is a compact open subgroup $P_1^{(2)} \subset P_+^{(2)}$ such that $(1 - P_1^{(2)})\ell(C_+)D_0 \subset (X^{(2)})^r D^{++}$. We may choose a compact open subgroup P_1 of P_0 such that $\ell(P_1) \subset P_1^{(2)}$, hence $\ell_M((1 - P_1)C_+ S) \subset (X^{(2)})^r D^{++}$. Relation (7.65) yields

$$(1 - P_1)C_+ S \subset J_\ell(N_0)M + X^r M^{++}.$$

Choosing k_0 such that $\varphi^k(J_\ell(N_0)) \subset \mathcal{M}_\ell(N_0)^r$ for $k \geq k_0$ (as we may by Lemma 7.114), the result follows from the φ-stability of $X^r M^{++}$(which follows from the fact that $\varphi(X^r) \in X^r \Lambda(N_0)$ and $\varphi(M^{++}) \subset M^{++}$). $\square$

Corollary 7.137. *Property $\mathfrak{T}(1)$ in Proposition 7.83 is satisfied.*

Proof. Let $C, C_+, \mathcal{M}$ as in Proposition 7.83 and choose r such that $\mathcal{M}_\ell(N_0)^r M + X^r M^{++} \subset \mathcal{M}$. Choose a lattice D_0 such that $C \subset M_s^{bd}(D_0)$. As $M_s^{bd}(D_0)$ is ψ-stable (Proposition 7.135), we can choose the subgroup P_1 and $k(C, \mathcal{M}, C_+) = k_0$ given by Proposition 7.136. $\square$

Recall that we defined (7.42) operators $s_g^{(k)} = \mathcal{H}_g^{(k+1)} - \mathcal{H}_g^{(k)}$ on $M_s^{bd} = M(\mathfrak{C}_s)$. From now on we fix a lattice D_0 in D and $g \in N_0 \overline{P} N_0$. We denote $S = M_s^{bd}(D_0)$.

Corollary 7.138. *There is $k_g \geq 0$ such that for all $x \in \mathbf{N}$ and $k \geq x + k_g$*

$$\psi^x \circ N_0 \circ s_g^{(k)}(S) \subset \ell_M^{-1}(D^{++}).$$

Proof. Let $r = 1$, $C_+ = \Lambda_g s$ and choose P_1 and k_0 as in Proposition 7.136, so that $s^k(1 - P_1)C_+ S \subset \ell_M^{-1}(D^{++})$ for $k \geq k_0$. Since S is $\Lambda(N_0)[\psi]$-stable and $\ell_M^{-1}(D^{++})$ is $o[N_0]$-stable, relation (7.43) and the inequality $k_g^{(2)}(P_1) \geq k_g^{(1)}$ yield $s_g^{(k)}(S) \subset N_0 \varphi^x \ell_M^{-1}(D^{++})$ for $k \geq x + k_0 + k_g^{(2)}(P_1)$. Applying $\psi^x \circ N_0$ yields the desired result, with $k_g = k_0 + k_g^{(2)}(P_1)$. $\square$

Lemma 7.139. *There is a lattice D_1 in D such that for all $u \in U_g$, $k \geq k_g \geq k_g^{(1)}$ and $x \geq k - k_g$*

$$\psi^x \circ N_0 \circ \alpha(g, x_u) \circ \mathrm{Res}(1_{uN_k})(S) \subset \ell_M^{-1}(D_1). \qquad (7.66)$$

Proof. Let $C'_+ = s^{k_g} t(g, U_g)$. This is a compact subset of L_+, since $k_g \geq k_g^{(1)}$. Since S is $\Lambda(N_0)[\psi]$-stable, we have $\mathrm{Res}(1_{uN_k})(S) \subset u \circ \varphi^k(S)$, hence

$$\alpha(g, x_u) \circ \mathrm{Res}(1_{uN_k})(S) \subset N_0 \circ t(g, u) \circ \varphi^k(S) \subset N_0 \varphi^{k-k_g}(C'_+ S).$$

Hence the left-hand side of (7.66) is contained in $\psi^{x-k+k_g}(\Lambda(N_0) C'_+ S)$, which is a subset of $\Lambda(N_0) C'_+ \ell_M^{-1}(D_0)$, because $S \subset \Lambda(N_0) \varphi^{x-k+k_g}(\ell_M^{-1}(D_0))$ and $C'_+ \Lambda(N_0) \subset \Lambda(N_0) C'_+$. Thus

$$\ell_M(\psi^x \circ N_0 \circ \alpha(g, x_u) \circ \mathrm{Res}(1_{uN_k})(S)) \subset \Lambda(N_0^{(2)}) \ell(C'_+)(D_0)$$

and the last subset of D is compact, hence contained in some lattice D_1. □

Corollary 7.140. *For all* $k \geq k_g$ *we have* $\mathcal{H}_g^{(k)}(S) \subset M_s^{bd}(D_1 + D^{++})$. *Moreover,* $\mathcal{H}_g(S) \subset M_s^{bd}(D_1 + D^{++})$.

Proof. The second assertion follows from the first by letting $k \to \infty$, since $M_s^{bd}(D_1 + D^{++})$ is closed in M. For the first assertion, we need to prove that $\psi^x(N_0 \mathcal{H}_g^{(k)}(S)) \subset \ell_M^{-1}(D_1 + D^{++})$ for all $x \geq 0$ and $k \geq k_g$. Fix $x \geq 0$. If $k \leq k_g + x$, simply add all relations (7.66) for $u \in J(U_g/N_k)$. If $k > k_g + x$, the equation $\mathcal{H}_g^{(k)} = \mathcal{H}_g^{(x+k_g)} + \sum_{j=x+k_g}^{k-1} s_g^{(j)}$ and Corollary 7.138 show that

$$\psi^x(N_0 \mathcal{H}_g^{(k)}(S)) \subset \psi^x(N_0 \mathcal{H}_g^{(x+k_g)}(S)) + \ell_M^{-1}(D^{++}).$$

But we have already seen that $\psi^x(N_0 \mathcal{H}_g^{(x+k_g)}(S)) \subset \ell_M^{-1}(D_1 + D^{++})$. □

Proposition 7.141. *All the assumptions of Proposition 7.84 are satisfied.*

Proof. Property $\mathfrak{T}(1)$ was checked in Corollary 7.137. Property $\mathfrak{T}(2)$ and the fact that $\mathcal{H}_g$ preserves M_s^{bd} (for $g \in N_0 \overline{P} N_0$) follow from Corollary 7.140 and the fact that any $m \in M_s^{bd}$ is in $S = M_s^{bd}(D_0)$ for some lattice D_0 in D. □

9.4. Functoriality and dependence on *s*

Let $Z(L)_{\dagger\dagger} \subset Z(L)$ be the subset of elements s such that $L = L_- s^{\mathbb{N}}$ and $(s^k N_0 s^{-k})_{k \in \mathbb{Z}}$ and $(s^{-k} w_0 N_0 w_0^{-1} s^k)_{k \in \mathbb{Z}}$ are decreasing sequences of trivial intersection and union N and $w_0 N w_0^{-1}$, respectively (see Section 6).

Let M be a topologically etale L_+-module over $\Lambda_\ell(N_0)$ and let $D := \mathbb{D}(M)$. We have $D/p^n D = \mathbb{D}(M/p^n M)$ for $n \geq 1$. By Lemma 7.104, M satisfies the properties a,b,c,d of Subsection 6.5 and is complete (the same is true for $M/p^n M$). The image $D_{0,n}$ in $D/p^n D$ of any lattice $D_{0,n+1}$ in $D/p^{n+1} D$ is a lattice and the maps ℓ and ψ commute with the reduction modulo p^n, hence $(M/p^{n+1} M)_s^{bd}(D_{0,n+1})$ maps into $(M/p^n M)_s^{bd}(D_{0,n})$. Therefore the special family $\mathfrak{C}_{s,n+1}$ in $M/p^{n+1} M$ maps to the special family $\mathfrak{C}_{s,n}$ in $M/p^n M$. As in Lemma 7.89 we define the special family $\mathfrak{C}_s$ in M to consist of all compact subsets $C \subset M$ such that $p_n(C) \in \mathfrak{C}_{s,n}$ for all $n \geq 1$. By Proposition 7.135 and Lemma 7.131 we have

$$M(\mathfrak{C}_s) = M_s^{bd}.$$

Theorem 7.142. *Let* $s \in Z(L)_{\dagger\dagger}$ *and* $M \in \mathcal{M}_{\Lambda\ell(N_0)}^{et}(L_+)$.

(i) *The* $(s, \text{res}, \mathfrak{C}_s)$*-integrals* $\mathcal{H}_{g,s}$ *of the functions* $\alpha_{g,0}|_{M_s^{bd}}$ *for* $g \in N_0\overline{P}N_0$ *exist, lie in* $\text{End}_o(M_s^{bd})$, *and satisfy the relations H1, H2, H3 of Proposition 7.72.*

(ii) *The map* $M \mapsto (M_s^{bd}, (\mathcal{H}_{g,s})_{g \in N_0\overline{P}N_0})$ *is functorial.*

Proof. (i) By Proposition 7.141 the assumptions of Proposition 7.91 are satisfied.

(ii) Let $f : M \to M'$ be a morphism in $\mathcal{M}_{\Lambda\ell(N_0)}^{et}(L_+)$. For $m \in M$ we denote $E_s(m) = \{\ell_M(\psi_s^k u^{-1}m) \text{ for } u \in N_0, k \in \mathbb{N}\}$. We have

$$\mathbb{D}(f)(E_s(m)) = E_s(f(m)) \text{ when } m \in M, \tag{7.67}$$

because the maps $\ell_M : M \to D$ and $\ell_{M'} : M' \to D'$ sending x to $1 \otimes x$ for $x \in M$ or $x \in M'$ satisfy $\ell_{M'} \circ f = \mathbb{D}(f) \circ \ell_M$, and f is P^--equivariant by Lemma 7.22. Any morphism between finitely generated modules on $\mathcal{O}_{\mathcal{E}}$ is continuous for the weak topology (cf. [12] Lemma 8.22). The image of a bounded subset by a continuous map is bounded. We deduce from (7.67) that $E_s(m)$ bounded implies $E_s(f(m))$ bounded, equivalently $m \in M_s^{bd}$ implies $f(m) \in M_s'^{bd}$. For $m \in M_s^{bd}$ we have $f(\mathcal{H}_{g,s}(m)) = \mathcal{H}_{g,s}(f(m))$ where

$$\mathcal{H}_{g,s}(.) = \lim_{k \to \infty} \sum_{u \in J(N_0/s^k N_0 s^{-k})} n(g, u)\varphi_{t(g,u)s^k} \psi_s^k u^{-1}(.),$$

because f is P_+ and P_--equivariant by Lemma 7.22.

$\square$

We investigate now the dependence on $s \in Z(L)_{\dagger\dagger}$ of the dense subset $M_s^{bd} \subseteq M$ and of the $(s, \text{res}, \mathfrak{C}_s)$-integrals $\mathcal{H}_{g,s}$.

Lemma 7.143. $Z(L)_{\dagger\dagger}$ *is stable by product.*

Proof. Let $s, s' \in Z(L)_{\dagger\dagger}$. Clearly $L_-s'^n = L_-s^{-n}s^n s'^n \subset L_-(ss')^n$ because L_- is a monoid and $s^{-1} \in Z(L)_- = Z(L) \cap L_-$. Therefore $L = L_-(ss')^{\mathbb{N}}$. The sequence $((ss')^k N_0(ss')^{-k})_{k \in \mathbb{Z}}$ is decreasing because

$$s'^{k+1}s^{k+1}N_0 s^{-k-1}s'^{-k-1} \subset s'^k s^{k+1}N_0 s^{-k-1}s'^{-k} \subset s'^k s^k N_0 s^{-k}s'^{-k}.$$

The intersection is trivial and the union is N because $s'^k s^k N_0 s^{-k} s'^{-k} \subset s^k N_0 s^{-k}$ when $k \in \mathbb{N}$ and $s'^k s^k N_0 s^{-k} s'^{-k} \supset s^k N_0 s^{-k}$ when $-k \in \mathbb{N}$. One makes the same argument with $w_0 N_0 w_0^{-1}$. $\qquad\qquad\qquad\qquad\qquad\qquad\qquad\qquad\qquad\quad\square$

Lemma 7.144. (i) *The action of* $t_0 \in \ell^{-1}(L_0^{(2)}) \cap L_+$ *on D is invertible.*

(ii) *There exists a treillis* D_0 *in D which is stable by* $\ell^{-1}(L_0^{(2)}) \cap L_+$.

Proof. (i) is true because the action of t_0 on D is étale and $N_0^{(2)} = \ell(t_0) N_0^{(2)} \ell(t_0)^{-1}$.

(ii) Let $s \in Z(L)_{\dagger\dagger}$ and let ψ_s be the canonical inverse of the étale action φ_s of s on D. We show that the minimal ψ_s-stable treillis $D^\natural$ of D (Proposition 7.93(iii)) is stable by $\ell^{-1}(L_0^{(2)}) \cap L_+$.

For $t_0 \in \ell^{-1}(L_0^{(2)}) \cap L_+$ we claim that $\varphi_{t_0}(D^\natural)$ is also a ψ_s-stable treillis in D. We have $\psi_s \psi_{t_0} = \psi_{t_0} \psi_s$ as $t_0 \in Z(L)$. Multiplying by φ_{t_0} on both sides, one gets $\varphi_{t_0} \psi_s \psi_{t_0} \varphi_{t_0} = \varphi_{t_0} \psi_{t_0} \psi_s \varphi_{t_0}$. Since ψ_{t_0} is the two-sided inverse of φ_{t_0} by (i) we get that φ_{t_0} and ψ_s commute. Hence $\varphi_{t_0}(D^\natural)$ is a compact o-module which is ψ_s-stable. It is a $\Lambda(N_0^{(2)})$-module because any $\lambda \in \Lambda(N_0^{(2)})$ is of the form $\lambda = \varphi_{\ell(t_0)}(\mu)$ for some $\mu \in \Lambda(N_0^{(2)})$ and $\lambda \varphi_{t_0}(d) = \varphi_{t_0}(\mu d)$ for all $d \in D$. As $D^\natural$ contains a lattice and φ_{t_0} is étale, we deduce that $\varphi_{t_0}(D^\natural)$ contains a lattice and therefore is a treillis. By the minimality of $D^\natural$ we must have

$$D^\natural \subset \varphi_{t_0}(D^\natural).$$

Similarly one checks that $\psi_{t_0}(D^\natural)$ is a treillis. It is ψ_s-stable because ψ_s and ψ_{t_0} commute. Hence

$$D^\natural \subset \psi_{t_0}(D^\natural).$$

Applying φ_{t_0} which is the two-sided inverse of ψ_{t_0} we obtain $\varphi_{t_0}(D^\natural) \subset D^\natural$ hence $D^\natural = \varphi_{t_0}(D^\natural)$.

$\qquad\qquad\qquad\qquad\qquad\qquad\qquad\qquad\qquad\qquad\qquad\qquad\qquad\quad\square$

We denote by $Z(L)_{\dagger} \subset Z(L)$ the monoid of $z \in Z(L)_+ = Z(L) \cap L_+$ such that $z^{-1} w_0 N_0 w_0^{-1} z \subset w_0 N_0 w_0^{-1}$. We have $Z(L)_{\dagger\dagger} Z(L)_{\dagger} \subset Z(L)_{\dagger\dagger}$.

Note that $L_0^{(2)}$ contains the center of $GL(2, \mathbb{Q}_p)$ and that $Z(L^{(2)})_{\dagger} = L_+^{(2)}$.

For $m \in M, t \in L_+, u \in U$, and a system of representatives $J(N_0/t N_0 t^{-1}) \subset N_0$ for the cosets in $N_0/t N_0 t^{-1}$ we have (7.12)

$$m = \sum_{u \in J(N_0/t N_0 t^{-1})} u \mu_{t,u}, \quad \mu_{t,u} := \varphi_t \psi_t(u^{-1} m). \qquad (7.68)$$

For $g \in N_0 \overline{P} N_0$ and $s \in Z(L)_{\dagger\dagger}$, we have the smallest positive integer $k_{g,s}^{(0)}$ as in (7.27). For $k \geq k_{g,s}^{(0)}$, we have $\mathcal{H}_{g,s,J(N_0/N_k)} \in \mathrm{End}_o^{cont}(M)$ where (compare with (7.28))

$$\mathcal{H}_{g,s,J(N_0/N_k)}(m) = \sum_{u \in J(U_g/N_k)} n(g,u)t(g,u)\mu_{s^k,u}. \tag{7.69}$$

When $m \in M_s^{bd}$, the integral $\mathcal{H}_{g,s}(m)$ is the limit of $\mathcal{H}_{g,s,J(N_0/N_k)}(m)$ by Theorem 7.142 and (7.29).

Proposition 7.145. *Let $s \in Z(L)_{\dagger\dagger}$, $t_0 \in \ell^{-1}(L_0^{(2)}) \cap Z(L)_{\dagger}$ and r a positive integer.*

(i) *We have $M_{st_0}^{bd} \subseteq M_s^{bd} = M_{s^r}^{bd}$.*
(ii) *For $g \in N_0 \overline{P} N_0$ we have $\mathcal{H}_{g,s} = \mathcal{H}_{g,st_0}$ on $M_{st_0}^{bd}$ and $\mathcal{H}_{g,s} = \mathcal{H}_{g,s^r}$ on M_s^{bd}.*

Proof. (a) Note that st_0 and s^r in proposition belong also to $Z(L)_{\dagger\dagger}$.
For a treillis D_0 in D which is stable by $\ell^{-1}(L_0^{(2)}) \cap L_+$ (Lemma 7.144), $(X^{(2)})^{-r}D_0$ is a treillis in D; it is also stable by $t_0 \in \ell^{-1}(L_0^{(2)}) \cap L_+$ because

$$\varphi_{\ell(t_0)}((X^{(2)})^{-r}\Lambda(N_0^{(2)})) = \varphi_{\ell(t_0)}((X^{(2)})^{-r})\varphi_{\ell(t_0)}(\Lambda(N_0^{(2)}))$$
$$= (X^{(2)})^{-r}\Lambda(N_0^{(2)}).$$

When M is killed by a power of p, this implies with Proposition 7.135 that M_s^{bd} is the union of $M_s^{bd}(D_0)$ when D_0 runs over the lattices of D which are stable by $\ell^{-1}(L_0^{(2)}) \cap L_+$.

(b) We suppose from now on, as we can by Lemma 7.131, that M is killed by a power of p to prove $M_{st_0}^{bd} \subset M_s^{bd} = M_{s^r}^{bd}$. Let $m \in M_{st_0}^{bd}(D_0)$ where D_0 is a $\ell^{-1}(L_0^{(2)}) \cap L_+$-stable lattice of D. For $u \in N_0$ and $k \in \mathbb{N}$, using (7.12) for $t = t_0^k$ we obtain that

$$\ell_M(\psi_s^k(u^{-1}m)) = \ell_M\Big(\sum_{v \in J(N_0/t_0^k N_0 t_0^{-k})} v \circ \varphi_{t_0}^k \circ \psi_{t_0}^k \circ v^{-1} \circ \psi_s^k(u^{-1}m)\Big)$$

$$= \sum_{v \in J(N_0/t_0^k N_0 t_0^{-k})} \ell(v)\varphi_{t_0}^k(\ell_M(\psi_{st_0}^k(\varphi_s^k(v^{-1})u^{-1}m)))$$

lies in D_0, since D_0 is both $N_0^{(2)}$- and φ_{t_0}-invariant and $\ell_M(\psi_{st_0}^k(u'm)) \in D_0$ for $u' \in N_0$. Therefore $M_{st_0}^{bd}(D_0) \subset M_s^{bd}(D_0)$ and by (a) we deduce $M_{st_0}^{bd} \subset M_s^{bd}$.

For any $m \in M$ we observe that

$$\{\ell_M(\psi_{sr}^k(u^{-1}m)) \text{ for } k \in \mathbb{N}, u \in N_0\} \subset \{\ell_M(\psi_s^k(u^{-1}m)) \text{ for } k \in \mathbb{N}, u \in N_0\},$$

as $\psi_{sr}^k = \psi_s^{rk}$. We deduce that $M_s^{bd}(D_0) \subset M_{sr}^{bd}(D_0)$ for any lattice D_0 of D hence $M_s^{bd} \subset M_{sr}^{bd}$. Conversely, for $k_1 \in \mathbb{N}$ we write $k_1 = rk - k_2$ with $k \in \mathbb{N}$ and $0 \le k_2 < r$ and we observe that

$$\ell_M(\psi_s^{k_1}(u^{-1}m)) = \ell_M\Big(\sum_{v \in J(N_0/s^{k_2} N_0 s^{-k_2})} v \circ \varphi_s^{k_2} \circ \psi_s^{rk}(\varphi_s^{k_1}(v^{-1})u^{-1}m))$$

$$= \sum_{v \in J(N_0/s^{k_2} N_0 s^{-k_2})} \ell(v)\varphi_s^{k_2}(\ell_M(\psi_{sr}^k(\varphi_s^{k_1}(v^{-1})u^{-1}m))).$$

The $\Lambda(N_0^{(2)})$-submodule D_r generated by $\sum_{i=1}^{r-1} \varphi_s^i(D_0)$ is a lattice because the action φ_s of s on D is étale. We deduce that $M_{sr}^{bd}(D_0) \subset M_s^{bd}(D_r)$ since $\ell_M(\psi_{sr}^k(u'm)) \in D_0$ for $u' \in N_0, m \in M_{sr}^{bd}(D_0)$. Therefore $M_{sr}^{bd}(D_0) \subset M_s^{bd}(D_r)$ hence $M_{sr}^{bd} \subset M_s^{bd}$. It is obvious that $\mathcal{H}_{g,s} = \mathcal{H}_{g,sr}$ on M_s^{bd}.

(c) Let $g \in N_0 \overline{P} N_0, k \ge k_{g,s}^{(0)}, t_0 \in \ell^{-1}(L_0^{(2)}) \cap Z(L)_+$ and $r \ge 1$. We have

$$k_{g,st_0}^{(0)} \le k_{g,s}^{(0)}, \quad k_{g,sr}^{(0)} \le k_{g,s}^{(0)}$$

because $(st_0)^k N_0 (st_0)^{-k} \subset N_k$ and $(s^r)^k N_0 (s^r)^{-k} = N_{kr} \subset N_k$. Let d in D and $v \in N_0$. By (7.12) we have

$$d = \sum_{u \in J(N_0^{(2)}/\ell(st_0)^k N_0^{(2)} \ell(st_0)^{-k})} u\varphi_{st_0}^k \circ \psi_{st_0}^k(u^{-1}d)$$

$$= \sum_{u \in J(N_0^{(2)}/\ell(s)^k N_0^{(2)} \ell(s)^{-k})} u\varphi_s^k \circ \psi_s^k(u^{-1}d),$$

with the second equality holding true summand per summand, because ψ_{t_0} is the left and right inverse of φ_{t_0} on D (Lemma 7.144 (i)) and $\ell(t_0)N_0^{(2)}\ell(t_0)^{-1} = N_0^{(2)}$. Since ι_D commutes with φ_t and ψ_t for $t \in L_+$, this implies

$$\iota_D(d) = \sum_{u \in J(N_0^{(2)}/\ell(st_0)^k N_0^{(2)} \ell(st_0)^{-k})} \iota(u)\varphi_{st_0}^k \circ \psi_{st_0}^k(\iota(u)^{-1}\iota_D(d))$$

$$= \sum_{u \in J(N_0^{(2)}/\ell(s)^k N_0^{(2)} \ell(s)^{-k})} \iota(u)\varphi_s^k \circ \psi_s^k(\iota(u)^{-1}\iota_D(d)),$$

again with the second equality holding true summand per summand. We choose, as we can, systems of representatives $J(N_0/(st_0)^k N_0(st_0)^{-k})$ and

$J(N_0/s^k N_0 s^{-k})$ containing $\iota(J(N_0^{(2)}/\ell(s)^k N_0^{(2)} \ell(s)^{-k}))$. For $k \geq k_{g,s}^{(0)} \geq k_{g,st_0}^{(0)}$, we obtain

$$\mathcal{H}_{g,st_0,vJ(N_0/(st_0)^k N_0(st_0)^{-k})}(\upsilon \iota_D(d)) = \mathcal{H}_{g,s,vJ(N_0/s^k N_0 s^{-k})}(\upsilon \iota_D(d)).$$

Passing to the limit when k goes to infinity, and using linearity we deduce that $\mathcal{H}_{g,st_0} = \mathcal{H}_{g,s}$ on the $o[N_0]$-submodule $< N_0 \iota_D(D) >_o$ generated by $\iota_D(D)$ in $M_{st_0}^{bd}$.

(d) Let $m \in M_s^{bd}(D_1)$ with $D_1 \subset D$ a ψ_s-stable lattice (Proposition 7.93 (iv)). For a positive integer k, and a set of representatives $J(N_0/s^k N_0 s^{-k})$, we write m in the form (7.12)

$$m = \sum_{u \in J(N_0/s^k N_0 s^{-k})} u\varphi_s^k(\iota_D(d(s,u)) + m(s,u))$$

with $m(s,u)$ in $J_\ell(N_0)M$ and $d(s,u) = \ell_M(\psi_s^k(u^{-1}m))$ in D_1. Then

$$m(s) := \sum_{u \in J(N_0/s^k N_0 s^{-k})} u\varphi_s^k(\iota_D(d(s,u))) \text{ lies in } < N_0 \iota_D(D) >_o$$

because ι_D is L_+-equivariant. Moreover $m - m(s)$ is contained in the $o[N_0]$-submodule $N_0 \varphi_s^k(J_\ell(N_0)M)$ generated by $\varphi_s^k(J_\ell(N_0)M)$. We show that

$$m(s) \in M_s^{bd}(D_1). \tag{7.70}$$

For $v \in N_0$ and $r \leq k$ we have

$$\psi_s^r(v^{-1}(m - m(s))) = \psi_s^r(v^{-1} \sum_{u \in J(N_0/s^k N_0 s^{-k})} u\varphi_s^k(m(s,u)))$$

$$= \sum_{u \in J(N_0/s^k N_0 s^{-k})} \psi_s^r(v^{-1}u)\varphi_s^{k-r}(m(s,u))$$

which lies in $J_\ell(N_0)M$ since $m(s,u)$ is in $J_\ell(N_0)M$ and $J_\ell(N_0)M$ is N_0 and φ_s-stable. This shows that $\ell_M(\psi_s^r(v^{-1}m(s))) = \ell_M(\psi_s^r(v^{-1}m))$ lies in D_1. On the other hand, for $r > k$ we have

$$\ell_M(\psi_s^r(v^{-1}m(s))) = \ell_M(\psi_s^r(v^{-1} \sum_{u \in J(N_0/s^k N_0 s^{-k})} u\varphi_s^k(\iota_D(d(s,u)))))$$

$$= \sum_{u \in J(N_0/s^k N_0 s^{-k})} \ell_M(\psi_s^{r-k}(\psi_s^k(v^{-1}u)\iota_D(d(s,u))))$$

which lies in D_1. Indeed, since D_1 is ψ_s-stable the formula in part (ii) of the proof of Proposition 7.132 implies that $\iota_D(D_1) \subseteq M_s^{bd}(D_1)$; hence the $\iota_D(d(s,u))$ lie in the ψ_s- and N_0-invariant subspace $M_s^{bd}(D_1)$. We conclude that $m(s) \in M_s^{bd}(D_1)$.

Therefore, for any ψ_{st_0}-stable lattice $D_1 \subset D$, any $k \geq 1$, and any set of representatives $J(N_0/(st_0)^k N_0(st_0)^{-k})$, we have defined an o-linear homomorphism

$$m \mapsto m(st_0) \quad M_{st_0}^{bd}(D_1) \to M_{st_0}^{bd}(D_1) \cap \; < N_0 \iota_D(D) >_o$$

such that

$$m - m(st_0) \in M_{st_0}^{bd}(D_1) \cap \varphi_{st_0}^k(J_\ell(N_0)M). \tag{7.71}$$

By (c) we have $\mathcal{H}_{g,st_0}(m(st_0)) = \mathcal{H}_{g,s}(m(st_0))$ for $m \in M_{st_0}^{bd}(D_1)$.

(e) To end the proof that $\mathcal{H}_{g,st_0} = \mathcal{H}_{g,s}$ on $M_{st_0}^{bd}(D_1)$ we use the $\mathfrak{C}_s$-uniform convergence of $(\mathcal{H}_{g,s,J(N_0/s^k N s^{-k})})_k$. We fix systems of representatives $J(N_0/(st_0)^k N_0(st_0)^{-k})$ and $J(N_0/s^k N s^{-k})$, for any $k \geq 1$. We also choose a lattice $D_0 \subset D$ which is stable by $\ell^{-1}(L_0^{(2)}) \cap L_+$ and such that $D_1 \subset D_0$. We recall that $M_{st_0}^{bd}(D_1)$ is compact (Proposition 7.135 i)) and that $M_{st_0}^{bd}(D_1) \subset M_{st_0}^{bd}(D_0) \subset M_s^{bd}(D_0)$ by (b). For any open $\Lambda(N_0)$-submodule in the weak topology $M_0 \subset M$, there exists a common constant $k_0 \geq k_{g,s}^{(0)} \geq k_{g,st_0}^{(0)}$ (by (c)) such that for $k \geq k_0$,

$$\mathcal{H}_{g,st_0,J(N_0/(st_0)^k N(st_0)^{-k})} \in \mathcal{H}_{g,st_0} + E(M_{st_0}^{bd}(D_1), M_0) \tag{7.72}$$

$$\mathcal{H}_{g,s,J(N_0/s^k N s^{-k})} \in \mathcal{H}_{g,s} + E(M_{st_0}^{bd}(D_1), M_0). \tag{7.73}$$

On the left-hand side of (7.72), (7.73), we have continuous endomorphisms of M. By Lemma 7.114, there exists an integer $k_1 \geq k_0$ such that they send $N_0 \varphi_{st_0}^{k_1}(J_\ell(N_0)M)$ into M_0. Therefore, for $m \in M_{st_0}^{bd}(D_1)$, they send the element $m - m(st_0)$ associated to k_1 and $J(N_0/((st_0)^{k_1} N_0(st_0)^{-k_1})$ as in (d) (7.71) into M_0 hence

$$\mathcal{H}_{g,st_0}(m - m(st_0)) \text{ and } \mathcal{H}_{g,s}(m - m(st_0)) \text{ lie in } M_0.$$

By (d) we obtain that $H_{g,st_0}(m) - H_{g,s_2}(m)$ lies in M_0 for $m \in M_{st_0}^{bd}(D_1)$. The statement follows since we chose M_0 to be an arbitrary open neighbourhood of zero in the weak topology of M.

$\square$

Definition 7.146. We define the transitive relation $s_1 \leq s_2$ on $Z(L)_{\dagger\dagger}$ generated by

$$s_1 = s_2 t_0 \text{ for } t_0 \in \ell^{-1}(L_0^{(2)}) \cap Z(L)_\dagger \quad \text{or} \quad s_1^{r_1} = s_2^{r_2} \text{ for positive integers } r_1, r_2.$$

Proposition 7.145 admits the following corollary.

Corollary 7.147. *Let $s_1, s_2 \in Z(L)_{\dagger\dagger}$.*

(i) When $s_1 \leq s_2$ we have $M_{s_1}^{bd} \subseteq M_{s_2}^{bd}$ and $\mathcal{H}_{g,s_1} = \mathcal{H}_{g,s_2}$ on $M_{s_1}^{bd}$.

(ii) *When the relation $\leq$ on $Z(L)_{\dagger\dagger}$ is right filtered, we have $\mathcal{H}_{g,s_1} = \mathcal{H}_{g,s_2}$ on* $M_{s_1}^{bd} \cap M_{s_2}^{bd}$.

Proof. (i) If $s_1 \leq s_2$ then there exists, by definition, a sequence $s_1 = s_1' \leq s_2' \leq \ldots \leq s_m' = s_2$ in $Z(L)_{\dagger\dagger}$ such that each pair s_i', s_{i+1}' satisfies one of the two conditions in Definition 7.146. Hence we may assume, by induction, that the pair s_1, s_2 satisfies one of these conditions, and we apply Proposition 7.145.

(ii) When there exists $s_3 \in Z(L)_{\dagger\dagger}$ such that $s_1 \leq s_3$ and $s_2 \leq s_3$, by (i) $M_{s_1}^{bd}$ and $M_{s_2}^{bd}$ are contained in $M_{s_3}^{bd}$ and $\mathcal{H}_{g,s_1} = \mathcal{H}_{g,s_2} = \mathcal{H}_{g,s_3}$ on $M_{s_1}^{bd} \cap M_{s_2}^{bd}$. □

Proposition 7.148. *We assume that the relation $\leq$ on $Z(L)_{\dagger\dagger}$ is right filtered. Then, the intersection and the union*

$$M_{\cap}^{bd} := \bigcap_{s \in Z(L)_{\dagger\dagger}} M_s^{bd} \subset M_{\cup}^{bd} := \bigcup_{s \in Z(L)_{\dagger\dagger}} M_s^{bd}$$

are dense étale L_+-submodules of M over $\Lambda(N_0)$.

For $g \in N_0 \overline{P} N_0$ the endomorphisms $\mathcal{H}_g \in \mathrm{End}_o(M_{\cup}^{bd})$ equal to $\mathcal{H}_{g,s}$ on M_s^{bd} for each $s \in Z(L)_{\dagger\dagger}$, are well defined, stabilize $M_{\cap}^{bd}$ and satisfy the relations H1, H2, H3 of Proposition 7.72.

Proof. $M_{\cap}^{bd}$ is an L_+-submodule of M over $\Lambda(N_0)$ by Proposition 7.130 and Remark 7.134. It is dense in M by Proposition 7.132 and Lemma 7.103. The action of L_+ on $M_{\cap}^{bd}$ is étale because $M_{\cap}^{bd}$ is L_--stable. When $\leq$ is right filtered, $M_{\cup}^{bd}$ is a $\Lambda_\ell(N_0)$-module by Corollary 7.147(i). For the same reasons as for $M_{\cap}^{bd}$, it is an étale L_+-submodule of M over $\Lambda(N_0)$.

By Corollary 7.147 the $\mathcal{H}_g$ are well defined and stabilize $M_{\cap}^{bd}$. They satisfy the relations H1, H2, H3 of Proposition 7.72 because the $\mathcal{H}_{g,s}$ satisfy them (Theorem 7.142). □

We summarize our results and give our main theorem.

Theorem 7.149. *For any $s \in Z(L)_{\dagger\dagger}$, we have a faithful functor*

$$\mathbb{Y}_s : \mathcal{M}_{\Lambda_\ell(N_0)}^{et}(L_+) \quad \to \quad \text{G-equivariant sheaves on } G/P,$$

which associates to $M \in \mathcal{M}_{\Lambda_\ell(N_0)}^{et}(L_+)$ the G-equivariant sheaf $\mathfrak{Y}_s$ on G/P such that $\mathfrak{Y}_s(\mathcal{C}_0) = M_s^{bd}$.

When the relation $\leq$ on $Z(L)_{\dagger\dagger}$ is right filtered, we have faithful functors

$$\mathbb{Y}_\cap, \mathbb{Y}_\cup : \mathcal{M}_{\Lambda_\ell(N_0)}^{et}(L_+) \quad \to \quad \text{G-equivariant sheaves on } G/P,$$

which associate to $M \in \mathcal{M}_{\Lambda_\ell(N_0)}^{et}(L_+)$ the G-equivariant sheaves $\mathfrak{Y}_\cap$ and $\mathfrak{Y}_\cup$ on G/P with sections on $\mathcal{C}_0$ equal to $\mathfrak{Y}_\cap(\mathcal{C}_0) = M_\cap^{bd}$ and $\mathfrak{Y}_\cup(\mathcal{C}_0) = M_\cup^{bd}$.

Proof. The existence of the functors results from Proposition 7.148, Theorem 7.142, Proposition 7.72, and Remark 7.69.

We show the faithfulness of the functors. For a non zero morphism $f : M \rightarrow M'$ in $\mathcal{M}^{et}_{\Lambda_{\ell}(N_0)}(L_+)$, we have $f(M^{bd}_{\cap}) \neq 0$ because f is continuous ([12] Lemma 8.22) and $M^{bd}_{\cap}$ containing $\Lambda(N_0)\iota_D(D)$ is dense (Proposition 7.148). We deduce $\mathbb{Y}_{\cap}(f) \neq 0$ since it is nonzero on sections on $\mathcal{C}_0$. A fortiori $\mathbb{Y}_s(f) \neq 0$, and $\mathbb{Y}_{\cup}(f) \neq 0$. $\qquad\square$

10. Connected reductive split group

We explain how our results apply to connected reductive groups.

(a) Let F be a locally compact non archimedean field of ring of integers o_F and uniformizer p_F. Let G be a connected reductive F-group, let S be a maximal F-split subtorus of G and let P be a parabolic F-subgroup of G with Levi component L containing S and unipotent radical N. Let $X^*(S)$ be the group of characters of S, let Φ_L, resp. Φ, be the subset of roots of S in L, resp. G, and let $\Phi_{+,N}$ be the subset of roots of S in N (we suppress the index N if P is a minimal parabolic F-subgroup of G).

Let s be any element of $S(F)$ such that $\alpha(s) = 1$ for $\alpha \in \Phi_L$ and the p-valuation of $\alpha(s) \in F^*$ is positive for all roots $\alpha \in \Phi_{+,N}$. *For any compact open subgroup N_0 of $N(F)$, the data $(P(F), L(F), N(F), N_0, s)$ satisfy all the conditions introduced in the section on étale P_+-modules, Section 3 and Subsection 3.2, the assumptions introduced in Section 6, and in Section 9.*

(b) We suppose that P is a minimal parabolic F-subgroup. Let $W \subset N_G(L)$ be a system of representatives of the Weyl group $N_G(L)/L$ and let w_0 be the longest element of the Weyl group. *The data $(G(F), P(F), W)$ satisfy the assumptions of Section 5 on G-equivariant sheaves on G/P.*

(c) We suppose until the end of this chapter that

$$F = \mathbb{Q}_p, \ G \text{ is } \mathbb{Q}_p\text{-split and } P \text{ is a Borel } \mathbb{Q}_p\text{-subgroup.}$$

The Levi subgroup $L = T$ of P is a split $\mathbb{Q}_p$-torus. The monoid of dominant elements and the submonoid without unit of strictly dominant elements are

$$T(\mathbb{Q}_p)_+ = \{t \in T(\mathbb{Q}_p), \ \alpha(t) \in \mathbb{Z}_p \text{ for all } \alpha \in \Delta\},$$
$$T(\mathbb{Q}_p)_{++} = \{t \in T(\mathbb{Q}_p), \ \alpha(t) \in p\mathbb{Z}_p - \{0\} \text{ for all } \alpha \in \Delta\}.$$

With our former notation $Z(L) = T(\mathbb{Q}_p)$, $Z(L)_{++} = T(\mathbb{Q}_p)_{++}$. For each root $\alpha \in \Phi$, let

$$u_\alpha : \mathbb{Q}_p \to N_\alpha(\mathbb{Q}_p), \ tu_\alpha(x)t^{-1} = u_\alpha(\alpha(t)x) \ \text{for} \ x \in \mathbb{Q}_p, t \in T(\mathbb{Q}_p), \tag{7.74}$$

be a continuous isomorphism from $\mathbb{Q}_p$ onto the root subgroup $N_\alpha(\mathbb{Q}_p)$ of $N(\mathbb{Q}_p)$ normalized by $T(\mathbb{Q}_p)$. We can write an element $u \in N(\mathbb{Q}_p)$ in the form

$$u = \prod_{\alpha \in \Phi_+} u_\alpha(x_\alpha)$$

for any ordering of Φ_+. The coordinates $x_\alpha = x_\alpha(u) \in \mathbb{Q}_p$ of u are determined by the ordering of the roots, but for a simple root α, the coordinate

$$x_\alpha : N(\mathbb{Q}_p) \to \mathbb{Q}_p \tag{7.75}$$

is independent of the choice of the ordering, and satisfies $u_\alpha \circ x_\alpha = 1$. We suppose, as we can, that the u_α have been chosen such that the product

$$N_0 = \prod_{\alpha \in \Phi_+} u_\alpha(\mathbb{Z}_p)$$

is a group for some ordering of Φ_+. Then N_0 is the product of the $u_\alpha(\mathbb{Z}_p) = N_\alpha(\mathbb{Z}_p)$ for any ordering of Φ_+.

We choose a simple root α. We consider the continuous homomorphisms

$$\ell_\alpha : P(\mathbb{Q}_p) \to P^{(2)}(\mathbb{Q}_p), \ \iota_\alpha : N(\mathbb{Q}_p)^{(2)} \to N(\mathbb{Q}_p), \ \ell_\alpha \circ \iota_\alpha = 1,$$

defined by

$$\ell_\alpha(ut) := \begin{pmatrix} \alpha(t) & x_\alpha(u) \\ 0 & 1 \end{pmatrix}, \ \iota_\alpha(u^{(2)}(x)) := u_\alpha(x) \ \text{for} \ u^{(2)}(x) := \begin{pmatrix} 1 & x \\ 0 & 1 \end{pmatrix},$$

for $t \in T(\mathbb{Q}_p), u \in N(\mathbb{Q}_p), x \in \mathbb{Q}_p$. They satisfy the functional equation

$$t\iota_\alpha(y)t^{-1} = \iota_\alpha(\ell_\alpha(t)y\ell_\alpha(t)^{-1})$$

for $y \in N(\mathbb{Q}_p)^{(2)}$ and $t \in T(\mathbb{Q}_p)$. *The data* $(N_0, \ell_\alpha, \iota_\alpha)$ *satisfies the assumptions introduced in Sections 8 and 9.*

We consider the binary relation $s_1 \le s_2$ on $T(\mathbb{Q}_p)_{++}$ generated by

$$s_1 = s_2 s_0 \ \text{with} \ s_0 \in T(\mathbb{Q}_p)_+, \alpha(s_0) \in \mathbb{Z}_p^*, \ \text{or} \ s_1^n = s_2^m \ \text{with} \ n, m \ge 1.$$

Lemma 7.150. *The relation* $s_1 \le s_2$ *on* $T(\mathbb{Q}_p)_{++}$ *is right filtered.*

Proof. Let $\Delta = \{\alpha = \alpha_1, \ldots, \alpha_n\}$. The image of $T(\mathbb{Q}_p)_{++}$ by $A = (\text{val}_p(\alpha_i(.)))_{\alpha_i \in \Delta}$ is contained in $(\mathbb{N} - \{0\})^n$ and $s_1 \leq s_2$ depends only on the cosets $s_1 T(\mathbb{Q}_p)_0$ and $s_1 T(\mathbb{Q}_p)_0$, where

$$T(\mathbb{Q}_p)_0 = \{t \in T(\mathbb{Q}_p), \alpha(t) \in \mathbb{Z}_p^* \text{ for all } \alpha \in \Delta\}.$$

(a) First we assume that, for any positive integer k, there exists $s_{[k]} \in T(\mathbb{Q}_p)$ such that $A(s_{[k]}) = (k, 1, \ldots, 1)$. Then we have $s_{[k]} \leq s_{[k+1]}$, and $s \leq s_{[k(s)]}$ for $s \in T(\mathbb{Q}_p)_{++}$ with $k(s) = \text{val}_p(\alpha(s))$. For any s_1, s_2 in $T(\mathbb{Q}_p)_{++}$ we deduce that $s_1 \leq s_{[k(s_1)+k(s_2)]}$ and $s_2 \leq s_{[k(s_1)+k(s_2)]}$. Hence the relation $\leq$ on $T(\mathbb{Q}_p)_{++}$ is right filtered.

(b) When G is semi-simple and adjoint the dominant coweights $\omega_{\alpha_1}, \ldots, \omega_{\alpha_n}$ for $\Delta = \{\alpha = \alpha_1, \ldots, \alpha_n\}$ form a basis of $Y = \text{Hom}(\mathbb{G}_m, T)$, and $A(T(\mathbb{Q}_p)_{++}) = (\mathbb{N} - \{0\})^n$. Hence $s_{[k]}$ exists for any $k \geq 1$.

(c) When G is semi-simple we consider the isogeny $\pi : G \to G_{ad}$ from G onto the adjoint group G_{ad} ([13] 16.3.5). The image T_{ad} of T is a maximal split $\mathbb{Q}_p$-torus in G_{ad}. The isogeny gives a homomorphism $T(\mathbb{Q}_p) \to T_{ad}(\mathbb{Q}_p)$, inducing an injective map between the cosets

$$T(\mathbb{Q}_p)_{++}/T(\mathbb{Q}_p)_0 \to T_{ad}(\mathbb{Q}_p)_{++}/T_{ad}(\mathbb{Q}_p)_0$$

respecting $\leq$, and such that for any $t_{ad} \in T_{ad}(\mathbb{Q}_p)$ there exists an integer $n \geq 1$ such that $t_{ad}^n \in \pi(T(\mathbb{Q}_p))$. Given $s_1, s_2 \in T(\mathbb{Q}_p)_{++}$ there exists $s_{ad} \in T_{ad}(\mathbb{Q}_p)_{++}$ such that $\pi(s_1), \pi(s_2) \leq s_{ad}$ by (b) and (a). Let $n \geq 1$ such that $s_{ad}^n = \pi(s_3)$ for $s_3 \in T(\mathbb{Q}_p)$. We have $s_{ad} \leq s_{ad}^n$ hence $\pi(s_1), \pi(s_2) \leq \pi(s_3)$. This is equivalent to $s_1, s_2 \leq s_3$.

(d) When G is reductive let $\pi : G \to G' = G/Z^0$ be the natural $\mathbb{Q}_p$-homomorphism from G to the quotient of G by its maximal split central torus Z^0. The group G' is semi-simple, $\pi(T) = T'$ is a maximal split $\mathbb{Q}_p$-torus in G', $\pi|_T$ gives an exact sequence

$$1 \to Z_0(\mathbb{Q}_p) \to T(\mathbb{Q}_p) \to T'(\mathbb{Q}_p) \to 1,$$

inducing a bijective map between the cosets

$$T(\mathbb{Q}_p)_{++}/T(\mathbb{Q}_p)_0 \to T'(\mathbb{Q}_p)_{++}/T'(\mathbb{Q}_p)_0$$

respecting $\leq$. By (c), $\leq$ is right filtered on $T'(\mathbb{Q}_p)_{++}$. We deduce that $\leq$ is right filtered on $T(\mathbb{Q}_p)_{++}$. $\square$

By Theorem 7.117 and Theorem 7.149, we can associate functorially to an étale T_+-module D over $\mathcal{O}_{\mathcal{E},\alpha}$ different sheaves:

- For any $s \in T_{++}$, a $G(\mathbb{Q}_p)$-equivariant sheaf $\mathfrak{Y}_s$ on $G(\mathbb{Q}_p)/P(\mathbb{Q}_p)$ with sections on $\mathcal{C}_0$ equal to $\mathbb{M}(D)_s^{bd}$.

- The $G(\mathbb{Q}_p)$-equivariant sheaves $\mathfrak{Y}_\cap$ and $\mathfrak{Y}_\cup$ on $G(\mathbb{Q}_p)/P(\mathbb{Q}_p)$ with sections on C_0 equal to $\cap_{s\in T_{++}}\mathbb{M}(D)_s^{bd}$ and $\cup_{s\in T_{++}}\mathbb{M}(D)_s^{bd}$.

In general $\mathbb{M}(D)$ is different from $\cup_{s\in T_{++}}\mathbb{M}(D)_s^{bd}$, by the following proposition.

Proposition 7.151. *Let M be an étale T_+-module M over $\Lambda_{\ell_\alpha}(N_0)$. When the root system of G is irreducible of positive rank $rk(G)$, we have:*

(i) *If $rk(G) = 1$, the $G(\mathbb{Q}_p)$-equivariant sheaf on $G(\mathbb{Q}_p)/P(\mathbb{Q}_p)$ with sections M_s^{bd} over C_0 does not depend on the choice of $s \in T_{++}$, and $M = M_s^{bd}$.*

(ii) *If $rk(G) > 1$, a $G(\mathbb{Q}_p)$-equivariant sheaf of o-modules $\mathfrak{Y}$ on $G(\mathbb{Q}_p)/P(\mathbb{Q}_p)$ such that $\mathfrak{Y}(C_0) \subset M$ and $(u_\alpha(1) - 1)$ is bijective on $\mathfrak{Y}(C_0)$, is zero.*

Proof. We prove (i). If $rk(G) = 1$, then $\mathcal{O}_{\mathcal{E}} = \Lambda_{\ell_\alpha}(N_0)$ and $M = D$ is an étale T_+-module over $\mathcal{O}_{\mathcal{E}}$. With the same proof as in Proposition 7.96, we have $M_s^{bd} = M$ for any $s \in T_{++}$ and the integrals $\mathcal{H}_g$ for $g \in N_0\overline{P}N_0$ do not depend on the choice of s.

(ii) is equivalent to the property: an étale $o[P_+]$-submodule M' of M which is also a $R = o[N_0][(u_\alpha(1) - 1)^{-1}]$-submodule of M, and is endowed with endomorphisms $\mathcal{H}_g \in \mathrm{End}_o(M)$, for all $g \in N_0\overline{P}(F)N_0$, satisfying the relations H1, H2, H3 (Proposition 7.72), is 0.

(a) Preliminaries. As $rk(G) \geq 2$ and the root system is irreducible, there exists a simple root β such that $\alpha + \beta$ is a root. The elements $n_\alpha := u_\alpha(1)$ and $n_\beta := u_\beta(1)$ do not commute. By the commutation formulas, $n_\alpha n_\beta = n_\beta n_\alpha h$ for some $h \neq 1$ in the group $H = \prod_\gamma N_\gamma(\mathbb{Z}_p)$ for all positive roots of the form $\gamma = i\alpha + j\beta \in \Phi_+$ with $i, j > 0$. Note that H is normalized by $N_\alpha(\mathbb{Z}_p)$. Let $s \in T_{++}$. We have the expansion (7.12)

$$(n_\alpha h - 1)^{-k} = \sum_{u\in J(N_\alpha(\mathbb{Z}_p)H/sN_\alpha(\mathbb{Z}_p)Hs^{-1})} u\varphi_s(\psi_s(u^{-1}(n_\alpha h - 1)^{-k})) \quad (7.76)$$

in R. We choose, as we can, a lift w_β of s_β in the normalizer of $T(\mathbb{Q}_p)$ such that

- $w_\beta n_\beta \in n_\beta \overline{P}(\mathbb{Q}_p)$
- w_β normalizes the group $N_{\Phi_+-\beta}(\mathbb{Z}_p) = \prod_\gamma N_\gamma(\mathbb{Z}_p)$ for all positive roots $\gamma \neq \beta$.

The subset $N'_\beta(\mathbb{Z}_p) \subset N_\beta(\mathbb{Z}_p)$ of $u_\beta(b)$ such that $w_\beta u_\beta(b) \in u_\beta(\mathbb{Z}_p)\overline{P}(\mathbb{Q}_p)$, contains n_β but does not contain 1. The subset $U_{w_\beta} \subset N_0$ of u such that $w_\beta u \in N_0\overline{P}(\mathbb{Q}_p)$ is equal to

$$U_{w\beta} = N'_\beta(\mathbb{Z}_p) N_{\Phi_+ - \beta}(\mathbb{Z}_p) = N_{\Phi_+ - \beta}(\mathbb{Z}_p) N'_\beta(\mathbb{Z}_p).$$

Hence $U_{w\beta} = u U_{w\beta}$, i.e. $w_\beta^{-1} C_0 \cap C_0 = u w_\beta^{-1} C_0 \cap C_0$, for any $u \in N_{\Phi_+ - \beta}(\mathbb{Z}_p)$.

(b) Let M' be an $R = o[N_0][(n_\alpha - 1)^{-1}]$-module of M, which is also an étale $o[P_+]$-submodule, and is endowed with endomorphisms $\mathcal{H}_g \in \mathrm{End}_o(M)$, for all $g \in N_0 \overline{P}(F) N_0$, satisfying the relations H1, H2, H3 (Proposition 7.72), and let $m \in M'$ be an arbitrary element. We want to prove that $m = 0$.

The idea of the proof is that, for $s \in T_{++}$, we have $m = 0$ if $\mathcal{H}_{w\beta}(n_\beta \varphi_s(m)) = 0$ and that $\mathcal{H}_{w\beta}(n_\beta \varphi_s(m)) = 0$ because it is infinitely divisible by $n_\gamma - 1$, where $\gamma = s_\beta(\alpha)$. An element in M with this property is 0 because $n_\gamma - 1$ lies in the maximal ideal of $\Lambda_{\ell_\alpha}(N_0)$.

Let $a \in \mathbb{Z}_p$. The product formula in Proposition 7.84ii implies

$$\mathcal{H}_{w\beta} \circ \mathcal{H}_{n_\alpha^a} \circ \mathrm{res}(1_{w_\beta^{-1} C_0 \cap C_0}) = \mathcal{H}_{w\beta n_\alpha^a} \circ \mathrm{res}(1_{w_\beta^{-1} C_0 \cap C_0}) =$$

$$\mathcal{H}_{n_\gamma^a w\beta} \circ \mathrm{res}(1_{w_\beta^{-1} C_0 \cap C_0}) = \mathcal{H}_{n_\gamma^a} \circ \mathcal{H}_{w\beta} \circ \mathrm{res}(1_{w_\beta^{-1} C_0 \cap C_0})$$

since $n_\alpha^{-a} w_\beta^{-1} C_0 \cap C_0 = w_\beta^{-1} C_0 \cap C_0 = w_\beta^{-1} n_\gamma^{-a} C_0 \cap C_0$. For all $k \in \mathbb{N}$, the elements

$$m_k := (n_\alpha - 1)^{-k} n_\beta \varphi_s(m) = n_\beta (n_\alpha h - 1)^{-k} \varphi_s(m) \tag{7.77}$$

lie in the image of the idempotent $\mathrm{res}(1_{w_\beta^{-1} C_0 \cap C_0}) \in \mathrm{End}_o(M)$, because

$$m_k = \sum_{u \in J(N_\alpha(\mathbb{Z}_p) H/s N_\alpha(\mathbb{Z}_p) H s^{-1})} n_\beta u \varphi_s(\psi_s(u^{-1}(n_\alpha h - 1)^{-k} m)) \tag{7.78}$$

by (7.76), (7.77). Therefore the product relations between $\mathcal{H}_{w\beta}$, $\mathcal{H}_{n_\alpha^a}$ and $\mathcal{H}_{n_\gamma^a}$ imply

$$\mathcal{H}_{w\beta}(n_\beta \varphi_s(m)) = \mathcal{H}_{w\beta}((n_\alpha - 1)^k m_k) = \sum_{a=0}^k (-1)^{k-a} \binom{k}{a} \mathcal{H}_{w\beta} \circ \mathcal{H}_{n_\alpha^a}(m_k)$$

$$= \sum_{a=0}^k (-1)^{k-a} \binom{k}{a} \mathcal{H}_{w\beta} \circ \mathcal{H}_{n_\alpha^a} \circ \mathrm{res}(1_{w_\beta^{-1} C_0 \cap C_0})(m_k)$$

$$= \sum_{a=0}^k (-1)^{k-a} \binom{k}{a} \mathcal{H}_{n_\gamma^a} \circ \mathcal{H}_{w\beta} \circ \mathrm{res}(1_{w_\beta^{-1} C_0 \cap C_0})(m_k)$$

$$= (n_\gamma - 1)^k \mathcal{H}_{w\beta}(m_k).$$

Hence $\mathcal{H}_{w\beta}(n_\beta \varphi_s(m)) = 0$ since it is infinitely divisible by $n_\gamma - 1$ which lies in the maximal ideal of $\Lambda_{\ell_\alpha}(N_0)$. We also have

$$n_\beta \varphi_s(m) = \mathcal{H}_1 \circ \text{res}(1_{w_\beta^{-1}\mathcal{C}_0 \cap \mathcal{C}_0})(n_\beta \varphi_s(m)) = \mathcal{H}_{w_\beta} \circ \mathcal{H}_{w_\beta}(n_\beta \varphi_s(m)) = 0.$$

As $n_\beta \circ \varphi_s \in \text{End}_o(M')$ is injective, we deduce $m = 0$. $\quad\square$

Corollary 7.152. *There exists a $G(\mathbb{Q}_p)$-equivariant sheaf on $G(\mathbb{Q}_p)/P(\mathbb{Q}_p)$ with sections M on $\mathcal{C}_0$ if and only if $\text{rk}(G) = 1$.*

References

[1] Bourbaki N., *Algèbre commutative.* Ch. 1 à 4. Masson 1985.

[2] Bourbaki N., *Topologie générale.* Ch. 1 à 4. Hermann 1971.

[3] Bourbaki N., *Topologie générale.* Ch. 5 à 10. Hermann 1974.

[4] Bosch S., Güntzer U., Remmert R., *Non-Archimedean analysis.* Springer 1984.

[5] Colmez P., (φ, Γ)-modules et représentations du mirabolique de $GL_2(\mathbb{Q}_p)$. *Astérisque* 330, 2010, 61–153.

[6] Colmez P., Représentations de $GL_2(\mathbb{Q}_p)$ et (φ, Γ)-modules. *Astérisque* 330, 2010, 281–509.

[7] Dixon J. D., du Sautoy M. P. F., Mann A., Segal D., *Analytic pro-p groups.* Second edition. Cambridge Studies in Advanced Mathematics, 61. Cambridge University Press, 1999.

[8] Ellis R., Locally compact transformation groups. *Duke Math. J.* 24, 1957 119–125.

[9] Gabriel P., Des catégories abéliennes. *Bull. Soc. Math. France* 90, 1962, 323–448.

[10] Kedlaya K., New methods for (φ, Γ)-modules, preprint (2011), http://math.mit.edu/kedlaya/papers/new-phigamma.pdf

[11] Schneider P., *p-Adic Lie groups.* Springer Grundlehren, Vol. 344, Springer, 2011.

[12] Schneider P., Vigneras M.-F., *A functor from smooth o-torsion representations to (φ, Γ)-modules. Volume in honour of F. Shahidi.* Clay Mathematics Proceedings, Volume 13, 525–601, 2011.

[13] Springer T. A., *Linear Algebraic Groups.* Second edition. Birkhäuser, 2009.

[14] Vigneras M.-F., *Représentations ℓ-modulaires d'un groupe réductif p-adique avec $\ell \neq p$.* PM 137, Birkhäuser, 1996.

[15] Warner S.: *Topological rings.* Elsevier, 1993.

[16] Zábrádi G., Exactness of the reduction of étale modules. *J. Algebra* 331, 2011, 400–415.

8

Intertwining of ramified and unramified zeros of Iwasawa modules

Chandrashekhar Khare and Jean-Pierre Wintenberger

1. Introduction

Let F be a totally real field, $p > 2$ a prime, $F_\infty \subset F(\mu_{p^\infty})$ the cyclotomic $\mathbb{Z}_p$-extension of F with Galois group $\Gamma = \mathbb{Z}_p = \langle \gamma \rangle$. This short note is a sequel to [5]. The sole aim is to point out the ubiquity of a phenomenon dicussed in a particular case in our previous paper. Namely, the (arithmetic) eigenvalues of γ acting on Galois groups of maximal p-abelian *unramified* extensions of $F(\mu_{p^\infty})$ intertwine with the eigenvalues acting on inertia subgroups of *ramified* p-abelian extensions of $F(\mu_{p^\infty})$. We make this vague philosophy precise in the text below after alluding for *mise-en-scène* to the p-adic L-functions that lurk suggestively in the wings, but do not play an explicit role in the algebraic computations of this note.

Let ψ be an even Dirichlet character of F. Consider the p-adic L function $\zeta_{F,p}(s, \psi) = L_p(s, \psi)$, $s \in \mathbb{Z}_p$, which is characterised by the interpolation property $L_p(1 - n, \psi) = L(1 - n, \psi\omega^{-n})\Pi_{v|p}(1 - \psi\omega^{-n}(v)N(v)^{n-1})$, for $n \geq 1$ a positive integer. When $\psi = 1$, we denote the corresponding L-function by $\zeta_{F,p}(s)$.

It is known that $L_p(s, \psi)$ is holomorphic outside $s = 1$, is holomorphic everywhere when $\psi \neq 1$, and otherwise has at most a simple pole at $s = 1$. This pole is predicted to exist by the conjecture of Leopoldt, which asserts the non-vanishing of the the p-adic regulator of units of F. The residue of $\zeta_{F,p}$ at $s = 1$ has been computed by Pierre Colmez:

$$\text{Res}_{s=1}\zeta_{F,p}(s) = \frac{2^d R_{F,p} h_F}{2\sqrt{D_F}},$$

CK was partially supported by NSF grants.
JPW is member of the Institut Universitaire de France.

Automorphic Forms and Galois Representations, ed. Fred Diamond, Payman L. Kassaei and Minhyong Kim. Published by Cambridge University Press. © Cambridge University Press 2014.

where $R_{F,p}$ is the p-adic regulator for F. The non-vanishing of $R_{F,p}$ is the Leopoldt conjecture for F and p.

We recall a folklore conjecture that is a very particular case of the general conjectures of Jannsen ([4]) about the non-vanishing of higher regulators.

Conjecture 8.1. *(Non-vanishing of higher p-adic regulators) For an integer $m \neq 0$, $L_{F,p}(m, \psi) \neq 0$ if either $m \neq 1$ or $\psi \neq 1$. Furthermore, $\zeta_{F,p}(s)$ has a pole at $s = 1$.*

As a supplement in the case $m = 0$, the case of "trivial zeros", the multiplicity of the zero at $s = 0$ of $L_p(s, \psi)$ is conjectured to be given by the number of $v|p$ such that $(1 - \psi\omega^{-1}(v)) = 0$.

We call the zeros of $\zeta_{F,p}(s)$ *unramified zeros* following a similar usage in [8].

The main conjecture of Iwasawa theory realises the zeros of the L-functions as roots of the characteristic polynomial of γ acting on "unramified arithmetic spaces". Nevertheless it gives no direct information about the zeros, belying the Hilbert–Polya philosophy in this case!

Our basic observation, coming from [5] that we reinforce here, is that the unramified zeros (of the p-adic L-function) always intertwine with (the eigenvalues of γ acting on) ramification at p. Further the Leopoldt zeros, i.e. eigenvalues of Γ that correspond on a finite index subgroup to its action on p-power roots of unity, intertwine with ramification at Q, for finite sets of primes away from p, for a *generic* choice of Q.

This shows that the non-vanishing of p-adic regulators is equivalent to splitting of ramification in naturally occuring exact sequences of Iwasawa modules. The reader is referred to Theorems 8.5, 8.9 and 8.10 for precise statements.

The proofs of these theorems rely on numerical coincidences between

– dimensions of certain Galois cohomology groups whose computation result from the work of Soulé and Poitou–Tate duality,

– dimensions of Iwasawa modules that follow from theorems of Iwasawa describing the structure of inertia at p (resp. at a set Q of auxiliary primes) in the Galois group of the maximal odd abelian p-extension of $F(\mu_{p^\infty})$ that is unramified outside p (resp. Q).

2. Galois cohomology

2.1.

In our previous paper [5] we paid attention to integral questions, while here we work over $\mathbb{Q}_p$ exclusively. We consider F a totally real field, an odd prime p.

We consider a sufficiently large finite extension K of $\mathbb{Q}_p$ with ring of integers $\mathcal{O}$ (that will contain values of the character under consideration).

Let S_p be the set of places of F above p and ∞ and let S be a finite subset of places of F containing S_p. Let G_S be a Galois group of the maximal algebraic extension of F unramified outside S. We consider a potentially crystalline, or arithmetic, character χ of G_S, of (parallel) weight m and thus of the form $\eta\chi_p^m$ where χ_p is the p-adic cyclotomic character, and η a finite order character. We impose that $\chi(c)$ is independent of the choice of complex conjugation $c \in G_F$, and call it odd or even according as this value is either -1 or 1. We denote by ω the Teichmüller character. We consider the cohomology subgroup $H^1_{(p)f}(G_S, K(\eta\chi_p^m))$ of $H^1(G_S, K(\eta\chi_p^m))$ defined by imposing for $v \in S$, $v \notin S_p$ the condition to be unramified (although χ might be ramified at these v). We denote by $H^1_f(G_S, K(\eta\chi_p^m))$ their Bloch–Kato subgroups, where for v primes over p, we impose the Bloch–Kato finiteness condition ([2]). We denote the dimensions over K of $H^1_{(p)f}(G_S, K(\eta\chi_p^m))$ and $H^1_f(G_S, K(\eta\chi_p^m))$ by $h^1(\chi)$ and $h^1_f(\chi)$. We use $h^1_{split}(\chi)$ to denote dimensions of cohomology groups where we ask that the classes are split locally at all places above p and unramified at other primes. We have the tables.

2.2. Odd χ

$\chi = \eta\chi_p^m$, χ odd	$m > 1$	$m = 1$	$m \leq 0$
$h^1(\chi)$	d	$d + \sum_{v\mid p} h^0(G_v, \eta^{-1}) - \delta_{\chi,\chi_p}$	d
$h^1_f(\chi)$	d	$d - \delta_{\chi,\chi_p}$	0

2.3. Even χ

$\chi = \eta\chi_p^m$, χ even	$m > 1$	$m = 1$	$m \leq 0$
$h^1(\chi)$	0	$\sum_{v\mid p} h^0(G_v, \eta^{-1})$	$\delta_{\chi,id} + h^1_{split}(\chi_p\chi^{-1})$
$h^1_f(\chi)$	0	0	0

Remark. The situation for even χ is thus not satisfactory as we do not have an explicit formula in all cases for $h^1(\chi)$. The conjectures in [4], concerning non-vanishing of higher p-adic regulators, predict the vanishing of $h^1_{split}(\chi_p\chi^{-1})$ for χ an even arithmetic character of G_F (see also [1] 5.2). This follows from the main conjecture of Iwasawa theory if the weight of χ is > 0.

2.4. Ingredients of the computation

We justify the values in the tables. We need the following ingredients:

- (Bloch–Kato) $h_f^1(\chi) = \dim_K K(\chi)^{G_F} + \mathrm{ord}_{s=1-m} L(\eta^{-1}, s)$.
- (global duality) χ even:

$$h^1(\chi) = h_{split}^1(\chi_p \chi^{-1}) + \delta_{\chi,id} + \sum_{v|p} h^0(G_v, \chi_p \chi^{-1}).$$

- (global duality) χ odd:

$$h^1(\chi) = d + h_{split}^1(\chi_p \chi^{-1}) - \delta_{\chi,\chi_p} + \sum_{v|p} h^0(G_v, \chi_p \chi^{-1}).$$

The (global duality) equalities follow from Theorem 8.7.9. of [6] and:

- (local Euler Poincaré characteristic)

$$h^1(G_v, V) = h^0(G_v, V) + h^2(G_v, V) + [F_v : \mathbb{Q}_p] \dim_K(V)$$

where $v|p$.
- (local duality) $h^0(G_v, V^*(1)) = h^2(G_v, V)$.

The Bloch–Kato formula, which directly implies the bottom row of both tables, follows from a theorem of Soulé (Theorem 1 of [7]) and duality as we justify.

- The theorem of Soulé directly implies Bloch–Kato formula for $h_f^1(\chi)$ for $m > 1$.
- The case of $m = 1$ for the bottom rows follows from Kummer theory.
- We have the equality for $m \leq 0$ and χ even:

$$h_f^1(\chi) = h_f^1(\chi_p \chi^{-1}) - d + \delta_{\chi,id},$$

and for $m \leq 0$ and χ odd:

$$h_f^1(\chi) = h_f^1(\chi_p \chi^{-1}).$$

These equalities follow from Theorem 8.7.9 of [6] and the fact that for $m \leq 0$, $H_f^1(G_v, \chi)$ coincides with the unramified cohomology $H_{ur}^1(G_v, \chi)$. These two equalities allow us to deduce the case $m \leq 0$ of the Bloch–Kato formula from the Theorem of Soulé (see also [1], §4.3.1).

This checks the second rows of both the tables.

Let us check the first row. For the first column in the case $m > 1$, we use the fact that $h^1(G_v, \chi) = h^1_f(G_v, \chi)$. The case $m \leq 0$ even χ follows from (global duality) as $h^0(G_v, \chi_p \chi^{-1}) = 0$ for all v above p. For the case $m = 1$, we use global duality and that $h^1_{split}(\eta^{-1}) = 0$ which follows easily from the fact that a $\mathbb{Z}_p$-extension of a number field has to be ramified at a place above p. For χ odd and $m \leq 0$, we use global duality and the fact that $h^1_{split}(\chi_p \chi^{-1}) \leq h^1_f(\chi_p \chi^{-1}) = 0$.

2.5. Galois groups

We consider ε totally odd and ψ totally even finite order characters of G_F, such that $\varepsilon \psi = \omega$. We set $\psi(n) = \psi(\omega^{-1} \chi_p)^n := \psi \kappa^n$, and likewise for $\varepsilon(n)$. We consider characters ψ_ζ of Γ that send a chosen generator γ of Γ to a p-power root of unity ζ and consider $\psi(n) \psi_\zeta$, $\varepsilon(n) \psi_\zeta$.

We assume that ψ is of type S, i.e. we assume that the field F_ψ cut out by ψ is linearly disjoint from the cyclotomic $\mathbb{Z}_p$-extension F_∞ of F. We denote by Γ the Galois group of F_∞/F with choice of generator γ, consider Λ the completed group algebra $\mathbb{Z}_p[[\Gamma]]$ that is isomorphic to $\mathbb{Z}_p[[T]]$ via the homomorphism which sends $\gamma \to 1 + T$.

We let $\Lambda_K = \Lambda \otimes K$. We consider the Galois group $\mathcal{G}_\psi = \mathcal{G}_\varepsilon = \mathrm{Gal}(F_\psi(\mu_p)/F) \times \mathrm{Gal}(F_\infty/F)$ of $F_\psi(\mu_{p^\infty})$ over F. A continuous character χ of $\mathcal{G}_\psi$ with values in $\mathcal{O}^*$ is the product of a finite character χ_ψ of $\mathrm{Gal}(F_\psi(\mu_p)/F)$ and a character χ_Γ of Γ. The character χ_Γ induces a K-algebra map $\Lambda_K \to K$. We denote by P_χ the corresponding prime ideal of Λ_K kernel of this map. It is generated by $\gamma - \chi(\gamma)$. The character χ_ψ induces a morphism $f_\chi : \mathbb{Z}_p[\mathrm{Gal}(F_\psi(\mu_p)/F)] \to \mathcal{O} \subset K$. The morphism $\mathbb{Z}_p[[\mathcal{G}_\psi]] \to K$ induced by χ is the composite of the map $\mathbb{Z}_p[[\mathcal{G}_\psi]] \to \Lambda_K$ induced by f_χ and the morphism $\Lambda_K \to K$.

We consider the maximal, abelian pro-p extension $\mathcal{L}_\infty$ of $F_\psi(\mu_{p^\infty})$ unramified everywhere: we denote its Galois group by X'_∞ and set $X_\infty = X'_\infty \otimes K$. We set $X_{\infty,\varepsilon}$ to be the maximal quotient of X_∞ on which $\mathrm{Gal}(F_\psi(\mu_p)/F)$ acts by ε.

We also consider the analogous extensions $\mathcal{L}_{\infty,\varepsilon}(Q)$ and corresponding Galois group $X_{\infty,\varepsilon,Q}$ when Q is any set of places of F and we allow ramification above Q. There are two natural cases to consider:

- Q is all the places above p, and then we replace Q by p in the notation;
- Q is a finite set of places disjoint from the places above p.

Let $F_{\psi,\infty}$ be the cyclotomic $\mathbb{Z}_p$-extension of the totally real field F_ψ. We denote by Y'_∞ the Galois group of the maximal abelian prop-p extension of

$F_{\psi,\infty}$ that is unramified outside p, and $Y_\infty = \mathbb{Q}_p \otimes Y'_\infty$. As above, we denote by $Y_{\infty,\psi}$ the related quotient on which $\mathrm{Gal}(F_\psi(\mu_p)/F)$ acs by ψ, and recall the perfect Γ-equivariant Iwasawa pairing

$$Y_{\infty,\psi} \times X_{\infty,\varepsilon} \to K(1).$$

Lemma 8.2. *Let ψ be an even character of G_F of type S and $n \in \mathbb{Z}$.*

(1) *We have $h^1(\psi_\zeta \psi(n)) = \dim_K(Y_{\infty,\psi}/P_{\psi_\zeta \psi(n)})$ if $\psi_\zeta \psi(n)$ is not trivial, and $h^1(\psi_\zeta \varepsilon(n)) = \dim_K(X_{\infty,\varepsilon,p}/P_{\psi_\zeta \varepsilon(n)})$.*
(2) *We have that*

$$\dim_K((X_{\infty,\varepsilon})_{P_{\psi_\zeta \varepsilon(-n)}}/P_{\psi_\zeta \varepsilon(-n)}) = \dim_K((Y_{\infty,\psi})_{P_{\psi_\zeta^{-1}\psi(n+1)}}/P_{\psi_\zeta^{-1}\psi(n+1)}).$$

Proof. For (i) we use the inflation-restriction sequence relative to $G_S \to \mathcal{G}_\psi$. Recall that the unramified condition for $\eta \in H^1_{(p)f}(G_S, K(\varepsilon \chi_p^m))$ at v not above p and such that ψ is ramified at v (Section 2.1). We check that this condition is equivalent to that the restriction of η to the kernel of the map $G_S \to \mathcal{G}_\psi$ is unramified at v.

For (ii), we invoke the pairing of Iwasawa. We use the cyclicity of Γ to identify dimensions of twisted invariants and covariants for Γ. $\qquad\square$

We rederive a standard result about trivial zeros of p-adic L-functions, usually proved using genus theory, which is a corollary to the lemma.

Proposition 8.3. *Suppose ε is an odd character of $\mathrm{Gal}(F_\psi(\mu_p)/F)$ as before. Then*

$$\dim_K((X_{\infty,\varepsilon})_{P_{\psi_\zeta \varepsilon}}/P_{\psi_\zeta \varepsilon}) = h^1(\chi_p \psi_\zeta^{-1} \varepsilon^{-1}) = \sum_{v|p} h^0(G_v, \psi_\zeta^{-1} \varepsilon^{-1}).$$

Proof. We deduce the first equality using (2) of Lemma 8.2 for $n = 0$ and the second equality using the above table for $m = 1$ even χ. $\qquad\square$

Note that $\sum_{v|p} h^0(G_v, \psi_\zeta^{-1} \varepsilon^{-1})$ is the number of places $v|p$ such that the Euler factor $(1 - \psi_\zeta^{-1} \varepsilon^{-1}(v))$ is 0. It is conjectured by Greenberg that the trivial zeros occur semisimply i.e. $\dim_K(X_{\infty,\varepsilon}/P_{\psi_\zeta \varepsilon}) = \dim_K(X_{\infty,\varepsilon})_{P_{\psi_\zeta \varepsilon}}$.

3. Main conjecture and higher regulators

We consider the Deligne–Ribet p-adic L-function $\zeta_{F,p}(s, \psi)$ for an even character ψ of F of type S. It is defined on $\mathbb{Z}_p$ when ψ is non-trivial, and on

$\mathbb{Z}_p \setminus \{1\}$ when ψ is trivial. It is characterised by the interpolation formula that for integers $n \geq 1$,

$$\zeta_{F,p}(1 - n, \psi) = L^p(1 - n, \psi\omega^{-n}),$$

where the superscript denotes that we have dropped the Euler factors at p. There is a power series $W_\psi(T)$ in $\Lambda_K = \mathbb{Z}_p[[\Gamma]] \otimes K$ (the latter by the isomorphism that sends a chosen generator γ of $\Gamma = \mathrm{Gal}(F_\infty/F)$ to $1 + T$ and $u := \chi_p(\gamma)$), with the property that:

$$\frac{W_\psi(u^s - 1)}{u^{1-s} - 1} = \zeta_{F,p}(s, \psi)$$

when ψ is trivial, and

$$W_\psi(u^s - 1) = \zeta_{F,p}(s, \psi)$$

otherwise. Furthermore we have:

$$W_{\psi\psi_\zeta}(T) = W_\psi(\zeta^{-1}(1 + T) - 1);$$

see the introduction of [8], where the notation is $G_\psi(T)$ for $W_\psi(u(1 + T)^{-1} - 1)$.

Then the main conjecture asserts for characters ψ of type S that

$$(W_\psi(T)) = \mathrm{char}_{\Lambda_K}(X_{\infty, \psi^{-1}\omega}),$$

i.e. the characteristic polynomial of the action of γ on the finite dimensional K-vector space $X_{\infty, \varepsilon}$ generates the same ideal as $W_\psi(T)$. (We ignore μ-invariants via this formulation.)

When m is an integer $\neq 0$, Conjecture 8.1 is equivalent via the main conjecture to the statement that the generalised ζu^m-eigenspace of $X_{\infty, \psi^{-1}\omega} \otimes \mathbb{Q}_p$ for the action of γ is trivial. When $m = 0$, via the main conjecture, we have that the generalised ζ-eigenspace is of dimension given by the number of $v | p$ of F such that $(1 - \psi_\zeta^{-1}\varepsilon^{-1}(v)) = 0$.

4. Intertwining of ramified and unramified zeros

4.1. Ramification at p

We consider the exact sequences:

$$0 \to I_Q \to X_{\infty, \varepsilon, Q} \to X_{\infty, \varepsilon} \to 0,$$

of finitely generated Λ_K-modules with $Q = p$ (see 2.5).

4.1.1. Ramification at p and intertwining with non-trivial unramified zeros

Next theorem follows from Th. 25 of [3] :

Theorem 8.4. *We have an ismorphism of Λ_K-modules*

$$I_p = \frac{\{\Lambda_K^d \oplus \oplus_{j=1}^s \mathrm{Ind}_{G_{\wp_j}}^\Gamma K(1)\}}{K(1)},$$

where $G_\wp$ are the decomposition groups of the places $\wp$ above p in Γ.

We deduce from this and the computations in Galois cohomology earlier:

Theorem 8.5. *Let $\chi = \varepsilon \psi_\zeta \kappa^m$ be an odd arithmetic character of $\mathcal{G}_\varepsilon$ of weight m, and let P_χ be the corresponding prime ideal of $\Lambda_\mathcal{O}$ for $\mathcal{O}$ such that χ is valued in $\mathcal{O}^*$. Then the exact sequence*

$$0 \to (I_p)_{P_\chi} \to (X_{\infty,\varepsilon,p})_{P_\chi} \to (X_{\infty,\varepsilon})_{P_\chi} \to 0,$$

of finitely generated Λ_K-modules splits if and only if $(X_{\infty,\varepsilon})_{P_\chi}$ vanishes. This is equivalent to that $\zeta_{F,p}(m, \psi_\zeta^{-1}\varepsilon^{-1}\omega) \neq 0$ when $\chi \neq \chi_p$, and when $\chi = \chi_p$, to that $\zeta_{F,p}(s)$ has a pole at $s = 1$.

Proof. One direction is trivial. For the other direction, one notes by the theorem of Iwasawa that $\dim_K((I_p)_{P_\chi}/P_\chi)$ is d if $m \neq 1$ and $d + \sum_{v|p} h^0(G_v, \varepsilon^{-1}) - \delta_{\chi,\chi_p}$ if $m = 1$. We have the numerical coincidence:

$$\dim_K((I_p)_{P_\chi}/P_\chi) = h^1(\chi).$$

Further one notes from Lemma 8.2 that $h^1(\chi) = \dim_K(X_{\infty,\varepsilon,p})_{P_\chi}/P_\chi$. Thus if the exact sequence in the theorem splits, we deduce that $(X_{\infty,\varepsilon})_{P_\chi}/P_\chi = 0$, which is equivalent to the vanishing of $(X_{\infty,\varepsilon})_{P_\chi}$.

By the main conjecture proved by Wiles, the vanishing of $(X_{\infty,\varepsilon})_{P_\chi}/P_\chi$ is equivalent to the vanishing of W_ψ at $\zeta u^m - 1$. It is equivalent to the vanishing of $\zeta_p(m, \psi\psi_\zeta^{-1}) = \zeta_{F,p}(m, \psi_\zeta^{-1}\varepsilon^{-1}\omega)$. $\qquad\square$

Remarks.

1. We also deduce that an equivalent formulation of the non-vanishing of the higher regulator conjecture is the conjecture that the exact sequence

$$0 \to (I_p)_{P_\chi} \to (X_{\infty,\varepsilon,p})_{P_\chi} \to (X_{\infty,\varepsilon})_{P_\chi} \to 0,$$

of finitely generated Λ_K-modules splits for *all* odd arithmetic characters χ of $\mathcal{G}$ of non-zero weight.

2. As trivial zeros (in weight 0) do occur we get examples of Iwasawa modules in which ramification is allowed at p such that γ acts non-semisimply.

We have a conditional result for any character $\chi = \varepsilon \kappa^s$ of $\mathcal{G}_\varepsilon$ not necessarily arithmetic, and that follows by similar arguments.

Proposition 8.6. *Let* $\chi = \varepsilon \kappa^s$ *be any odd character of* $\mathcal{G}_\varepsilon$. *Assume further that* $H^1_{split}(G_{F,p}, \chi_p \chi^{-1})$ *is trivial. Then the exact sequence*

$$0 \to (I_p)_{P_\chi} \to (X_{\infty,\varepsilon,p})_{P_\chi} \to (X_{\infty,\varepsilon})_{P_\chi} \to 0,$$

of compact finitely generated Λ_K-*modules splits if and only if* $(X_{\infty,\varepsilon})_{P_\chi} = 0$.

Remark. Note that the vanishing of $H^1_{split}(G_{F,p}, \chi_p \chi^{-1}) = 0$ is predicted by Greenberg's conjecture that the p-part of the class group of the cyclotomic $\mathbb{Z}_p$-extension F_∞ of F is finite. The above proposition suggests that there may be a formulation of the main conjecture using $\mathrm{Ext}_{\Lambda_K}(X_{\infty,\varepsilon}, I_p)$.

4.2. Ramification away from p

4.2.1. ψ and ψ_ζ trivial

Consider the maximal abelian p-extension L_∞ of F, and denote its $\mathbb{Z}_p$-rank by $1 + \delta$. The Leopoldt conjecture asserts that $\delta = 0$.

Definition 8.7. (generic sets Q) We say that a finite set of primes Q of cardinality r away from p is generic if the rank r_Q of the subgroup generated by the Frob_q's for $q \in Q$ in $\mathrm{Gal}(L_\infty/F)$ is $\min(r, 1 + \delta)$.

The terminology is meant to reflect the fact that when $\delta > 0$, the $\mathrm{Frob}_{q_1}, \mathrm{Frob}_{q_2}$ will be linearly independent in $\mathbb{Z}_p^{1+\delta} = \mathrm{Gal}(L_\infty/F)$ for most choices of q_1, q_2. If $r = 2$ and we choose a prime q_1 freely, then for a density one set of primes q_2, the set $Q = \{q_1, q_2\}$ is generic.

We now show the intertwining of the Leopoldt zero $u = u_\gamma$ with the ramification at Q provided Q is a generic set of primes with $|Q| > 1$.

Proposition 8.8. *Let* Q *be a finite set of primes of* F *away from* p. *Then the subgroup* I_Q *of* $X_{\infty,\omega,Q}$ *generated by the conjugacy class of the inertia groups* I_q *for* $q \in Q$ *is isomorphic as a* $\mathrm{Gal}(\mathcal{F}_\infty/F) = \mathrm{Gal}(F(\mu_p)/F) \times \Gamma$-*module to*

$$\frac{(\prod_{q \in Q} \mathrm{Ind}^\Gamma_{G_q} K(1))}{K(1)}$$

where G_q *is the decomposition subgroup at* q *in* $\Gamma = \mathrm{Gal}(\mathcal{F}_\infty/F)$, *and where we declare that* $\mathrm{Gal}(F(\mu_p)/F)$ *acts by* ω.

Proof. This follows from class field theory.   □

Note that when the primes q are inert in F_∞/F then $I_Q = K(1)^{r-1}$.

Theorem 8.9. *Let Q be generic set of primes Q of cardinality $r \geq 2$. Then the exact sequence*

$$0 \to (I_Q)_{P_{\chi_p}} \to (X_{\infty,\omega,Q})_{P_{\chi_p}} \to (X_{\infty,\omega})_{P_{\chi_p}} \to 0$$

splits if and only if $(X_{\infty,\omega})_{P_{\chi_p}} = 0$, i.e., if and only if the Leopoldt conjecture is true.

Thus if there is a Leopoldt zero, then it intertwines with the ramified zeros at Q for a generic set of primes (away from p) with $|Q| \geq 2$.

Proof. If the Leopoldt conjecture is true, $(X_{\infty,\omega})_{P_{\chi_p}} = 0$ and the exact sequence splits.

Let us prove the converse.

For a finite dimensional vector space V over K endowed with a continuous action of G_F that is unramified outside a finite set of places, and a set of Selmer conditions $\mathcal{L} = \{\mathcal{L}_v\}$ for $\mathcal{L}_v \subset H^1(F_v, V)$ where $\mathcal{L}_v$ is outside a finite set of places the unramified subgroup, we have the formula :

$$h^1_{\mathcal{L}}(F, V) - h^1_{\mathcal{L}^\perp}(F, V^*(1)) = h^0(F, V) - h^0(F, V^*(1))$$
$$+ \sum_v (\dim_K \mathcal{L}_v - h^0(F_v, V)).$$

We apply this formula for $V = K(1)$, and with the Selmer conditions $\mathcal{L}$ to be unramified everywhere. In particular the Selmer condition is trivial at places v above p as $V^{I_v} = 0$ for v above p. We get:

$$h^1_{\mathcal{L}}(F, V) - h^1_{\mathcal{L}^\perp}(F, V^*(1)) = -1.$$

Furthermore, we have $h^1_{\mathcal{L}^\perp}(F, K) = 1 + \delta$.

Consider the Selmer conditions $\mathcal{L}_Q$ that arise when we allow ramification at Q, i.e., $(\mathcal{L}_Q)_v = \mathcal{L}_v$ for $v \notin Q$ and $(\mathcal{L}_Q)_v = H^1(G_v, K(1))$ for $v \in Q$. We get:

$$h^1_{\mathcal{L}_Q}(F, V) - h^1_{\mathcal{L}_Q^\perp}(F, V^*(1)) = -1 + r.$$

Furthermore, we have $h^1_{\mathcal{L}_Q^\perp}(F, K) = 1 + \delta - r_Q$. We see that:

$$h^1_{\mathcal{L}_Q}(F, K(1)) = h^1_{\mathcal{L}}(F, K(1)) + r - r_Q.$$

If the exact sequence splits, it remains exact after reduction modulo P_{χ_p}, hence we have:

$$h^1_{\mathcal{L}_Q}(F, K(1)) = h^1_{\mathcal{L}}(F, K(1)) + r - 1.$$

Thus we get $r_Q = 1$. As Q is generic and $\mid Q \mid \geq 2$, i.e., $r_Q = \min(r, 1 + \delta)$ with $r \geq 2$, we get that $1 + \delta = 1$, thus $\delta = 0$ and the Leopoldt conjecture is true. $\qquad\square$

4.2.2. Weight 1, ψ or ψ_ζ non-trivial

Theorem 8.10. *Consider the exact sequence*

$$0 \to (I_q)_{P_{\psi_\zeta \varepsilon\kappa}} \to (X_{\infty,\varepsilon,q})_{P_{\psi_\zeta \varepsilon\kappa}} \to (X_{\infty,\varepsilon})_{P_{\psi_\zeta \varepsilon\kappa}} \to 0.$$

Assume $\varepsilon \neq \omega$ or that $\zeta \neq 1$. Then the sequence splits for all choices of primes q of F away from p, if and only if $(X_{\infty,\varepsilon})_{P_{\psi_\zeta \varepsilon\kappa}} = 0$.

Proof. We only sketch the proof as its very similar to the proof of Theorem 8.9.

By (2) of Lemma 8.2, if $(X_{\infty,\varepsilon})_{P_{\psi_\zeta \varepsilon\kappa}} \neq 0$, the maximal abelian p-extension L of $F_{\varepsilon^{-1}\psi_\zeta^{-1}\omega} = F_{\psi\psi_\zeta^{-1}}$ on which $\mathrm{Gal}(F_{\psi\psi_\zeta^{-1}}/F)$ acts by $\psi\psi_\zeta^{-1}$ and which is unramified outside p, has Galois group that is of positive rank as a $\mathbb{Z}_p$-module. Let q be a prime of F away from p, that splits in $F_{\psi\psi_\zeta^{-1}}$, such that the Frobenius at a prime above q of $F_{\psi\psi_\zeta^{-1}}$ is a non-torsion element in $\mathrm{Gal}(L/F_{\psi\psi_\zeta^{-1}})$. With $V = K(\psi_\zeta \varepsilon\kappa)$ and the Selmer conditions as above, it follows that $h^1_{\mathcal{L}^\perp} - h^1_{\mathcal{L}_q^\perp} = 1$.

Using the above formula, we get $h^1_{\mathcal{L}_q} - h^1_{\mathcal{L}_q^\perp} - h^1_{\mathcal{L}} + h^1_{\mathcal{L}^\perp} = 0$. If the exact sequence were to split, we would have $h^1_{\mathcal{L}_q} = h^1_{\mathcal{L}}$, which is a contradiction. $\qquad\square$

References

[1] Joël Bellaiche. An introduction to the conjecture of Bloch and Kato. Lectures at the Clay Mathematical Institute summer School, Honolulu, Hawaii , 2009.

[2] Spencer Bloch, Kazuya Kato. L-functions and Tamagawa numbers of motives. *The Grotendieck Festschrift*, vol. 1, 333-400, Prog. Math. 86, Birkhäuser Boston, Boston, MA, 1990.

[3] Kenkichi Iwasawa. On $\mathbb{Z}_\ell$-extensions of algebraic number fields. *Ann. of Math.* (2) 98 (1973), 246–326.

[4] Uwe Jannsen. On the ℓ-adic cohomology of varieties over number fields and its Galois cohomology, in *Galois groups over* $\mathbb{Q}$ (Berkeley, CA, 1987), 315–360, Math. Sci. Res. Inst. Publ., 16, Springer, New York, 1989.

[5] Chandrashekhar Khare, Jean-Pierre Wintenberger. Ramification in Iwasawa modules and splitting conjectures. To appear in International Mathematics Research Notices.

[6] Jürgen Neukirch. *Cohomology of number fields*. 2nd edition. Grundlehren des mathematischen Wissenschaften, 323, Springer, 2008.

[7] Christophe Soulé. On higher p-adic regulators, in *Algebraic K-theory*. Evanston 1980 Lecture Notes in Math., 854, Springer, 1981.

[8] Andrew Wiles. The Iwasawa conjecture for totally real fields. *Ann. of Math.* (2) 131 (1990), no. 3, 493–540.

Printed in the United States
by Bookmasters.

Printed in the United States
By Bookmasters